Cold War, Hot Science

Studies in the History of Science, Technology and Medicine
Edited by John Krige, CRHST, Paris, France

Studies in the History of Science, Technology and Medicine aims to stimulate research in the field, concentrating on the twentieth century. It seeks to contribute to our understanding of science, technology and medicine as they are embedded in society, exploring the links between the subjects on the one hand and the cultural, economic, political and institutional contexts of their genesis and development on the other. Within this framework, and while not favouring any particular methodological approach, the series welcomes studies which examine relations between science, technology, medicine and society in new ways e.g. the social construction of technologies, large technical systems.

Other Titles in the Series

Volume 1
Technological Change: Methods and Themes in the History of Technology
Edited by Robert Fox

Volume 2
Technology Transfer out of Germany after 1945
Edited by Matthias Judt and Burghard Ciesla

Volume 3
Entomology, Ecology and Agriculture: The Making of Scientific Careers in North America, 1885–1985
Paolo Palladino

Volume 4
The Historiography of Contemporary Science and Technology
Edited by Thomas Söderqvist

Volume 5
Science and Spectacle: The Work of Jodrell Bank in Post-war British Culture
Jon Agar

Volume 6
Molecularizing Biology and Medicine: New Practices and Alliances, 1910s–1970s
Edited by Soraya de Chadarevian and Harmke Kamminga

Volume 7
Cold War, Hot Science: Applied Research in Britain's Defence Laboratories, 1945–1990
Edited by Robert Bud and Philip Gummett

Other Volumes in Preparation
Making Isotopes Matter: F.W. Aston and the Culture of Physics
Jeff Hughes

This book is part of a series. The publisher will accept continuation orders which may be cancelled at any time and which provide for automatic billing and shipping of each title in the series upon publication. Please write for details.

Cold War, Hot Science

APPLIED RESEARCH IN BRITAIN'S DEFENCE LABORATORIES, 1945–1990

Edited by

Robert Bud
Science Museum, London, UK

and

Philip Gummett
University of Manchester, UK

Published in association with the Science Museum

harwood academic publishers
Australia • Canada • China • France • Germany • India • Japan
Luxembourg • Malaysia • The Netherlands • Russia • Singapore
Switzerland

Amsteldijk 166
1st Floor
1079 LH Amsterdam
The Netherlands

British Library Cataloguing in Publication Data

Cold war, hot science : applied research in Britain's
defence laboratories. – (Studies in the history of science,
technology and medicine ; v 7)
1. Military research – Great Britain
I. Bud, Robert II. Gummett, Philip
355'.07'0941

ISBN: 90-5702-481-0
ISSN: 1024-8048

Contents

List of Illustrations

Foreword

In the search for intelligent life elsewhere in the universe it is widely thought that such life is likely to generate a technological spacefaring civilisation. It seems to me that this likelihood may in fact be very low. Human involvement in space is based not only on our intelligence and the resources the Earth provides, but also on some peculiarities of our emotional make-up. One of these is our curiosity, without which science is unlikely to have developed. However, it may well be argued that there is little evolutionary advantage in intelligence if it is not coupled with curiosity, so it is perhaps not so surprising a feature.

However, surely it is most odd that our loyalties and feelings of belonging are very strong in relation to a nation state of some tens of millions of people but weak in relation to mankind as a whole. Only this emotional peculiarity can account for the occurrence of war between huge well-organised states. Group loyalty to those we know personally or perhaps at one remove is readily explicable in evolutionary terms. It is known to occur in other animals and presumably leads to the distrust of others, which lies at the root of gang conflict and border raids. However, warfare on a national scale seems very odd indeed and may well not be a usual feature of evolved intelligent life elsewhere in the universe.

This is so relevant because military needs have been a major driving force for all of science and technology throughout history, but nowhere have they been as dominant as in space technology. The human ability to operate in space is a by-product of the Cold War. Thus we should realise that it may be rare in the universe for a highly-technological civilisation to be spacefaring. Our attachment to the nation state and so to military technology may be much less common than the application of intelligence within the universe at large to general technology.

In the UK much of the defence research and development effort has been led by a group of government establishments recently brought together as the Defence Evaluation and Research Agency. To study, as this volume does, how these organisations interacted with a wide range of industries is an important contribution to an understanding of the last 50 years. The establishments produced a few outstanding successes and made many useful contributions to meeting the needs of the armed forces. Alas, there were also cases where a considerable development effort led to nothing, however successful the preceding research had been. Sometimes the military requirement had then to be met by foreign purchase, while on some other occasions external happenings had made the requirement obsolete. The successful developments therefore were necessarily interspersed with less fruitful ones. About these one sometimes heard the sad, proud comment that with 10 per cent of the outlay of the Americans one had got to 90 per cent of project completion. Unhappily, 90 per cent of a project is generally not much more useful than 0 per cent. However much brainpower and dedication is devoted to the development of a project, success cannot be taken for granted.

The history described in these pages is therefore sometimes cheering, sometimes saddening, but always worth recounting.

Sir Hermann Bondi FRS
formerly Director-General, ESRO;
Master, Churchill College, Cambridge;
Chief Scientific Adviser, Ministry of Defence;
Chairman and Chief Executive, NERC

Steering Committee

General Sir Hugh Beach

Professor Hans-Joachim Braun
Universität der Bundeswehr, Hamburg, Germany

Dr Robert Bud
Science Museum, London, UK

Dr Martin Edmonds
University of Lancaster, UK

Professor Philip Gummett
University of Manchester, UK

Dr Eric Grove
University of Hull, UK

Dr John Krige
CRHST, la Villette, France

Dr Adrian Mears
DERA, Porton, UK

Mr Andrew Nahum
Science Museum, London, UK

Mrs Elizabeth Peace
DERA, Porton, UK

Notes on Contributors

Jon Agar is a lecturer at the Centre for the History of Science, Technology and Medicine, University of Manchester. He is the author of *Science and Spectacle: the Work of Jodrell Bank in Post-war British Culture*, and co-editor of *Making Space for Science.* He is working on a cultural history of the government machine, and jointly on a history of radar development, and a history and anthropology of information technologies in Manchester.

Brian Balmer is a lecturer in the Department of Science and Technology Studies, University College London. He was previously a research fellow at the Science Policy Research Unit, University of Sussex, where he started his research on the history of biological weapons research. He has also published several articles on the sociology and politics of research in human genetics.

Robert Bud is head of Research (Collections) at the Science Museum, London and is also head of Life and Communications Technologies. His work concentrates on the history of modern applied science. Recent books include *Instruments of Science: An Historical Encyclopedia* which was edited with Deborah Warner, and *Manifesting Medicine: Bodies and Machines* which was edited with Bernard Finn and Helmuth Trischler.

Gradon Carter is a consultant at the CBD Sector of DERA at Porton. He served from 1948 to1995 in both Porton Establishments and on the Defence Intelligence Staff. Earlier publications reflect his microbiological work, but from 1989 some 21 papers, a book and book chapters on historical aspects of chemical weapons and biological weapons reflect his wider interests.

John Ernsting joined the Medical Branch of the Royal Air Force in 1954 and was posted to the RAF Institute of Aviation where he conducted and later directed research in aviation medicine over a period of 39 years, during the last five of which he was Commandant. After retiring from the Royal Air Force, having attained the rank of Air Vice-Marshal, he undertook teaching and research as Visiting Professor in Human and Applied Physiology at King's College London. He is now Head of Human Physiology and Aerospace Medicine within the GKT School of Biomedical Sciences at King's College London.

Eric Grove is deputy director of the Centre for Security Studies at the University of Hull. Formerly a civilian lecturer at both the Royal Naval Colleges at Dartmouth and Greenwich, he has written widely on twentieth-century naval history and maritime strategy. His books include *Vanguard to Trident*, a history of post-1945 British naval policy, and *The Future of Sea Power.*

Philip Gummett is professor of government and technology policy, and Pro-Vice-Chancellor, at the University of Manchester. He has written widely on science and technology policy, with particular reference to defence, technological and industrial affairs in Britain and Europe, and advised on these subjects in the UK and European Parliaments and the European Commission. He is academic convenor of the nine-nation CREDIT (Capacity for Research into European Defence and Industrial Technology) network.

Jeff Hughes studied at the Universities of Oxford and Cambridge. Since 1993 he has been a lecturer at the Centre for the History of Science, Technology and Medicine, University of Manchester. His book *Making Isotopes Matter* will be published by Harwood Academic Press. With Jon Agar he is working on a history of radar development in the UK, 1935–50.

Richard Mills has a first degree in chemical physics, a postgraduate diploma in politics and international relations and a PhD in politics. He has held several posts in the defence industry and academia before taking up his current position as research fellow with the Defence Engineering Group at University College London. His doctoral thesis investigated the influence of defence expenditure on the early development of laser technology in the UK. Dr Mills' main research interests lie in the areas of technology and research management.

Andrew Nahum is senior curator of the National Aeronautical Collection at the Science Museum, London and is currently directing the creation of a major new synoptic gallery covering the history of technology, entitled *Making the Modern World*. He has written extensively on the history of technology and aviation for both scholarly and popular journals; his books include *The Rotary Aero Engine* and *Alec Issigonis*. He is currently completing a technological and economic study of the British aircraft industry in the years following the Second World War.

Richard Ogorkiewicz is a graduate in mechanical engineering of London University and Fellow of the Institution of Mechanical Engineers, and currently visiting professor at the Royal Military College of Science where he has been lecturing for 20 years on armoured fighting vehicles. For the past 25 years he has also been a consultant on armoured vehicles to industrial and government organisations in the UK, USA, Brazil, Italy, South Korea and Turkey, and has lectured on the subject in these and many other countries, as well as being the author of three books and numerous articles and papers.

Ernest Putley joined TRE in 1942 and began working on infrared in 1948. With W.D. Lawson he began research on the preparation and properties of single crystals of infrared-sensitive semiconductors. He was a member of the team which discovered cadmium mercury telluride. He was the first to observe free electron photoconductivity in indium antimonide and also initiated the British development of pyroelectric devices, used in the fire brigades' thermal imagers, and the widely used infrared intruder alarms.

Stephen Robinson was a fighter control officer in the Royal Air Force in 1950 and studied physics at Cambridge. He then joined the Philips Laboratories to carry out microwave defence research. He became product director of MEL Ltd and managing director of PYE TVT Ltd before transferring to the Royal Signals and Radar Establishment in 1985 and retiring as director in 1991. He received an OBE in 1971 for work with the Navy on electronic warfare.

Graham Spinardi is senior research fellow in the Research Centre for Social Sciences at the University of Edinburgh, currently looking at the exploitation of inventions. He has worked on a number of research projects investigating the relationship between technology and society, looking in particular at nuclear weapons systems, information technology and environmental technologies. He is the author of *From Polaris to Trident: the Department of US Fleet Ballistic Missile Technology* and joint editor of *Defence Science and Technology: Adjusting to Change.*

Stephen Twigge works in the Records Management Department at the Public Record Office, Kew. He was formerly lecturer in contemporary history at Salford University and research associate in the Department of International Politics, University of Wales, Aberystwyth. His book *Planning Armageddon: Britain, the United States and the Command of Western Nuclear Forces 1945–1964*, co-written with Len Scott, is due for publication with Harwood in 1999.

Matthew Uttley is senior lecturer in the Defence Studies Department, Joint Services Command and Staff College (JSCSC) and director of the JSCSC Defence Technology Research Group. He was formerly senior lecturer in the Department of History and International Affairs at the Royal Naval College, Greenwich, and has been a research fellow at the Centre for Defence Economics at the University of York and the Centre for Defence and International Security Studies, Lancaster University. During 1987 he was the Ford Foundation Scholar in Science, Technology and International Affairs at Lancaster University and was elected Fellow of the Royal Society of Arts in 1994. Dr Uttley has published widely on defence procurement, technology transfer and West European defence integration. He is joint editor of *Defence Science and Technology: Adjusting to Change* (1993).

Tom Wright is a senior research fellow of the Science Museum, London. He was formerly an assistant director of the Museum, responsible for the Collections Division. In the past he has served as an officer in the merchant marine.

Acknowledgements

This project has benefited from the commitment of a large number of people whose support over the years has converted a vision into a book.

The project has been guided by a steering committee whose commitment to analysing draft after draft has been key to the building of a very happy team. Dr John Becklake has driven through the implementation of their decisions as project manager and we are most grateful to him. Anna Hodson and then Ela Ginalska, the Science Museum's publications managers, ensured that the book would find a publisher.

Once texts had been delivered, several people worked very hard to convert manuscripts into a book. We are grateful to Shanika Cruz, Giskin Day and Jennifer Bew who worked at various stages copyediting a most complex text and even more complex notes. Caroline Thompson has worked hard with us and with staff at DERA to research pictures. Stephen Twigge took responsibility for the index. Colleagues at the Science Museum have provided support and encouragement.

Finally we should like to record our gratitude to DERA and to Elizabeth Peace, Corporate Affairs Director. She has given her personal enthusiasm as an historian and as a member of DERA's Board, to ensure that DERA has been a wonderful patron for this project. The agency has made its papers and staff available, whilst respecting the independence of authors and editors.

ACKNOWLEDGEMENTS

CHAPTER 1

Introduction: Don't You Know There's a War On?

ROBERT BUD AND PHILIP GUMMETT

'Don't you know there's a war on?' This Second World War catchphrase could as easily have been invoked during the 45 subsequent years of the Cold War. To many in the West's military establishment, hot war with the Soviet Union often really did seem imminent. Preparation was deemed both the best means of prevention and urgently required. Almost ignored today, the cover of the *Bulletin of the Atomic Scientists* has featured a clock whose hands indicate the countdown to Armageddon: they rarely moved back from ten minutes to midnight during the period of this book.

The Cold War is central to an understanding of science in the second half of the twentieth century. Stretching across technological eras, it began with the weapons of the Second World War. Then, computers, software and microchips, as well as thermonuclear capability, transformed the context of weapons development. Military expenditure drove the early development of the information age, paying for the solution of countless scientific problems.

In countries such as the USA and the UK, more than half of government-funded research and development, and something like a quarter of the national total, was funded out of defence budgets. At the same time it was recognised that throughout history great empires had exhausted themselves through their investment in military might.[1] The new investment in science and high technology, however, provided some hope that these enormous expenditures might yield economic as well as security benefits.

It was, nonetheless, unclear how this translation between the military and the economic spheres could best be encouraged. Often it seemed that most of the scientific knowledge going into weapons development had been generated and then retained within a closed military research world. The benefits did not appear to be diffusing into the civilian economy.[2] At a central moment of the confrontation, 1967, the US National Academy of Sciences reported to the US House of Representatives on 'applied science and technological progress'. The Academy was concerned to ensure that US applied science would continue to be healthy at a time when the economies of other countries seemed to be growing faster than their own. Among its key conclusions were that 'studies of the history and sociology of applied science are important'.[3] The late 1960s did see a number of other surveys, investigating the

FIGURE 1.1: APPLIED SCIENCE IN ACTION: DETERMINATION OF THE RATE AT WHICH A CHEMICAL WARFARE AGENT PENETRATED ARMY BOOTS, CHEMICAL DEFENCE EXPERIMENTAL ESTABLISHMENT, PORTON, IN THE 1960S (© CROWN COPYRIGHT, DERA)

passage of ideas from basic science through applied science into technology, but as this process was found to be most complex, and rarely obeyed the ideal trajectory, attention rapidly diminished. Such investigations also proved difficult because the study of recent applied science, conducted with commercial or security ends in mind, has usually been impeded by secrecy. Thus, despite the importance of applied research in our culture and economy, we still know little about its practice. With a few honourable exceptions the history of the process of applied—and especially industrial—research since the Second World War (Fig. 1.1) has remained untold.[4]

Conducted, like pure science, in laboratories, applied research is none the less managed and funded explicitly because of its place in the development of identifiable products and processes. Sometimes it is seen as the 'application' of pure science, at other times as an activity in itself. 'Engineering science' is a term sometimes used, particularly in the US.[5] Whatever the precise interpretation, it has seemed to be a key link between pure science and its ultimate application. In scale too applied science lies between the two. Whereas, to some, the word 'laboratory' might conjure up the Hollywood image of a single room inhabited by a 'mad inventor', many industrial and military research laboratories better resemble large university campuses or even industrial zones, comprising many buildings and employing thousands of workers. The category 'R&D' also includes development which costs far more than research and commonly requires factory-scale production.

If industrial research has been hidden from sight, even more so has been work conducted to support national security, even though it has constituted such a large proportion of all national research and development. Only now, at the beginning of a new geopolitical age, are scholars looking back in an attempt to better understand the last half-century of our science. In the USA, France and elsewhere, historians of technology are examining national endeavours.[6]

In this book, the British defence R&D effort has at last begun to be explored. We make a start here on recording one of the largest R&D campaigns ever mounted in the UK, and we aim also to illuminate the more general nature of Cold War science. Britain's strategy was particularly interesting in the history of applied science because, behind a strong sense of the traditional, there was another, determinedly modernist, face. Faith has repeatedly been placed in science and technology as the solution to economic woes, and as a new focus for national self-esteem in a country retreating from empire. There has been enormous pride in technological breakthroughs, which might be seen to leapfrog the efforts of those nations more ploddingly productive, such as the Germans, or irritatingly wealthy, such as the Americans.[7] The Spitfire, penicillin, the jet engine and radar were wartime examples. The development of computing at that time also became a matter of immense national pride. In the 1950s, Christopher Cockerell's hovercraft, Issigonis' Mini car and civil nuclear power provided new examples.

This enthusiasm for the technological leap declined in the late 1960s as the costs of such once-beloved breakthroughs as the world's first supersonic jetliner, Concorde, escalated and commercial potential withered without apparent benefit. National champions of high technology, such as the computer company ICL, remained minnows compared to US competitors on the international scene. Rolls-Royce, the aeroengine manufacturer, went bankrupt attempting to maintain world-class status. The country was spending a higher proportion of its national income on scientific research than any other except the USA, but its economic growth was lower than that of most of its competitors. However, one cannot overestimate the importance of the preceding love affair with science and technology. The Cold War was an R&D war, as Agar and Balmer have argued elsewhere.[8] In such a confrontation the very names 'Farnborough', 'Malvern' and 'Porton Down' were felt to be superweapons.

The British effort was larger than any European counterpart until the 1960s expansion of the French endeavour; a decade earlier it had been several times the scale of French defence research. It was, however, always much smaller than the US effort which, broadly speaking, was about ten times larger. Another comparison would be between the in-house British conventional R&D effort in 1960 and the R&D of the largest US chemical company, Du Pont. Both in expenditure and in numbers of scientists employed, these two were quite comparable, although the balance between research and development probably differed.[9]

Unlike its larger US counterpart, British defence research (as distinct from development, where the picture changed over time) was conducted largely within dedicated civil service establishments, employing at its height over 30,000 scientists and administrators. In 1959 the non-nuclear military research establishments were spending about £2 million a year on basic research, £45 million a year on applied research, and £15 million on development.[10] About 3000 technicians and 2000 professional scientists were employed.[11]

Many of the people working in these establishments have been scientifically distinguished, even if they did not become household names. Their output, as explored in this book, ranged from the now-familiar angled flight deck of the modern aircraft carrier, to Concorde, the still-secret composite tank armour, to the infrared detector used in modern burglar alarms, liquid crystal displays and carbon fibre. It is, however, important to appreciate that the establishments were not simply centres of research: they were (and are) also a reservoir of technical advice available to government, whether on the feasibility of proposals from would-be contractors, on the seriousness with which technical dimensions of intelligence information should be viewed, or on the future of technology as a factor in strategic planning. In times of crisis they also proved to be a valuable asset, able to mobilise effort rapidly to address novel problems, or to work out how to adapt existing equipment, which has so often found itself being used in circumstances different from those for which it had been intended. Finally, civilian spinoff has also been an important aspect of their work, even though, except for a short period in the late 1960s, this has been only a secondary concern.

Although London's Science Museum—perhaps the most important public place for the exploration, exposition and experience of the history of science and technology in the UK—is not focused principally upon military affairs, it has been enthusiastic about exploring this 'dark matter' of postwar science. Over many years, the Museum's collections have been enriched by material transferred from defence research establishments. In 1995 it was fortunate to reach an agreement with the Defence Research Agency, (shortly to become the Defence Evaluation and Research Agency) that, aided by a distinguished steering committee, it would orchestrate the writing of a history of key parts of the postwar military research enterprise. This history would be focused upon the so-called conventional weapons, and not deal with the more high-profile and already better-studied nuclear arsenal. Because of this support, some of the intense secrecy in which papers have been kept was lifted. It was, moreover, agreed that there should be no DERA control over what was published, except on grounds of national security—a condition that has not been invoked.

The authors have drawn on expertise accumulated within the establishments and within the academic (including military academic) historical community. Their detailed chapters deal with the broad scope of publicly funded conventional military research in the UK over half a century. This introduction seeks to put this in the context of general shifts in the equipment strategy over that time, to introduce the role of the establishments, and to identify issues which the authors have together highlighted.

BEFORE THE SECOND WORLD WAR

'Command technology', as the historian William McNeill calls the attempt deliberately to create new weapons systems that surpass existing capabilities, was an invention of the eighteenth century.[12] Before 1914 it was mainly the world's leading navies that patronised and funded deliberate invention. Their warships were the most expensive and complex weapons systems of their day.[13]

Not until the First World War did deliberate invention cease to be a predominantly naval matter and come ashore. Faced with the embarrassment (and worse) of dependence on Germany for such essential items as optical glass, magnetos, and even the khaki dye for uniforms, the UK set up the Department of Scientific and Industrial Research and began the systematic incorporation of science and technology into government.[14] It also set about building up—without any overarching plan—a set of research establishments devoted to defence matters, which developed in various ways up to and after the pre-Second World War rearmament period.

A number of them were long established, having grown by adventitious extension of testing facilities to incorporate experimental work.[15] Others grew out of the slightly more systematic approach that followed the First World War: for example, the Admiralty Research Laboratory and the Armament Research Establishment were set up to fill gaps identified during the war. This approach continued during the Second World War with, for example, the establishment in 1943 of the Naval Construction Research Establishment to carry out experimental work on underwater explosions and their effects upon ships' structures. Yet others owed their existence to the foresight and vigour of particular individuals. One such was the Admiralty Experiment Works, set up in 1870 at the instigation of the distinguished naval architect William Froude, in order to conduct tests of model ships. Another was the pair of organisations that were the forerunners of what later became the National Gas Turbine Establishment. The private firm of Power Jets Ltd was set up in the 1930s under the influence of Air Commodore Whittle, and, as the nationalised Power Jets (R&D) Ltd and later the National Gas Turbine Establishment became the main centre for work on jet engines.[16] Some, finally, took shape under political pressure. During the First World War, as aircraft manufacture became a commercial proposition, industrial disapproval of government involvement in this business had led to a decision to discontinue the design and manufacture of aircraft at the Royal Aircraft Factory, and to reconstitute the factory as the Royal Aircraft Establishment. This was given the dual role of technical adviser to the commercial designers and actual designer of aircraft equipment. A reverse example was perhaps the case of the Admiralty Engineering Laboratory, which was set up in 1915 to break the monopoly held by the Vickers company in the design and development of submarine engines.

It was with the Second World War, however, that the watershed was finally reached. Not only were radically new weapons technologies developed—or at least presaged—during the war (radar, the atomic bomb, jet engines, guided missiles), but the very scale of effort and complexity of organisation was revolutionised. A large armaments company such as Vickers could count itself the equal of any government agency in its sphere in the interwar years. By 1945, however, the government organisation had grown enormously. In the words of Vickers' historian, Scott:

> There was now a whole complex of naval and military research, development and design establishments, employing a vastly greater number of scientists, engineers, administrators and specialist naval and military officers than had been the case before the war. There had been, too, a vital change of balance. In the comparatively small pre-war organisation the important people had been the serving officers and the practical engineers. In the war, however, the scientist had achieved a new standing, and generally speaking the research and development establishments at the end of the war were more 'research minded', more disposed to seek a radical solution of their problems, bolder and more imaginative, but less 'practical' in their approach. Could Vickers keep up with them in design?[17]

The military effort was divided into three ministries: the Ministry of Supply responsible for the army; the Ministry of Aircraft Production responsible for aircraft development and production; and the Admiralty. New senior posts were created, such as that of Controller General of Research and Development in the Ministry of Supply in 1941.[18] Scientists came to play new roles as advisers at the highest levels: Tizard and Lindemann, and at a slightly less exalted level Blackett and Jones, Zuckerman and Bernal, became familiar names. Their impact was felt not only on equipment (from the dramatic level of the atomic bomb to the mundane), but also in the development of tactical and strategic ideas, as the 'Sunday Soviets', in which the radar scientists brainstormed freely with military officers, or the debates about the appropriate targets for strategic bombing, demonstrated superbly.[19] In short, scientists came to play a new role as colleagues and advisers of those men in Whitehall and in the field.

THE POSTWAR POSITION

Things were different at many levels in 1945 compared with before the war. The number of research establishments had risen substantially,[20] and manufacturing industry was on a war production footing. Those parts of it that intended to continue in armaments production found, as in the case of Vickers, that the government machine, with its new scientific strength, had to be dealt with on a different basis from before. The scale and complexity of the weapons systems that were now wanted also led to a decline in the capacity of industry to develop them as private ventures. By 1953, the Chairman of the Fairey Aviation Company was to report to his shareholders that 'Prior to the war over 50 per cent of the aircraft in use by the Services were the result of private ventures. Since the war the financing of complete aircraft is becoming rare.'[21]

Companies such as Fairey still continued to spend heavily on development work, however. Overall, Britain remained a world leader in many fields of defence technology, and was well placed in many others. On the other hand, there was much to be done to rebuild the economy and, presented with both the need and the opportunity for exports in the global postwar economy, many companies saw their future in civil, not defence, markets. Resources, of both capital and skilled people, were in short supply, necessitating sharp choices over priorities. Industry was concerned that an excessively military focus would detract from civilian competence. Thus the great electrical company GEC was concerned that its wartime emphasis on radar had meant that it had fallen behind US competitors in telecommunications. Therefore, when in 1949 it was asked on take on a contract for the development of air interception radar, GEC was loath to divert staff from civil work. After finding no success with other firms, the government eventually prevailed upon GEC to create a new, government-financed organisation for military electronics R&D, with 70 scientific staff, 280 auxiliary staff, and a running cost of over £1 million per annum.[22] Even then, concern was expressed at Cabinet level about the need to guard against the drain of staff from other defence research establishments, or from other key industrial projects, into this new organisation. Preparation for warfare was also seen to have become even more challenging. As a 1958 Ministry of Supply memorandum explained, 'the aim must also be to ensure. . . a fund of scientific knowledge and resources is built up which will be adequate for the development of future generations of weapons, even though their precise nature cannot yet be foreseen'.[23]

Militarily, the UK remained a great power, albeit one that was in the process of being eclipsed by the emergent superpowers. Demobilisation could not be carried out overnight; in 1946, defence expenditure still stood at one-fifth of the GNP; military commitments were extensive; British forces were serving in the armies of occupation in Europe, and others were tied down by disorder in Greece. They were also serving in ex-Italian colonies in Africa, and in Burma, Malaya, Singapore, Hong Kong, Gibraltar, Malta, Iraq, and from Suez to Aden. Mounting political unrest in India and Pakistan created additional burdens. By the beginning of 1947 the extra-European responsibilities alone commanded some 300,000 Service personnel. It was only with the coincidence of the 1947 economic crisis and the approaching independence of India, Pakistan, Burma and Ceylon that Attlee, the Labour Prime Minister elected in 1945, announced in August 1947 a proposal to reduce the number of Service personnel overseas from half a million to 300,000.[24]

The cost of this commitment was considerable, and particularly burdensome for an economy ravaged by war. For six years civilian industry had been starved of investment, three million houses had been damaged and, in the cold winter of 1947, coal shortages closed factories. Bread rationing, avoided during the war, was imposed in a dollar-starved postwar austerity. All production had to be focused upon export rather than consumption.[25]

Circumstances did not ease for several years. Extra demands arose from the Berlin blockade of June 1948–May 1949. The pound was devalued in September 1949 (on the day before the USA announced its detection of the first Soviet atomic weapons test in August). And, just as the economy took a marked turn for the better in 1950, the Korean War prompted another large rearmament programme.

The history of British defence policy over the next 40 years, until the late 1980s, could be summed up as reflecting, on the one hand, the withdrawal from Empire and the definition of a new world role in the face of growing superpower strength, and on the other the search for a defence policy that could be economically sustainable. An interwoven theme has been the role to be played by nuclear weapons, both tactical and strategic. Britain announced its possession of thermonuclear weapons in 1957 and through the post-war years has boasted of a deterrent employable, if necessary, 'independently' of the USA.[26] Even with conventional weapons the main focus of policy was a potential war against the Soviet Union, which would be fought on the north German plain. Meanwhile, British arms were tested in numerous police actions in quite different terrains, ranging from the jungles of Malaya and Belize to the deserts of South Arabia. Despite US pressure, the country did not join the Vietnam War and thus watched from the sidelines the defining experience of its closest ally and competitor in military sales. Not until the 1982 Falklands conflict was a major conventional campaign waged.

The period saw a succession of defence reviews. These aimed first at contraction then, with Korea, at expansion, followed, once Churchill regained office, by 'stretching' of the defence programme inherited from Attlee. Even so, the Conservative government considered the balance of payments to be critical, and observed that, even with the reduced expenditure, 10 per cent of GNP was being absorbed by 900,000 Service personnel, and that by the end of 1953 the civilian workforce engaged on defence work would be 2 million.[27] There followed doctrinal revision, in particular the argument, dating from 1952, that much greater reliance should be placed in future upon atomic weapons, both as a deterrent at the strategic level and as a substitute for the manpower requirements laid down by the NATO Council meeting in Lisbon that year.[28]

Bartlett observes that the average life expectancy of every defence review between 1945 and the mid-1960s had been a mere two and a half years. As the usual gestation period for new weapons systems was from two to four times that figure, this was clearly too short. Things were to get worse in the mid-1960s, 'with doubts and dissension developing at almost every level of British defence policy, with strategy, fundamental weapons and administrative methods all being thrown into the melting pot'.[29] At the same time, equipment development programmes begun at the end of the Second World War would show remarkable resilience. The failed expedition to retake the Suez Canal of 1956, when the British and French governments found their intervention condemned by the USA as well as by the Soviet Union, caused a profound shock. It led to further reappraisal, both of Britain's world role and of her defence arrangements. The new Prime Minister, Harold Macmillan, appointed as Minister of Defence Duncan Sandys. His tenure of office was marked by a reform of the central organisation for defence, and by controversial emphases upon the strategic nuclear deterrent and missiles (both as delivery vehicles for nuclear weapons and to replace fighter aircraft in air defence). This therefore entailed a sharply reduced emphasis on manned aircraft. Sandys also sought substantial cuts in military manpower, mainly through the ending of National Service. There followed the ambitious programme to develop Britain's own intermediate-range ballistic missile, known as Blue Streak. When this programme was cancelled in 1960 it was to be replaced by the US air-launched missile Skybolt. This was cancelled in turn by the USA, and instead the UK finally purchased, from the USA, Polaris submarine-launched ballistic missiles.[30]

The radical tone of the 1957 Sandys review was not translated into correspondingly radical action. Instead, the late 1950s and early 1960s saw reversal of policy and a build-up in military activity and an increase in commitments to new development and procurement programmes. British troops went into action in the Protectorate of Southern Arabia and in Borneo, as well as in Cyprus. A promise was given in 1964 to increase the strength of the British Army of the Rhine. Work also proceeded on an extensive and expensive equipment programme, including a number of aircraft projects. There was to be a new world-class strike aircraft known as TSR-2, and a revolutionary supersonic aircraft which would for the first time take off and land vertically (VTOL), with the code name P-1154. Until its cancellation in 1960 there was also the cost of the Blue Streak system. Later, there arose the question of maintaining the ability of the planes developed after the Second World War to deliver atomic bombs, the so-called V-bombers, to penetrate the increasingly sophisticated Soviet air defences, and the proposal, mooted in the Defence White Paper of 1964, of converting some of the V-bombers to low-level flight so as to maintain the credibility of the deterrent until the Polaris force was operational.[31]

There developed, however, a growing unease with the managerial control of the equipment programme. Much of the blame was attributed to inter-Service rivalry and the incapacity of the central Ministry of Defence to control the Service ministries and supply agencies. A review of the administrative machinery led to the creation of a new, more unified, Ministry of Defence from April 1964. This instrument was to be used to powerful effect after the election of a Labour government in October of that year and the appointment of Denis Healey as Secretary of State for Defence.

The new government was determined to hold down defence spending. Healey reported to the House of Commons in November 1964 that Britain was spending more on defence than any other country of her size and wealth, and instituted yet another wide-ranging defence review. One result was a significant alteration to the shape and timing of the RAF re-equipment programme. Three major projects—(the TSR-2, the P-1154 and the Hawker-Siddeley HS-681 transport plane—were cancelled. They were replaced with one national programme (the P-1127, which became the Harrier), American purchases (the Phantom, the Hercules transport and the F-111), and a collaborative programme with France to develop a tactical reconnaissance-strike aircraft for the longer term (the Anglo-French variable geometry (AFVG) plane, from which the French were, however, soon to withdraw. The Navy's programme of large aircraft carriers was to be scrapped; the Polaris programme was to proceed, but with only four of the five submarines that had been ordered. A year later, in February 1967, came another major decision: Britain was to restrict its overseas role by evacuating Aden by the end of the year. The import of this decision was emphasised in July 1967 by the further decision to withdraw from mainland bases east of Suez, and from the commitment to maintain forces outside Europe.[32] By January 1968 this decision had sharpened. The Prime Minister, Harold Wilson, was able to announce that Britain was to withdraw from Malaysia, the

Persian Gulf and Singapore (but not Hong Kong) by the end of 1971, and that therefore the order for the F-111s could be cancelled.[33] Britain was pulling in its military tentacles: henceforth, the emphasis was to be on NATO and Europe, both operationally and in terms of the procurement of equipment.

Even with the Healey cuts, however, the problem of living within a realistic equipment budget was not solved. Further programmes, many of which exhibited considerable cost overruns, were begun. Thus, the abandonment of the AFVG in July 1967 opened the way for what became the Anglo-German-Italian multirole combat aircraft (MRCA), later called Tornado, which would see service in the Gulf War. Two other examples of new programmes begun in the late 1960s were the enormously expensive programme to upgrade the single nuclear warheads on Polaris missiles to multiple (though not independently targetable) warheads, known as Chevaline, and a new high-technology torpedo, known as Stingray.[34]

By the mid-1970s, and despite the establishment within the Ministry of Defence of a new, more powerful procurement executive, the need for further defence cuts had become imperative. More followed in 1977, leading the Chiefs of Staff, according to press reports, to protest directly to the Prime Minister, James Callaghan, in mid-January. Further reappraisals followed, driven by the rising costs of such programmes as Chevaline and Stingray, along with, by now, the Nimrod airborne early warning aircraft and the MRCA programme. Also coming down the line was the need to make provision for a new fighter for the RAF and, following its announcement by Mrs Thatcher's Conservative government in 1980, for the purchase of US Trident missiles and new British submarines to replace the Polaris system.

There was an increasing sense that the ever more costly equipment requirements could be met only by striking a new balance between the military roles Britain was trying to play. Even in 1980, with the Empire long gone, Britain's Armed Forces were still performing a greater range of missions than any others in the world, except those of the superpowers and, possibly, France. The UK was maintaining independent strategic and theatre nuclear forces, which were committed to NATO. It was providing for the direct defence of the UK while also maintaining a major Army and Air Force contribution on the European mainland. It was also deploying a major surface and submarine force in the eastern Atlantic and the Channel. In addition, there continued to be a requirement for a limited capability for operations outside the NATO area, including the defence of such territories as Hong Kong and the Falkland Islands.

An attempt to deal with these problems radically was made in 1981 by Mr (now Sir) John Nott, the Secretary of State for Defence. In a White Paper entitled *The United Kingdom Defence Programme: The Way Forward*, Mr Nott argued that changes in the defence programme were required for two main reasons.

First, even the increased resources which the government planned to allocate could not fund all the force structures and the plans for their improvement that were in hand. A particular cause of this problem was cost growth, especially in equipment, as illustrated by the following example from the *Statement on the Defence Estimates* that Mr Nott had presented to Parliament in April 1981 (Table 1.1).

TABLE 1.1: RELATIVE REAL PRODUCTION COSTS OF SUCCESSIVE GENERATIONS OF EQUIPMENT (*SOURCE*: CMND 8212-I, FIG. 14)

New weapon weapon	Cost multiplier	Old
Type 22 frigate	3×	Leander
Harrier GR1	4×	Hunter F6
Lynx Mk 2 helicopter	2.5×	Wasp Mk 1
Sea Wolf guided missile	3.5×	Sea Cat

The second of John Nott's reasons for change in the defence programme was, as the June White Paper observed, that technological advance was dramatically changing the defence environment.

> The fast-growing power of modern weapons to find targets accurately and hit them hard at long ranges is increasing the vulnerability of major platforms such as aircraft and surface ships. To meet this, and indeed to exploit it, the balance of our investment between platforms and weapons needs to be altered so as to maximise real combat capability.[35]

This passage implicitly alluded to the threat to surface ships, aircraft and tanks raised by new precision-guided weapons, and to the substantial imbalance between the costs of a major weapons platform and the missile, that might be used to destroy it. This had been a subject of much discussion (in the specialist literature at least) since the 1973 Middle East war. Its introduction into the White Paper indicated that those members of the Ministry of Defence who were concerned about this question (these included the Chief Scientific Advisor) had gained the Secretary of State's ear.

The conclusion had been reached, not that any defence roles should be abandoned, but that a new and sustainable force structure was needed. The White Paper acknowledged that this would be less glamorous than maximising the number of large and costly platforms in the armoury, and that moving in this direction would mean 'substantial and uncomfortable' change in some fields.

The impact of the Nott decisions was limited in some areas. Trident was to go ahead, together with various programmes concerned with direct defence of the UK. The British Army of the Rhine was to remain at its current size, rather than be increased as previously planned, but planned programmes (including the Challenger tank, improved night-fighting equipment, and new equipment for automated data handling) were to proceed. The RAF was to be equipped with the AV8B development of the Harrier, produced jointly by McDonnell Douglas and British Aerospace, and with the strike version of the Tornado. More advanced weapons for attacking armoured forces and suppressing air defences were to be developed, so as further to exploit the potential of Tornado. Early replacement of the Jaguar force would not be possible, but work and discussions with potential partners was continuing on future combat aircraft.

The important changes came in the maritime sector. Arguing that it was impossible to sustain current levels of equipment in all three dimensions of naval warfare, the White Paper announced the controversial decision to enhance the maritime air and submarine effort, but to reduce below current plans the size of the surface fleet and the scale and sophistication of new shipbuilding. The current numbers of frigates and destroyers was 59, to be reduced to about 50. For advocates of the surface Navy the downgrading of this element of the force structure was almost tantamount to the sinking of the fleet, and Britain's naval heritage with it. Soon, however, the Navy was to be given an unexpected chance to demonstrate the value of a powerful surface fleet.

THE FALKLANDS AND AFTER

In April 1982, Argentine troops invaded the Falkland Islands (Malvinas). The British government promptly despatched a large naval task force, including carriers, frigates, destroyers, troopships and at least one nuclear-propelled submarine; a successful landing was made on the Islands, and in June the Argentine forces surrendered.

It would be inappropriate here to dwell on the details of the conflict,[36] but some aspects are worth recalling. First, and despite the major role played by the Navy, the outcome was very much a product of combined operations between the three Services. Second, British industry and the defence research establishments played an important behind-the-scenes part. For example, the RAE rapidly conducted about 90 trials into ways of decoying Exocet missiles. It also worked on new combinations of aircraft and weapons, such as arming Harriers with Sidewinder missiles supplied by the USA, and it used its wind tunnels to explore ways of erecting tents in high winds. It also accelerated the development and testing of new night-vision and radiolocation navigation systems.

A number of lessons were learned for equipment policy. The versatility and reliability of the Harrier, and of the Rapier antiaircraft guided-missile system, were well demonstrated. Inadequacies in army clothing, especially boots, were exposed. Inadequacies were also revealed in methods for controlling fires on ships. Perhaps the most dramatic lessons, and certainly the most costly in terms of UK Servicemen's lives, concerned the defence of ships against aircraft armed with missiles, or even old-fashioned iron bombs. These lessons bore directly upon the debate, which had been continuing since Mr Nott's White Paper, on the value of surface ships in the conditions of modern warfare.

On the one hand, it was clear that without a substantial surface fleet, including an aircraft carrier, the Falklands Campaign could never have been mounted. The carrier was the platform from which Harriers could attack targets on the islands and defend the troopships against air attack. The support ships in turn defended the carrier against submarine and air attack. It was argued in some quarters that, had the fleet already been cut to the numbers proposed by Mr Nott, the task force could never have succeeded. Certainly, it was operating at the limit of its capacity and benefited considerably from good luck and Argentine incompetence.[37]

On the other hand, it was also clear that single strikes with guided missiles could indeed do what the proponents of precision-guided munitions had claimed. The vulnerability of surface ships was further demonstrated by a number of successful strikes upon them with iron bombs. That many of these bombs failed to explode on contact was partly luck, but was mainly a product of the aircraft being forced to fly low by the effectiveness of British air defences at higher levels. Those air defences were, however, stretched thin, and the general problem of air defence was compounded by the lack of long-range airborne early warning of attack which, had the task force been operating in the North Atlantic, would have been provided by land-based aircraft.

The outcome was a spate of new or accelerated equipment programmes and some revision of the 1981 White Paper proposals. All the ships, aircraft and helicopters lost in the Falklands were to be replaced, usually with the most up-to-date designs. The Navy was also to receive additional Harriers and helicopters over and above those lost. The short-range air defence of ships was to be improved. Better army equipment and an improved air tanker fleet were also to be acquired, and more effort was to be put into electronic warfare equipment. The carrier *Invincible* was to be retained in the fleet, the new plan being that two of the three carriers in this class should be operational at any time.[38]

In addition, the other programmes that had been referred to in the 1981 White Paper were now moving further through their lifecycles, and new programmes were embarking upon theirs. By 1983 new programmes were under way on antitank missiles for the 1990s, on an experimental aircraft programme, and on antiradiation missiles for use against radar transmitters. Plans were proceeding to develop a new generation of short and medium-range air-to-air missiles in cooperation with the USA, West Germany and France. Research was also starting into an advanced airborne antiarmour weapon for the l990s and a long-range standoff air-to-ground missile carrying conventional munitions. For the Navy, design work was proceeding on the Type 23 frigate, and longer-term plans were being laid with seven other countries for a NATO frigate for the 1990s. The Stingray torpedo was about to enter service, and work was proceeding on a new heavyweight torpedo. Better radars were being developed and fitted to enhance the ability of ships to detect low-flying missiles. Better electronic warfare systems were also on their way, and the first of a new set of military communication satellites was expected to be launched in 1985.[39]

By 1985, all these items remained on the agenda and some new ones had been added. The Ministry of Defence, under Michael Heseltine (who replaced John Nott in early 1983) underwent a major reorganisation. The new Secretary of State also reacted positively to new opportunities for international collaboration in defence equipment development and production, seeing here a route to economies. Such economies were in any case needed more than ever as, in a reaction against the dominant role played by nuclear weapons in western defence policy, sources inside and outside government increasingly pressed the case for a greater emphasis to be placed upon conventional weapons.

This new emphasis sprang in part from the political strength of the antinuclear movement in various parts of Europe, which had been revived by the controversies over the deployment of Cruise missiles in Europe, and Trident in Britain, in the late 1970s and early 1980s. It also emerged from orthodox military circles, most notably from the office of NATO's Supreme Allied Commander in Europe, General Bernard Rogers, with his concept of deep strikes against the follow-on forces of the Warsaw Pact simultaneously with the near battle at the front.[40] These concepts became bound up with an initiative launched in 1982 by the US Secretary of Defense Caspar Weinberger. He developed the notion of using the so-called emerging technologies (such as microprocessors, new materials and sensors, advanced conventional munitions and guidance systems, and vastly improved command, control and communications facilities) to improve the effectiveness of NATO forces. For the Europeans, fearful that this 'ET' initiative might result in massive imports of American equipment to the detriment of the European defence industries, and concerned also at the expected cost of developing these technologies, a high premium was put upon a collaborative response to the US.[41] Britain was particularly active in trying to coordinate a European response to this new challenge.

Concern about US initiatives continued when, in March 1983, President Reagan launched the Strategic Defense Initiative (SDI). This raised many questions, political, strategic, economic and technological.[42] Suffice it here to note that the immediate reaction of Margaret Thatcher's government was an uncomfortable mixture of loyalty to the US government, doubts about the technical feasibility and strategic desirability of SDI (and, in particular, the twin worries of its effect upon the US commitment to Europe and the effect of a USSR response upon the value of Britain's Trident missile force), and concern to obtain a share of the money that was to be spent on the technologies that made up the SDI.

The SDI also raised issues of cooperation within Europe. At one level attempts were made (unsuccessfully), to orchestrate a European response to the initiative. At another, France proposed an alternative programme, called Eureka, which grew out of earlier French proposals for European technological cooperation but was accelerated by the SDI.[43] The French argued that SDI, with its emphasis on technologies that were crucial not only to the military but also to the civil sector (such as optoelectronics, computing, advanced communication, software design, new materials and space technology), threatened to drain skilled people to the USA. This would have serious consequences for European technological development. They therefore proposed European cooperation in a programme of high-technology development linked to promising consumer markets. For Britain, this counterproposal raised familiar tensions between transatlantic and cross-channel loyalties and interests.

The overall picture in 1985, then, was one of a holding operation. The guidelines of the Nott review were still being adhered to in formal terms, but defence equipment policy-makers faced grave problems. The post-Falklands equipment replacement programme (admittedly funded largely by a supplement to the defence budget) and rising equipment prices was coupled with the need to interact somehow with the opportunities presented in such initiatives as ET, SDI and Eureka. Controversial decisions, such as that over the future of the Westland helicopter company (with its transatlantic versus European dimension), leading to the resignation of Mr Heseltine in 1986 and the decision in the same year to abandon the British Nimrod AEW system and buy American instead, continued to highlight what by now were familiar dilemmas.[44]

This book takes as its end-point the end of the Cold War in 1990. Suddenly, the enemy on whom the entire military enterprise, procurement and research programme had been focused, disappeared. Within a year, however, the Gulf War served notice that war itself had not disappeared. It also proved a test of British weapons designed in the Cold War. By means of enormous exertion the country had retained the capacity for the (relatively) autonomous development and production of a wide range of weapons systems, including nuclear weapons and all the major weapons platforms except long-range ballistic missiles, heavy bombers and large aircraft carriers. In some fields (notably VTOL, communications equipment, night vision, display systems, short-range antiaircraft missiles, and computer software, to name but some), Britain's capability could be seen to be in the front rank by world standards.

At the same time, it must be acknowledged that British weapons have rarely been tested in the kind of confrontation for which they were designed: nuclear weapons have never been used in anger by the British. Despite a fortune spent on the development of torpedoes, only one single use was made in war in half a century. The criteria for judging success cannot therefore be the same as those for technologies tested in action. In general the capacity of the potential opposition was calculated and the performance of British weapons evaluated against this estimated challenge. Whereas in the Second World War it could be said of the Spitfire that it was a fine weapon that outperformed its opponents in many circumstances, or of the Cromwell tank that it proved inferior to its German opposition, such comparisons have rarely been possible in peacetime. Of course British equipment has been sold in the world market, and indeed the UK has been one of the four major arms exporters since the Second World War. Customers such as the Israelis have had the opportunity to test weapons such as the Centurion tank, but in contexts and against oppositions different from those intended by their designers. In this sense, therefore—and this should not be a cause for complaint—the jury remains out on just how effective the products of the Cold War defence equipment programmes might have been.

OVERVIEW OF R&D EFFORT

In this context, some key features of defence R&D in Britain will be reviewed. The UK has by international standards traditionally been a high spender on military R&D. It was the only major European state to maintain a high and unbroken commitment to military R&D throughout the postwar decade. The Attlee government made R&D a major element of its defence planning, taking a deliberate decision not to re-equip at the end of the war with state-of-the-art technology, but to delay and then leap a generation. Accordingly, it began major programmes in atomic weapons, new aircraft, guided missiles and so on. In contrast, Germany and Italy (and Japan) were limited by treaty (as well as by inclination) from engaging in certain defence activities, and the French recovery in military R&D began only in the 1960s.

Thus in spending terms Britain remained, at least until the late 1980s, the clear leader of the second division, although an order of magnitude behind the superpowers.[45] From the late 1960s until well into the 1980s, half of government expenditure on R&D went on defence. Three concerns would be repeated throughout the postwar half-century: How could the research establishments best focus on priorities, select key affordable programmes that would benefit the military most efficiently, and maximise spinoff to the civilian economy? These would be expressed through frequent reports and ministerial and institutional reorganisations which reflected both the fluctuations of policy and the consequences of an unprecedented technological change. Thus in May 1958, under pressure from the Treasury, the Committee on Management and Control of Research and Development was set up.[46] Its progress was slow but it finally reported in 1961, under the chairmanship of Sir Solly Zuckerman, with the urgent recommendation that there should be a seamless web between the definition of an operational requirement and monitoring of a final manufacturing process. By contrast, under the succeeding Labour government, which came into power in 1964, the priority was made the successful use of research expertise outside the military. The establishments of the Ministry of Supply and the Ministry of Aviation would be incorporated within the new Ministry of Technology, whose remit was increasingly to be a great national research network. It would perhaps emulate the success of the USA's ARPA, which would later father the Internet.[47] However, the Ministry of Technology was disbanded under the Conservative administration of the early 1970s.

Today, the government research establishments are all under the aegis of the unified Ministry of Defence. In the past, however, they were scattered across a number of its precursor ministries, notably the Ministry of Supply, the Admiralty, the Ministry of Aviation and, later, the Ministry of Technology. Thus the current defence research establishments are the products of a lengthy process of rationalisation. From the 30 or so of the early postwar years, the numbers of both institutions and their staffs have been steadily cut. This has been partly because of restructuring, but also because of a gradual transition from the early postwar position, in which nearly all military R&D and design work was done in the establishments, to one in which two-thirds is done in industry, including nearly all the final, full-scale design and development work on non-nuclear systems. The last field to be so transferred—torpedo design—moved into industry in the late 1970s.

The roles of the establishments in the 1980s and 1990s, (nuclear work again excepted) have been restricted to those that fall most naturally to government, and without which government would find it hard to act as an informed customer of the defence industries and to stay abreast of international trends. Thus, the establishments assist with the setting of specifications, perform longer-range research, run certain expensive central facilities, support industrial contractors, engage in evalu-

TABLE 1.2: THE DEFENCE RESEARCH ESTABLISHMENTS, 1989[48]

Name	Locations	Staff	Activities
Royal Aerospace Establishment (RAE)	Farnborough, Hants; & Bedford	5200	Research covering all aerospace activities, including airframes, engines, weapons & systems; facilities included airfields, ranges and wind tunnels
Royal Armament Research & Development Establishment (RARDE)	Fort Halstead, Kent; & Chertsey, Surrey	2262	Land systems research, including guns and ammunition, combat and logistics vehicles, engineering and bridging equipment
Admiralty Research Establishment (ARE)	Portsdown, Hants; Portland, Dorset; & Teddington, Greater London	2931	Sea systems research, including marine technology, weapons and sensors for above and below surface warfare
Chemical Defence Establishment (CDE)	Porton Down, Wilts	560	Research into defence against chemical and biological attack, including physical protection and medical treatment
Royal Signals and Radar Establishment (RSRE)	Malvern, Worcs	1586	Research into fundamental and applied electronics, including displays, sensors and computing
Aeroplane & Armament Experimental Establishment (A&AEE)	Boscombe Down, Hants	1101	Aeroplane and armament testing, particularly trials for aircraft clearance, equipment evaluation and aerial delivery
Atomic Weapons Establishment (AWE)	Aldermaston and Burghfield, Berks	Not known	Atomic weapons research and production

ation and acceptance work, and generally act as a reservoir of advice for government. As recently as 1989 they were organised into seven main units (Table 1.2) distributed across about 20 sites, many of them containing special facilities. Staff numbers were cut back from 34,000 staff in 1971 to 23,000 in 1985!

Going beyond the time-frame of this book, in 1991 the main non-nuclear defence research establishments were moved into a more contractual relationship with the Ministry of Defence in the form of the Defence Research Agency (DRA), which was put under considerable pressure to generate a specified annual return on the capital invested in it, and to run, in effect, like a business, while still being owned by the ministry and with its staff still regarded as public officials. At the same time, this new status has freed it from various aspects of Treasury control as applied to conventional civil service organisations. This has made it more possible for the agency to consider ways of making money from non-defence customers, which in the past might have been constrained by such factors as a prohibition on taking on the necessary extra staff. Moreover, the pressure to make the required return on investment added further incentive to seek a wider customer base for its scientific and technological services.[49]

In 1995 DRA underwent a 60 per cent expansion, with the addition of test and operational analysis facilities, plus the Chemical and Biological Defence Establishment, and was renamed the Defence Evaluation and Research Agency (DERA). Hence, by mid-1996, and the publication of what by then was its annual report for 1995–96, DERA was the largest science and technology organisation in Europe, with a turnover of more than £1 billion. Staff (excluding contract staff), numbered 11,700, compared with 8600 in 1995, though it was projected to fall to about 10,000 by 1997–98. Its revenues from government customers outside the MOD had increased by 5 per cent compared with the previous year, and from industry by 72 per cent. As this book is being completed in 1998 there are plans to establish a Defence Diversification Agency associated with DERA. The enduring issues of coping with strategic change, and maximising civilian benefit, continue.

The organisational history of the British defence research establishments is, then, in a certain sense, a history of concentration into smaller, more multifunctional units. What initially were single-Service-oriented establishments gradually began to serve wider constituencies, though still, in most cases, with a strong single-Service emphasis. This process was not much accelerated by the formation of the unitary Ministry of Defence in 1964, but began to gather pace with the formation of the Procurement Executive (PE) in 1970. All the establishments except those concerned with atomic energy were moved under central control within the PE, and AWRE itself followed within a few years. The formation of DERA has taken the process to its limit, AWRE apart. Table 1.2 illustrates this history well. To give a slightly fuller sense of some of the other developments referred to above, let us look more closely at two of the establishments.

The Royal Signals and Radar Establishment (today the electronics division of DERA) is located principally around Malvern, near Worcester. Historically, it and its precursors have been the main centre for work on electronic systems and devices and their application in electronic equipment for all the Services. Because full exploitation of new hardware techniques requires parallel advances in computer software, RSRE has also acted as a focus for defence work on computer languages and support software. This 'technology base' work has fed into work on complete systems, not only in RSRE but in most other defence research establishments and in industry. Indeed, in the past much of it was supported by the Department of Trade and Industry, and before that the Ministry of Technology. The main systems activities at RSRE have been airborne, battlefield and ground radar, optoelectronics, guided weapons, ground-based communications and air traffic control, the last of these being funded by the Civil Aviation Authority.

The history of RSRE illustrates well the process of fusion of different elements over time that is so characteristic of the evolution of the establishments. RSRE's history can be traced to 1914, when the Army set up an Experimental Wireless Telegraphy Section of the Royal Engineers. In 1916 it moved to Woolwich as the Signals Experimental Establishment. The Second World War brought a major expansion and another transfer, this time to Christchurch, Hampshire, where work continued under the title of the Signals Research and Development Establishment (SRDE), and where important advances were made in mine detection techniques and in radiocommunications.

A second strand emerged during the rearmament period, when Robert (late, Sir Robert) Watson Watt suggested that radio might be capable of detecting aircraft in flight. After preliminary investigations the Air Ministry founded a research establishment, first at Orford Ness and later at Bawdsey Manor in Suffolk, charged with developing and installing a chain of early-warning radar stations around the coast of Britain. The War Office soon established a parallel organisation at Bawdsey to develop techniques for controlling searchlights and antiaircraft guns. Later, the Air Ministry group moved to Swanage, where it formed the Telecommunications Research Establishment (TRE). The War Office team moved to Christchurch to join an Air Defence Research and Development Establishment (ADRDE), which later became the Radar Research and Development Establishment (RRDE). In 1942, both TRE and ADRDE moved to Malvern, although retaining their separate identities until 1953, when they were amalgamated to form the Radar Research Establishment, becoming the Royal Radar Establishment (RRE) in 1957 (Fig. 1.2). Chapter 8 explores this amalgamation in more detail.

A third strand emerged in 1945. The Services Electronics Research Laboratory (SERL), a descendant of the Admiralty Signals School set up in the nineteenth

FIGURE 1.2: ROYAL RADAR ESTABLISHMENT (RRE), MALVERN, FROM THE AIR, 1972 (© CROWN COPYRIGHT, DERA)

century, was founded at Baldock to develop the production of electronic devices. It was originally staffed by teams of scientists and engineers who had worked during the war at various Admiralty establishments. Most of its early work concerned microwave valves for radar and guided-weapons systems, special valves for shell fuses and infrared detection (often in loose collaboration with TRE and RRE), dealt with in detail in Chapter 7. The programme expanded to include semiconductor devices, lasers and night-vision devices, explored in Chapters 13, 10 and 7 respectively. Although primarily a defence establishment, SERL also worked on civil projects, such as cancer therapy, satellite communications, metal cutting and welding.

The three establishments—SRDE, RRE and SERL—were amalgamated early in 1976 as the Royal Signals and Radar Establishment, in Malvern. The resulting establishment was not only large, with several thousand staff, but also covered almost every area of electronic component and systems technology of military importance. RSRE has long enjoyed a reputation as an outstanding research centre, in recognition of which it has since 1979 received numerous Queen's Awards for Technological Achievement. One of these was for the well-known work, carried out with Hull University and BDH Chemicals Ltd, on light-modulating liquid crystal displays, which became a worldwide commercial success. Chapter 13 looks at this as a case of 'spinoff'. Another was for the 'Malvern puller', a technique for growing large single crystals of electronically important materials.

RSRE remains today a world centre of excellence in microwave and millimetric systems, display devices, signal processing, infrared detectors, displays, electronic materials, laser devices, computing and associated systems, such as all forms of radar and optoelectronics, communications and electronic warfare. It also plays a major role both nationally and internationally in the field of standards, measuring and calibration, especially with respect to microwave standards. It has for many years been one of the more open of the defence research establishments, partly owing to moves made during the days of the Ministry of Technology in the 1960s, and subsequently under the Department of Trade and Industry (until the 1990s funding cuts) to develop its role as a *de facto* national electronics laboratory; and partly to the relatively long-range nature of much of the research.

Although Malvern has done a great deal of systems work, it has been known mainly for its work in developing enabling technologies that can be applied to underpin systems development. In contrast, the Royal Aircraft (later Aerospace) Establishment, Farnborough, has been better known for its work on systems development, although it also has done a good deal to enlarge the technology base.

At its peak, RAE may well have been the largest research establishment in Europe. It has been the principal air systems establishment of the Ministry of Defence;[50] it has had overall responsibility for the conduct and coordination of R&D into all aerospace activities, with the exception of radar, and has provided support for civil aviation. Many developments in aviation and rocketry (such as the English Electric Lightning, TSR-2, Blue Streak, Tornado, Concorde, Harrier, lithium–aluminium alloys, and carbon-fibre materials) are associated with the establishment, and its activities warrant an entire volume. Here Chapter 2 deals with its most influential years before the 1960s, Chapter 13 discusses the development of carbon fibre, and Chapter 4 explores RAE's work on missiles.

RAE began to grow on the Farnborough site in 1905, first as part of the War Office, and later of the Air Ministry, the Ministry of Aircraft Production, the Ministry of Supply, the Ministry of Aviation and the Ministry of Technology, before in 1970 becoming part of the Procurement Executive. From the outset, RAE's work included both advanced research and practical development and testing in-flight and ground facilities. The design and construction of prototype aircraft was important in the early years, although this ended in 1916 with the transfer of airframe work to industry. Similar transfers of design and construction responsibility to industry occurred at later dates for successive categories of aircraft equipment and missiles, with RAE continuing to provide component development and testing, and taking up design and construction work in the newer technologies, such as satellites.

As with RSRE, RAE has undergone an elaborate process of fission and fusion between establishments. By 1962, the organisation was both very large and highly complex, with 8500 staff, including 1500 qualified scientists and engineers employed in eight research departments—as many as in a fair-sized university. A wide range of engineering, physical science and mathematical disciplines were represented, organised partly for disciplinary support and partly to enable the assembly of multidisciplinary teams for specific tasks—a common problem in R&D management. It had several wind tunnels; engineering and carpentry workshops with over 1300 employees; a central administrative, security, stores and transport staff of 1200; and airfields with maintenance and air traffic control staff at Farnborough, Bedford and Llanbedr, as well as the missile test ranges at Aberporth, West Freugh, Salisbury Plain (Larkhill) and Woomera in Australia. To call all this a 'research establishment' is to understate its complexity.

Although RAE was subsequently slimmed down, in 1983 it also assumed new responsibilities when the National Gas Turbine Establishment at Pyestock was amalgamated with it. NGTE had been the main government centre for R&D on gas turbine engines and their related systems. Its main emphasis had been on military and civil aircraft engines, but it was also actively concerned with engines for warship propulsion. As with the rest of RAE, therefore, this work required a combination of research, design, development and especially testing of components and entire engines in large and elaborate facilities, which have been used not only by MoD but also, on a repayment basis, by industry.

The RAE in the postwar period was very much in advance of the aircraft industry in aerodynamics and in structural analysis. The power of the jet engine was making higher aircraft speeds possible and aircraft were entering a new aerodynamic regime. This made the RAE contribution enormously important to the development of the postwar generation of British aircraft. It should be noted that in the early period the RAE had research facilities, such as high-speed wind tunnels, that did not exist elsewhere, and the institution can perhaps be regarded as a device for 'technologising' the British aircraft industry.

Chapter 2 shows how the RAE's role varied in different projects. In the case of the supersonic fighter the RAE led, adopting a role almost like that of the advanced projects office in an aircraft company. It analysed the strategic importance (in the emerging atomic age) of a supersonic interceptor with a very high rate of climb, and adopted a propagandist posture which helped to sell the concept to policy makers. It also produced a preliminary design which showed a possible solution in terms of aerodynamic layout and engine installation. This was then adopted and developed by the chosen contractor firm, English Electric, and developed into a practical aeroplane, the P.1 Lightning.

In the case of the V-force bombers RAE performed a role that was more conventionally like that of a research establishment. In this case, the requirement for the aircraft came clearly from the Air Staff, who defined the need for a bomber able to carry a 10,000 lb bomb load over a range of approximately 3500 nautical miles at 500 knots and an altitude of 50,000 feet. Widely different solutions were developed by the aircraft companies, and RAE was involved in the appraisal of these. In this case it was more cautious than with the supersonic fighter, and it was partly RAE advice about the uncertainties of design for high subsonic speeds that led to the expensive decision to procure three separate types: the Vickers Valiant, the Avro Vulcan and the Handley Page Victor.

Concorde is again an example of a high level of RAE innovation and intervention. All the preliminary aerodynamic work on a supersonic civil transport, to solve both the economics of high-speed cruising, and the slow speed flight for landing and takeoff, was done at the RAE. These results were then exposed to the firms in conferences and meetings to persuade them of the practicality of the concept. The RAE also worked at a political level, through the agency of the Supersonic Transport Aircraft Committee, which had been established largely through the efforts of Morien Morgan, Deputy Director of the RAE.

ISSUES

This book cannot be a comprehensive survey of all the activities of thousands of scientists during the period of half a century: rather, the authors have examined major activities at key research sites. Their work will, we are sure, provide both food for thought and stimulus to those who would do better. Two themes emerge: relations between science, technology and the military environment, in a historical context; and issues in applied science and technology.

A constant theme of the book is the way in which needs identified by the military fed through into the cultural environment of the establishments. This can be seen at both macro and micro levels. Until the late 1950s, the risk of war was seen by many as high. Sandys was actually physically assaulted by the Chief of the Imperial General Staff when he suggested his reformation of the Armed Services and the end of conscription.[51] That war was widely envisaged as a continuation of the Second World War. This book illustrates a point made by Zuckerman, in his autobiography, that many of the projects started at the end of the war were still in being in 1960: this shows how much the work of both British and German scientists during the war was being drawn upon more than a decade later. Aircraft interceptors against incoming nuclear bombers were designed to be faster than their wartime ancestors, flying supersonically with jet engines, but still fulfilling a Spitfire-like role. Sonar was developed for the Navy to facilitate Second-World-War-like convoys.

From the early 1960s, with the growth of computing, technology changed and concepts of war evolved further. On the one hand, competition between the Services came increasingly under MoD control, and research into certain weapons, such as the helicopter, was facilitated (see Chapter 3). On the other hand there were the many cancellations, such as the TSR-2 strike aircraft, which have already been referred to. The knock-on effects are explored throughout this book. They were, however, not always obvious. The radar for TSR-2 came to be used in the Tornado, and its engines were those employed for Concorde. Increasingly the challenge moved to the development of high-quality radars. Agar and Hughes (Chapter 8) point out that TSR-2 was to have been the UK's first computerised aeroplane. Putley (Chapter 7) shows how the emergence of large-scale integrated circuits transformed the problem of seeing in the dark. In the naval context, Wright (Chapter 6) shows how the emergence of nuclear submarines and the new emphasis on the submarine-based missiles for the UK's defence transformed the problems of submarine finding. The type 2001 sonar was the solution (this was the weapon the Kruger spy ring was famously guilty of giving away to the Soviet Union).

Whereas the growth of the Soviet Navy was a general concern, the particular stimulus for the investment in the 2001 was a US development, the Nautilus nuclear submarine. This illustrates a remarkable feature of the postwar weapons development process explored in this book. Time and time again we see the leapfrogging of offence and defence, but often all on the side of the West. New tank armour and new forms of explosive (see Chapter 5), submarines and sonar (Chapter 6), radar and jamming, chemical warfare and protection therefrom, biological weapons and antibiotics, the race between the development of aircraft described by Nahum (Chapter 2) and the missiles development described by Twigge (Chapter 4) all show that very often the countermeasure that stimulated further research was developed by the West itself. The carcinotron jamming device which rendered obsolete earlier radar was a French development. The nuclear submarines that required the new sonars were American. In the absence of reliable intelligence, the assumption was made that the USSR would either already have or would soon develop the like, and therefore pre-emptive steps should be taken. We are in no position to comment on the validity of that analysis, but simply note that the high level of investment in western military research and development undoubtedly stimulated further research to nullify its consequences.

After 1960 collaboration in weapons R&D also grew in importance, as economic pressures mounted. This was not always a route to stability, as Ogorkiewicz shows in the development of tanks (Chapter 5) However, as Wright (Chapter 6) explores in the treatment of the angled-deck aircraft carrier, it could catalyse the acceptance in the US of a British development. Collaboration, principally with the USA but increasingly with European states also, is a feature of several other chapters, as in

Ernsting's discussion of joint studies of diving physiology with the US and French Navies (Chapter 12), and Uttley's discussion of helicopter development (Chapter 3).

The second theme in this book is the nature of applied science and technology. The focus of research in the military field could, of necessity, rarely be on single devices alone: rather, it was upon entire systems, hence the importance of having access within the establishments to a wide range of cognate technologies and—critically—the capacity to integrate them. This did not just mean complexes of receivers and components as in radar, but the capacity to integrate all the elements required for a particular military task. To take three widely differing examples: the growth of automation and the performance of land, sea and air vehicles ironically placed more, not less stress, on the human operator: hence research could not just be into weapons, but must also be into the physical limits and clothing of the warriors. The development of nerve gas required not just the exploration of chemistry, but an understanding of delivery and protection as well. The most extreme claim to be made in this respect here is in relation to the development of naval warfare. Grove (Chapter 9) shows how British skill in the computer integration of relatively modest components and subsystems resulted in a naval air defence system that, in 1958, amazed the USA by enabling a British aircraft carrier to inflict a crushing defeat on an apparently better equipped, heavily coordinated attack by US Navy aircraft. As he argues, if there really has been an information-led revolution in military affairs, together with movement towards the much-vaunted 'system of systems', then its foundations were laid half a century ago in the UK by the Admiralty research establishments.

As technologies developed away from wartime practice the scientific phenomena exploited changed too. Solid-state physics had been in its infancy during the war; thereafter, it became a major focus for an enormous range of weapons-related developments, such as computing, infrared detection, and lasers. This affected the organisation of establishments set up in previous eras, as Agar and Hughes (Chapter 8) show in their study of electronic-related organisations. The development of new synthetic and composite materials affected technologies from aircraft wings to tank armour. Above all, computing changed the nature of weapons development. Computer languages suitable for the specialised needs of the military became a major challenge. Wright (Chapter 6) points out how between the 1960s and 1990s the language of naval research changed from 'asdic' and 'torpedo' to more general categories such as 'sensor', 'integration' and 'platform'.

Of course the weapons developed in such a rapidly evolving context were rarely tested in action, so what could be the criteria for success? Effectiveness at meeting Service requirements could be self-serving. A widely used, if informal, measure was the takeup of UK developments by the USA. The angled-flight-deck aircraft carrier was such a success that the US used it even before the UK. Protection against chemical warfare could be counted as a success because it was used by allies. From this perspective the role of UK military R&D, which in any case was but a small fraction of its US counterpart, was that of a consultant engineer. In some cases, such as helicopters, the research establishments' main role was precisely to monitor US developments. In other cases, as in low-light detection, developments shadowed US advances.

Here was, however, a rather unusual consulting engineering business, raising complex problems of management, as Robinson discusses in Chapter 14. In peacetime especially, the requirements of public accountability and the need to meet budget cuts put a premium on increasingly tighter specifications and controls. Yet this has also been an area where the direction of work has frequently been driven by the contractor rather than the customer. Even Zuckerman, the arch policy maker, wondered at the end of his life whether his procedures for seeking to drive R&D from the requirement end really did much good. He concluded that efficient financial management of military R&D mattered far less than choosing the right projects on which to work and the right people to do the job.[52] But here the circle closes on the military R&D dilemma, for only with use in war (and maybe not even then) can the correctness of earlier decisions, made under conditions of peacetime government, be finally judged.

As for the other available test—contribution to the wider science, technology and innovation base—this is of course secondary to the defence objective. Nevertheless, the management of military research has indeed been conducted to some extent with regard to commercial exploitation—more so at some times than others. Spinardi and others record the technological achievements that did lead to civilian exploitation, but this frequently occurred outside the UK: it was Japanese companies that reaped the benefits of carbon-fibre and liquid-crystal developments. Moreover, unlike the defence customers, commercial corporations were inevitably offered not systems, but rather isolated technologies which, in many cases, British companies were not well-placed to use, nor need they necessarily have been. The function of the establishments within 'UK plc' has therefore been ambivalent: pressed to contribute, yet in the nature of the innovation process not best placed to do so.

CONCLUSION

This history of military science and technology can be seen as recounting campaigns of the Cold War, much as actual military engagements constitute the campaigns of hot war. In broad outline it follows the requirements of the evolving geopolitical context, but not in a simple way, as often the work of the establishments themselves altered the strategic possibilities.

At one level we see an attempt to shape and steer an element of a defence system that was itself undergoing profound change during the period under review. From the end of the Second World War to the end of the Cold War, Britain's place in the world, its relations with the USA and Europe, and—at least as crucially—its perceptions of those changes (which often lagged behind the reality), altered enormously. Uncertainty about—and change in—foreign and security policy fed through in complex ways into requirements for defence science and technology, but echoes of the wider geopolitical scene are heard at every corner.

At another level we see an attempt to manage (in the broadest sense) an exceedingly large applied science and technology activity. The difficulties of the customer in specifying requirements, and keeping them stable for long, coupled with periodically changing customer organisation, created an unusual managerial challenge. The absence of a clear 'bottom line' in government, analogous to that with which a major industrial corporation would have judged its success, and its corporate research laboratory, contributed more managerial complexity. Contributing in some way to 'UK plc', but not to the detriment of the primary role, was necessarily a secondary role which made the managerial and political decisions even more complex. Frequently, to national surprise and dismay, the outcome of Britain's investment in military R&D was that technological leadership did not lead to civilian industrial spinoff.

Of course the experience of the establishments in negotiating the nature of applied science in an era of rapid change and multiple objectives would not be unique. Indeed, it was destined to become typical of the experience of industrial and even academic establishments. Today's circumstances, both geopolitical and organisational, are radically different from those of 1945. DERA today is a different kind of organisation, operating in a different world, from that of the early postwar establishments, but the essential dilemmas remain: how to maintain technological capabilities against uncertain and shifting requirements; how to organise projects so as to achieve success; how to recruit and retain the necessary staff; how to cope within the budget set; and how to contribute effectively to the wider science, technology and innovation base. These are questions that have been endemic throughout the period covered in this book, and are, it seems, inherent to the activity itself.

1. Paul Kennedy, *The Rise and Fall of the Great Powers: economic change and military conflict from 1500 to 2000* (Unwin Hyman, 1988).
2. C.W. Sherwin and R.S. Isenson, *First Interim Report on Project Hindsight: summary*, Report AD-642–200 (Office of the Director of Defense Research and Engineering, Washington D.C., 1966).
3. US National Academy of Sciences, *Applied Science and Technological Progress. A Report to the Committee on Science and Astronautics, US House of Representatives* (Washington DC, 1967): 22.
4. David Hounshell and John Kenly Smith, *Science and Corporate Strategy: Du Pont R&D, 1902–1980* (Cambridge University Press, 1988).
5. On 'Engineering Science' see Walter Vincenti, *What Engineers Know and How They Know It. Analytical Studies from Aeronautical History* (Johns Hopkins University Press, 1990); David Francis Channell, *The History of Engineering Science: an annotated bibliography* (Garland, 1989). On the development of the concept of applied science see Robert Bud and Gerrylynn K. Roberts 'Thinking about science and practice in British education: the Victorian roots of a modern dichotomy', in *Industry and Higher Education*, edited by P.W.G. Wright (Open University Press, 1990): 18–30.
6. See, for instance, the Cold War Science and Technology Studies project at Carnegie Mellon University and work carried out by Dominique Pestre in France. Current British work includes David Edgerton, *England and the Aeroplane: an essay on a militant and technological nation* (Macmillan, 1991).
7. For the German style of post-war technology, see Joachim Radkau, '"Wirtschaftswunder" ohne technologische Innovation? Technische Modernität in den 50er Jahren', in *Modernisierung im Wiederaufbau. Die westdeutsche Gesellschaft der 50er Jahre*, edited by Axel Schildt and Arnold Sywottek (Dietz, 1993): 129–54.
8. Jon Agar and Brian Balmer, 'British scientists and the Cold War: the Defence Research Policy Committee and information networks, 1947–1963', *Historical Studies in the Physical and Biological Sciences*, 28, 2 (1998): 209–52. See also Daniel Kevles, 'Cold-War and hot physics—science, security, and the American state, 1945–56', *Historical Studies in the Physical and Biological Sciences*, 20, pt 2 (1990): 239–64.
9. In 1961 Du Pont Research and Development Department employed 3787 professional staff and had a budget of $160.8 million or about £50 million. See David Hounshell and John Kenly Smith, *Science and Corporate Strategy* (n. 4 above): 328.

10. Great Britain Minister for Science, Committee on the Management and Control of Research and Development (Chairman Solly Zuckerman), *Report of the Committee on the Management and Control of Research and Development* (HMSO, 1961): 51.
11. *Report of the Committee on the Management and Control of Research and Development* (n. 10 above): table II, p. 20.
12. William H. McNeill, *The Pursuit of Power: technology, armed force and society since A.D.1000* (Basil Blackwell, 1983).
13. William H. McNeill, *The Pursuit of Power* (n. 12 above): 331.
14. Philip Gummett, *Scientists in Whitehall*, chap. 2 (Manchester University Press, 1980).
15. M.M. Postan, D. Hay and J.D. Scott, *Design and Development of Weapons. Studies in Government and Industrial Organisation*, History of the Second World War, United Kingdom Civil Series, chap. 16 (HMSO & Longman, 1964).
16. M.M. Postan, D. Hay and J.D. Scott, *Design and Development of Weapons*, chap. 16 (n. 15 above); David Edgerton, *England and the Aeroplane*, chap. 5 (n. 6 above).
17. J.D. Scott, *Vickers, A History* (Weidenfeld and Nicolson, 1962): 310–12.
18. M.M. Postan, D. Hay and J.D. Scott, *Design and Development of Weapons* (n. 15 above): 474.
19. M.M. Postan, D. Hay and J.D. Scott, *Design and Development of Weapons* (n. 15 above): 478–83.
20. House of Commons, Select Committee on Estimates, *Third Report, Session 1946–47, Expenditure on Research and Development*, HC 132 (HMSO, 1947).
21. *The Times* (13 November 1953).
22. Cabinet Defence Committee, *Expansion of Capacity in Industry for Electronic Research and Development*, Memorandum by the Minister of Supply, PRO CAB 131/7, DO (49) 28 (25 March 1949).
23. O.H. Wansborough Jones to D. Neville-Jones, 12 September 1951, enclosure, PRO CAB 124/1/422.
24. Christopher Bartlett, *The Long Retreat: a short history of British defence policy, 1945–70* (Macmillan, 1972): *passim*.
25. Margaret Gowing, with the assistance of Lorna Arnold, *Independence and Deterrence: Britain and atomic energy, 1945–1952*, Vol. 2: Policy Execution (Macmillan, 1974): 36.
26. John Simpson, *The Independent Nuclear State: the United States, Britain and the military atom* (Macmillan, 1983).
27. Colin Gordon, 'Duncan Sandys and the independent nuclear deterrent', in *Politicians and Defence: studies in the formulation of British defence policy*, edited by I.F.W. Beckett and J. Gooch (Manchester University Press, 1981): 138.
28. Colin Gordon, 'Duncan Sandys' (n. 27 above): 138–9.
29. Christopher Bartlett, *The Long Retreat* (n. 24 above): 170.
30. Lawrence Freedman, *Britain and Nuclear Weapons* (Macmillan, 1980).
31. Christopher Bartlett, *The Long Retreat* (n. 24 above): 172–4.
32. Peter Nailor, 'Denis Healey and rational decision making in defence' in *Politicians and Defence* (n. 27 above): 169, 170–1.
33. Christopher Bartlett, *The Long Retreat* (n. 24 above): 224–5.
34. Lawrence Freedman, *Britain and Nuclear Weapons* (Macmillan, 1980); Mary Kaldor, *The Baroque Arsenal* (Deutsch, 1982); Dan Smith, *The Defence of the Realm in the 1980s* (Croom Helm, 1980).
35. Ministry of Defence, *The United Kingdom Defence Programme: the way forward* Cmnd 8288, par. 5 (HMSO, 1981).
36. Lawrence Freedman and Virginia Gamba-Stonehouse, *Signals of War* (Faber, 1990).
37. International Institute for Strategic Studies (IISS), *Strategic Survey 1982–83* (IISS, 1983): 121–2.
38. Ministry of Defence, *Statement on the Defence Estimates 1983*, Cmnd 8951, par. 308–12 (HMSO, 1983): 308–12.
39. Ministry of Defence, *Statement on the Defence Estimates 1983* (n. 39 above), par. 313–34.
40. Philip Gummett, 'New conventional technology and alternative defence: help or hindrance?', in *Alternative Defence Policy*, edited by G. Burt (Croom Helm, 1988).
41. D. Holloway, 'The Strategic Defense Initiative and the Soviet Union', *Daedalus*, 114 (1985): 257–8
42. Judith Reppy and Philip Gummett, 'Economic and technological issues in the NATO alliance', in *Evolving European Defense Policies*, edited by C.M. Kelleher and G.A. Maddox (Lexington Books, 1987): 17–38.
43. Albert C. Pierce, 'The Strategic Defense Initiative: European perspectives', in *Evolving European Defense Policies*, edited by C. M. Kelleher and G. A. Maddox (Lexington Books, 1987): 145–65.
44. Lawrence Freedman, 'The case of Westland and the bias to Europe', *International Affairs*, 63 (1987): 1–19.
45. Council for Science and Society, *UK Military R&D* (Oxford University Press, 1986).
46. [Solly Zuckerman] *Report of the Committee on the Management and Control of Research and Development*; Zuckerman described his experience of the committee in his autobiography, *Monkeys, Men and Missiles: an autobiography 1946–88* (Collins, 1988).
47. See Richard Coopey, 'Industrial policy in the white heat of the scientific revolution', in *The Wilson Governments 1964–1970*, edited by R. Coopey, S. Fielding and N. Tiratsoo (Pinter, 1993): 102–22.

48. International Security Information Service, *Defence Research Establishments* (London: ISIS Briefing no. 12, July 1990).
49. House of Lords, Select Committee on Science and Technology, Defence Research Agency, Session 1993–94, HL 24 (HMSO, 1994).
50. M.J. Lighthill (subsequently Sir James Lighthill), 'The Royal Aircraft Establishment', in Sir John Cockroft, *The Organisation of Research Establishments* (Cambridge University Press, 1965): 28–54.
51. Martin S. Navias, '"Vested Interests and Vanished Dreams": Duncan Sandys, the Chiefs of Staff and the 1957 White Paper', in *Government and the Armed Forces in Britain 1856–1900*, edited by Paul Smith (Hambledon Press, 1996): 217–34 (see 223).
52. Solly Zuckerman, *Monkeys, Men and Missiles* (n. 46 above): 164.

CHAPTER 2

The Royal Aircraft Establishment From 1945 To Concorde

ANDREW NAHUM

INTRODUCTION

Britain ended the Second World War with aircraft manufacturing forming the largest single industrial sector and with its designers enjoying enormous prestige. However, in spite of the great achievements of the industry, it was becoming apparent to perceptive observers that the methods that had produced outstanding British aircraft such as the Spitfire and Lancaster might not be adequate for the increasingly complex types that were emerging. There was, in fact, among those with a thoughtful overview, an apprehension that the industry was not as advanced as it should be, and that 'so many of our wonderful aircraft and their engines owe their origins to a few men of genius'.[1] The procurement of aircraft for British forces in the USA had also taken British industrialists, civil servants and Service officers into US firms, often surprising them with the high overall technical and scientific level of their employees and the very high number of university-educated aeronautical engineers, compared with 'the large number of apprentice-trained artisans in England given limited theoretical training'. In contrast to the US pattern, one critic complained that British aircraft design was 'of necessity ... an art, rather than an applied science'.[2]

The state of the industry in 1945 has been touched on because it is crucial to an understanding of the extraordinary influence of the Royal Aircraft Establishment (RAE), for it can be regarded as an institution that was, in part, devoted to 'technologising' the British aircraft industry and, in the postwar years, there is no doubt that the advanced projects were critically reliant on the RAE. In fact, it would have been impossible for the firms to have completed the design, testing and evaluation of the new postwar generation of high-speed jet aircraft without the RAE, as they lacked the research equipment and wind tunnels required. In 1951, for example, the English Electric company was one of the first firms to build a transsonic wind tunnel as the design for its P.1 supersonic fighter started to advance. The firms also lacked specialist staff: 'they always had their chief aerodynamicist', one RAE scientist recalled, 'but to be honest, he was the one aerodynamicist'.[3] Another RAE member was more severe, recalling in the mid-1950s 'some twenty design departments, all more or less inadequate'.[4]

From the time of the wartime coalition government it had been accepted that these deficiencies would have to be addressed by government action, largely through the research establishments, and the early postwar years saw what was almost an inversion of the capitalist ethos, with government scientists and civil servants often adopting a more 'progressive' and explicitly technocratic posture than the firms were prepared to do, recalling Lord Beaverbrook's quip as the war ended that the aircraft industry 'was a hotbed of cold feet'.[5] However, it was also a period in which the wartime consensus still prevailed, and in which collaboration between the Services, the firms and the government research establishments in pursuit of a common national purpose was the norm.

The example of German aeronautical research, as discovered when the Allies began to enter Germany, also proved influential. Some British analysts initially tended to regard the German work as almost too advanced and divorced from the needs of industry and the Services to have been of proper use.[6] However, the scale and sophistication of the facilities soon made a deep impression on British visitors, tending to reinforce the effects of exposure to US practice and helping to underpin the arguments then being made for an even greater role for science in future defence work. Thus W.S. Farren, as Director of the RAE, visited the vast secret camouflaged Luftfahrt Forschungs Anstalt at Völkenrode, near Brunswick, shortly after it was discovered by Allied forces, and commented ruefully that it had 'a magnificence ... that beggars the imagination of anyone who has seen similar institutions in the UK'.[7] Exposure to German facilities and work acted as a stimulus, and in the postwar period the RAE became an extraordinarily powerful establishment which could do both long-range work at the highest theoretical level and also cover practical applied aircraft problems of weapons and equipment installation, flying behaviour (handling) and so on.[8]

By 1962 the RAE was employing 8500 people (of whom 1500 were 'qualified scientists and engineers') at Farnborough and its other sites, making it probably the largest research establishment in Europe. The organisation covered the complete spectrum of aeronautical technology, including guided weapons research, armaments, radio, instrumentation and navigation, as well as maintaining its own chemistry, physics and materials section, and all this expertise contributed to the emerging postwar generation of British jet aircraft. It was, wrote M.J. Lighthill, the then Director of the RAE, 'a central reservoir of scientific and technical knowledge ... on which all British producers and users of aircraft are able to call'.[9]

The account here cannot cover all the areas in which the RAE was active and therefore deals mainly with the aerodynamic and 'project work' that contributed to specific aircraft or aircraft proposals. However, the main focus of this study is the role of the RAE in the two decades from the end of the Second World War, during which, it is argued, its influence on national defence was at its peak. Certainly by 1970 the firms had become more independent and were forced into competing groups. They were, moreover, required to pay for government facilities and the establishments were themselves being encouraged to become more accountable, and 'the freedom was going out of research'.[10] The industrial situation also changed radically during the postwar years, and during the 1960s the firms were becoming transformed, bringing in more analytical talent and investing in research facilities.

During the Second World War the RAE had been largely occupied with the immediate problems of aircraft entering or already in service. This necessity for Farnborough to deal with pressing ad hoc problems was partly responsible for an ambitious plan for a new, independent research establishment, with eminent aerodynamicists such as Sir Brian Melvill Jones, Chairman of the Aeronautical Research Committee, arguing that 'aviation was entering a new era in which revolutionary changes are inevitable', and that there was a need for a great increase in wind tunnels and experimental equipment.[11] Sir Stafford Cripps, Minister of Aircraft Production, independently held a visionary and strategic view of aviation, urging Cabinet colleagues that it was to 'industries such as this, and no longer to coal and cotton' that Britain must look in the postwar period for the essential expansion of exports. These interests coincided to help the creation of the National Aeronautical Establishment at Bedford, with several important new wind tunnels of higher speeds, size and power, but although it was conceived as a separate establishment it never achieved independence, and came to form an additional arm of the RAE.

The work of the RAE in this period underlines the thesis that the Cold War was, in a sense, a war—one of competitive innovation in weapons utilising the most advanced technology that the country could deploy and some of its best scientific and engineering personnel.[12] This was clearly a hugely expensive undertaking, and has been analysed from one perspective as a waste of human assets and national treasure, as well as a dangerous game of escalation that carried the risk of triggering conflict, particularly at crucial periods when new systems were about to be deployed. However, an alternative strategic and utilitarian view, which has been defended in the USA rather more robustly than in Britain, is that the effort was worthwhile because the threats were met. According to this view, World War III was averted not only by the balance of terror but, in the early period, when atomic war fighting was considered a possibility, by a balance of capability. The technical success of Western defence science in countering and meeting Soviet threats can therefore, on this alternative view, be seen as a contribution to stability.

Perhaps the surprising feature of the work at the RAE in this period is that, in spite of all the pressing needs and perceived dangers implicit in this world order, the direction of the research performed there by its highly intelligent and motivated scientists was done with a remarkably light touch. Indeed, much influential work was done because RAE scientists themselves felt that it would be strategically significant, or that it would be scientifically interesting.

There were various reasons for this comparative freedom. One was the prevailing sense throughout universities and research institutes, which lasted until the Thatcher era, that advanced research should not be shackled to closely too specific ends. Useful results would certainly accrue but, the argument went, who could say where? No-one had foreseen, it used to be said, that Fleming's stray mould would lead to penicillin, and this perspective was underpinned by the administration of the establishment, as the director was generally a government aeronautical scientist who had grown up in this system with its tradition of intellectual independence. Another reason for this comparative independence was that aircraft were entering a new aerodynamic realm as they approached the speed of sound. Up to 1945, for example, the Air Staff had expected incremental progress in engine performance, leading to improvements in the speed and height ceiling of aircraft; the main question was how rapidly these would come along in comparison to the progress of the enemy. With the approach to the speed of sound there was a genuine discontinuity of knowledge and the solutions were in the hands of the scientists, whereas the experience within the aircraft companies did not extend to trans-sonic and supersonic flow, or to the associated problems of stability, control, structural integrity and the prediction of performance in this regime.

Thus postwar British aeronautical work can be seen as being led as much by scientific research as by military requirements and military assessments of threat. Furthermore, because of the centrality of new science to aviation, the perception of threat would be very likely to derive from the aeronautical scientists' own appreciation of what the potential enemy might be able to achieve. The pre-eminence of science in this process is conveyed, for example, by the Ministry of Supply statement that 'the Scientific Officer Class is the initiating, directing and inventive brain' of the scientific civil service.[13] However, it is important to note that in this new, uncertain postwar era (uncertain both politically and aerodynamically) the development of military aircraft specifications was a much more fluid process than during the Second World War. Then, the mission could be defined reasonably closely on the basis of actual experience with losses, evaluation of success, and continual knowledge of enemy capabilities. In the Cold War era the nature of the threat was speculative, as were the potential capabilities of the defensive and offensive aircraft that might be produced to operate in the new trans-sonic and supersonic regimes.

The complement of aerodynamic staff at the RAE was drawn from the rather few universities which had aerodynamics courses (that run by Sir Brian Melvill Jones at Cambridge was one), as well as academically trained engineers and a considerable number of mathematicians of exceptionally high attainment, who frequently came directly on completion of their undergraduate degrees. Thus, to a considerable degree the RAE developed its own cadre of aerodynamicists and its own style of research.[14] However, the work of the theoreticians, it must be noted, was supported by a large number of talented practical engineers and technicians, and by well-equipped workshops in which prototypes, wind-tunnel models and instrumentation of the highest quality were made. To support this the RAE ran its own craft apprentice scheme, although exceptional individuals recruited in this way could move into more directing positions in research programmes. Thus the early work on jet lift and the control of vertical takeoff craft was done by a former craft apprentice, Dennis Higton, and his moving platform rig was the conceptual predecessor to the Rolls-Royce Flying Bedstead and, eventually, to the reaction-control system of the Harrier.

At the theoretical centre of the innovative work of the RAE was the Aerodynamics or 'Aero' Department, whose members, according to one recollection, led 'a god-like separate existence'. Although the level of mathematical analysis required in, say, the Structures Department can certainly have been no less taxing, it was within Aero Department that new concepts for aircraft and new directions for theoretical research tended naturally to arise.[15] Within this group there was a strong team spirit, perhaps engendered by the closed nature of the establishment and the consequent absence of an incentive to publish work competitively.[16] In this atmosphere scientists within the group were, to a very high degree, self-directing and, in the Cold War climate, there was a good chance that many of the ideas and suggestions emanating from this group could be tested in wind-tunnel work and with research aircraft. Thus aerodynamicists have recalled being 'in the fortunate position of never having to think about money for the things we were doing' and that 'we never thought about cost—we never asked'. It must certainly have been an almost idyllic time for aeronautical scientists with high theoretical capabilities and self-motivation. The perception of some of those who worked at the RAE in this period was that 'it was absolutely marvellous', and that 'we thought of the topics ourselves and found the answers; we had the perfect jobs'. The RAE was also a highly integrated community in which the various groups knew what each was working on and collaborated. 'We really were a team here. People had ideas, but passed them into the team; everyone was very co-operative'.[17]

The Aerodynamics Department also contained an inner circle of perhaps six people, the Projects Division (also referred to as Aero Projects), which played a leading part in devising proposals and considering what new types of aircraft could result from the latest aerodynamic research. Tom Sommerville, who was head of the group for a time, considered that the job was 'to look one step ahead', because 'aerodynamics always played a leading part' in advanced projects and that 'we were able to integrate the thinking of all the [other] departments'.[18] Within the Aero Department internal contact was ensured by bimonthly meetings where the theoreticians, and perhaps flight test practitioners, would stimulate each other with problems, queries and proposals, often in the absence of any top-down directive or requirement. On these days the internal morning discussion was followed by an afternoon meeting with senior government and RAF personnel, and, for example, the Vice Chief of Air Staff (VCAS) of the RAF would come to survey the progress of practical short-term experimental work on service types and also to meet the Director and senior staff and to learn what advanced ideas were in the air.'

In this period the Air Staff targets for future aircraft were frequently 'not in black and white', and were evolved and modified in discussion with Aero Projects or with specially formed multidisciplinary teams, such as the Advanced Bomber Project Group. In addition, the Director of the RAE always took a close interest in the work of these groups and was able to represent it at higher policy levels, both informally and through bodies such as the Air Warfare Committee at the Air Ministry.[19]

In the context of so much open-ended and self-directed research it is often hard to disentangle where aircraft developments actually originated, and it is important to see Farnborough as part of a wider aeronautical community involving the RAF, the Air Staff and civil servants, in which ideas and possibilities could circulate and resonate. In this milieu aircraft requirements and projects developed in a 'social' and even an informal way, so that Operational Requirements (ORs), which were issued to the industry to invite design submissions for future aircraft, had been refined by an interchange, which helped to define what might be attainable, between RAE experts, Ministry of Supply officials and RAF officers.

Of course the Air Staff could originate Operational Requirements for aircraft on the basis of strategic and combat needs in advance of the firms' ability to meet them, and even in advance of the power of the RAE to predict whether the performance was attainable. This was certainly the case with the medium bomber requirement which was to result in the V-force, issued in 1946. However, in other cases, such as that of the supersonic fighter and also, later, with Concorde, it seems clear that the impulse to initiate the projects came from Farnborough.

It is also noteworthy that RAE staff in this period recall considerable contact with RAF personnel at a practical and an experimental level, with, on occasion, uniformed RAF officers going into the wind tunnels to collaborate on programmes. There was also a very high degree of collaboration with the firms at all levels, from the Director and Deputy Director at the RAE to the research scientists on the particular projects, who all liaised in a personal way with the appropriate contacts in the aircraft firms. Through this, the research flying and wind tunnel programmes at the RAE were completely integrated into the development programmes for the forthcoming service aircraft. Indeed, this had to be so, for as the case studies here will show, the companies in this period were simply not equipped to carry out the fundamental research to support the advanced programmes.

This disparity between the capability of the RAE in advanced areas and that of the industry led, for a short time, to a proposal from the Minister of Supply that the RAE might be made responsible for the direct design and manufacture of military aircraft, 'with the object of producing better types' and 'for the benefits of research to be passed on to operational aircraft as early as possible'.[20] The main objective (which is significant in the light of the discussion below on the supersonic fighter) appears to have been to obtain an advanced interceptor. The RAE discussed this proposal internally but, despite reflecting on 'the problems of getting the firms to do advanced work'—particularly the problem of aeroelastic distortion on highly swept-back wings—cautioned strongly against the proposal. The establishment view was that the existing cooperation between RAE specialists and the firms' designers had only been achieved 'after years of striving to gain the industry's confidence'. If the RAE were to become a competitor, the effective partnership with the firms would be dissolved and 'the whole structure of the aircraft industry would then lose its main scientific support—a disaster for which a single national factory, however efficient, would be poor recompense'.[21]

In pursuit of its aim to nurture the capability of the industry, the RAE mounted numerous conferences and meetings in this period to pass on the implications of its long-range research to the designers and the aerodynamicists from the aircraft firms. These would take the form of presentations, followed by discussion, from a range of RAE personnel, including aerodynamicists, control and stability experts, structural and 'flutter' analysts and so on, to acquaint the firms of 'what it would be like to design a supersonic aeroplane'. Flutter was a particular province of the RAE, as the mathematics of the interaction between aerodynamic flow and the elastic or vibratory movements of the aircraft structure is particularly demanding. It was an area in which, during the 1950s, the firms appeared unable to cope, often leaving the final validation of a design, or the cure of a fault, to the RAE while the groundwork for the theoretical solutions that came to be adopted by the industry was established at Farnborough. The theoretical excellence of the RAE at this time was accompanied by a laconic, almost self-deprecating style, and one mathematician, describing his presentation on the expected solutions for supersonic wing design to the industrial audience, recalled mildly that 'I'd just done some sums really ... and drawn some graphs'.[22]

THE P.1 LIGHTNING SUPERSONIC FIGHTER

In 1945 a study of the 'interception problem', prepared by Don Hallowes, an Aero Department mathematician, analysed a new type of threat—the high-altitude high-speed bomber made possible by the new and rapidly developing jet engine. Hallowes assessed the chances of intercepting hostile incoming bombers at a range of speeds from 300 mph up to 500 mph and various heights up to 50,000 feet. He hypothesised as the defence 'a good conventional jet-propelled fighter such as might be flying in a few years' time' powered by the Rolls-Royce Nene, which was then, at 5000 lb thrust, the most powerful jet engine in the world, providing a top speed of 595 mph. Hallowes treated the problem as a series of exercises in three-dimensional geometry, calculating the results of a large range of different defence conditions, such as radar range and fighter readiness, and their interaction with bombers at varying speeds and heights. From this he arrived at what he called 'a pseudo-statistical approach' to answer the question 'how fast and high must a bomber fly to avoid interception?', and reached the important conclusion that 'in a large number of cases (70–90 per cent) a bomber flying at 500 mph at 50,000 ft would be free from the possibility of interception'.[23]

This study is interesting because the performance parameters of such a bomber mirror almost exactly those selected by the Air Staff in the following year for the high-altitude high-speed medium bomber force called for in its OR 229, issued in November 1946, that led to the Vulcan, Victor and Valiant 'V-force' aircraft.[24] However, the RAE, as we have seen, nurtured a culture in which the various groups knew what lines the others were working on and where ideas were shared. The RAE was essential to the development of the nuclear V-force but, as Sir John Charnley recalled, 'at the same time that you'd be thinking about that, you'd be thinking about how to beat it'.[25]

To combat the high-altitude, high-subsonic-speed bomber and make interception possible, it was essential to restore the speed advantage of the fighter over the bomber, but this would require 'a step into the unknown'—the design and production of a supersonic fighter.

Speeds approaching that of sound had been encountered in dives by late Second World War combat aircraft, often with disconcerting loss of control, buffeting and even structural failure. Partly to analyse these problems, which came to be known as 'compressibility effects', the RAE had, during the war, pioneered high-Mach-number trials in Britain by diving a Spitfire at increasing speeds up to Mach 0.88 while simultaneously conducting wind-tunnel model experiments to help predict the control effects that would be encountered and to analyse the pilots' experiences.[26]

Bullets and artillery shells were, of course, supersonic and stable in flight, and the mathematics of fully supersonic flight was fairly simple and reasonably well known. The greatest uncertainty was in the trans-sonic region, where the aircraft had not quite reached the speed of sound (a Mach number of 1), but where the local flow over convex shapes on the aircraft accelerated the flow to become supersonic in parts. This mixed flow was extremely complex aerodynamically and mathematically, and although some aircraft, such as the Spitfire, had approached the speed of sound with reasonably progressive control changes that a prepared pilot could cope with, other types had shown severe problems. No body of theory existed to explain this.

In the latter stages of the war attention had been given in Britain to developing an experimental supersonic aircraft—the Miles M52, using a special development of the Whittle W2/700 engine. This programme was cancelled, partly through doubts about the ability of the small Miles company to sustain an advanced programme, but mainly on advice from the RAE that the M52 wing was not thin enough to allow supersonic flight. Furthermore, the Allied missions that entered Germany as the war ended found that extensive work had been conducted on the swept wing in the trans-sonic regime, which appeared to suggest that the straight-wing approach adopted for the M52 was mistaken.[27] A number of German aerodynamicists concerned with high-speed flight came to Farnborough, and in 1948 M Winter and Hans Multhopp (formerly an aerodynamicist with Kurt Tank in the Focke-Wulf design office) proposed an experimental swept-wing trans-sonic research aircraft. They calculated that the new Rolls-Royce Avon engine was just sufficient to give the aircraft a supersonic performance of Mach 1.24 at 36,000 ft 'if equipment and instrumentation are restricted to only the most essential items'. The other restriction was to keep the diameter of the fuselage to the absolute minimum dictated by the Avon engine, and to this end the pilot was to be located in a prone position in a compartment located, in effect, within the inlet duct (Fig. 2.1). The wing was to be swept back at an angle of 55° to delay compressibility effects, and the tailplane ('a scheme which was developed some years ago for the Focke Wulf 183 fighter') was to be mounted high on the fin to keep it clear of the wing wake and to avoid the loss or alteration of pitch control which had been encountered in trans-sonic flight.[28]

This aircraft, it was suggested by the authors, could be regarded as 'an experimental reduced scale model ... of an aircraft having possible practical applications' (i.e. a fighter) and was not built, but later, in 1948, another RAE paper by Owen, Nonweiler and Warren proposed a larger supersonic fighter which was clearly developed from this suggested trans-sonic research proposal.[29] In general layout and wing plan form the proposed fighter followed closely the Winter/Multhopp design, including a version with a prone pilot position, although an alternative layout was

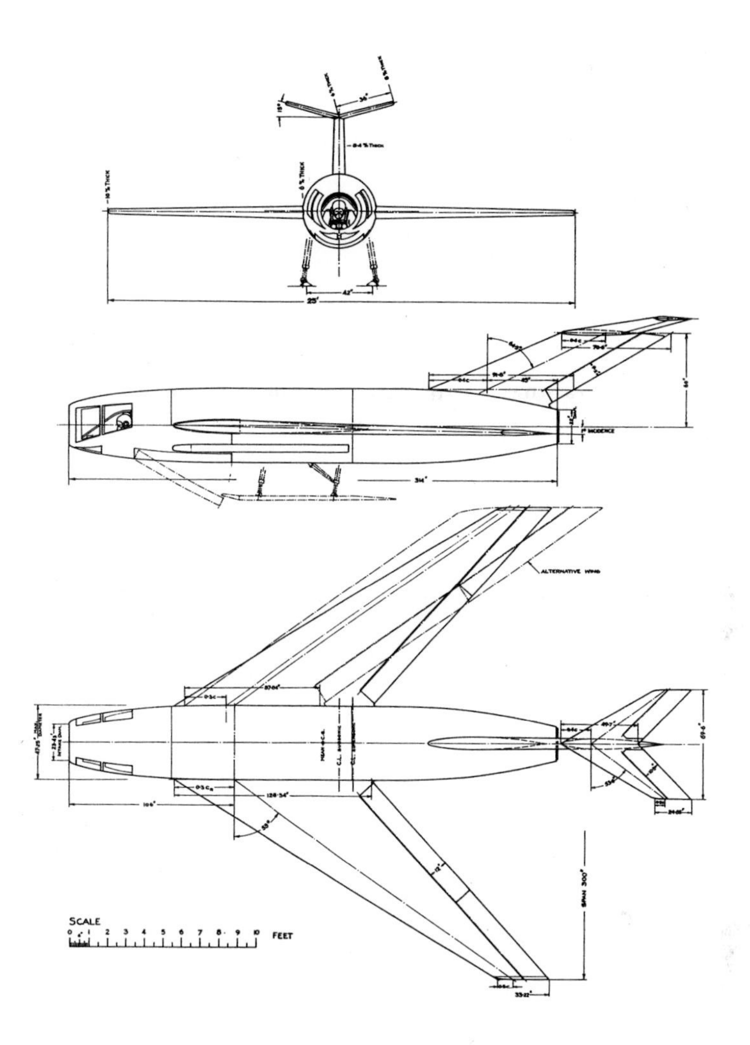

FIGURE 2.1: DRAWING OF THE SUPERSONIC RESEARCH AIRCRAFT PROPOSED BY M. WINTER AND HANS MULTHOPP AT THE RAE IN FEBRUARY 1948 (© CROWN COPYRIGHT, DERA)

sketched with a conventional pilot position above the intake and a radar scanner dish faired into the centre of the intake duct (Fig. 2.2). However, the Winter/Multhopp aircraft was only supersonic by dint of scrupulous streamlining and the avoidance of all unnecessary structure. A practical fighter would need much more power to attain this performance, and the new feature of this June 1948 proposal was the use of multiple engines, staggered so that the thickest part of one lay over the thinnest part of the other—the so-called 'hip and waist' arrangement. The earlier conclusions of Hallowes on interception are implicit in the argument of the paper, which noted that the success of bomber interception depended largely on the margin of speed of the fighter over the bomber, and that, as bomber speeds increased, this

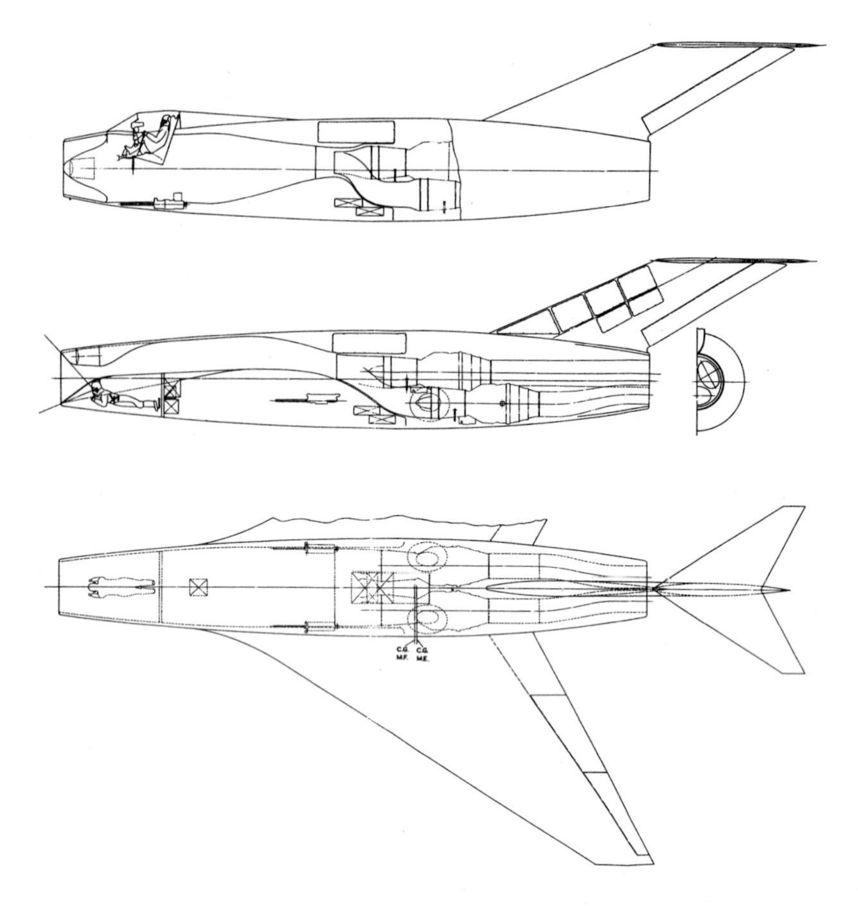

Figure 2.2: Project study for a supersonic fighter developed in June 1948 by Owen, Nonweiler and Warren at the RAE from the Winter/Multhopp proposal with conventional or prone position for pilot. The design accommodates multiple engines by an ingenious 'hip and waist' layout which was carried forward with other elements of the design to the English Electric P.1 Lightning interceptor. (© Crown copyright, DERA)

margin was becoming progressively harder to maintain. The authors noted that instability in the trans-sonic regime was expected from work to date, including disturbance in pitch and a severe loss, or even reversal, of aileron effectiveness (control in roll). However, these problems, which might be tolerated in a research aircraft and mitigated by accelerating rapidly through the 'sound barrier' (not the RAE term), had to be solved for a practical fighter to give progressive control at all speeds.

By November 1948 the Advanced Fighter Project Group, which had been set up at the RAE, reported on work to date, stressing the difficulty of predicting the nature of the threat (in terms of speed and altitude) for which 'the fighter which must stop the bomber' should be designed. The task, they suggested, was that of defending 'this island against the attacks of enemy bombers similar to the long-range high altitude bombers we ourselves are developing'—aircraft capable of delivering bombs at 500 knots and from 50,000 ft.[30] Although the study noted that this was a 'restrictive and possibly unrealistic assumption', these were also the performance characteristics of the hypothetical intruder aircraft for the earlier RAE interception analysis.

The group considered that, although the state of knowledge on aerodynamics, stability and control was still developing, the main uncertainty centred on the structure. The operational supersonic fighter was required to be a large and complex aircraft weighing perhaps 30,000 lb (at a time when the relatively simple 'first-generation' jet fighters, such as the De Havilland Vampire and the Supermarine Swift, weighed only 8000–10,000 lb). The gamble of estimating strength and weights closely in the absence of 'real guiding experience' or established design principles is shown by the structural challenge of providing enough stiffness to wings and tail surfaces to prevent flutter and aileron control reversal.[31] The catch here was that the forces would be higher than those met hitherto, although the surfaces were required to be much thinner and, for geometric reasons, the high degree of sweepback also would tend to compound the problems of twist and aeroelastic distortion. However, the price of a slightly 'safer' and more conservative design, increasing the structure weight by a factor of only 3 per cent, would reduce flight endurance from 55 minutes to 29 minutes—scarcely a useful fighter.[32]

The RAE admitted that the new fighter would be 'unlike anything designed or built in this country before', and that there would be many entirely new design problems to solve, but it argued that 'the requisite information, experience and design skill can only come from a direct attack on the problem'. Indeed, in the light of these technical reservations the RAE took a bold and even propagandist role in weapons development policy, arguing that 'a fully operational supersonic fighter would be an immeasurably valuable asset to the defences of this country', and actively promoted work on it in spite of the many uncertainties. It noted:

> The unknown factors are many and frightening but the prize may be immense. ... If we had unlimited time ... the obvious way to achieve this prize would be to tackle the problem slowly. ... It would, however, be a lengthy process. In view of this we would like to suggest, in all seriousness, that we take a short cut by proceeding forthwith with the design of a fully operational supersonic fighter on the lines sketched [out]. ... This would admittedly be an appalling gamble.[33]

The RAE suggested that 'a first class design team from the Industry' be asked to proceed with the design of an operational supersonic fighter. In August 1948 the Ministry of Supply (MoS) issued OR F.23/49 based on this RAE thinking which asked for 'a minimum top speed of Mach = 1.2 or higher' and a climb from the moment the pilot presses 'the first button' to 50,000 ft in not more than six minutes. The MoS then began to pursue discussions with English Electric as the most likely company to build the aircraft and, by March 1949, confirmed to the company that it was to develop the concept as the English Electric P.1—the prototype that was to lead to the Lightning fighter.[34] Thus the project, it should be noted, was set in train at virtually the same time as the trans-sonic Hawker Hunter (and long before the Hunter flew), with the intention of leapfrogging a generation of fighters.

It is important in studies of Cold War work to bear in mind that nuclear strategic thinking evolved continuously during the period, and the P.1 interceptor programme must be understood in the context of its times. At the outset of the programme the atomic bomb constituted a new and terrifying weapon, but in the minds of planners there was still some hope for defence. It appeared that technical constraints meant that the actual rate of production of bombs was quite low and the relative scarcity of atomic weapons suggested that an attack could be defeated by a really exceptional technological effort.[35] A glimpse of this thinking can be gathered from a paper on 'The Air Defence Problem' given to aircraft industry designers and RAE representatives in December 1953, which hypothesised what soon came to be seen as an unrealistically limited and strategic attack on Britain, assuming that 'the elimination of the United Kingdom capital, ports and bomber bases would be a strong factor towards the success of Russian aims in Europe in the next war'.[36] The initial development of the Lightning therefore took place in the context of a range of suggestions for fast-climbing manned rocket- or hybrid rocket-and-gas-turbine-powered fighters. Sir Charles Gardner (as Director of Guided Weapons Development) also gave a glimpse of a certain optimism for defence when he noted that 'the Million-fold increase in striking power of a single aircraft has transformed the defence problem from one in which an attrition of 5 or 10 per cent could be worthwhile ... to one in which it is necessary to achieve an annihilation defence in which virtually every aircraft must be destroyed'.[37]

One of the challenges for this study is to elucidate the contribution of government scientists at the establishments to the aircraft and equipment that were actually supplied. This has traditionally been difficult because of the mythologising tendency in British historical accounts, which conventionally emphasise individual contributions and efforts. This can be seen, for example, in the story of the Whittle jet, where the contribution of government scientists from the Engine Department of the RAE is completely discounted.

Much aviation history has been written by enthusiasts with a particular regard for the firms. Thus, in the case of the English Electric P.1 Lightning, Bill Gunston, one of the best-known British aviation historians, opined that the English Electric designer Teddy Petter and his team 'needed all their skill to create a supersonic aircraft which ... could form the basis for an operational fighter, in an environment devoid of practically any supersonic experience and with the most meagre facilities'. Their success, he suggests, 'was to some extent despite the advice of officials who caused endless trouble trying to make English Electric adopt a T-tail'.[38] These 'officials' were actually the most talented and experienced aerodynamicists in the country, feeling their way, like those in the companies, into the new trans-sonic and supersonic regimes, and it is clear that, rather than keeping them at arm's length, the companies were most eager to have RAE advice on their designs. Of course, both government and company scientists made both good choices and bad. The initial English Electric 1948 project drawings mirrored closely the planform of the RAE study, including the ingenious 'hip and waist' engine arrangement. This became a distinctive and successful feature of the production aircraft, although in the case of the T-tail English

Electric became convinced that RAE advice was wrong. In this they proved to be correct, and the low tail position eventually adopted proved far more effective in the nose-high landing attitude.

This study demonstrates that the aerodynamic and structural facilities open to the industry were far from meagre; they were, in fact, enormous, although largely concentrated at the Farnborough and Bedford sites of the RAE, and any account of this period must stress the tremendous integration between the companies and the government establishments. Mike Dobson (subsequently Superintendent of RAE Bedford), who joined in 1954 as an aerodynamicist and worked initially on the P.1A, recalled a colleague who worked 'for months and months in the "three foot" wind tunnel. It was a long time before I realised that he was English Electric and not RAE. The attitude was ... RAE was part of UK Limited'.[39] In this period there was a huge amount of work in the RAE tunnels on stability and control for the supersonic fighter, and also on the aerodynamics of the engine intake. There was also deep involvement of the RAE in the flight test programme and the Short S.B.5, built as a smaller model of the P.1 to evaluate its low-speed handling, was flown extensively by Aero Flight, Bedford, from 1954 to 1960 in the service of the development of the interceptor.

The RAE also served as the point at which the contributions of the other government research establishments, and RAF fighter development experience, were integrated into the aircraft. This included, for example, tactical ideas as well as radar and infrared research. The effect of these important elements of the interceptor can be seen in the arguments that began to surface from about 1953 for a collision-course interception, rather than the usual turn by the fighter on to the tail of the bomber, as a way of getting the interception point back from the coast of Britain.[40]

Thus at the Sixth Fighter Tactical Convention in 1952, described as 'a family gathering of the Fighter World, for the free exchange of views on current and future problems of common interest', the view was advanced that 'from the Communist point of view Korea can be classed as a second Spain ... a testing and proving ground for their equipment and crews against Western powers', and it was suggested that the 'immediate threat was ... of atomic attack on the UK by bombers at the beginning of a war'.[41] This marked an early interest in collision-course attack and, a little later, Dr F.E. Jones of the RAE noted explicitly that 'if the over-riding requirement was for interception at the earliest possible moment, and certainly before the coast, we may be forced to collision course intercept using guided weapons'.[42]

This represented a change in the way in which the supersonic fighter would be used, but the RAE studies in terminal dynamics showed that the advantage of collision-course interception was strikingly great. The need to destroy a bomber carrying an atomic weapon as far out from the coast as possible is clear, but there were also aerodynamic and operational considerations. First, the timing and positioning of the approaching fighter for the turn on to the tail of the bomber for a pursuit-course attack were critical. Furthermore, making this turn at a high rate caused the drag to build up and the fighter speed to decay. It was also a manoeuvre which, at supersonic speed, used a very large proportion of the fighter's fuel load and limited the number of targets that could be engaged.

The requirements of fighter interception also acted as a driver for all the establishments contributing to airborne interception (AI) radar, as well as to guided missile development and the guidance components for them. This extended to infrared work, as the collision-course attack placed far higher demands on the heat-sensing cells in the missiles, and in 1955 English Electric observed that:

> the present lead sulphide cell in Blue Jay Mk.I has to rely for most of its radiation [on] the hot exhaust pipes. ... Before firing such weapons the fighter must often close to short range behind the target and within a narrow cone around the axis of its jet exhausts. ... Improved guidance range can result from the introduction of lead telluride cells in Blue Jay Mk.II in place of the lead sulphide cells in Mk.I which will allow lock-on to the jet plumes. Unlike the jet pipe emission on which Blue Jay Mk.I depends these plumes cannot be shielded or cooled as a countermeasure to attacks from the side. Moreover, the jet plumes are visible over a wide range of directions round from the side towards the front, provided the fighter attacks from *below* the bomber.[43]

The Lightning, when it entered service in 1960, certainly vindicated the early RAE advocacy of the supersonic interceptor, and although, like so many British aircraft, it arrived awfully late, its performance substantially exceeded initial RAE predictions. It was, however, an aircraft that was predicated on the special air-defence and quick-reaction needs of Britain. In this role it was probably the most potent interceptor in the world at the time, but this specificity of role denied it really substantial export sales, although 40 were sold to Saudi Arabia and a further 14 to Kuwait.

The performance of the Lightning perhaps also says something about the complexion of British technology in this period. Lightning pilots recall a real thoroughbred with an astonishing rate of climb (2.5 minutes to reach 40,000 ft), a speed, in later versions, of more than Mach 2, and flying qualities that were 'beautiful' and 'perfect'. However, the promised integration of the computational aids and electronics—the aircraft radar, navigation system and attack sight—never reached the level of contemporary US semiautomatic systems. Interception was, a pilot recalled, 'a real one-armed paper-hanging operation' imposing an extremely high workload.[44] This deficiency was spotted early on by the RAE, and it is noteworthy that, during the programme, the instrument experts at Farnborough reflected that the electronics industry was overloaded and that 'the effects of lagging equipment development are becoming very apparent in current fighter development'.[45] In terms of maintenance, too, the aircraft throughout its operational life reflected the initial RAE view of it as an experimental aircraft which could be converted to an 'operational prototype' if flight development was a success, and it demanded a huge and continuous effort on the part of the RAF engineering ground services.

In strategic terms the fairly simple scenario for the overall air defence of Great Britain which had spawned the supersonic fighter also changed. The planning for the interception of a rather limited atomic attack proved unrealistic in the thermonuclear age, and it became tacitly accepted that the UK would not be able to build and man enough supersonic fighters to defend the country against a concentrated

Soviet bomber attack. In the light of this perception Duncan Sandys' 1957 Defence White Paper, surprisingly, served to preserve the Lightning programme. Sandys is mainly remembered for prematurely anticipating the age of the missile and the end of manned military flying, but this view overlooks the main objective behind his strategic thinking. Sandys, with a rather remorseless pragmatism, held the position that Britain could not withstand a nuclear attack, and that both conventional and nuclear war-fighting strategies were unrealistic. The overwhelming need therefore was to secure the credibility of the British deterrent, and he saw that this would necessitate the move away from the manned nuclear V-force to intercontinental missiles.[46] In the interim, while the V-force continued to pose a viable threat, the Lightning force that was coming into service was to be devoted largely to bolstering this credibility, and to 'the defence of the deterrent' against a Soviet counterforce strike.[47]

The Lightning was certainly the most complex and potent fighter ever built by Britain alone and, from the perspective of Cold War history, it was a response, at the limits of what was technologically achievable, to the nuclear threat faced by specifically by the UK, which for geographical reasons was different from and probably much more difficult technically than either the Soviet or the US air defence problems. The programme was extraordinarily ambitious for a country the size of the UK, and it could be argued that the influence of the RAE was to nudge defence policy towards an excessively technological aircraft, rather than to the more flexible fighters exemplified by, say, the Mirage series, with its great possibilities for extended and incremental development. Backing the Lightning to leapfrog a whole generation of fighters perhaps also caused the UK to forgo interim supersonic aircraft such as the American F-100 Super Sabre or the F-104 Starfighter.[48]

Nevertheless, in the long term the Lightning was the learning project that really established the English Electric team at Warton as the leading military aircraft group in the UK. Indeed, it is clear that the development of technological and analytical power in the company, looked for in the postwar years by defence planners and civil servants, did in fact occur. When C.H.E. Warren returned from the USA in 1960 to head the Flutter, Vibration and Noise unit in the Structures Department at the RAE, he found the firms much more capable. 'In the old days the RAE had done the sums which worked out whether the aircraft would flutter—now the firms were doing the work and we were just monitoring it'.[49] The technical expertise of the group has been maintained during the mergers that established the British Aircraft Corporation and then British Aerospace, and through the collaborative Jaguar, Tornado and now Eurofighter projects. Today, as the Military Aircraft and Structures Department of British Aerospace, the group that grew out of English Electric has become established as one of the largest and most capable military aircraft organisations in Europe.

THE V-FORCE BOMBERS

Advice from the RAE was also crucial in the development of the British postwar V-bomber force, which formed the initial delivery system for British nuclear weapons and which remained in service until 1970. However, its role in this programme was rather different from that in the case of the supersonic fighter discussed above. With the fighter the RAE took a proactive and propagandist role, and could even be said to have originated the concept, whereas the requirement for a long-range, high-speed bomber derived from strategic thinking within the Royal Air Force.

The experience of the Second World War had shown that massed formations of bombers, even with substantial capability for defensive fire, were highly vulnerable to fighter attack. By contrast, the survivability of unarmed, high-speed high-altitude aircraft, such as photoreconnaissance Spitfires and the unarmed Mosquito bombers, had proved good. This practical experience complemented the mathematical analysis of interception at RAE, discussed above, and the provisional medium bomber Operational Requirement, as we have seen, asked for an aircraft with speeds in excess of 500 mph at 50,000 feet. It asked, moreover, for a radius of action of 2000 nautical miles (subsequently reduced in the final specification to 1500 nm).

The major factor driving the creation of a new type of aircraft at the limits of aeronautical science was knowledge of the atomic bomb, and it is interesting to note that the Air Staff 'jumped the gun' by issuing a specification for aircraft capable of delivering the bomb before the British government had made a commitment to the production of atomic weapons. However, the GEN 75 committee of ministers, convened to consider atomic matters, had already received unequivocal advice that 'the United Kingdom should undertake the production of atomic bombs as soon as possible', and the Chiefs of Staff (Lords Alanbrooke, Cunningham and Portal) minuted the Prime Minister directly on 1 January 1946 that 'we are convinced that the best method of defence against the atomic bomb is likely to be the deterrent effect', and that 'we must have a considerable number of bombs at our disposal'.[50]

The Attlee government formally decided to proceed with the development of the atomic bomb in late July 1946. Thereafter, the principle of deterrence for the United Kingdom by atomic bomb-carrying high-speed high-altitude aircraft was firmly established by Lord Tedder, who followed Lord Portal as Chief of Air Staff in 1946, and was continued by his successor, Sir John Slessor (1950–54).

The initial aircraft requirement, despite making no reference to an atomic weapon, was clearly drafted with one in mind, as it called for an aircraft capable of carrying a single 10,000 lb bomb, some 60 inches in diameter and 24 ft long.[51] These parameters effectively defined the characteristics of the plutonium (Nagasaki) weapon (with its ballistic case) that William Penney's team was to replicate in Britain.[52]

The process of procuring these high-performance aircraft was inevitably complex. A.V. Roe (Avro) and Handley Page, having built large numbers of heavy bombers during the war, were clearly front-runners, and Shorts (which had been nationalised during the war) and Vickers all became involved in production. Handley Page had, in fact, anticipated the RAF's need for a high-speed jet bomber and had started an ambitious project with slender swept wings almost a year earlier, partly using German design experience which had been acquired by Handley Page designer Godfrey Lee during his membership of an Allied technical intelligence mission to Germany.

The Avro team also understood that sweepback was needed for the high transsonic speeds required, but did not consider it was possible to provide adequate strength for a wing of the span required without excessive structural weight. The Avro solution was to maintain the sweepback and reduce the span, but to restore the wing area by filling in the space between the swept-back wing and the fuselage, thus independently reinventing the delta wing. In this period, incidentally, the aerodynamic orientation of Avro was strengthened by the departure of W.S. Farren, then Director of the RAE, to become chief designer at Avro.

Both these proposals were highly ambitious, unlike any preceding aircraft, and carried major risks. The RAE, functioning as independent design advisers for the Air Staff, formed the Advanced Bomber Project Group from the Aero and Structures Departments, and noted that the very long range required and 'operation at a Mach number of 0.87 and at a height of 50,000 ft means that designers are being asked to go right outside the realm of past experience into a region bristling with aerodynamic and structural unknowns'. The group reflected that the drag coefficient of an aerofoil did not vary appreciably until a critical Mach number was reached (and thereafter rose steeply), but considered that experimental information on whether this critical Mach number could be deferred to a value of at least 0.87 was 'as yet scanty and conflicting'. Unless this could be achieved the requirement for range could not be met, and the group noted that 'present knowledge is ... grossly inadequate for the safe design of an advanced bomber'.[53]

The RAE suggested that 'we cannot put all our eggs in one basket', and advised pursuing several designs in order to spread the risk. Out of the large number of notional designs for tailed and tailless bombers of varying sweepback considered, it suggested work on a tailless delta and a tailed aircraft of about 45° sweep. These were theoretical or 'paper' aircraft, but they corresponded broadly to the Avro Vulcan and Handley Page Victor proposals. The RAE also proposed an intermediate design with less sweepback, which presumably gave comfort to the Air Staff, who were clearly nervous about the progress of procurement. In fact, a more conventional intermediate or 'insurance' type—the Short Sperrin—had been ordered by the Ministry of Supply in case the firms were not able to solve the aerodynamic problems of the advanced types, but the Air Staff considered this aircraft to be 'unimaginative' and took a gloomy view of its likely performance 'from our knowledge of the work of the firm'. This hiatus was exploited by George Edwards, chief designer of Vickers, who was able to reassure officials that a bomber that almost approached the original performance requirements could be built to a quicker timescale, and the Vickers proposal (subsequently named the Valiant) was ordered to a relaxed specification which called for a speed of 465 knots at 45,000 ft. Even so, the production of this aircraft was an industrial *tour de force* and the first arrived for squadron service in February 1955, whereas the Vulcan and Victor, which started almost a year earlier, did not reach operational squadrons until May 1957 and April 1958, respectively. Thus the development of the V-force operational procedures and the airborne dropping trials of inert versions of the first British atomic weapon (Blue Danube) took place with Valiants in 1955. These showed that weapon release was also an aerodynamic problem and that the streamlined shape of the bomb could allow it to 'fly', keeping station beneath the tail of the bomber, until strakes were fitted below the fuselage to create turbulence. The first live air drops of British nuclear weapons were also performed by Valiants with the test of Blue Danube, the fission weapon dropped at Maralinga in south Australia in October 1956, and with the first Bristish thermonuclear weapon dropped at Christmas Island in the South Pacific in May 1957. However, the Valiants were grounded as their fatigue life approached in 1964.

The decision to order RAF nuclear bombers in triplicate was one of the most notorious cases of the multiplication of British aviation projects, which imposed a heavy financial burden and cost Britain very dear in the Cold War years. It also represented a profligate use of British aircraft design skills, squeezing out, among other possibilities, better British civil airliners. The puzzle is that the RAE had shown itself to be remarkably single-minded about the supersonic interceptor (asking for only one basic layout and manufacturer), although this may have been because of the consciousness that it was proposing a large, risky and expensive step which was not so dear to the hearts of the Air Staff as the bomber, and which therefore could not be duplicated. There was also some latitude in the interceptor performance: it was a 'hot-rod' whose mission was short. By contrast, the bomber, to constitute a viable deterrent with a reasonable chance of survival, could not compromise on range, speed or height over the target.

This habit of multiple procurement was also due in part to Second World War experience, where it had proved impossible to predict which designs would be superior until they reached service (the superiority of the Avro Lancaster over the Handley Page Halifax and Short Stirling was advanced as proof of this), creating a doctrine which was held dearly by the RAF. The RAF, furthermore, had ended the war with immense prestige and was used to getting what it wanted. It was not, of course, the role of the RAE to comment on procurement policy, but in the light of genuine uncertainty about the right structural and aerodynamic solutions for new aircraft the RAE was perhaps too ready to see a large number of alternative aircraft projects as useful full-sized experiments. Handel Davies, referring to the trans-sonic fighter and bomber programmes, suggested that 'many of us thought at the time that it would be better to do a bit more experimental work on prototypes than to go ahead with three of each. ... I think we dithered a great deal on the merits of delta or sweepback and shared the blame for the multiplication of projects'.[54]

Throughout the development programme of the three V-bombers there was again a major contribution of RAE aerodynamic work, and this also extended to the aerodynamics of the carriage of nuclear weapons and their release (Fig. 2.3). Thus in February 1958 the Ministry of Supply noted that Vulcan B. Mk 1 'had now been completed by the firm and handed over to the RAE. ... Trials are now in progress to clear Yellow Sun for carriage'.[55] RAE work also contributed to the development of the Vulcans and Victors to carry the Blue Steel powered 'standoff' missile, and the finding that the partially recessed missile 'has little effect on the drag coefficient of the aircraft' was important to establishing the viability of the development.[56]

RAE expertise of a different kind was needed in the service of the V-force when

Figure 2.3: RAE wind tunnel model of the Avro Vulcan (Science Museum inv. no. 1993–2287, Science Museum/SSPL)

a Handley Page Victor on trials from Boscombe Down in August 1959 crashed unaccountably into the sea off the Pembrokeshire coast. More than 600,000 pieces of debris were recovered by trawling the sea bed (16 trawlers were involved), and the investigation was conducted by the Structures Department at Farnborough, where the fragments were reassembled on a wooden framework. The investigation concluded that the loss of a Pitot head through vibration had given a false indication of low airspeed, causing the automatic system which protected against stall to push the aircraft nose down into a catastrophic supersonic dive.

This investigation, conducted under the direction of P.B. Walker, Head of the Structures Department, recalled the earlier investigation at Farnborough following the unexplained crashes of two de Havilland Comet airliners in 1954. As the Comet was the first jet airliner in the world to enter service, the losses were seen as entailing a grave loss of national prestige and the RAE was asked to take charge of the investigation, underlining the unitary nature, at the time, of both British military and civil aviation effort. Although de Havilland had previously been less inclined than other aircraft makers to consult the RAE, only Farnborough at the time had the resources and expertise to construct a rig to repeatedly pressurise an entire Comet fuselage to simulate the cycle of takeoff, climb and descent, while hydraulic jacks simulated the flight loads on the wings. After some 3000 'flights' a fatigue crack, originating at the corner of a window, pushed out a section of the fuselage. This pattern of failure was confirmed by debris then being recovered from the sea bed off Rome (Fig. 2.4). The extraordinary efforts made to elucidate the cause of the

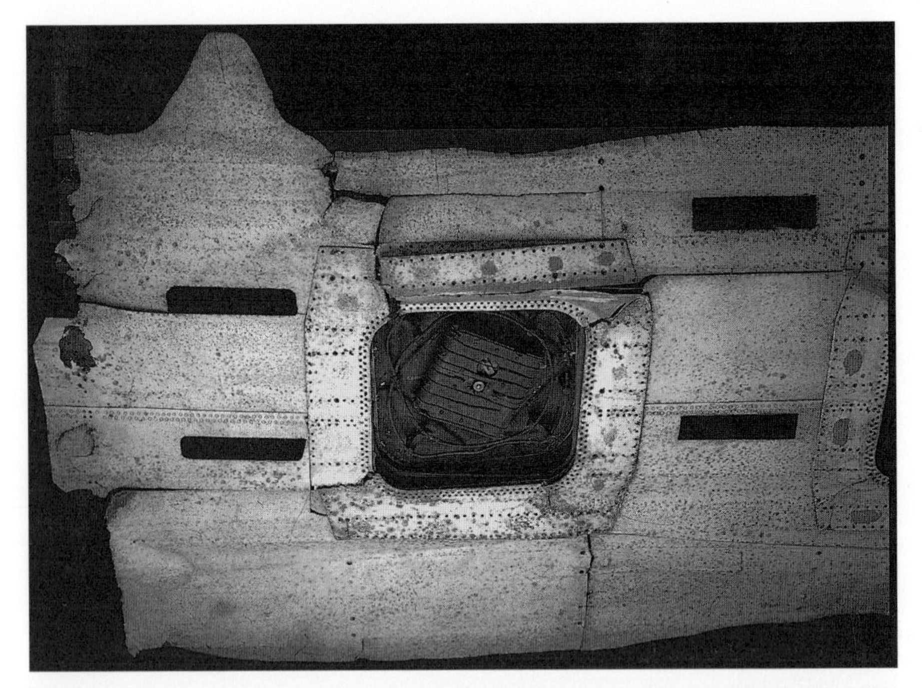

FIGURE 2.4: SECTION OF THE UPPER FUSELAGE OF DE HAVILLAND COMET G-ALYP RECOVERED FROM THE SEA BED OFF ELBA AFTER CRASHING JANUARY 1954. THROUGH THE STUDY OF THIS WRECKAGE, TOGETHER WITH STATIC TESTS ON ANOTHER AIRCRAFT AT FARNBOROUGH, THE RAE SHOWED THAT THE COMET CRASHES HAD BEEN CAUSED BY FATIGUE CRACKS STARTING AT THE CORNERS OF THE PASSENGER WINDOWS OR THE STRUCTURALLY SIMILAR HATCH USED FOR THE RADIO COMPASS, SEEN HERE (SCIENCE MUSEUM INV. NO. 1999–2155, SCIENCE MUSEUM/SSPL)

failures should not, however, be seen merely as industrial support for a flagship civil aircraft, as the perception of the technical capability of the British aircraft industry internationally also had a defence dimension. Writing in April 1954, Air Chief Marshall Sir John Baker (Controller of Aircraft, MoS) noted 'the most urgent need ... to resolve and rectify the technical cause of the accidents ... [in the] interests of the integrity of the aircraft industry from both the national and strategic viewpoints'.[57]

THE MACH 2 FIGHTER

The supersonic interceptor and the V-bomber force show two ways in which the RAE contributed to operational aircraft. However, this reasonably smooth progression from a project study, whether initiated by the firms or by the RAE, did not always occur. For example, following the initiation of the supersonic interceptor as a counter to high-subsonic-speed bombers, the RAE looked ahead to the possibility of the Soviet Union developing bombers with low supersonic performance (Mach 1.3), and proposed, as a countermeasure, a new design of fighter based on an RAE project for an aircraft capable of Mach 2.[58]

The RAE conducted an analysis of the utility of such a defensive fighter together with the Radar Research Establishment (RRE).[59] Their report suggested that the weapon was worth developing, in conjunction with more advanced airborne radar systems, and that such an aircraft 'can achieve a kill line 30 miles out to sea' against a Mach 1.3 target. The system, it noted, could also cope with a Mach 2 target, but 'the kill line would shrink in very close to our coast line'.[60]

However, the main argument for a Mach 2 fighter—that the Soviet Union might supplement their high-subsonic-speed bombers with some Mach 1.3 aircraft—seemed rather implausible, even to its proponents, since it began to appear that designing a long-range Mach 1.3 bomber was no easier than designing a Mach 2 bomber. As the missile age dawned the perceived threat of such high-altitude supersonic manned bombers receded, whereas the Lightning proved capable of development to Mach 2 performance, and some elements of the AI and ground control electronics permeated into the Lightning programme. The RAE Mach 2 fighter proposal proved to be a dead end, with only two prototypes of the Bristol 188 built to test the wing layout.

However, the work was influential at least in that, to test the wing layout, the Bristol 188 was built in stainless steel to cope with the expected kinetic heating. The expense and difficulty of this construction was to play an important part in the later decision at RAE to restrict the design speed of Concorde to about Mach 2 (where aluminium alloy structure would just suffice), and not to aim for a speed of Mach 3 or more, as was being contemplated in the USA.

The project also throws further light on the role of the RAE, both as a kind of 'advanced projects office' for the industry, and as a 'booster' in policy-making circles for advanced aviation projects. Thus the same RAE Mach 2 thinking also gave rise to a research contract to Avro to develop a supersonic high-altitude reconnaissance and bomber aircraft (the Avro 730), and the RAE helped to mediate between the company and the Ministry of Supply in the development of the new type.

Indeed, RAE protagonists of advanced aircraft, such as Morien Morgan, cooperated (one could almost say colluded) with the companies to promote advances in aviation technology, and when Avro sought government funding for the design and construction of a small-scale flight research variant of the supersonic bomber Morgan, as Deputy Director, wrote that the RAE was preparing an appreciation of the Avro design 'for official consumption, which I hope will help'. He wrote also to the Ministry of Supply that 'we are in full agreement with the reasons the firm give ... and consider that they have made an excellent case for this development'.[61] A supersonic airliner version of the Avro bomber was also proposed but made little progress, owing to its marginal payload, although Morien Morgan's real powers of advocacy were shortly to bear fruit with another supersonic airliner proposal—the narrow delta layout that led to Concorde.

CONCORDE

The RAE's contribution to V-bomber development corresponded to a conventional sense of the work of a research establishment, but its role in initiating and promoting the supersonic airliner proposal that led to Concorde followed a more inspirational and propagandist pattern. In some respects this activity—relating the latest theo-

FIGURE 2.5: FAMILY OF DELTA SHAPES DEVELOPED FOR WIND TUNNEL TESTING AT RAE AS PART OF THE SUPERSONIC AIRLINER (CONCORDE) PROGRAMME (SCIENCE MUSEUM INV. NO. 1993–2996, SCIENCE MUSEUM/SSPL)

retical aerodynamics to an advanced aircraft—recalled the role the RAE had adopted in the supersonic fighter, when it urged a development that was well in advance of what the aircraft industry, and even perhaps the Air Staff, was contemplating. However, in the case of Concorde, RAE influence spread more widely than this and into the political domain, largely through the efforts of Morien Morgan and his efforts in the creation of the Supersonic Transport Aircraft Committee.

Supersonic passenger flight had been considered at the RAE from the mid-1950s and, as we have seen, had been linked to the Avro 730 supersonic bomber project. From 1955, the highly persuasive Morgan 'did the missionary work to get the companies to take an interest in it', and persuaded all the chief designers who were prepared to cooperate to join the Supersonic Transport Aircraft Committee (STAC), which first met in November 1956.[62] However, as interest began to develop, RAE aerodynamics research began to suggest a completely different supersonic configuration for the aircraft—the narrow delta (Fig. 2.5).

The understanding of the narrow delta is associated particularly with the RAE aerodynamicist Dietrich Küchemann, who had studied at Göttingen with Ludvig Prandtl and was one of the German scientists offered employment in England by British investigators in Germany at the close of the war. He came to Farnborough in 1946, and stayed to become the highly respected head of the Aero Department from 1966 to 1971. His associate, the mathematician and aerodynamicist Johanna Weber, came, at his instigation, some months later in 1947.

Weber initiated the first interest in the aerodynamics of the narrow delta at the RAE, with a survey paper in 1955 on all the available information on this type of wing, and the configuration was first considered for its low 'wave drag' in supersonic flight.[63] Indeed, it appeared to be the only shape that would allow the carriage of a reasonable civil payload. The solution to the supersonic end of the speed range thus began to become emerge, but it was still unclear whether it would it be possible for this supersonic, low-drag shape to queue and land at normal airliner speeds, using available runways. 'The problems were the low speed problems of the high speed aeroplane—it seemed hopeless', C.H.E. Warren recalled.[64]

The aerodynamicists considered that 'we'd solved the classical aeroplane', but the narrow delta entailed a new type of flow and although, with its echoes of the schoolboy paper dart, the configuration seems with hindsight an almost unsurprising choice for a high-speed aircraft, the way in which it works amounted to a real paradigm shift in aerodynamic thinking.[65] In essence, at high angles of incidence the airflow over each wing rolls up into two huge stable vortices. This contrasts with classic aerodynamics on straight wings, where orderly flow front to back was the desired state, and wandering unstable vortices over the upper surface of the wing were associated with stalling and loss of lift.

It soon began to emerge from this research that the vortex flow of the narrow delta could offer the combination of very high speeds and the possibility of flying slowly (and landing) in a 'nose-high' flight attitude. Thus speed could be reduced but lift could be maintained, it seemed, without the complicated flaps and leading-edge devices on which conventional airliners depended, by progressively raising the nose and increasing the angle of incidence of the wing. This discovery of the greatly extended range of 'non-linear lift' was at the core of the RAE advocacy of the narrow delta.

Even once it had been shown theoretically that this slender supersonic shape should also be able to fly slowly, doubts still existed. Although low-speed lift was assured by the pair of large stable vortices rotating over the upper wing, how stable were they in the ultraslow landing regime? In a side gust, might they not be liable to slide off one wing or the other and take some seconds to regenerate? American wind-tunnel studies had predicted that this form of aircraft might be prey to 'Dutch roll', a spiral instability with the aircraft rolling and swinging in a corkscrew motion about the direction of flight.

However, as the theoretical solution of the narrow delta began to emerge for the civil supersonic mission, it was adopted with tremendous enthusiasm by the Farnborough aerodynamicists. One in particular, W.E. Gray, believed that Farnborough and Britain had a mission to develop the supersonic aircraft, and he also considered that the danger of Dutch roll was exaggerated. Gray was an extraordinary character whose research effort was entirely 'self-propelled'. He had been a pilot, awarded the DFC in the First World War, and was often seen cycling round Farnborough in his old leather flying cap. He had a talent for cheap pragmatic experiments, and made it his mission to investigate this roll phenomenon experimentally. Gray's private programmes often cut across the intentions and plans of other departments, and sometimes section heads and senior officers would refuse to see him to avoid being badgered for research facilities, flight time or resources. Gray, who was an expert rose grower, usually managed to detain his quarry by finding a moment to present a bunch of his own blooms.

Gray initiated one of his typically ingenious and cheap research programmes to test the claimed instability of narrow deltas, building and launching simple balsa models with a variety of shapes across the 24-ft wind tunnel (Fig. 2.6). The models

FIGURE 2.6: ROUGH-AND-READY MODELS CONSTRUCTED BY THE RAE RESEARCHER W.E. GRAY IN BALSA AND FOAM TO INVESTIGATE THE STABILITY OF DELTA SHAPES FOR THE PROPOSED SUPERSONIC AIRLINER (SCIENCE MUSEUM INV. NO. 1993–2473, SCIENCE MUSEUM/SSPL)

began their glide in still air and, as they crossed the tunnel nozzle, encountered the tunnel flow from the side, simulating the kind of side gust the airliner might encounter on a landing approach. The aim was to see whether the disturbance damped itself out or was self-sustaining. The mathematician Jean Ross, who had just joined the RAE, was assigned to help with these trials and recalled: 'I was his ball girl', retrieving the models from across the wind tunnel.[66]

However, Gray went almost too far with his next sally. He argued that 'the national good [and] the RAE's good name' rested on flying a narrow delta supersonic shape without delay, and advocated building a range of cheap wooden manned gliders which he volunteered to fly 'and so permit the Captains of Industry to keep their white robes unblemished'.[67] His criticism of RAE direction and 'the massive slow-motion approach' of the Ministry of Supply got him into trouble, but the view that Jean Ross held of his particular style was that 'you could be amused or you could be cross, or you could respect him. I think most people respected him'. Indeed, the conjunction of German theoretical and mathematical training, as exemplified by Küchemann and Weber, with Gray's pragmatic English empiricism, which bordered on the eccentric, is one of the most fascinating episodes in the generation of the Concorde project.

As RAE confidence in the solution increased the establishment again mounted a series of conferences and 'selling' visits to the firms, explaining the thinking behind the narrow delta, its aerodynamics, performance and control characteristics. Thus the initial project design for the SST can certainly be attributed to the RAE and not to any particular firm. This parentage is revealed by the remarkable similarity of the shapes proposed independently for the aircraft by separate design teams at Bristol, Handley Page and English Electric, all of which showed a similar planform and the distinctive ogee (wineglass) shape that came to typify Concorde.

CONCLUSION

A central point about the RAE which needs to be touched on again concerns its crucial role in the integration of defence science and technology, for it was largely through the work at Farnborough and Bedford that all the then new and developing technologies for trans-sonic and supersonic flight came to be integrated into military aircraft. Many specialisms were the province of separate establishments, and are referred to in other chapters. However, the incorporation of such work into the aeroplane had to take place in the early years at the RAE. The pressure suits mentioned by John Ernsting in Chapter 12, which were essential to flight at high altitude and high *g* are a good example of the physiological work of the RAF Institute of Aviation Medicine; the suit itself had to interface with the aircraft systems, and therefore the Mechanical Engineering Division of the RAE assessed techniques for providing the suit with the compressed air it needed from the compressor stage of the jet engine.[68] To take another example, the work on weapons, missiles and guidance done elsewhere also had to be integrated into the aircraft. This included the aerodynamic effects on performance and handling of the aircraft while the ordnance was still attached, as well as establishing safe separation on firing. Guidance and the integration of missile and aircraft electronics also fell to the RAE in this period, and 'terminal dynamics'—the complex three-dimensional theoretical treatment of interception, pitting the fighter/missile ensemble against the hypothetical enemy bomber—was very much an RAE speciality. The strength of the RAE was that, 'presented with any aircraft-like problem, the RAE could pull together a high quality group over a range of subjects'.[69] Even as late as 1963, the RAE was itself pioneering a new integrated navigation and attack (nav-attack) system for the TSR-2 tactical strike aircraft without any lead contractor from industry.[70]

The work cited here is necessarily only a sample of the huge output of research work carried out by the RAE in the period under review. Nonetheless, it shows the Establishment's far-reaching influence in British defence and weapons systems. Even an RAE programme with the apparently innocuous purpose of developing a blind landing system for poor visibility conditions was tied in to the British posture on nuclear deterrence, as 'the point about the blind landing was to get the V force from their base airfields to their dispersal airfields where their weapons were stored'.[71] This was a system that the V-force needed to have to be credible as a deterrent, and it is an intriguing thought that part of the effect of the experimental transmissions from the RAE Blind Landing Unit at Martlesham in Suffolk was to acquaint Soviet signals analysts with the capability of the force for all-weather operation.

The competitive development of aircraft at the frontiers of aerodynamic knowledge illustrates the special Cold War phenomenon of science at war, which is an overriding theme of this book. However, the point should be made that this chapter is not a critique of the procurement policy of the UK for aircraft, and it has tried to show that in policy matters the RAE was part of a larger political and decision-making milieu. In addition, the concentration on the preliminary aerodynamic design of new aircraft should not be taken to imply that the firms did not contribute a huge amount to the designs as they began to take shape. The RAE aerodynamicists would themselves be at pains to point this out, being essentially modest, cautioning that 'a lot of things go into an aeroplane' and stressing the contribution to the actual aircraft of 'the immensely practical people from industry'.[72]

Finally, we should perhaps remember the large national and international role that the RAE played in aeronautics. The work done there circulated within the aeronautical community of the NATO countries, but the RAE also fuelled the postwar British university expansion in aerodynamics, with high-level teaching staff for aerodynamics and aviation subjects.[73] Internationally it also contributed, for example, chief scientists to Lockheed and Martin-Marietta. It is also interesting to note that from 1971 Dietrich Küchemann, then head of the Aero Department, promoted the initiative to set up powerful new trans-sonic wind tunnels with NATO partners through his promotion of the European Large Wind Tunnels Group. Thus through the Cold War the RAE served as the main and enduring core of European aerodynamic knowledge, and was instrumental in returning some of this expertise to Germany as part of the larger Europe.

1. Lord Hankey to Sir Stafford Cripps, 26 August 1943, in file 'College of Aeronautics—Proposal for Establishment', PRO AIR 19/389.
2. Sir Roy Fedden in *The Fedden Mission to America*, Final Report, Ministry of Aircraft Production (June 1943). Although opinionated and 'difficult' Fedden can certainly be relied upon for this relative assessment of US and UK capabilities. W.G.A. Perring, Deputy Director of the RAE also took a mission to the USA to review research facilities there (see W.G.A. Perring, *The Perring Aeronautical Research Mission to the USA and Canada, 1945*, RAE Report Aero 2140).
3. C.H.E. Warren, interview with the author (1 June 1998). As an influential member of the Aero Department 'Chew' Warren was involved in many of the advanced RAE aircraft proposals in the period covered. From 1960 he headed the Noise, Flutter and Vibration section in the Structures Department where he made a special study of the effects of the sonic boom over land, as part of the Concorde programme.
4. John Bagley, conversations with the author (1985–97). John Bagley was a member of the Aero Department and worked for a long period with Dietrich Kuchemann. He preceded the author at the Science Museum as Curator of Aeronautics.
5. Sir Peter Masefield, *The Story of the Brabazon Committees*, Lecture at the Royal Aeronautical Society (28 October 1995). Beaverbrook had, of course, been Minister of Aircraft Production in 1940 but was then negotiating civil air transport agreements with the USA.
6. R. Smelt of the Aero Department, RAE, in 'A critical review of German research on high-speed airflow', *Journal of the Royal Aeronautical Society*, 50 (1946): 900, noted that 'the great increase in aircraft speeds and, in particular, the arrival of jet propulsion changed our attitude to the subject [of high-speed flow]'. What had been 'an absorbing fundamental study became, quite suddenly, an important operational problem, looming large in every new fighter design'. The discussion after the paper as well as commenting on the 'pilgrimage from this country' by aerodynamicists and aeronautical engineers to the German research institutes noted the lack of contact between scientists, aircraft designers and the services—'the fault of the German general staff, who had a poor idea of the way scientists could be used in war-time'. Smelt also contrasted 'the imposing array of high speed wind tunnels with the 'ridiculously small amount of German flight research at high speeds' (compared to the RAE).
7. *Report on Visit to Messerschmitt Aircraft Factory, 1945*, PRO AVIA 10/113 (1945).
8. A major programme for the exploitation of German aeronautical facilities in the British sector under the codename 'Operation Surgeon' was set in train in January 1946 which contributed much research equipment

and plant to Farnborough, Bedford, and the new College of Aeronautics at Cranfield. Selected German scientists were also brought to Britain to work in the establishments and at one time some fifty were engaged at Farnborough. This episode has been dealt with in some detail by the author in a forthcoming paper.

9. M.J. Lighthill (subsequently Sir James Lighthill), 'The Royal Aircraft Establishment' in *The Organisation of Research Establishments*, edited by Sir John Cockroft (Cambridge University Press, 1965): 28–54.
10. Mike Dobson, personal communication (28 March 1998).
11. Aeronautical Research Committee, *National Requirements for Aeronautical Research and Development*, PRO DSIR 23/13337 (23 March 1944).
12. This point is well made by Peter Nailor, 'The Ministry of Defence 1959–1970' in *Government and the Armed Forces in Britain 1856–1990*, edited by Paul Smith (Hambledon Press, 1996): 235–48.
13. Booklet: *Ministry of Supply Exhibits at the Society of British Aircraft Constructors' Exhibition, 1957*, Science Museum Archives.
14. On completion of his degree during the Second World War, Sir John Charnley recalls being interviewed by C.P. Snow who was involved in the allocation of scientific manpower for the war effort and sent to Farnborough because of his interest in hydraulics and structures. Sir John Charnley, interview with the author (28 May 1998). Sir John Charnley was, among other posts, Superintendent of the RAE Blind Landing Experimental Unit, Head of the Weapons Department, RAE and Chief Scientist to the RAF. C.H.E. Warren recalls being allocated to the RAE on completion of his mathematics study at Cambridge with the winning proposition that 'in the last war you'd have been sent to the trenches but we've been thrown out of France now. Have you heard of a place called Farnborough?' (n. 3 above).
15. Brian Kervell, personal communication (3 February 1997).
16. An ingenious mechanism did exist for peer review of secret work through the Aeronautical Research Council which had a fluid dynamics committee composed of security cleared aerodynamicists from the universities and other establishments at which, C.H.E. Warren considered, the ARC would put forward ideas and comments on the work at RAE 'but in a generous way' (n. 3 above).
17. Johanna Weber and C.H.E. Warren, interview with the author (1 June 1998).
18. Tom Sommerville, interview with the author (16 October 1997). The group also acted as arbiters when several competing designs were produced by the industry.
19. For an understanding of British defence policy it seems important to give attention to the role of informal and social contacts between the protagonists. I have attempted to show how essential these contacts were in a slightly earlier period to the establishment of the Second World War radar chain and the concurrent development of high power interceptor engines in 'Two-stroke or Turbine: the Aeronautical Research Committee and British Aero Engine Developments in World War II', *Technology and Culture*, 38 (1997): 312–54.
20. *Proposals for the RAE to Take Direct Responsibility for the Design and Production of Aircraft*, PRO AVIA 13/666 (March 1947). This proposal of course recalls the role of Farnborough in the First World War, when as the Royal Aircraft Factory, it had a contentious role as both a government manufacturer and as a design and research authority.
21. *Proposals for the RAE to Take Direct Responsibility for the Design and Production of Aircraft* (n. 20 above). These opinions were expressed by Morien Morgan, then Head of Aero Flight, and P.B. Walker, Head of Structures, who became well known for his investigation of the Comet crashes.
22. C.H.E. Warren (n. 3 above). The major theoretical insights on flutter analysis at Farnborough were ascribed to E.G. (Ted) Broadbent, currently Visiting Professor in Mathematics at Imperial College.
23. D.M. Hallowes, *An Examination of the Interception Problem*, RAE Report No. Aero 2035 (April 1945). The analysis included a range of figures for radar detection range, the distance between defending airfields, and the 'lost time'—the period elapsing between the first detection of the bomber and fighter take-off. Even taking an optimistic figure for radar detection range (150 miles, at a time when the more likely range was 100 to 120 miles) it concluded that 'in a large number of cases (70–90 per cent) a bomber flying at 500 mph. at 50,000 ft. would be free from the possibility of interception by the fighter'. Clearly all these variables interact and bombers far slower than 500 mph can be immune from interception if the radar warning distance and the 'lost time' are degraded. Thus a bomber flying at just over 300 mph at 40,000 feet—by no means an unlikely performance, even with piston engines—could escape interception if only detected 100 miles away and if the 'lost time' was 10 minutes or more.
24. Probing the influence of work such as this is always difficult. However the circulation of the paper is recorded and included a large part of the scientific direction in the Ministry of Supply including the Controller of Research and Development (CRD) and the Director of Scientific Research (DSR). The operational research sections at the Air Ministry and in Fighter Command are also included. On the basis of the performance targets for the V-bomber force and the British supersonic interceptor it appears that Hallowes' study was highly influential.
25. Sir John Charnley (n. 14 above).
26. ARC Monograph, *Research on High Speed Aerodynamics at the Royal Aircraft Establishment from 1942 to 1945*, edited by W.A. Mair, R&M No. 2222, 1946.

27. The cancellation of the Miles M52 has always been a contentious issue with British aircraft enthusiasts who have personalised the issue by blaming the then Director of Scientific Research at MAP, Sir Ben Lockspeiser, for setting back the British supersonic programme by years. However the interviews conducted by this author make it clear that the advice came from RAE aerodynamicists who also calculated that the thrust from the special Whittle W2/700 was inadequate. The critics point out that the American rocket-powered Bell X-1, the first supersonic aircraft, used straight wings but the clear flying weather of Nevada and the almost limitless desert runways provided a far better environment for early supersonic trials.
28. M Winter and H Multhopp, *Transonic Research Aircraft with "Avon" Turbine Jet Engine (A.J.65)*, RAE Technical Note Aero. 1928 (February 1948).
29. P.R. Owen, T.R. Nonweiler and C.H.E. Warren, *Preliminary Note on the Design and Performance of a Possible Supersonic Fighter Aircraft*, RAE Technical Note Aero 1960 (June 1948). It is significant that Winter and Multhopp, as German scientists, were able to work on the research aircraft but not, at that time, on the fighter derivative.
30. *Report of RAE Advanced Fighter Project Group*, RAE Report Aero 2300 (November 1948).
31. One problem encountered as aircraft approached the speed of sound was that of the much higher aerodynamic loads encountered. Control reversal could occur because movement of the control surface could produce a force adequate to twist the wing or tailplane supporting it, giving a control effect in the opposite sense to that applied. Thus a major problem in the development of supersonic aircraft was to design in adequate stiffness and strength at acceptable weight.
32. *Report of RAE Advanced Fighter Group* (n. 30 above): 11. The RAE noted that 'it may be extremely difficult, or even impossible, to keep the structure weight down to that assumed. . . . It is hardly to be expected that on a type of structure having many new features such as very thick skin, less than half the usual ratio of thickness to chord, and large sweepback angle, a firm will achieve first time the same high structural efficiency [as for subsonic types]. . . . This difference . . . is unfortunate but time does not permit the repetition of the work.'.
33. *Report of RAE Advanced Fighter Group* (n. 30 above): 11.
34. An 'Experimental Requirement' ER 103, issued in late 1947, preceded this. The advantages of English Electric were an industrial management that was more solid than that usually found in the aircraft companies and a design team led by Teddy Petter that had, largely through its own initiative, launched the Canberra jet bomber.
35. The initial post-war assumptions were that Britain would be able to make bombs at a rate of 15 a year from 1951 while the USSR was considered to be behind Britain in the essential technology, science and industrial resources (although P.M.S. Blackett independently estimated that the USSR would be able to make 40 bombs a year from 1952). Also, in 1949, there was said to be 'desperate anxiety' in the USA over the shortage of uranium which was a limiting factor in their production programme. Margaret Gowing, *Independence and Deterrence, Britain and Atomic Energy 1945–1952*, Vol. I, (Macmillan, 1974): 167–8, 360. The 'Rotor' programme to upgrade the radar defences and place the operations rooms underground (described in Chapter 8) should also be seen as part of this British technological effort for atomic defence.
36. Paper by Air Commodore Grandy at the Designers' Convention, Ministry of Supply, 15 and 16 December 1953, PRO AIR 64/167.
37. Sir Charles Gardner, DGWRD, *The Future of Military Aviation*, lecture to the Imperial Defence College (19 June 1957). The hybrid rocket/gas turbine interceptors developed to prototype stage as the Saunders-Roe SR.53 and Avro 177 were cancelled in the aftermath of the Sandys 1957 Defence White Paper. However it was by then becoming clear that the gas turbine engine, with reheat, as in the Lightning, could give impressive climb performance without the complication of mixed powerplants and dangerous chemical fuels.
38. Bill Gunston, *Fighters of the Fifties* (Stephens, 1981): 74.
39. Mike Dobson (n. 10 above).
40. *Report of the Designers' Convention*, Ministry of Supply, 15 and 16 December 1953, PRO AIR 64/167.
41. *Report of the Sixth Fighter Tactical Convention*, 1–4 July 1952 held at Central Fighter Establishment, West Raynham, PRO AIR 64/180 (1952).
42. *Report of the Designers' Convention* (n. 40 above).
43. Project brochure: *Aircraft Project P 8. Fighter Weapon System to Specification F.155.T. (OR 329)*, English Electric Company Ltd, ex-RAE Museum now RAF Museum Archives AC 94/27/227 (4 October 1955). Although this 2-seat development of the Lightning was not built the brochure sets out well the details of the development of guidance and the thinking on interception technique and flight profile which was applied to the Lightning.
44. Brian Carroll, personal communication (25 June 1998).
45. A. Stratton, Memo: Instruments and Photography Department, RAE, PRO AVIA 65/369 (5 December 1956).
46. T.C.G. James, *Defence Policy and the Royal Air Force 1956–1963*, Air Historical Branch Archives (Ministry of Defence, 1987). One senior civil servant likened Sandys, in pursuit of these policies, to 'a programmed tank'.
47. In fact the 1956 Defence White Paper (Command 969) had initiated the policy that close defence of vulnerable areas was outmoded and that the aim was to concentrate on defending the V-bomber bases using a guided weapon

(Bloodhound) to break up enemy formations before engagement by UK fighters. Discussed in Jack Gough, *Watching the Skies* (HMSO, 1993): 178.

48. 'By 1950 we knew enough to design a supersonic aircraft as good or better than the F104. I've never understood why we were so slow to do it.' Handel Davies, interview with the author (15 April 1998). Handel Davies was Head of Aero Flight (J) in the immediate post-war period and, among other posts, was Scientific Adviser at the Air Ministry (1955–6), Deputy Director of the RAE (1959–63) and British chairman of the management boards for Concorde and Tornado.
49. C.H.E. Warren (n. 3 above).
50. Humphrey Wynn, *The RAF Strategic Nuclear Deterrent Forces: their origins, roles and deployment, 1946–1969* (HMSO, 1994): 7–43.
51. 'The RAF did not wait to hear whether Attlee's inner Cabinet had ratified the decision to "go nuclear", before it started to lay plans as "delivery vehicles" for nuclear bombs. As a result, three new V-bombers were being readied for deployment by the end of the 1950s.' Solly Zuckerman, *Monkeys, Men and Missiles* (Collins, 1988) quoted in Wynn, *The RAF Strategic Nuclear Deterrent Forces* (n. 50 above).
52. For an account of the British replication of the 'Fat Man' plutonium weapon, see Brian Cathcart, *Test of Greatness* (Murray, 1994). OR1001, the procurement document for the atomic bomb, was issued on 9 August 1946.
53. *Report of the RAE Advanced Bomber Project Group*, RAE Report Aero 2246 (February 1948).
54. Handel Davies (n. 48 above). Apart from the single supersonic Lightning fighter programme there were several trans-sonic fighter programmes including the Supermarine Swift and Scimitar, the Hawker P1081 and Hunter, the Gloster Javelin and the De Havilland Sea Vixen.
55. *Speeding up Development, Production and Introduction into Service of Military Aircraft—Policy*, PRO AVIA 65/30 (1954).
56. PRO AVIA 65/369 includes P. Lee and K.W. Newby, *Tests in the RAE 10 ft by 7 ft High Speed Tunnel of a Winged Bomb mounted under the Fuselage of the Avro Vulcan*, RAE Tech Note 2365 (May 1955).
57. CA to DCAS, *Comet Aircraft Production Policy*, PRO AVIA 65/59 (21 April 1954).
58. C.H.E. Warren, J. Poole and D.C. Appleyard, *An Investigation into an Aircraft to Fly at a Mach Number of 2*, RAE Report No. Aero 2462 (June 1952).
59. *Defence against High Altitude Bombers by Mach 2 Fighters—a Joint RAE/RRE Study*, RAE Reports Aero 2513 and 2513A (June 1954).
60. *Joint RAE/RRE Supersonic Fighter Working Party*, PRO AVIA 65/369 (1954–1956). The putative fighter was a straight wing Mach 2 aircraft to OR 329.
61. J.R. Ewans, Chief Designer, A.V. Roe and Co. Limited, to Morien Morgan (5 January 1956, and his reply of 13 January); Morien Morgan to Secretary, MoS, FAO PDSB(A). Letters with Avro company project brochures on Avro 730 Supersonic High Altitude Aircraft to Specification R.156.T (July 1955), and Avro 731 Research Model of Avro 730, ex-RAE Museum now Archives Department, RAF Museum AC/94/27, 39, 40 and 41.
62. It was some hint of this, presumably, that led Reginald Maudling, as Minister of Supply, to make the first government reference to a supersonic airliner in October 1955. Reviewing the reasons for the cancellation of the Vickers V1000 (the RAF transport and airliner version of the Valiant bomber) Maudling questioned whether Britain could produce a jet airliner that would be competitive with American types like the Boeing 707 and whether 'we have got to surrender this particular field to the Americans until we can re-enter it with a true supersonic aeroplane, not before 1970'. Paper to Cabinet on cancellation of the Vickers V1000 by Reginald Maudling, Minister of Supply, PRO AVIA 19/811 (1955).
63. J. Weber, *Some Effects of Flow Separation on Slender Delta Wings*, RAE Technical Note Aero 2425 (1955).
64. C.H.E. Warren, Handel Davies and Sir John Charnley, meeting with the author (14 May 1998).
65. The concept of paradigm shift in technological development, explicitly borrowed from Thomas Kuhn, has been explored by Edward W. Constant in *The Origins of the Turbojet Revolution* (Baltimore, 1980). Interestingly, Constant in the main attributes paradigm shifts to individuals like Frank Whittle outside the mainstream profession but, in the case of the vortex flow for Concorde-like wings, the new theory arose and was promoted among the established practitioners of the RAE.
66. Jean Ross, conversation with the author (23 May 1994).
67. W.E. Gray, *The Case for Flying a Narrow Delta Glider in April 1959*, RAE Technical Memo Aero 618. The model experiments are described in *Dynamic Tests on Free-Flying Models of Slender Wings Subjected to Side-Gusts*, RAE Technical Note Aero 2983 (1958).
68. A typical example of this kind of work is represented by G.R. Allen and M.D. Chamberlain, *The Operation of Anti-G Suits with Air taken from the Turbine Engine Compressor*, RAE Technical Note Mech. Eng. 128 (July 1952).
69. Sir John Charnley (n. 14 above).
70. Handel Davies and Sir John Charnley (n. 64 above).
71. Sir John Charnley, personal communication (20 May 1998).

72. C.H.E. Warren (n. 3 above).
73. Austin Mair became Director of the Fluid Motion Laboratory, University of Manchester, and then Professor of Aeronautical Engineering at Cambridge (1952); Paul Owen, Director of the Engineering Research Laboratory at Manchester (1963) and then held the Zaharoff Chair of Aviation, Imperial College (1984); Alec Young held the Chair of Aeronautical Engineering at Queen Mary College, London (1954–78); Alan Thom at Oxford and A.G. Pugsley, Professor of Engineering Science at Bristol.

CHAPTER 3

Rotary-Wing Aircraft

MATTHEW UTTLEY

INTRODUCTION

In 1945 the helicopter was a marginal branch of aviation technology in the UK. The Armed Services had 25 operational helicopters and the industry comprised two teams, each employing fewer than ten designers. In contrast, by 1990 helicopters had become a key element in the UK's force posture, accounting for 40 per cent (by number) of the Armed Services' aircraft inventory.[1] They represented a significant segment of British aircraft manufacturing, with a turnover of £447 million and a research, development (R&D) and production workforce of approximately 9000.[2]

The role and impact of UK government research establishments throughout the period was varied. Between 1945 and 1965, despite some notable establishment projects, intramural[3] helicopter efforts were marginal, underfunded, and had only a limited impact on development projects in industry. After 1965, however, intramural helicopter research priority and government investment increased significantly, which was reflected in important practical achievements. This chapter analyses these trends and tentatively evaluates the impact of establishment helicopter research on technological innovation in British industry.

The chapter contains four main sections. The first outlines the wartime legacy in 1945 for subsequent intramural helicopter research. The second main section evaluates the major themes characterising establishment helicopter activities between 1945 and 1965. The third develops these themes for the period between 1965 and 1990. The final section discusses the findings in terms of the generic themes of this volume, introduced in Chapter 1. Since 1945 government research establishments have undertaken a spectrum of roles, ranging from advanced research aimed at supporting specific projects in industry, to providing technical advice to assist procurement ministries to act as 'intelligent customers'.[4] To keep the length of this chapter reasonable the focus is primarily on the first of these roles, namely the factors affecting the impact of intramural rotary research on industrial project developments.

POSTWAR ROTARY-WING RESEARCH—THE 1945 LEGACY

Before 1938, the British Air Ministry took a 'technical interest'[5] in rotary-wing aircraft. After 1919, for example, at the Royal Aircraft Establishment (RAE), Brennan received Air Ministry funding to develop a multirotor helicopter design.[6] During the 1920s and 1930s, the RAE evaluated autogyro prototypes under development in industry on behalf of the Air Ministry. However, the major factor precluding concerted intra- or extramural government-sponsored rotorcraft research was the rudimentary technological status of the helicopter in relation to fixed-wing aviation. By the late 1930s, whereas fixed-wing aircraft were in scale manufacture and operational in a range of civil and military roles, technical and development difficulties[7] encountered by helicopter designers meant that no rotorcraft prototypes had entered production.

During 1938, the Air Ministry adopted a more systematic helicopter industrial policy when:

> it became clear that, like every other important project in aviation, it [helicopter development] could not be run as a commercial undertaking and that Air Ministry control was essential if this country were to retain its position in relation to other countries.[8]

Stimulated by concerns that the German firm Focke Achgelis was threatening to establish a technological lead[9] by placing the FW 61 helicopter in production, the Air Ministry response was to sponsor three helicopter projects: the 'Gyroplane' AR.5 at AR III (Hafner Gyroplane) Company and the W.7 at the G. & J. Weir firm, both to meet Specification No. 22/38; and a single-seater single-rotor experimental helicopter at AR III to meet Specification No. S.10/39. By 1940, achievements on these contracts meant that British designers 'were on the verge of leading the world in helicopter development'.[10]

In July 1940 the government policy of supporting indigenous helicopter R&D was abandoned because of the higher priority of fixed-wing aircraft production[11] and the RAF's 'unwillingness to divert production manpower engaged in making normal RAF types of aircraft'.[12] The government's revised policy, formulated after secret intergovernmental aircraft manufacturing agreements reached with Washington in December 1940,[13] was that the UK would rely for any Service helicopter requirements on allocations from US helicopter production,[14] and that for the duration of the war it would place all 'its knowledge on helicopters' at the disposal of US industry.[15] Although Ministry of Aircraft Production (MAP) officials recognised that the policy risked placing the UK 'in an inferior position to the Americans commercially if and when we restart helicopter construction', it was accepted as 'the price we [the UK] pay for placing the war first and post war commerce second'.[16]

Somewhat paradoxically, given the 1940 industrial policy shift, intramural rotorcraft research was conducted on a previously unprecedented scale throughout the war. In late 1940, the MAP made a concession in its blanket policy on indigenous helicopter R&D when it sponsored work at the Central Landing Establishment (CLE) on a rotary-wing glider under the technical direction of Raoul Hafner, designated the 'Rotorchute'[17] project (Fig. 3.1). The objective was to develop a one-man rotary-wing Airborne Forces parachute. In 1942 this work was transferred to the Airborne Forces Experimental Establishment (AFEE) and extended to include the development of larger rotary-wing gliders.[18] Although rotary-wing gliders were never produced,

FIGURE 3.1: THE ROTORCHUTE ROTARY-WING GLIDER WAS DEVELOPED AT THE CENTRAL LANDING ESTABLISHMENT/AIRBORNE FORCES EXPERIMENTAL ESTABLISHMENT AFTER 1940 UNDER THE TECHNICAL DIRECTION OF RAOUL HAFNER (AIRBORNE FORCES MUSEUM ALDERSHOT)

CLE/AFEE activities were significant because most British prewar helicopter designers, including Hafner (former Chief Designer, AR III Construction) and J.A.J. Bennett (former Chief Designer, Cierva), were engaged in intramural research. AFEE work continued until October 1944, when it became apparent that rotary-wing gliders had limited operational utility, after which Hafner and other rotorcraft designers were recruited for helicopter development by industry.

An important consequence of the 1940 MAP policy shift was that it allowed the US helicopter industry to acquire a clear technological lead over the UK design teams that were emerging in 1945. Between 1940 and 1945, in the USA, the combination of specific military requirements, operational urgency and large-scale government finance and technical resources enabled US firms to transform the helicopter from an experimental to a production technology. In 1945, the US helicopter industry comprised 21 firms that had initiated approximately 80 different wartime helicopter programmes,[19] of which 16 types had undergone flight testing. Moreover, two US companies—Sikorsky and Nash-Kelvinator—commanded design and manufacturing staffs of over 1000 and 700, respectively.[20] By comparison, during 1945 two design teams, at the Bristol Aeroplane Company and G & J Weir, each comprising fewer than ten designers, represented the extent of the UK helicopter industry. Most significant, however, was that the MAP's 1940 decision to opt out of helicopters allowed the US firm Sikorsky to acquire a production lead over British firms by 1945, when it began scale manufacture of the R-4 helicopter.

By 1945, three important factors that were to influence postwar intramural helicopter research until the mid-1960s were in place. First, wartime CLE/AFEE activities established the concept of intramural helicopter research at the MAP and its successor procurement ministries. Second, the wartime period was important because US innovations transformed the helicopter from an experimental to a production technology with defined military and civil roles. And, third, the UK's 1940 decision to abandon indigenous helicopter development had enabled US helicopter firms to acquire a comprehensive manufacturing lead.

INTRAMURAL HELICOPTER RESEARCH, 1945–65

On 1 May 1945, the MAP adopted the approach that endured until 1965:

> The policy of this Ministry is to undertake research at our Establishments and stimulate interest of aircraft firms on Helicopters as we believe that there are potential fields, civil and military, for this type of aircraft.[21]

What emerged was a twin-track strategy. Government R&D funding was to be channelled to industry in an attempt to 'establish and foster a strong [indigenous] helicopter industry with design and production capacity to cater for all home and [potential future] export markets'.[22] In parallel, to support industrial development, the MAP planned 'a comprehensive programme including the setting up of a research group at RAE to deal with the difficult and complex aerodynamic problems associated with rotary wing aircraft'.[23] Policy, therefore, was to encourage aircraft firms to develop and produce 'all-British' helicopter technology, with the establishments providing research assistance.

Four factors underpinned the new policy. The first was UK government aircraft industrial policy *in toto*. During 1945, the impending decline in demand for fixed-wing production created pressures for government intervention 'to maintain employment and, hopefully, the profitability of a strategically and economically vital sector'.[24] Supporting helicopter R&D in the aircraft industry provided a mechanism for government to encourage aircraft firms to diversify away from fixed-wing manufacture. Second, wartime production priorities had meant that, by 1945, the UK was 'falling behind in the science of aerodynamics'.[25] Enunciated formally in February 1946, the first 'principle' of defence technology policy was to be 'concentration on research', as:

> scientific progress ... is so rapid that safety lies far more in the maintenance of an adequate organisation for pure and applied research than in the building up of stocks of obsolescent equipment.[26]

Against this imperative, MAP officials were keen to sponsor helicopter research because 'basic knowledge on helicopters [was] more limited in the United Kingdom than elsewhere at the end of the war and much ground had to be made up'.[27]

The third factor was MAP estimates of civil and military helicopter market potential. By 1945, imported Sikorsky R.4, R.5 and R.6 helicopters obtained under wartime agreements were entering UK military service,[28] and their performance indicated that future military demand would arise for replacement types. At the same time, the Brabazon Committee, tasked by the Cabinet with planning a programme of aircraft designs that would form the basis of postwar civil aircraft production, had produced preliminary specifications which included three civil helicopter types.[29] On this basis, MAP perceptions about future helicopter markets meant that officials were eager to encourage British firms to exploit this potential and provide supporting intramural research.[30] Finally, although recognising the US helicopter production lead,[31] MAP estimates in 1945 were that 'generally speaking, the design fundamental knowledge on rotating wing aircraft [in the UK] is at least as good if not better, than that in the USA'.[32] This assumption that British designers could easily 'catch up' with US firms provided an additional stimulus for industrial and intramural R&D helicopter sponsorship.

Sharing MAP market predictions, and eager to diversify, six UK aircraft manufacturers entered the helicopter field between 1945 and 1960 (Table 3.1). With the exception of Cierva, where rotorcraft development was a continuation of the firm's prewar activities, Fairey, Bristol, Saunders-Roe (Saro), Percival and Westland saw helicopter development as a mechanism for adjusting to declining fixed-wing demand.[33] In each case, the company Boards shared the perception that:

TABLE 3.1: MAIN UK HELICOPTER MANUFACTURING COMPANIES, 1945–60

Company	1945	1946	1947	1948	1949	1950	1951	1952	1953	1954	1955	1956	1957	1958	1959	1960
Westland Aircraft																
Cierva Autogiro Co.								Helicopter interests merged								
Saunders-Roe								as Saunders-Roe Helicopter								
Bristol Aeroplane Co.								Division in 1950								WESTLAND
Percival																AIRCRAFT
Fairey Aviation								Bristol Aircraft								

TABLE 3.2: UK HELICOPTER R&D AND PRODUCTION PROJECTS, 1945–66

Project	Company	Comments
W.9	Cierva	Research project initiated in 1944 and abandoned in 1946
Air Horse	Cierva	Research project abandoned in 1951
W.14	Cierva/Saro	Entered production as the 'Skeeter' after 1948
P.531	Saro/Westland	Entered service as the 'Scout' and 'Wasp' after 1960
Gyrodyne	Fairey	Commenced in 1946 and abandoned in 1949
Jet Gyrodyne	Fairey	Experimental compound helicopter
Rotodyne	Fairey	Commenced in 1947 and abandoned in 1962
Ultra-Light	Fairey	Commenced in 1954 and abandoned in 1956
P.74	Percival	Commenced in 1952 and abandoned in 1956
Type 171	Bristol	Produced as the 'Sycamore'
Type 173	Bristol	Produced as the 'Belvedere'
WS.51	Westland	Produced as the 'Dragonfly'
WS.55	Westland	Produced as the 'Whirlwind'
WS.58	Westland	Produced as the 'Wessex'
Widgeon	Westland	First flight in 1955
Westminster	Westland	Abandoned in 1960

> By the time the war had ended, the brief period of service of these aircraft [helicopters] had given an indication of the usefulness of this type of flying machine. There was good reason to believe that the helicopter would be needed on an ever-increasing scale for military applications, and for numerous jobs in the civil field.[34]

Between 1945 and 1965 these firms undertook a range of R&D and production projects (Table 3.2). Within the parameters of the 1945 'twin-track' policy, the research establishments' primary remit was to provide scientific advice and research data to enhance innovation on these projects.

Intramural Helicopter Research Achievements, 1945–65

After 1945, the main body responsible for drawing up and assigning relative priorities for intramural helicopter research was the Ministry of Supply (MoS) Rotating Wing Committee (RWC), which comprised MoS officials and, *inter alia*, the directors of establishments conducting helicopter-related research.[35] With industry's needs in mind, the RWC formulated the intramural helicopter programme in conjunction with the MoS Assistant Directorate of Research and Development (Helicopters), a specialist branch with terms of reference that included 'supervision and co-ordination of research work at Establishments'.[36] Additional inputs were provided by the Helicopter Committee of the Aeronautical Research Council (ARC), which included MoS officials as well as representatives from industry.[37] Together, these bodies framed overall intramural helicopter research strategy and policy.

Helicopter research was located principally at the RAE and the Aeroplane and Armaments Experimental Establishment (A&AEE). RAE teams focused on helicopter aerodynamics and structures, performance estimation and flight control. The A&AEE Rotary Wing Section specialised in performance and handling trials for all helicopters, and assessment trials of modifications and new equipment.[38] Both establishments investigated instrumentation and night flying, engine-off techniques and performance analysis. In addition, the National Gas Turbine Establishment (NGTE) had a more limited role which included analysing the general problems of applying jet and turbine power units to helicopters.[39]

In the immediate postwar period noteworthy RAE and A&AEE research included basic theoretical work and experimental data collection on rotor characteristics, including model tests in the RAE's 24ft wind tunnel.[40] Other activities included flight evaluation and experiments at Farnborough on imported American Sikorsky R.4 helicopters.[41] In addition, a tower for full-scale rotor testing was constructed under MoS contract at the Bristol Aeroplane Company in 1946, which proved 'useful in the elimination of rotor flutter'[42] on a range of prototypes under development in industry.

During February 1951, in an effort to rationalise and augment establishment research, the RWC formulated the Helicopter Research Programme, which identified work necessary to 'ensure development of satisfactory helicopters', and prioritised items considered 'of greatest importance'.[43] Compiled under 73 detailed headings,[44] the programme advocated urgent intramural research in ten key areas: stability and control; performance; project work; rotor aerodynamics; tunnel, tower and ground testing; vibration; flight testing techniques; operating techniques; instrumentation; and structural and mechanical engineering. Within the programme, projects included longitudinal stability studies (RAE), tests of performance after engine failure (A&AEE), jet propulsion systems analysis (RAE), rigid rotor helicopters studies (RAE), and blade motion (RAE) research.[45]

In July 1958 the RWC advocated that the 1951 programme 'should be revised and re-cast in the light of general progress and the facilities currently available',[46] and a second Rotary Wing Research Programme was formulated. The 1958 programme was drawn up under three headings:[47] 'Engineering and Structural' (general structural design criteria, fatigue aspects, improved methods of component construction, vibration aspects, methods of propulsion, operational equipment and mechanical reliability); 'Flight and Operation' (work necessary to meet various operational requirements, including all-weather, blind and night flying, and deck landing); and 'Aerodynamics', or the bulk of basic research work, the main items being concerned with aerodynamic problems (e.g. higher forward speeds), together with control and stability. Despite minor revision,[48] the 1958 programme remained the basic template for intramural helicopter research until 1965.

Under the 1958 programme, establishment scientists and engineers pursued various research projects. In the 'Engineering and Structural' field, the RAE Structures Department collected data on helicopter vertical landing. Under the 'Flight and Operation' heading the A&AEE (DA Nav.) and RAE (Instruments Department) investigated the general problems of blind/night flying and restricted-site operations, and the RAE (Naval Aircraft Department) investigated deck landings to satisfy Royal Navy (RN) requirements. Under the 'Aerodynamics' heading the A&AEE developed flight-testing techniques and performance reduction methods, and RAE teams gave some consideration to high-speed flight, largely to meet RN requirements for higher cruising speeds in the antisubmarine warfare (ASW) role. Important work on stability theory was also initiated at the RAE (Naval Aircraft Department) to the extent that 'the successful development of the British flight control system owed a great deal to the pioneering work of Mr Oliver St. John, Head of the Helicopter Section of Avionics, at RAE Farnborough, from 1957–1963.'[49]

Marshalled under the strategic direction of the MoS RWC, therefore, a number of intramural projects were initiated after 1945 that were 'for the most part, linked to the development of a specific helicopter type against its Operational Requirement'.[50] According to one assessment, these projects had important impacts on helicopter development in UK firms: 'it was from these foundations that the British helicopter industry grew'.[51]

Constraints Affecting Intramural Helicopter Research, 1945–60

Between 1945 and 1965, official recognition of 'high-quality'[52] achievements on individual intramural helicopter projects was only part of a more complex picture. By contrast, taken in aggregate, intramural helicopter research received only limited resources and it remained a major source of concern to government officials and committees responsible for framing helicopter R&D policy.

The marginal status of intramural helicopter research compared to other aviation branches is evident in two statistics. First, between 1945 and 1965 rotary-wing projects attracted only between 0.1 and 0.6 per cent[53] of total annual MoS intramural aircraft research budgets. Second, intramural helicopter research was assigned a negligible proportion of establishment personnel: in 1959, of the 1515 qualified scientists and engineers (QSEs) at the RAE,[54] only 1.3 per cent were employed on helicopter projects, and the entire helicopter team comprised just 15/20 full-time staff.[55] More revealing, where RAE helicopter work was concentrated—the Naval Aircraft Department, Instruments and Photographic Department, Mechanical Engineering Department and the Structures Department—of some 417 QSEs[56] only 4 per cent were employed on helicopter-related research.

In October 1946, when the MoS conducted its first intramural helicopter research review,[57] officials noted that practical progress had been limited because qualified 'personnel available [were] very limited in number'. This reflected:

> (a) the early development state of the helicopter; (b) the scarcity of technical specialists, pilots, ground crews and aircraft; (c) the fact that the helicopter is not yet generally accepted as the coming thing.[58]

However, after 1946 the limited allocation of financial and labour resources became an evident source of frustration for those officials responsible for framing helicopter research policy. Despite the 1951 RWC Helicopter Research Programme, in March 1954 the ARC Helicopter Committee felt it necessary to highlight:

> the dire need of more rapid progress in this work to provide a rational backing for design work, both that now in progress and that which will be required for the new service and civil requirements. At no time since the birth of the A.R.C. was the situation of fixed wing aircraft design in relation to fundamental work so lamentable as that of the helicopter is now.[59]

The ARC recommended an immediate increase in MoS intramural investment and pointed to the detrimental effects of existing expenditure levels for industrial development.[60] Again, during March 1954 claims of 'underfunding' were reiterated by MoS officials:

> the expansion of effort by transferring half-a-dozen White Paper grades at R.A.E. to full-time work in this field, and if possible the appointment of one officer at R.A.E. as a co-ordinator of all fundamental work on rotating wing aircraft would pay dividends quite out of proportion to the additional expenditure (?£50,000 p.a.) involved. Some action of this kind is indeed essential.[61]

Similarly, in April 1959 an MoS review of the 1958 RWC Research Programme's implementation revealed similar anxieties: 'in general ... progress was slow and effort available very short'.[62] The review concluded that 'few items on the Programme are receiving any worthwhile attention on the [intramural] R&D side';[63] 'little basic research [was] under way' in the 'Rotor Performance' category, despite the fact that it was 'frequently stated ... by designers [in industry] that there [was] a pressing need to augment our present knowledge'; establishment 'airframe aerodynamics' research was non-existent, despite its 'increasing importance as higher speeds are demanded'; and there was no establishment wind tunnel experimentation, despite pressing requirements from industry for rotor/tunnel size and tunnel interference data.[64]

Rather than specific intramural helicopter projects themselves, therefore, the central issue for the period 1945–65 is why establishment helicopter efforts were so limited and controversial compared to other branches of aviation. Over the period, British helicopter firms initiated 12 major R&D projects involving prototype construction and flight testing,[65] and succeeded in manufacturing 1022 helicopter units.[66] Moreover, British-built military helicopters had been extensively deployed in low-intensity conflicts including, *inter alia*, the Malayan and Suez operations.[67] Consequently, why did an aircraft type that had entered scale production, and which became a significant component of the UK force posture, attract so little intramural effort compared to other aviation technologies?[68] And what factors explained the evident frustration of officials and committees responsible for framing intramural helicopter research policy? The answers to these key questions lay in the interplay of the following three important variables that diverted UK government resources and priority away from establishment helicopter research.

The first factor that diverted resources was that helicopter firms, rather than the establishments, were the primary locus of government-sponsored research. This is evident in two statistics. First, almost all government helicopter research expenditure was channelled to industry: in FY 1954/55, for example, the establishments received just 4.1 per cent of the total MoS helicopter research budget.[69] Second, helicopter firms contained practically all the qualified labour: during 1958, for example, helicopter-specific QSEs in industry[70] numbered 628, compared to the 15/20 at the establishments.[71]

One reason for this pattern was UK government priorities. Under the MAP's 1945 'twin-track' helicopter approach, considerations of industrial policy and market potential dictated that industrial R&D took priority over intramural research in the allocation of scarce technical resources. MAP priorities were manifest between 1944 and 1946 when, with government sanction,[72] the helicopter experts coopted to the AFEE during the war were encouraged to seek employment in the newly formed industrial design teams. Consequently, after 'wartime occupations' were removed, key helicopter designers including, among others, R. Hafner (Bristol), O. Fitzwilliams (Westland) and J.A.J. Bennett (Fairey),[73] were recruited by firms rather than the establishments.

A further reason was the nature of basic helicopter research *per se*. By 1945, in the USA Sikorsky had established a production lead in tail-rotor-configured gear-driven helicopters. In an attempt to 'leapfrog' the US manufacturing lead, MoS policy was to sponsor the development of highly speculative helicopter configurations. In the light of existing establishment labour constraints, MoS officials concluded that industry would have to conduct the extensive fundamental research required to fulfil the new policy. In this regard, after MoS endorsement of the RWC's March 1951 Helicopter Research Programme,[74] procurement officials had conceded:

> that we [the MoS] shall be unable appreciably to increase work on helicopter problems at Establishments until priorities are revised and more staff is available. The only way we can start to meet the various recommendations for increase of effort is to make more use of firms.[75]

The result was that industry, rather than the establishments, undertook extensive theoretical and applied research on tandem-rotor (Bristol Type 173), three-rotor (Cierva 'Air Horse'), compound (Fairey Gyrodyne), compound tip-jet (Fairey Jet-Gyrodyne, Fairey Ultra-Light, Fairey Rotodyne) and pure tip-jet (Percival P.74) designs. This locus of effort in industry led the MoS to allocate the bulk of its helicopter research budget to firms rather than the establishments. Illustrative here was MoS research funding for the Fairey Rotodyne project which, in FY 1954/55 alone, exceeded cumulative total intramural helicopter expenditure between FY 1950/51 and FY 1954/55.[76]

The second factor diverting government resources away from intramural projects was technology transfer. Between 1946 and 1960, MoS policy was explicitly to foster 'all-British' helicopter technology, namely indigenous R&D and manufacture. In reality, however, UK production was dominated by Westland—a licensed producer of US Sikorsky technology—to the extent that of the 1022 UK helicopters manufactured by 1960, approximately 75 per cent (767) were Westland variants of Sikorsky designs (Table 3.3).

Westland's commercial success, achieved through licensed production of the Sikorsky S.51, S.55 and S.58 helicopters, reflected a two-stage process. The first stage was that a series of factors in the USA, including extensive government R&D funding and large-scale postwar orders, enabled Sikorsky to maintain its 1945 production lead over 'all-British' helicopter capabilities. The second stage was that, through its licensed production strategy, Westland 'anglicised' this lead and exploited the competitive cost, lead-time and performance edge that the licences provided over comparable 'all-British' projects.

TABLE 3.3: UK HELICOPTER PRODUCTION, 1945–60[77]

Firm/helicopter	First service	Role	Total military	Total civil	Total export
Westland Dragonfly[a]	Apr. 1950	Rescue	88	8	41
Bristol Sycamore	Feb. 1953	Rescue	109	2	66
Westland Whirlwind[b]	June 1954	Rescue/ASW	308	20	90
Saunders-Roe Skeeter	Jan. 1957	Army light	66	2	10
Westland Widgeon	Sept. 1957	Civil	—	6	6
Westland Wessex[c]	Apr. 1960	ASW/assault	*c.*200		
Total					1022
Total Westland					767

[a] Westland-licensed variant of the Sikorsky S.51 helicopter.
[b] Westland-licensed variant of the Sikorsky S.55 helicopter.
[c] Westland-licensed variant of the Sikorsky S.58 helicopter. The 1960 figures for the Wessex are approximate, based on 1960 production estimates.

The significance of Westland's licensed production was that it enabled the British Armed Services to 'free-ride' on US Sikorsky technology. However, UK Service access to US technology via Westland simultaneously reduced the requirement and priority for MoS-sponsored 'all-British' helicopter projects. An indicator of the extent to which this process undermined resource allocation to 'all-British' research was the degree of MoS antagonism towards Westland's approach. During 1953, for example, after Westland had won S.51 and S.55 orders at the expense of 'all-British' alternatives, the MoS Assistant Secretary (Air) argued that Westland/Sikorsky licensed agreements raised 'a real problem', as 'any further extension of US design in [the UK] would make a nonsense of our [MoS] efforts to stimulate a positive policy into British helicopter efforts'.[78]

The third factor militating against MoS resource allocation to 'all-British' helicopter research, both intra- and extramural, was the 'priority problem'.[79] The source of the 'priority problem' was that, despite the MoS commitment to sponsor 'all-British' helicopter R&D, under extant aircraft procurement arrangements the *scale* of MoS intra- and extramural R&D funding and the *extent* of military production orders were dictated by other government departments. Decisions by these bureaucracies—primarily the Admiralty and the Air Ministry—created the 'priority problem' by diverting R&D resources and manufacturing orders away from 'all-British' helicopter programmes.

The first contributor to the 'priority problem' was the Admiralty, the purchaser of the majority of UK helicopter production output between 1945 and 1960. By 1958, 97.3 per cent (by number) of cumulative RN procurement was of Westland licensed variants of Sikorsky designs.[80] Admiralty preferences for proven Sikorsky/Westland technology, however, meant that the RN had little interest in pressing for scarce MoS aircraft R&D resources to be spent on intra- or extramural 'all-British' helicopter projects, as this expenditure would have competed with other MoS-sponsored naval aviation programmes. The second contributor was inter-Service rivalry between the War Office and the Air Ministry. After 1945, although the War Office was quick to recognise the battlefield potential of the helicopter,[81] extant procurement procedures were that the Air Ministry was responsible for ordering and paying for all aircraft destined for Army use. Air Ministry priorities, however, were for fixed-wing strategic aircraft, and the RAF thus had little interest in helicopter R&D and orders to support Army requirements.[82] The result was that the Air Ministry's objectives further undermined MoS helicopter R&D priority and expenditure. Without military backing and orders for 'all-British' projects, therefore, the MoS was unable to release large-scale helicopter R&D funding for intra- and extramural helicopter research.

The key indicator of the 'priority problem' is relative MoS helicopter and non-helicopter extramural R&D expenditure. Had helicopter and non-helicopter projects received equal priority then, *ceteris paribus*, they could have been expected to have attracted a pro rata share of MoS R&D expenditure. Between 1945 and 1955, the MoS sponsored 26 aircraft R&D projects in industry: ten helicopter (38 per cent by number) and 16 non-helicopter (62 per cent by number). Given their numerical importance, however, helicopter projects received a negligible proportion of MoS aircraft R&D funding (Table 3.4): despite accounting for nearly 40 per cent of the total *number* of projects, they attracted less than 6 per cent of total *expenditure* in any one year. These statistics suggest that helicopter R&D in industry was substantially 'underfunded' by the MoS compared to expenditure on fixed-wing projects.[84]

TABLE 3.4: MOS EXTRAMURAL AIRCRAFT R&D EXPENDITURE, 1950–55 (ACTUAL EXPENDITURE IN CURRENT PRICES) (£000)[83]

Description	1950/51	1951/52	1952/53	1953/54	1954/55
Airframes:					
Service helicopter prototypes	0	0	0	2	269
Civil helicopter prototypes	0	0	0	0	200
Research helicopter prototypes	192	246	423	440	820
Total of helicopter airframes	192	246	423	442	1,289
Total of all airframes (helicopters + non-helicopters)	12,351	12,898	17,700	20,527	22,550
Helicopter airframes as a proportion of all airframes (%)	1.5	1.9	2.3	2.2	5.7
Engines:					
Total helicopter engines	0	43	150	664	1,710
Total of all engines	16,508	15,580	22,700	25,974	33,307
Helicopters as a proportion of all engines (%)	0	0.27	0.66	2.5	5.0
Total of all helicopter work	192	289	573	1,106	2,800
Total of all airframe and engine work	28,859	28,478	40,400	46,501	55,857
All helicopter work as a proportion of all work	0.66	1.0	1.4	2.4	5.0

Moreover, relative MoS intra- and extramural helicopter project expenditure levels discussed previously indicate that 'underfunding' of establishment helicopter projects was even more acute than that encountered by industry.

In summary, between 1945 and 1965 the performance of intramural helicopter research was mixed. On the individual research project level, establishments made notable contributions to applied helicopter R&D in industry. In aggregate terms, however, the combination of industry's role in basic helicopter research, the 'Westland effect' and the 'priority problem' all reduced the scale and scope of intramural investment and effort. In short, the position after 1945 was aptly summed up in the MoA's 1965 submissions to the Committee of Inquiry into the Aircraft Industry: 'intramural research on helicopters has never exceeded 2 per cent of total effort. This stems from the low priority given to helicopters and the lack of specialist staff'.[85]

INTRAMURAL HELICOPTER RESEARCH, 1965–90

By early 1964, important policy developments had led to an increase in the priority of intramural helicopter research. A key stimulus was the MoD Defence Research Committee (DRC), which noted on 20 February 1964 that Ministers 'were likely to take a keen interest in helicopter problems',[86] and at which the MoA decided to establish a working party 'to examine from a technical point of view helicopter research'.[87] The new working party had wide-reaching terms of reference:[88]

(1) To consider the present state of knowledge and technique at home and abroad, in respect of:
 (a) Rotor aerodynamics
 (b) Power plants and rotor drive systems
 (c) Control and guidance systems
 (d) Mechanical and structural engineering
(2) To consider the general level and content of the current Ministry of Aviation rotary-wing research programme.
(3) To examine the scope for improved design capability in the UK and for new applications and methods.
(4) To investigate the part that future research could play in raising the standards of rotary-wing technology and extending the usefulness of rotorcraft.
(5) To report to the Defence Research Committee.[89]

Consequently, by mid-1964 intramural helicopter research had finally attracted high-level policy interest and priority.

Three factors led to the 1964 policy shift. The first was MoD recognition that, in numerical terms, the helicopter was becoming a core military aviation technology: MoA estimates indicated that, by 1966, the ratio of helicopters to fixed-wing aircraft would be 1: 1 for the RN, 5: 1 for the Army and 1: 10 for the RAF, and that relatively, 'over the following years these [helicopter] numbers are likely to increase'.[90] An increase in basic research effort was thus considered essential to support ongoing development of what was becoming a major technology in the Services' aircraft inventories. The second factor was events in industry. In 1960, through a process of government-induced mergers in the aircraft industry, Westland had acquired the helicopter interests of Fairey, Saunders-Roe and Bristol to become the UK's sole helicopter manufacturer. The emergence of a monopoly supplier had focused intramural research on supporting fewer helicopter projects, increased the clarity of the industry/establishment interface, and generated incentives for additional intramural support.[91] The final factor was the spur of emerging Anglo-French collaboration, and Westland's negotiations with Sud Aviation for production of the SA.330 helicopter.[92] Impending international collaboration elevated the political priority of intramural research in support of broader technical negotiations with the French government.

Between 1965 and 1990 the RAE remained the primary focus of intramural helicopter research. RAE helicopter work was located in several departments, but the main efforts were concentrated in the Materials and Structures Department, the Flight Management Department and the Mission Management Department. In its specialist rotorcraft division, the Materials and Structures Department focused on rotor aerodynamics and the dynamics of rotors and fuselage response. In several RAE divisions at Bedford, the Flight Management Department covered research into flight mechanics, handling and control, simulation studies and cockpit displays. The Mission Management Department specialised in avionic systems with a particular emphasis on night vision aids. On a lesser scale, other establishments had a role in the development of generic technologies or innovations with helicopter applications. Illustrative here was infrared (IR) detector work conducted in the 1970s by the Royal Signals and Radar Establishment (RSRE), Malvern, which was carried out in parallel with RAE studies into low-light TV (LLTV) for helicopter applications.[93] Another example was A&AEE work at Boscombe Down in the mid-1980s that played a vital role in investigating the effects of icing on helicopters.[94]

The scale and dispersal of helicopter research after 1965, coupled with data limitations,[95] preclude an analysis of intramural projects *in toto*. As a result, short case studies are adopted to highlight how important characteristics of the establishments' activities before 1965 evolved subsequently. Two caveats accompany this approach. The first is that the cases selected do not necessarily constitute a representative sample of the establishments' work or *modus operandi*. For example, 'successful projects' are highlighted rather than those that failed, partly because the former have understandably been described more extensively in technical journals. The second concerns the concept of 'successful project'. As MacKenzie and Spinardi point out, major methodological and empirical constraints hamper measurement of the impact of establishment research. These include, among others, the problem that:

> It is not enough simply to note the involvement of a defence organisation or applications to defence purposes. Rather, the crucial issue is whether the history of the technology in question would have been different in some fundamental technical respect had the defence organisation or application not existed. Because that necessitates hypothesising a counterfactual alternative history, it involves delicate questions of judgement rather than empirical investigation alone.[96]

With this in mind, the case studies are intended to provide a snapshot rather than a comprehensive assessment of how the pre-1965 constraints affecting intramural helicopter activities evolved subsequently.

Case Study 1: Amelioration of the 'Priority Problem' and Rotary Avionics Research

Until the mid-1960s, the 'priority problem' detracted from UK intra- and extramural helicopter R&D expenditure both in absolute terms and relative to fixed-wing aviation. Subsequently, the combination of increased helicopter reliability and their adoption for an increasing array of military roles meant that by the early 1990s helicopters accounted for approximately half of the Services' aircraft. Throughout the intervening period, operational priority was accompanied by a commensurate increase in the priority for intramural helicopter-related research.

Illustrative of this trend was establishment helicopter avionics research. Before the mid-1970s, the RAE (Farnborough) had undertaken some studies into the use of electronic aids for helicopters, but on a much smaller scale than for conventional aircraft. Instead, helicopter requirements were met by the 'simple modification of "off-the-shelf" equipment', in part because there were major reservations in the avionics industry as to whether helicopters posed problems that could not readily be solved 'by a re-application of fixed wing equipments'.[97] However, after the mid-1970s increasing Service priority for helicopters led to funding for extensive rotary-specific avionics research at the RAE (Fig. 3.2).

A systematic research programme that covered helicopter flight control systems, navigation and approach guidance, visual displays and pilot's TV displays was initiated.[98] Throughout the late 1970s and 1980s, RAE work in these and related fields increased considerably, to the extent that, by 1990, helicopter-specific projects included the integrated flight management system which covered, *inter alia*, the development of flight aids to ensure that aircrews were presented with essential information in a readily assimilable manner.[99] Equipment installed in the system included colour CRT displays, a speech recogniser, synthetic speech output, and interfaces with most of the aircraft's systems and instruments, which were designed 'to form an easy-to-use, pilot-friendly system'.[100]

In addition to mission management and task guidance research, a major area of RAE work was in the field of vision sensors. With the development of night vision goggles (NVGs), important areas of the RAE's activities included studies to improve cockpit lighting, as part of a wider research programme to facilitate 24-hour helicopter operations.[101] In terms of current avionics system research, a measure of the extent to which the 'priority problem' that afflicted establishment research in the 1950s and 1960s has been remedied is the close and extensive collaboration between Westland and DERA on 'carefree handling' concepts.[102]

Case Study 2: Establishment Research and the Technology Transfer Dimension

Before 1965, a primary characteristic of British helicopter procurement was reliance on US technology via Westland licensed production. After 1965, technology transfer remained a constant feature in both industry and government research establish-

FIGURE 3.2: THE COMPLETE WESTLAND WESSEX COCKPIT INCORPORATING RAE-DEVELOPED AVIONICS SYSTEMS (© CROWN COPYRIGHT, DERA)

ments. However, in contrast to the UK governments' earlier antagonism, after 1965 the approach towards technology transfer was more systematic and positively policy driven.

After 1965, in industry, Westland relied on two types of technology transfer. On the one hand, the firm consolidated its US connection. In 1967 it had acquired a further Sikorsky licence to manufacture the SH.3, designated the Sea King. Following the 1985 'Westland Affair'[103] and the United Technologies (UTC)—Fiat bid, Westland acquired permission to produce the Sikorsky Black Hawk helicopter under licence. Most recently, since the UK government's July 1995 decision to purchase the Apache attack helicopter, Westland is assembling the US system and manufacturing the transmission units.[104] On the other hand, since the mid-1960s Westland has established extensive collaborative links within Europe. The Memorandum of Understanding (MoU) signed in February 1967 between the French and UK governments committed both to the joint purchase of three different helicopter types developed by Westland and Aerospatiale: the SA.330 (Puma), the SA.340 (Gazelle) and the WG13 (Lynx).[105] In 1979, under an Anglo-Italian MoU, Agusta and Westland formed a joint company—European Helicopter Industries (EHI)—to design, develop and manufacture the EH101 Merlin helicopter.

Since 1965, intramural helicopter research has also been characterised by extensive collaboration and technology transfer with both US and European partners. By the

TABLE 3.5: NATIONS INVOLVED IN SOME MULTILATERAL COLLABORATIVE AEROSPACE RESEARCH AGREEMENTS[106]

Nation	NATO–AGARD	TTCP	CAARC	GARTEUR
India			+	
Australia		+	+	
New Zealand		+	+	
Canada	+	+	+	
United States	+	+		
France	+			+
Germany	+			+
Netherlands	+			+
UK	+	+	+	+

NATO–AGARD: North Atlantic Treaty Organisation—Advisory Group for Aerospace Research and Development (includes all 15 NATO members)
TTCP: The Technical Cooperation Program
CAARC: Commonwealth Advisory Aeronautical Research Council
GARTEUR: Group for Aeronautical Research and Technology in Europe

early 1980s the RAE was participating in a number of multilateral aerospace research agreements, the most important of which are summarised in Table 3.5. Within the terms of these agreements research arrangements have varied: some have been longer term whereas other have been applied; some have involved the exchange of information, whereas others have involved cooperative or complementary research; some have been intergovernmental agreements implemented by their respective R&D establishments, and others (e.g. GARTEUR) have included defence contractors.[107]

By 1990 the RAE had participated in a spectrum of collaborative helicopter research programmes, a sample of which is shown in Table 3.6. RAE helicopter collaboration highlights two trends: first, like the UK helicopter industry, basic research has been highly internationalised, with establishment scientists involved in extensive technology transfer; and second, intramural projects have involved both European and US partnerships.

Case Study 3: The RAE/Westland Connection, the British Experimental Rotor Programme (BERP), and 'World Class' Technology

The evidence suggests that until the mid-1960s the lack of priority for intramural helicopter research had adverse implications for the RAE's contribution to development and production in industry. Since the 1970s, however, and the increased priority for intramural helicopter research, a major achievement of the RAE Materials and Structures Department has been in the fields of rotor-blade aerodynamics and composite materials. Under the umbrella of the British Experimental Rotor Programme (BERP), joint RAE/Westland research led to fundamental improvements in rotor design and fabrication. Establishment innovations contributed directly to British helicopter performance enhancements, and provided UK industry with a world lead in a key subsystem niche.

TABLE 3.6: EXAMPLES OF RAE INVOLVEMENT IN COLLABORATIVE HELICOPTER RESEARCH UNDER MULTILATERAL AGREEMENTS, 1990[108]

Agreement	Research programme	Nation
TTCP	Tail rotor flight dynamics Work on MilSpec 8501 (Helicopter Handling Qualities)	US
AFARP	Blade research	UK/France
GARTEUR	Assessment of mathematical models covering helicopter flight and aerodynamics Rotorcraft concepts	UK, France, Germany, Netherlands
AGARD	Involvement in Working Group on System Identification Techniques	NATO states
Non-specific	Cabin vibration	US Army
	Sensor images/data overlays	US Army
	Icing	France/US
	Blade research (pressure distribution measurement)	France/UK/Australia

The BERP has addressed two interrelated problems associated with helicopter rotors. The first is 'retreating blade stall', i.e. degradation of helicopter handling qualities and increased fuel consumption caused by the blade and control vibration which is brought about by the difference in velocity between advancing and retreating main rotor blades. The second is the 'advancing blade tip compressibility', encountered as the advancing rotor blade velocity approaches Mach 1. The major breakthrough of the BERP was the development of a new rotor blade that ameliorated these constraints.

The BERP originated in the mid-1970s. Westland Helicopters Limited launched a project to assess the utility of RAE composite material innovations for rotor blade manufacture.[109] Sea King tail rotors in composite materials were compared with orthodox metal units, and the firm found that:

> On two counts substantial advantages emerged, first, the fatigue life of the composite rotor was seen to increase dramatically and, second, the methods of production were economical and capable of achieving fully reproducible quality in components difficult to manufacture in metal.[110]

The results were encouraging enough for Westland to initiate composite Sea King main rotor development, which proved satisfactory in prototype flight trials.[111] At this stage the RAE suggested a joint project marrying RAE aerodynamic design expertise with Westland's practical experience of blade applications using composite materials. The primary RAE inputs to the project, now designated the BERP, were the development of a new range of aerofoils and fundamental swept-tip blade research, which was then to be fed into the final blade design.[112]

What emerged from the BERP was an entirely novel rotor design. Before the 1980s most helicopter rotor blades had employed a constant chord length and aerofoil section along the entire blade span. In contrast, the BERP rotor (Fig. 3.3) had a

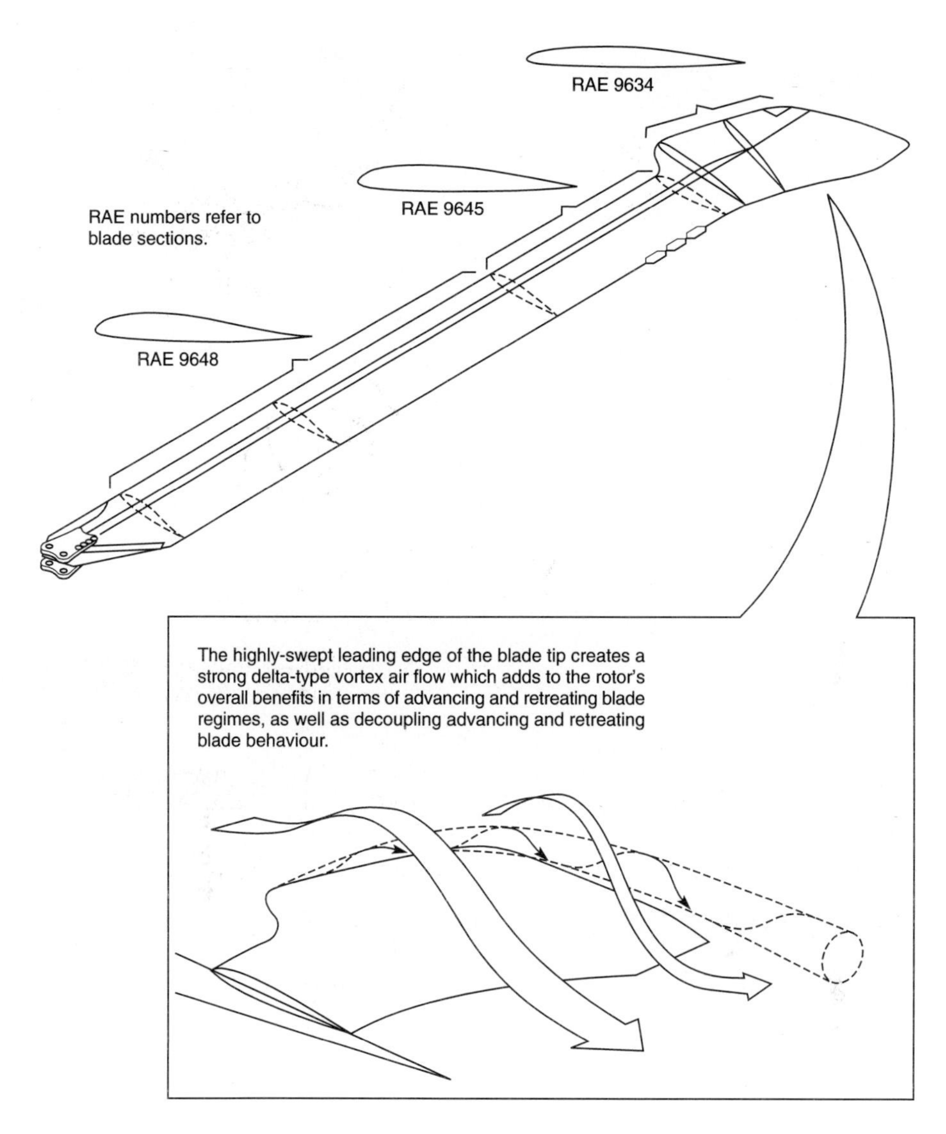

FIGURE 3.3: THE BERP BLADE CONFIGURATION (REDRAWN FROM D.J. TYLER AND A.H. VINCENT, 'FUTURE ROTORCRAFT TECHNOLOGY DEVELOPMENT', *THE AERONAUTICAL JOURNAL*, 100, NO. 1000 (1996): 457)

new blade tip design, a cambered aerofoil section, and was fabricated in advanced composite materials.

In a number of vital respects the BERP was a major achievement by RAE scientists and industry. The first indicator of project success was rotor performance. The new blade tip provided substantial benefits in advancing and retreating blade regimes while decoupling advancing and retreating blade behaviour. The tip, in conjunction with distributed cambered aerofoil sections (Fig. 3.3), provided substantially greater thrust (>30 per cent) than contemporary designs, and extended the forward speed potential of the edgewise rotor in excess of 200 knots. As White points out, 'the

FIGURE 3.4: WESTLAND LYNX HMA MK 8 FITTED WITH BERP ROTORS (© CROWN COPYRIGHT, DERA)

key to these performance increases is the combination of tip and distributed cambered sections; each is required to fully exploit the potential of the other'.[113] The thrust increases achieved have been reflected in a commensurate increase in helicopter top speed: when BERP rotors were applied to a Westland Lynx (Fig. 3.4) it established a new world helicopter speed record of just over 216 knots in August 1986.[114]

Second, industry officials have been the first to acknowledge that RAE teams 'played a vital role in the development and characterisation of [the] composite systems'[115] that were essential for rotor blade fabrication. Third, quality control procedures were critical to composite blade development. The collection of vital quality control data covering material characteristics, both physical and chemical, over a wide range of environmental conditions, was 'strongly led'[116] by RAE, Farnborough, in topics including artificial and natural ageing. Finally, a measure of the longer-term success of the RAE contribution is the contemporary application of BERP results to composite rotor blade development for the

EH101 Merlin project.[117] Currently, to retain their technological edge, GKN Westland Helicopters and DERA are collaborating on BERP-related initiatives which include:[118]

- Design of new high-performance rotor aerofoils.
- Work to improve the lift capability of thin blade tip sections.
- Use of computational fluid dynamics (CFD) methods[119] for aerodynamic aerofoil design, and the calculation of complete rotor aerodynamic behaviour.
- Investigation of 'aeroelastic tailoring' and the deliberate use of aerodynamic couplings to control blade twist in different regions of the helicopter flight envelope.

DISCUSSION

A central contention in this chapter is that the history of postwar establishment helicopter research falls into two discrete periods: the first between 1945 and 1965, and the second between 1965 and 1990. This section briefly discusses these findings in relation to some of the generic themes introduced in Chapter 1.

In important respects, the history of intramural helicopter research between 1945 and the mid-1960s corresponds broadly to the orthodox historiography of British aircraft projects *in toto* for this period. The dominant interpretation—the 'Project Cancelled'[120] paradigm—is alluded to by Sir Roy Fedden:

> When history comes to be written it will be found that the years 1946—56 have played an enormously important part in aeronautical advance, but in spite of this the period has not been a vintage one for Britain.[121]

The central tenets of the orthodoxy are that UK government procurement policies sustained 'too many' aircraft firms and 'too many' projects. In turn, it is argued that this approach was 'detrimental ... in spreading the country's industrial and design effort too thinly',[122] leading to a 'dispersion of effort'[123] and the 'crowding out' of promising development lines. Two arguments subsumed under the 'Project Cancelled' paradigm are relevant to the case of intramural helicopter research. On the one hand, the dispersion of government R&D resources across a spectrum of aircraft development meant that *aggregate* funds assigned to helicopter work were limited. On the other hand, the scarce resources that were allocated to helicopter research were then dissipated in support of the 16 projects under development (Table 3.2) at the UK's six helicopter firms (Table 3.1).

In other respects, the broader process and policy environments militated against 'all-British' technology in establishments and industry. First, under extant procurement procedures, purchasing and funding authority was dictated in part by individual Service ministries. Examples of the negative effects of inter-Service rivalry on fixed-wing aircraft projects have been documented elsewhere; in the helicopter case, its primary manifestation was in the 'priority problem'. Finally, despite deliberate attempts on the part of successive procurement ministries to protect the industrial base and remain independent from the USA, the reality was that service preferences for

Westland licensed production further detracted from intra- and extramural 'all-British' helicopter development and expenditure.

After 1965, however, new policies and processes gradually ameliorated these constraints. First, increased Service requirements for operational helicopters were reflected in a growth in helicopter-specific establishment research, as exemplified in the avionics field. Second, as government policies emphasised international collaboration rather than protectionism, the establishments' international collaborative links were increasingly organised on a systematic basis. Finally, the government-induced industrial rationalisation that resulted in Westland's emergence as the sole UK helicopter company had a range of positive effects. In contrast to the earlier period, intramural research was subsequently targeted at supporting a small number of helicopter projects in industry. Moreover, industrial rationalisation facilitated a close and continuous relationship between establishment and industry teams, exemplified in the BERP. In short, as contemporary technologists from industry have described it, the close partnership between the establishments and Westland that has emerged over several decades:

> has prospered despite many changes within both organisations. It has already yielded a number of examples of world class rotorcraft technology and it is fundamental to the future exploitation of many ... potential improvements.[124]

Throughout the late 1990s, since the conversion to agency status, the primary challenge for establishment scientists working in the rotary field remains one of consolidating this link with industry and, as with their achievements on the BERP, contributing to the development of new world-beating innovations.

1. Figures calculated from *The Military Balance 1994–1995* (Brassey's, 1994).
2. Figures for 1994 calculated from: European Commission, *The European Aerospace Industry: trading position and figures 1994* (European Commission, 1994).
3. The terms 'intramural' and 'extramural' research are used subsequently in accordance with Ministry of Supply practice. 'Intramural' research refers to research activities conducted in government research establishments. 'Extramural' research refers to research conducted in British industry, either under government contract or on a private venture basis.
4. For a general discussion of the roles undertaken by government research establishments, see M.J. Lighthill, 'The Royal Aircraft Establishment', in *The Organization of Research Establishments*, edited by J. Cockcroft (Cambridge University Press, 1965): 28–9.
5. J.A.J. Bennett, *Rotary Wing Development*, PRO AVIA 15/954 (23 December 1940).
6. For more detail, see R. Graham, 'The Brennan Helicopter', in *A History of British Rotorcraft 1866–1965*, edited by R.A.C. Brie (Westland Helicopters Ltd., 1968): 35.
7. For more detail, see M.R.H. Uttley, *Licensed Production Versus Indigenous Innovation: A Case Study of Helicopter Research, Development and Production in Britain, 1945–1960* (Ph.D. thesis, Lancaster University, 1997), chap. 2.
8. J.A.J. Bennett, *Rotary Wing Development* (n. 5 above).
9. M.R.H. Uttley, *Licensed Production Versus Indigenous Innovation* (n. 7 above).
10. E.J. Everett-Heath, *British Military Helicopters* (Arms and Armour Press, 1986): 27.
11. Minute from the Deputy Director of Research, Development and Technology, PRO AVIA 15/954 (14 January 1943).
12. Minute from MO9, PRO AVIA 15/954 (9 December 1940).
13. United States National Archives (USNA), State Department Collection, *Statement on UK Government Assistance to Commercial Aircraft Industry*, File 841.3333 (16 April 1954).
14. Minute from the Chief Executive, PRO AVIA 15/954 (4 February 1943).

15. Minute by MAP Director of Scientific Research, PRO AVIA 15/954 (2 February 1943).
16. Minute by MAP Director of Scientific Research (n. 15 above).
17. A number of analyses have emerged on the 'Rotorchute' project and subsequent rotary-gliders under development at the CLE and AFEE between 1940 and 1944. For more detail see E.J. Everett-Heath, *British Military Helicopters* (n. 10 above): 21–4; R. Sturtivant, *British Research and Development Aircraft: seventy years at the leading edge* (Haynes, 1990): 69–71 and R. Hafner, 'Four decades of rotating wing activity 1928–1965', in *A History of British Rotorcraft 1866–1965* (see n. 6 above): 38–9. The relevant MAP file covering developments is PRO AVIA 44/536, Work of the A.F.E.E. 1940–5.
18. See R. Hafner, 'Four decades of rotating wing activity' (n. 17 above): 39.
19. USNA, Grover Loening (Chairman of the Helicopter Committee of the National Advisory Committee for Aeronautics), *The Helicopter and Its Future*, File No. 312.248 (United States War Production Board Collection, 3 December 1943).
20. Memorandum from Major Tappan to General Chidlaw, File No. 421.12, Air Corps and Army Air Forces Library Collection (8 May 1943).
21. Minute by MAP Director of Scientific Research, Ministry of Aircraft Production, PRO AVIA 15/954 (1 May 1945).
22. Note on the MoS Helicopter R&D Programme with Special Emphasis on the Civil Transport Role, MoS Air 2(a), PRO AVIA 65/6 (17 October 1953).
23. Minute by MAP Director of Scientific Research, Ministry of Aircraft Production, PRO AVIA 15/954 (1 May 1945).
24. K. Hayward, *Government and British Civil Aerospace: A Case Study in Post-War Technology Policy* (Manchester University Press, 1983): 13.
25. *The Supply of Military Aircraft*, Cmd 9388 (HMSO, 1955): 3.
26. *Statement Relating to Defence, 1947*, Cmd 7042 (HMSO, 1947).
27. Second Report of the Select Committee on Estimates, 'The supply of military aircraft', *House of Commons Proceedings*, 34, session 1956–7 (1956): vii.
28. See D.W. Wragg, *Helicopters at War: A Pictorial History* (Robert Hale, 1983): 47.
29. For more detail, see H. Self, 'The status of civil aviation in 1946', *Journal of the Royal Aeronautical Society* (September 1946): 759.
30. Note on the MoS Helicopter R&D Programme with Special Emphasis on the Civil Transport Role (n. 22 above).
31. A. McClements, 'Review of British rotorcraft activities 1943–46', in *A History of British Rotorcraft 1866–1965* (n. 6 above): 103.
32. A. McClements, 'Review of British rotorcraft activities' (n. 31 above).
33. See, for example, J.W.R. Taylor and M.F. Allward, *Westland 50* (Ian Allen, 1965): 158; C.H. Barnes, *Bristol Aircraft Since 1910* (Putnam, 1988): 361 and H.A. Taylor, *Fairey Aircraft Since 1915* (Putnam, 1974): 390.
34. J.W.R. Taylor and M.F. Allward, *Westland 50* (n. 33 above): 106.
35. For more details about the structure, composition and activities of the RWC, see *MoS Rotating-Wing Committee, 1951–1958*, PRO AVIA 65/107.
36. *The Work of AR/H*, PRO AVIA 65/723 (17 July 1959). This document describes the various internal reorganisations of the branch between its formation in 1947 and 1959.
37. The Helicopter Committee of the ARC was appointed in 1946 to encourage helicopter research. By July 1959, the Helicopter Committee was disbanded, and the RWC remained the only official committee devoted to rotary-wing R&D.
38. For a general overview of A&AEE work, see J. Poole, 'Rotorcraft work at the Aeroplane and Armament Experimental Establishment', *Journal of the Royal Aeronautical Society* 67 (1963): 501–13.
39. A. McClements, 'Review of British rotorcraft activities' (n. 31 above): 112.
40. A. McClements, 'Review of British rotorcraft activities' (n. 31 above): 112.
41. P. Wilby, 'Helicopter research at the RAE', *Air Clues* (February 1990): 67.
42. Ministry of Supply Aeronautical Research Council, *Review for the Years 1939–1941* (HMSO, 1950): 25.
43. *Proposed Helicopter Research Programme*, PRO AVIA 65/107 (23 February 1951).
44. *The Work of AR/H*, PRO AVIA 65/723 (17 July 1959).
45. *Proposed Helicopter Research Programme* (n. 43 above).
46. *A Note on the Rotary Wing Research and Development Programme Prepared by the Ministry of Supply Rotating Wing Committee*, PRO AVIA 65/723 (April 1959).
47. *A Note on the Rotary Wing Research and Development Programme* (n. 46 above).
48. See, for example, *Ministry of Aviation Deputy Controller of Aircraft (Research and Development) Helicopter Research Programme*, PRO AVIA 65/1642 (17 October 1961).
49. R. Gardner and R. Longstaff, *British Service Helicopters: A Pictorial History* (Robert Hale, 1985): 172–3.
50. *A Note on the Rotary Wing Research and Development Programme* (n. 46 above).

51. R. Gardner and R. Longstaff, *British Service Helicopters* (n. 49 above): 42.
52. R. Hafner, 'British rotorcraft', *Journal of the Royal Aeronautical Society* 70, pt. 1 (1966): 241.
53. Percentages calculated from: PRO AVIA 65/6, MoS Air 2(a) (23 October 1953) and *A Note on the Rotary Wing Research and Development Programme* (n. 46 above).
54. Aggregate employment figures derived from M. J. Lighthill, 'The Royal Aircraft Establishment' (n. 4 above): 30.
55. *A Note on the Rotary Wing Research and Development Programme* (n. 46 above).
56. *A Note on the Rotary Wing Research and Development Programme* (n. 46 above).
57. A. McClements, 'Review of British Helicopter Activities', PRO AVIA 15/2245 (RDL2L, 7 December 1946).
58. A. McClements, 'Review of British Helicopter Activities' (n. 57 above).
59. *Rotating Wing Aircraft*, MoS Air 2(a), PRO AVIA 65/6 (5 March 1954).
60. *Rotating Wing Aircraft* (n. 59 above).
61. *Rotating Wing Aircraft* (n. 59 above).
62. Note of Meeting on Wednesday, 22nd April, 1959, to Consider Future Helicopter Work and the Best Organisation to Cope with It, PRO AVIA 65/723 (24 April 1959).
63. Note of Meeting on Wednesday, 22nd April, 1959 (n. 62 above).
64. Note of Meeting on Wednesday, 22nd April, 1959 (n. 62 above).
65. *Statement of Expenditure on Rotary Wing Aircraft*, PRO AVIA 65/6 (10 May 1954).
66. Figures compiled from *Report of the Committee of Inquiry into the Aircraft Industry, 1964–65*, Cmnd 2853 (HMSO, 1965): 121–2 and R. Hafner, 'British rotorcraft' (n. 52 above): 241.
67. See, for example, J. Dowling, *RAF Helicopters: the first twenty years* (HMSO, 1992).
68. Compare, for example, the much greater allocation of intramural resources to guided missile technology as discussed in S. R. Twigge, *The Early Development of Guided Weapons in the United Kingdom, 1940–1960* (Harwood, 1993), and this volume.
69. Calculated from: *Extra-Mural Expenditure on Rotating-Wing Aircraft*, PRO AVIA 65/6, 1954; figures for Parliamentary Question 10264 (1954) and PRO AVIA 65/6, MoS Air 2(a) (23 October 1953).
70. *Helicopter Research and Development*, PRO AVIA 65/723 (14 March 1958).
71. *A Note on the Rotary Wing Research and Development Programme Prepared by the Ministry of Supply Rotating Wing Committee*, PRO AVIA 65/723 (April 1959).
72. See, for example, *A Review of British, American, French and Russian Helicopters*, ADI (Tech.) Paper 3/47, PRO AVIA 55/9029 (April 1947).
73. See, for example, G.E. Walker, 'Activities of Weir, Hafner and Bristol 1933–1960', in *A History of British Rotorcraft 1866–1965* (see n. 6 above): 74.
74. RAE Aero. Dept., *Proposed Helicopter Research Programme—Summary*, PRO AVIA 65/107 (23 February 1951).
75. Minute by H.F. Vessey to DMARD, *Increase of Effort on Helicopter Work*, PRO AVIA 65/107 (9 March 1951).
76. Figures calculated from *Statement of Expenditure on Rotary Wing Aircraft*, Air 2a (n. 65 above); and, intramural estimates in PRO AVIA 65/6, MoS Air 2a (23 October 1953).
77. Figures for Table 3.3 from *Report of the Committee of Inquiry into the Aircraft Industry* (n. 66 above): 121–2; R. Hafner, 'British rotorcraft' (n. 52 above): 24.
78. Memorandum for AS/Air 2 to US (Air), *Extension of Manufacturing License Agreement by Westland Aircraft Ltd.: financial and policy considerations*, PRO AVIA 65/6 (7 October 1953).
79. See M.R.H. Uttley, 'British helicopter developments 1945–1960: government technology policy in a changing defence environment', in *Defence Science and Technology: adjusting to change*, edited by R. Coopey, G. Spinardi and M.R.H. Uttley (Harwood, 1993): 125–41.
80. Statistics calculated from *Helicopter Research and Development*, PRO AVIA 65/723 (14 March 1958). According to MoS production figures, cumulative Admiralty helicopter orders by 1958 totalled 336 units, of which 327 were contracts with Westland. The remaining nine purchases were contracts for the Mk 50 and Mk 51 variants of the Bristol Type 171.
81. See, for example, *A Study of the Potential Applications of the Helicopter to Army Roles*, PRO AVIA 55/90 (11 August 1945).
82. For a discussion see: P.A. Towle, *Pilots and Rebels: the use of aircraft in unconventional warfare, 1918–1988* (Brassey's, 1989): 126; M.R.H. Uttley, *Licensed Production Versus Indigenous Innovation* (n. 7 above): chap. 4–7.
83. Figures for Table 3. 4 from 'Extra mural research and development expenditure on rotating wing aircraft' (n. 69 above).
84. A contrasting interpretation of the statistics is that helicopter projects were inherently less resource-intensive than fixed-wing development. However, this hypothesis is undermined by MoS assessments of helicopter project resource requirements during the period:
'. . . basic knowledge on helicopters had been more limited in the United Kingdom than elsewhere at the end of the war, and much ground had to be made up. Furthermore, . . . [helicopter projects initiated] . . . concerned

radically different types of helicopters; some research was concerned with machines designed to carry light loads; some with medium and heavy loads and others with investigating new principles of helicopter flight' (Second Report of the Select Committee on Estimates, n. 27 above). Instead, the figures imply a 'priority problem' that was militating against large-scale government R&D support for 'all-British' helicopter development in industry.

85. Note by the Ministry of Supply, *Committee of Inquiry into the Aircraft Industry—Helicopters*, PRO AVIA 65/1642 (29 April 1965).
86. Ministry of Defence Defence Research Committee, *Minutes of a Meeting*, PRO DEFE 10/570 (20 February 1964).
87. Ministry of Defence Defence Research Committee, *Minutes of a Meeting* (n. 86 above).
88. Ministry of Defence Defence Research Committee, *Note by the Secretaries, Helicopter Working Party*, PRO DEFE 10/571 (10 June 1964).
89. Ministry of Defence Defence Research Committee, *Note by the Secretaries, Helicopter Working Party* (n. 88 above).
90. Note by the Ministry of Supply (n. 85 above).
91. P. Wilby, 'Helicopter research at the RAE' (n. 41 above): 67.
92. Ministry of Defence Defence Research Committee, *Minutes of a Meeting* (n. 86 above).
93. For more information, see P.J. Cooper, *Forever Farnborough: Flying the Limits 1904–1996* (Hikoki, 1997): 153–4.
94. See, for example, M.H. Battersby, P.A. Knowles, 'Icing clearance criteria for UK military helicopters', *Helicopter Rotor Icing Symposium Proceedings* (Royal Aeronautical Society, January 1986): 64.
95. Official papers relating to intramural research for the period prior to 1965 are in the public domain, and are housed in the Public Record Office, Kew. However, much of the establishments' subsequent work is covered by the Official Secrets Act or commercial-in-confidence considerations.
96. D. MacKenzie and G. Spinardi, 'The technological impact of a Defence Research Establishment', in *Defence Science and Technology: adjusting to change* (n. 79 above): 86.
97. H.B. Johnson, 'Helicopter avionics—UK research programme', *AGARD Conference Proceedings CP-148 on the Guidance and Control of V/STOL Aircraft and Helicopters at Night and in Poor Visibility* (AGARD, 1975): R-I-1.
98. H.B. Johnson, 'Helicopter avionics' (n. 97 above): R-I-3–R-I-4.
99. P. Wilby, 'Helicopter research at the RAE' (n. 41 above): 70.
100. P. Wilby, 'Helicopter research at the RAE' (n. 41 above): 70.
101. P. Wilby, 'Helicopter research at the RAE' (n. 41 above): 70.
102. D.J. Tyler and A.H. Vincent, 'Future rotorcraft technology development', *The Aeronautical Journal*, 100, no. 1000 (December 1996): 457.
103. For more details, see S. Croft, 'The Westland Helicopter crisis: implications for the British defence industry', *Defense Analysis* 3, no. 4 (1987): 291–303.
104. *Statement on the Defence Estimates*, Cmd 3223 (HMSO, 1996): 59.
105. For a general discussion of Anglo-French helicopter collaboration, see A.G. Draper, *European Defence Equipment Collaboration: Britain's involvement 1957–87* (RUSI, 1990), 79–85.
106. From T.H. Kerr, 'The role of the Research Establishments in the developing world of aerospace', *Aeronautical Journal* (December 1982): 362.
107. T.H. Kerr, 'The role of the Research Establishments' (n. 106 above).
108. From: P. Wilby, 'Helicopter research at the RAE' (n. 41 above): 67–72.
109. R.W. White, 'Development in UK Rotor Blade Technology', *Future* (Spring 1985): 31.
110. *Composite Materials in Aircraft Structures*, edited by D.H. Middleton (Longman, 1990): 305.
111. *Composite Materials in Aircraft Structures* (n. 110 above).
112. P. Wilby, 'Helicopter research at the RAE' (n. 41 above): 68.
113. R.W. White, 'Development in UK Rotor Blade Technology' (n. 109 above): 32.
114. D.H. Middleton, *Composite Materials in Aircraft Structures* (n. 110 above): 305.
115. D.H. Middleton, *Composite Materials in Aircraft Structures* (n. 110 above): 306.
116. R.W. White, 'Development in UK Rotor Blade Technology' (n. 109 above): 39.
117. For more details on the EH101 project see, for example, P. Beaver, *Modern Military Helicopters* (Patrick Stephens, 1987): 70–2.
118. D.J. Tyler and A.H. Vincent, 'Future rotorcraft technology development' (n. 102 above): 453–4.
119. For more general discussion of CFD activities in the RAE, see: Royal Aerospace Establishment (Farnborough), 'Royal Aerospace Establishment', in *World Aerospace Technology '90*, edited by S.R. Broadbent (Sterling, 1990): 207–8.
120. The expression 'Project Cancelled' is derived from D. Wood, *Project Cancelled: the disaster of Britain's abandoned aircraft projects* (Tri-Service Press, 1990). It is adopted as a convenient shorthand for the dominant interpretation

of the performance of British aircraft R&D, production and procurement between 1945 and 1965.

121. R. Fedden, *Britain's Air Survival: an appraisement and strategy for success* (Cassell, 1957): 11.
122. Second Report of the Select Committee on Estimates (n. 27 above): xvi.
123. E. Devons, 'The aircraft industry', in *The Structure of British Industry*, edited by D. Burns, Vol. II (Cambridge University Press, 1964): 59.
124. D.J. Tyler and A.H. Vincent, 'Future rotorcraft technology development' (n. 102 above): 460.

CHAPTER 4

Ground-Based Air Defence and ABM Systems

STEPHEN TWIGGE

INTRODUCTION

The development of guided weapons in the UK represents a significant aspect of British defence strategy. Second only to the military aircraft sector in terms of government R&D defence expenditure, guided weapons are central both to the work of the research establishments and the output of the British defence industry (see Table 4.1 for details of spending in year 1967–68). This chapter explores the role undertaken by the research establishments in this process. Attention is focused first on the organisation of the establishments and the difficulty of coordinating the activities of the various establishments concerned with guided weapons. The events of the early postwar period are then explored, with emphasis on the expansion of the guided weapons programme and the introduction of industry. The remaining sections deal with the development of Bloodhound, early research into ABM defence and the Rapier air defence system. These projects are chosen as they both highlight a number of themes common to all guided weapons projects, and provide a framework against which the often fragmented process of guided weapons development can be viewed.

GUIDED WEAPONS DEVELOPMENT AND THE ROLE OF THE RESEARCH ESTABLISHMENTS

In the guided weapons field the defence R&D establishments are responsible for performing two major functions. First, to provide a focus of scientific expertise allowing the government and Services to formulate new operational requirements, and second, to facilitate a close working relationship between the military and industry without which the transition from operational requirement to industrial production is impossible to manage efficiently. In both these tasks the establishments fulfilled a critical role in assessing technological risk and in judging the cost-effectiveness of major investments in the development and production of new guided weapons systems. The work undertaken within the establishments can broadly be divided between three main activities: the direct support of projects under development; the management of ranges and testing equipment to meet the needs of both Services and industry; and exploratory research aimed at improving existing

TABLE 4.1: ESTIMATED DEFENCE R&D EXPENDITURE (£M) 1967–68, MINISTRY OF DEFENCE AND MINISTRY OF TECHNOLOGY[1]

	Extramural		Intramural		Total
	Research	Development	Research	Development	
Aircraft	6.4	82.9	5.6	12.6	107.5
Guided weapons	5.0	39.1	3.8	4.7	52.6
Military space	0.2	2.0	0.6	0.5	3.3
Ship construction, underwater weapons and equipment	2.3	5.5	2.7	3.3	13.8
Ordnance, military ground vehicles, engineering stores, biological and chemical defence	0.8	4.4	5.0	12.8	23.0
Other electronic	4.6	13.8	4.8	4.4	27.6
Other (net)*					32.9
Total	19.3	147.7	22.5	28.3	260.7

*Includes works expenditure charged to the Ministry of Public Buildings and Works.

equipment or establishing the next generation of technology. In this latter capacity a major part of the research was carried out extramurally under contracts with industry for prefeasibility work and the design and manufacture of experimental hardware generated by the R&D activities of the establishments.[2]

As with much other military equipment, the development of guided weapons was not the sole preserve of any single establishment. Indeed, given the complexity of guided weapons and the requirement to integrate many different technologies and operating environments, the concentration of research into a single establishment was not considered feasible. As a result, since 1945 a variety of different establishments have been associated with various aspects of guided weapons development. In 1950, for example, the number of establishments undertaking work on guided weapons totalled at least 13. Details of these establishments are given in Table 4.2. In general, however, the majority of work was concentrated at four main establishments.

1. The Royal Aircraft Establishment (RAE). Retitled in 1918 and located at Farnborough, the RAE served as the main centre within the UK for the development of air-to-air, air-to-ground and surface-to-air guided weapons.

Research conducted at the establishment was undertaken within the Guided Weapons Department and concentrated primarily on improving fair-weather guidance using electro-optical (TV, infrared, laser) techniques, exploratory work on aerodynamic control, and feasibility studies into antiballistic missile (ABM) development. To support these various research programmes the RAE maintained three weapons ranges, at Aberporth, Larkhill and West Freugh. In addition, the RAE operated an experimental range at Woomera, in Australia, which was used for prototype development, weapons evaluation and service acceptance trials.[4]

2. The Royal Armament Research and Development Establishment (RARDE). Operating from two sites, at Fort Halstead and Langhurst, RARDE was responsible for the development of ordnance, including antitank guided missiles, warhead design and safety, and arming systems. The guided weapons work undertaken at RARDE was conducted primarily into guidance and control techniques (manual, semiautomatic and automatic), including the development of weapons used from rotary-wing aircraft. In addition to its R&D activities, RARDE was also the main UK centre for detailed wargame studies of land warfare to examine the tactical impact of the various weapons systems under development.
3. The Royal Signals and Radar Establishment (RSRE). Located at Malvern in Worcestershire, RSRE, and its forerunners, was the main UK centre responsible for research into all aspects of radar. In relation to missile development, the work undertaken at RSRE included the integration of completed radar units into ground-to-air guided weapons, research into ground and missile-borne radar homing and command and surveillance systems, portable surface-to-air (SAM) missiles, and close air defence systems. To manage these activities, a Guided Weapons Department with three divisions was established within RSRE.
4. The Admiralty Surface Weapons Establishment (ASWE). Based at Portsdown, ASWE was responsible for integrating guided weapons into a shipborne environment, with particular emphasis directed towards ship-to-ship and ship-to-air guided weapons, including target surveillance, tracking and missile guidance.

Given the large number of research establishments connected with missile development, overall coordination of the guided weapons programme within the UK remained largely fragmentary. In 1951, to redress this deficiency, the position of Controller of Guided Weapons and Electronics (CGWL) was created to serve as the focus for all guided weapons research undertaken in the UK. The role of CGWL was to act as a link between government departments, industry and the various research establishments.[5] In August 1973, however, following publication of the Rayner Report, the position of CGWL was abolished. As a direct result, responsibility for guided weapons and electronic systems was transferred to the Sea, Land and Air Controllerates, with overall coordination the responsibility of the Defence Ministry's Procurement Executive Management Board, under the direction of the Chief Executive (Procurement Executive).[6]

TABLE 4.2: MAIN ESTABLISHMENTS ENGAGED ON GUIDED WEAPONS.[3]

Establishment	Type of work
Royal Aircraft Establishment (RAE), Farnborough	Technical consideration of projects and trials
RAE / Rocket Propulsion Department (RAE/RPD), Westcott	Solid and liquid rocket fuels; motor design and testing
National Gas Turbine Establishment (NGTE), Pyestock Cove	Ramjet research and development
Telecommunications Research Establishment (TRE), Gt Malvern	Radar homing and infra-red guidance for air-to-air missiles. Investigations into semi-active CW homing techniques.
Radar Research and Development Establishment (RRDE), Malvern	Research and development of ground- based radar equipment and tracking radar; instrumentation for ranges
Signals Research and Development Establishment (SRDE), Christchurch	Research and development into telemetry
Explosives Research and Development Establishment (ERDE), Waltham Abbey	Research on liquid and solid fuel rocket motors and explosives
High Explosives Research Establishment (HERE), Fort Halstead, Sevenoaks	Research in collaboration with RAE into fragmentation and warhead design
Armament Research Establishment (ARE), Fort Halstead, Sevenoaks	Research into fragmentation
Armament Design Establishment, (ADE) Fort Halstead, Sevenoaks	Warhead design
Admiralty Signals & Radar Establishment (ASRE), Halsmere	Research and development of type 901 beam-riding radar set and type 902 radar gathering set
Admiralty Gunnery Establishment (AGE), Teddington	Research and development on servo-mechanisms and ship directors
Joint Services Experimental Research Laboratory (SERL), Baldock	Work on valve reliability and valve design

To manage the various projects and maintain a central focus for research, the Policy Co-ordinating Committee of Guided Weapons and Electronics was established. The Committee, which met on a monthly basis, was chaired by the Deputy Under-Secretary (Policy) Procurement Executive and was charged with keeping overall guided weapons policy under review and to provide a forum for the discussion of any matter of general concern. Owing to the lack of a clear focus, however, the organisational structure came under criticism from industry. To address these concerns, the post of Director General Guided Weapons and Electronics was later established within the MoD to act as a sponsor for the guided weapons industry. To provide coordination at the research level, the Controller R&D Establishments and Research (CER) was provided with a Director of Research Guided Weapons (DRGW) to exercise control and coordination over the work undertaken at the various research establishments over the whole guided weapons field.[7] The organisational changes affecting the research establishments during this period were mirrored within central government, with responsibility for conducting guided weapons research residing in turn with the Ministry of Supply, the Ministry of Aviation, the Ministry of Technology, the Department of Trade and Industry and the Ministry of Defence. At the departmental level similar changes were also evident, with authority for approving guided weapons projects given to a number of different bodies, including the Defence Research Policy Committee (DRPC) (1946–63), the Weapons Development Committee (WDC) (1963–71), the Defence Equipment Policy Committee (DEPC) (1971–84) and the Equipment Policy Committee (EPC) (1984–90).

POSTWAR DEVELOPMENTS

In contrast to Nazi Germany, the development of guided weapons in wartime Britain was not given a high priority. Characterised by small-scale studies undertaken at the Projectile Development Establishment (Aberporth), the Air Defence Research and Development Establishment (Malvern) and the Royal Aircraft Establishment (Farnborough), the work was largely experimental and lacked overall coordination. The outcome of these studies resulted in two projects. The first was named Spaniel and consisted of a projectile, a few inches in diameter and weighing 130 lb, which was controlled along a radio beam before reaching its target. The second was known as Brakemine, a ground-launched missile using radar as its means of guidance. In 1944 the two projects were combined to meet a requirement issued by the Ministry of Supply for a medium-range missile with a ceiling of 40,000 feet for air defence. Initially the project was given the code name LOPGAP (liquid oxygen and petrol guided antiaircraft projectile), but with the transfer of the project to the RAE in 1947 this was later amended to RTV 1 (Rocket Test Vehicle 1).[8]

Given the psychological impact of the German V-weapons during the war, the development of rocket technology was widely assumed to have a significant role in any future conflict.[9] Despite this expectation, the Tizard Committee (tasked with evaluating scientific advances made during the war) remained equivocal, concluding that although experimental work on guided weapons was important and should continue, 'we do not think that too much faith should be put on them.'[10] This view was challenged by the military. A particularly vocal critic was Sir John Slessor, a future Chief of the Air Staff, who questioned both the validity of the report and the role of scientists in the formulation of postwar strategy:

> I have the greatest respect for those members of the Committee whom I know, especially Sir Henry Tizard and Mr Blackett, as scientists. But I have no regard for their views on strategy, which is not their business. I think experience in this war has shown that scientists are very often quite unsound as tacticians; as strategists I see no reason why they should be expected to have opinions worth consideration by the Chiefs of Staff. I have a superficial knowledge of the application of some forms of science to some forms of war; but I should not expect the Royal Society to take the slightest notice of my opinions on the possible influence of warfare on the future of science.[11]

Given these reservations, and denied access to atomic information, the report's value as a planning document proved limited. To address the perceived deficiencies, a further report was compiled by the Joint Technical Warfare Committee (a sub-committee of the Chiefs of Staff). Completed in July 1946, the report's conclusions on guided weapons represented a significant reversal of previous policy: 'The most promising form of defence so far conceived is the Guided Anti-aircraft Projectile; and the importance and urgency cannot be over-emphasised'.[12] This view was later endorsed by the Chiefs of Staff, who agreed that absolute priority should be given to research on guided weapons over all other methods of antiaircraft defence.[13] To consolidate the various research programmes, a Directorate of Guided Weapons Research and Development was created within the Ministry of Supply under Dr George Gardner DGW(R&D), with all work on guided weapons transferred to either the Guided Weapons Department, RAE, or the Rocket Propulsion Department, Westcott. The responsibility for coordinating the programme was given to the Defence Research Policy Committee (DRPC), with overall priority determined by the Cabinet's Defence Committee.[14]

In its first meeting to discuss progress in the guided weapons field, the DRPC discovered that 20 projects were currently under investigation. To rationalise the programme, four were accorded the highest priority:

Sea Slug: a ship-to-air guided weapon
Red Heathen: a ground-to-air guided weapon
Blue Boar: a controlled bomb
Red Hawk: an air-to-air guided weapon.[15]

Despite shifting the focus of research to only four projects, progress remained slow. To address the problem the DRPC agreed to relax performance characteristics, with effort now concentrated on three distinct areas: a surface-to-air weapon (i.e. for use from land or sea) with a maximum range of not less than 30,000 yards; an air-to-air weapon for use from a limited cone astern of the target; and a controlled bomb which need not necessarily be capable of use in fully blind conditions. In approving the programme, the DRPC expressed concern over the lack of progress and voiced its disquiet over the degree of priority accorded to guided weapons research.[16] A particular area of disagreement centred on the atomic weapons programme, which was considered by many to divert resources away from work on guided weapons. The matter became particularly salient after the Soviet Union exploded its first nuclear weapon in August 1949.

Before this event, research priorities between guided weapons and atomic energy had been determined on the belief that the Soviets would not possess a nuclear weapon until the mid-1950s. This assumption was no longer valid, and the relative priorities of the two programmes were re-examined. A firm proponent of increasing the resources allocated to guided weapons was Sir Henry Tizard, Chairman of the DRPC. Presenting his case to the Minister of Defence, Tizard contended that to defend Britain against atomic attack, the guided weapons programme should be given the maximum boost possible.[17] Tizard's position was largely endorsed by the Chiefs of Staff. Outlining their concerns in the 1950 Global Strategy Paper, the Chiefs made the following recommendation: 'There are a limited number of projects which should receive the highest priority for research and development. Among these are guided missiles of all types. It is of the greatest importance that work on these projects should proceed with the utmost energy'.[18] Meeting in July, the Defence Committee agreed to accelerate the guided weapons programme to the maximum possible extent. To sanction the decision, the Prime Minister issued a directive which stated that guided weapons 'must be accorded the same special degree of priority that has been given to the atomic energy programme'.[19]

To augment the work undertaken by the research establishments (and to provide the necessary production capability) a number of development contracts were placed with industry. To facilitate the exchange of ideas, technical staff were seconded to RAE and other government establishments. The result was an increase in the number of development projects, with a variety of propulsion and guidance techniques under active examination. The content of the programme in 1952 was as follows:

- Red Shoes (Thunderbird), a ground-to-air weapon with a range of 40,000 yards. Developed by English Electric to meet the original Red Heathen specification, the missile was rocket propelled and guided by semiactive pulse radar homing from launch to the target.
- Red Duster (Bloodhound), a ground-to-air weapon with a range of 50,000 yards. Developed by Bristol Aircraft Co. Ltd in conjunction with Ferranti, the missile was also designed to meet the Red Heathen specification and was powered by two ram-jet engines with a semiactive pulsed radar guidance system similar to Red Shoes.
- Blue Sky (Fireflash), an air-to-air weapon employing beam riding as means of guidance. Manufactured by Fairey and designed for use against Soviet Tu-4 bombers, the missile was unpowered and relied on two solid-fuel boosters for propulsion.
- Blue Jay (Firestreak), an air-to-air weapon using infrared as a means of guidance. Developed by De Havilland and employing an internal solid-fuel motor, the missile had a maximum range of 6 km.
- Red Dean, an air-to-air weapon employing active pulse radar as a means of guidance. Under development by Vickers and powered by an internal solid-fuel motor, the project was eventually cancelled in 1956.

- Blue Boar, a controlled bomb using TV and radar as a means of guidance, eventually cancelled as AST in 1954.
- Sea Slug, a ship-to-air weapon with a range of 30,000 yards. Developed by Armstrong-Whitworth, the missile was powered by four solid-fuel booster motors and an internal sustainer, and used beam riding as a means of guidance.[20]

The collaboration between industry and the research establishments on guided weapons yielded a number of valuable results, including advances in guidance techniques, propulsion and aerodynamics. The resultant activity was described to the Minister of Supply, Duncan Sandys, in the following terms:

> The work has been undertaken both in the Ministry's Experimental Establishments, practically all of which have made some contribution although the greater part has been carried out at RAE (G.W. Dept.), the Rocket Propulsion Dept. and at RRE and in industry, where, in addition to about 12 major contractors there are some 200 others covering about 450/500 contracts. Contractors engaged on major projects employ over 1600 White Paper staff.[21]

The substantial progress achieved in the guided weapons field was noted in the 1954 Defence White Paper, which stated that 'guided weapons have reached an advanced stage of development: an air-to-air weapon will be the first to come into service and surface-to-air weapons will follow'.[22] Continued expansion of the programme, however, proved difficult. A significant factor was the rising cost of research and the government's desire to reduce the proportion of GDP spent on defence (Fig. 4.1)[23]. The position was later summarised by Sir Robert Cockburn, Controller of Guided Weapons and Electronics, Ministry of Supply, as follows:

> There was tremendous argument about the technology of guided weapons, whether it should be semi-active, active, passive, long-range, short-range etc, it went on and on. It was mainly at the DRPC that the answer was decided. Soon after I arrived at the Committee, Tizard put his foot down, we were going to concentrate on two ground-to-air missiles and one air-to-air missile. Mitchell, my predecessor, placed many contracts anywhere. My job in 1955 was to cancel about half as absolutely useless.[24]

The guided weapons programme was also affected by a reappraisal of strategic priorities initiated by the Chiefs of Staff in the mid-1950s. A significant outcome of the review was increased reliance on nuclear deterrence, with the role of surface-to-air guided weapons (SAGW) now centred on the defence of the V-bomber airfields and Thor IRBM missile sites. Prior to this decision, plans for the deployment of SAGW had been based on a requirement to locate sites near the coast, so that attacking bombers could be destroyed as far out to sea as possible. However, the requirement for a coastal deployment made it uneconomic to use short-range SAGW (stage 1) for the full-scale defence of the UK, as considerable numbers of missile sites would be required to provide an effective defence. The decision to limit the deployment of stage 1 SAGW was also based on the very low estimates of the system's operational performance. As a result of these factors, the Chiefs recommended that, in addition to the Service Trials Station at North Coates, only six stage 1 stations should be deployed operationally. In short, the primary purpose of the first-genera-

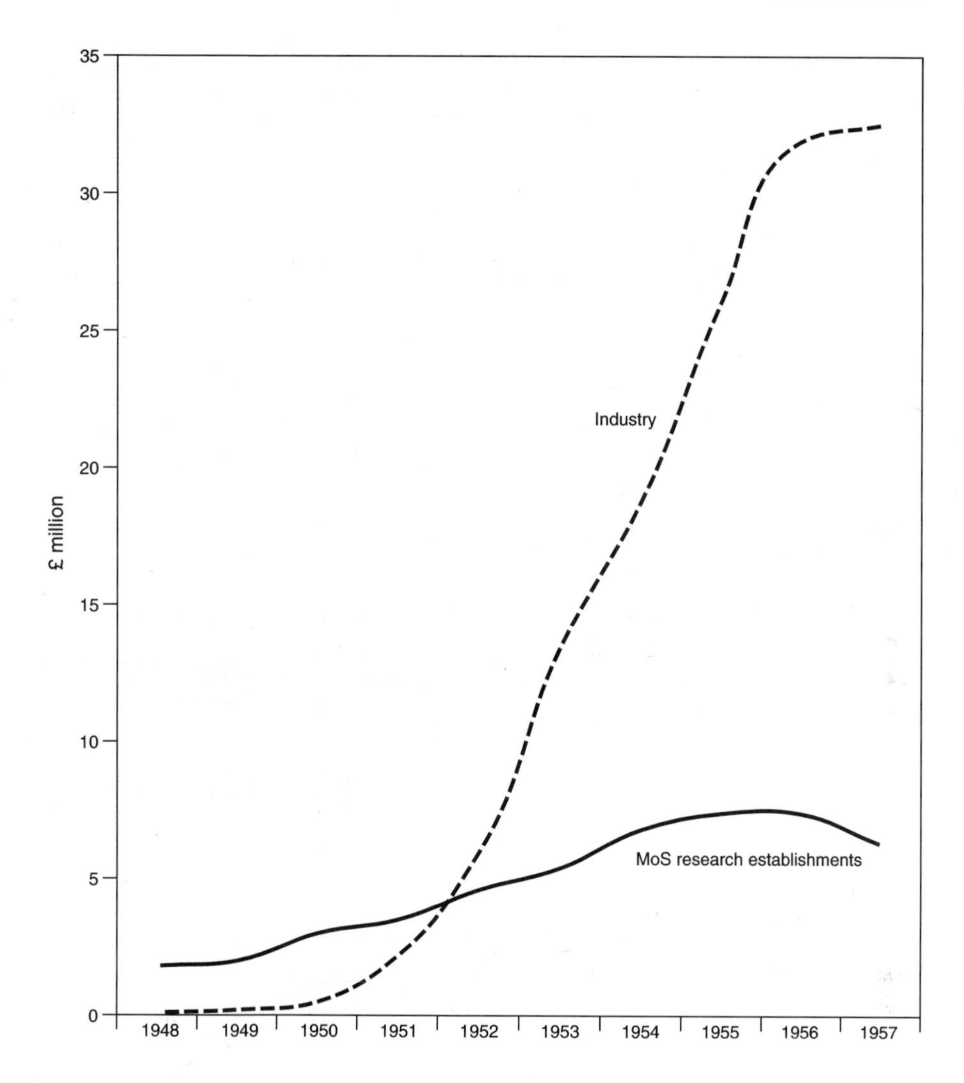

FIGURE 4.1: TOTAL EXPENDITURE OF INDUSTRY AND MINISTRY OF SUPPLY RESEARCH ESTABLISHMENTS ON GUIDED WEAPONS R&D, INCLUDING ASSOCIATED ELECTRONIC EQUIPMENT, 1948–57 (© CROWN COPYRIGHT, REDRAWN FROM AVIA 54/1907, PRO)

tion systems was 'to get into the SAGW business as quickly as possible and to provide training and experience in readiness for the introduction of improved types of weapon systems at a future date.'[25]

BLOODHOUND

Of all the first-generation guided weapons systems developed in the UK, the most significant in terms of length of deployment and export sales was Bloodhound. A member of the 'heavy trio' of guided weapons which included the Army's Thunderbird and the Navy's Sea Slug, Bloodhound was a semiactive surface-to-air homing missile developed jointly by Bristol Aircraft Ltd and Ferranti.[26] Launched by four solid-fuel motors which were jettisoned at burnout, the missile was sustained in flight by two ram-jet engines mounted externally above and below the main body. In its early development stage, the Bloodhound programme benefited considerably from work undertaken at RAE. This was particularly important in relation to test vehicles, a field in which RAE had unrivalled experience. Comprising four main components, designated RTV (guidance and control), CTV (aerodynamics), GPV (general purpose) and JTV (propulsion), and employing the ranges at Larkhill, Aberporth and Woomera, the test vehicle programme conducted at RAE 'laid the foundations of knowledge and experience on which guided weapons work in the United Kingdom was built'.[27]

To consolidate the work undertaken at RAE, further research was conducted at the Radar Research and Development Establishment (RRDE, later to become RSRE) on both tracking and illuminating radars. This work was essential, as the two land-based first-generation SAGWs developed in the UK were designed to operate using semiactive homing, in which each missile contained a scanning receiver but with the radar transmitter located on the ground. To operate the system, the ground-based tracking radar first 'illuminated' the target, allowing the missile's onboard receiver to home on to the reflected signal. This process was continued throughout flight until impact, usually 30 seconds after launch. The radar system adopted for both Mk 1 Bloodhound and Thunderbird was a gun-laying pulse radar known as Yellow River, which required only minor modifications to become an effective SAGW illuminator.[28]

The results obtained from both RAE and RRDE fed directly into the development of Bloodhound. This was achieved by a series of regular meetings and via the secondment of personnel to Farnborough and Malvern. In this fashion the Bloodhound design programme proceeded through a number of discrete evolutionary stages, as follows:

XTV 1 One-quarter-scale model
XTV 2 One-third-scale controlled model
XTV 3 Full-scale unpowered missile
XTV 4 Full-scale ram-jet propulsion test vehicle
XTV 5 Complete experimental missile, with homing head.[29]

The development contract for Bloodhound Mk 1 was awarded in mid-1951, and the first successful homing missile was flown in March 1956. In its final form the Bloodhound Mk 1 missile was 6.7 metres long, 55 centimetres in diameter and travelled at a speed of Mach 2.2 to a range of 50 kilometres. Service acceptance trials began in February 1958 and resulted in a 66 per cent success rate, including five direct hits against Canberra aircraft flying at 50,000 feet.[30] These trials were also notable for the first use of mathematical modelling techniques pioneered by the Weapons Research Establishment (Salisbury) using the tridimensional analogue computer (TRIDAC) developed at RAE.[31]

As shown above, the deployment of the Mk 1 Bloodhound was intended primarily to gain experience of handling guided weapons, the first operational squadron being formed in October 1958 at North Coates RAF base in Lincolnshire. Indeed, prior to the squadron's formation development work on the next generation of SAGW was already under way. Initially known as Bloodhound 1 3/4, or Blue Envoy, the project was technically overambitious and was cancelled in late 1957. In its place, two complementary development programmes were initiated. The first, Bloodhound Mk 2 (QF 169), employed a continuous-wave (CW) semiactive radar homing system with increased protection against electronic countermeasures (ECM); the second, Bloodhound Mk 3 (RO 166), was a command guidance system to be fitted with a nuclear warhead.[32] Both projects were to be developed jointly by Bristols and Ferranti, with deployment scheduled for the mid-1960s.

The development programme, however, became embroiled in the protracted debate surrounding British air defence policy in the 1960s. The crux of the debate centred on deterrence or, more accurately, its implementation. In short, air defence could only be justified if it contributed to the credibility of the British nuclear deterrent. The nature and scale of that contribution (if any) was therefore a key question that required clarification. A major factor which directly affected this decision was the anticipated deployment of Soviet ballistic missiles, against which SAGWs (with conventional warheads) would provide only marginal defence. To compound the problem, the Air Defence Committee, with the support of Sir Fredrick Brundrett, recommended the abandonment of Bloodhound in favour of an improved version of Thunderbird for both mobile and static defence, a position wholly rejected by the Air Ministry.[33] To help resolve the issue, the Joint Planning Staff were directed to prepare a paper outlining the various options. Presented to the Chiefs of Staff in January 1960, the paper recommended the cancellation of Bloodhound Mk 3 but confirmed a need for the Mk 2 version, both for the overseas defence role and for the export market. Future research into command guidance techniques and ABM requirements was left for further study (see below). The final decision on the future of Bloodhound was taken by the Defence Committee in October 1960; its main recommendations were as follows:

> Bloodhound Mark 2 uses continuous wave radar and semi-active guidance; originally developed as a static single shot SAGW replacement for Bloodhound Mark 1 for the defence of the United Kingdom in global war, it is being re-designed as a transportable system for the air defence of overseas bases. All elements of the system are now being made air transportable to enable them to be moved wherever they are needed.[34]

To fulfil this role, lightweight equipment was required which entailed a complete redesign of the target-illuminating radar (Blue Anchor) developed by the Royal Radar Establishment (RRE), a modified launcher capable of reload in less than five minutes, a portable electric power supply and a launch-control post housed in an air-transportable container. To coordinate work on the various elements a three-tiered management structure was established. At the bottom level each component of the system was the responsibility of a single R&D establishment. These establishments were then grouped together under appropriate coordinating R&D authorities. In relation to Bloodhound three such groupings were established, concerned with the missile (airframe, propulsion, guidance, fuse and warhead); the electronic ground

FIGURE 4.2: A LIGHTNING MK 6 OF NO. 56 SQUADRON FLIES OVER A BLOODHOUND MK 2 MISSILE OF NO. 112 SQUADRON AT PARAMALI, CYPRUS (© CROWN COPYRIGHT, MOD)

equipment (tracking radar, launch-control post and missile test equipment); and the mechanical ground equipment (launcher, assembly stands and missile transporters). Finally, to manage the development of the entire missile system, a system coordinating R&D authority was established which, in the case of Bloodhound, was the responsibility of RRE.[35]

The original deployment plan for Bloodhound Mk 2, approved by the Defence Committee in October 1960, consisted of three missile squadrons to be stationed in the UK, two each in Cyprus and Singapore, and a single squadron situated at Butterworth in Malaya. Each squadron was to consist of 16 missiles and launchers, plus a further 16 missiles for reloading. In January 1962, however, 'as a direct result of pressure to reduce expenditure overseas', these plans were revised.[36] The outcome was a reduction in deployment to only five squadrons, with two stationed in the UK and one each in Cyprus, Singapore and Malaya (Fig. 4.2).

Accepted into RAF service in 1964, the Bloodhound Mk 2 system was eventually exported to Australia, Sweden, Switzerland and Singapore. Further overseas sales of Bloodhound, however, were effectively curtailed by the US administration, which blocked potential markets in either of two ways. First, those NATO countries that showed an interest in acquiring air defence systems were offered the American Nike missile 'on generous terms, if not free', making it 'virtually impossible' to sell Bloodhound within NATO.[37] Second, potential markets outside NATO were often denied by legal interpretations concerning the export of sensitive technology developed jointly under the terms of Anglo-American mutual defence agreements. One such example was the possible sale of Bloodhound to Israel, where the British government was first required 'to consult with the US not only on general policy issues but also because of the American techniques embodied in Bloodhound.'[38]

To test the reliability of production models, Bloodhound evaluation trials continued in Australia until 1966. Conducted by No. 15 JSTU (Joint Service Trials Unit) under the control of RRE, the trials were designed to provide further technical details and demonstrate the mutual compatibility of all the components under a range of operational conditions.[39] With Britain's decision to withdraw from east of Suez in 1966, the Bloodhound missile squadrons stationed in the Far East were redeployed to Europe to provide air defence cover for RAF bases in West Germany, and were later withdrawn to provide a similar role in East Anglia (Fig. 4.3). In frontline service with the RAF for over 20 years, Bloodhound was eventually withdrawn from active service in 1991. In announcing its decision to Parliament to seek a replacement for Bloodhound, the government paid tribute to all those involved in its development, with particular praise given to the research establishments both for providing protection for British forces overseas and for enabling the British defence industry to remain a world leader in the development and production of high-technology equipment.[40]

PROJECT R-1

A significant factor in the development of Bloodhound—the details of which are only now slowly emerging—was the decision taken in 1956 to initiate the devel-

FIGURE 4.3: A ROYAL AIR FORCE POLICE HANDLER ON BLOODHOUND GUARD DUTY AT WEST RAYNHAM (© CROWN COPYRIGHT, MOD)

opment of a nuclear-armed SAGW for use in air defence. The practicalities of designing such a system had first been studied in the early 1950s, but because of the scarcity of fissile material and the difficulty of producing a lightweight warhead, no large-scale development programme was initiated.[41] Small-scale studies were nevertheless continued, with a small team at RAE investigating the practicality of using nuclear-tipped SAGWs against both bombers and ballistic missiles.[42]

In conjunction with this work, research conducted at AWRE indicated that the destructive power of a hydrogen bomb could be largely nullified by subjecting the weapon to a high flux of neutrons, a phenomenon known as the R-1 effect.[43] It was further realised that a comparable neutron source could be produced from the detonation of a small atomic weapon in close proximity to a nuclear-armed bomber. These results were far reaching: a defensive nuclear explosion in the vicinity of a megaton bomb would either physically destroy the bomb or considerably reduce its yield. Further studies indicated that a 1 kiloton nuclear weapon detonated at 50,000 ft would produce blast and thermal effects which would destroy the aircraft at a range of over 600 ft. Moreover, if a high-neutron flux weapon was used the effective range could be increased by a factor of 4.[44] These early studies resulted in a long-term research programme leading to the development of two lightweight kiloton warheads (Blue Fox and Indigo Hammer) capable of delivery by a modified Bloodhound missile.[45]

The discovery of the R-1 effect had important implications for the deployment of SAGWs in the UK.[46] As described above, the original plans approved by the Chiefs of Staff in October 1955 called for the introduction of SAGWs at the earliest practical date, so that any problems could immediately be fed back into the development programme. In light of the R-1 effect, the assumptions underpinning these plans were no longer valid. The policy implications for the future of air defence in the UK were described by the Chief of the Air Staff, as follows:

> The development during recent months of nuclear warheads for use with SAGW has increased considerably the operational potential of these weapons. Firstly, SAGW fitted with nuclear warheads will have a much greater lethality than weapons fitted with High Explosive warheads. Secondly, as a bonus, they may also be capable of "poisoning" thermonuclear bombs to the extent that even if these bombs were to explode subsequently, their yield could be reduced to more tolerable proportions.[47]

A particular aspect of SAGW deployment which now required clarification was the location of the six stage 1 missile sites. Two alternatives were available: to retain the present policy of coastal deployment, or to relocate the stations to defend either the deterrent or areas of population. In assessing the various options, the Chiefs of Staff relied on intelligence estimates of the probable scale of a Soviet nuclear attack. These indicated that for an attack on the UK in 1958, the Soviets would allocate 15 bombs of 1–5-megaton yield and 200 bombs, or ballistic missile warheads, of 10–100-kiloton yield. The primary objective for these weapons was the elimination of SAC and V-force airfields. Secondary targets included components of the air defence system, the centre of government, ports, population centres, war industries, electrical power and the oil industry. It was further estimated that by 1960 the Soviets would possess sufficient megaton weapons to attack the majority of intended targets within the UK.[48]

Given these assessments, the Chiefs of Staff recommended the deployment of nuclear-headed SAGWs on sites near the V-bomber and Thor IRBM bases, and high-explosive SAGWs to guard against jamming and aerial reconnaissance, on coastal sites forward of the deterrent bases. To meet this requirement, the Chiefs agreed that 'if an improved defence against the manned threat is to be developed, it should be based on command guidance and should use the Red Duster [Bloodhound] airframe with an Indigo Hammer warhead'.[49] It was further proposed that research on both continuous-wave and command guidance techniques should be intensified, a decision that ultimately led to the development of Bloodhound Marks 2 and 3. In approving the development of both types of guidance system, the Chiefs sought to provide an air defence system capable of meeting a variety of threat. As they later explained, 'the combination of command guidance Bloodhound and C.W. Bloodhound would force the enemy to spread his jamming effort and enhance the effectiveness of the defence'.[50] The decision that Britain was to develop a nuclear-capable SAGW was contained (albeit implicitly) in the 1957 Defence White Paper:

> A central feature of the defence plan is the maintenance of an effective deterrent. High priority will therefore continue to be given to the development of British nuclear weapons suitable for delivery by manned bombers and ballistic rockets. Nuclear warheads are also being evolved for defensive guided missiles.[51]

Scheduled for initial deployment in 1961, a nuclear-armed SAGW system was expected to be fully operational by 1964–65. A significant problem in achieving this date was the provision of warheads. Provisional estimates for the production of fissile material placed Britain's military requirements at 100 megaton weapons for strategic use and at least 1500 kiloton weapons for defensive purposes. Deployment on this scale, however, could not be met from Britain's own resources: existing capacity would provide only 250 defensive weapons, in addition to 100 megaton bombs.[52] To make good the shortfall, informal discussions were initiated with the US administration to determine whether American lightweight nuclear warheads could be provided for the air defence of the UK under the terms of the recently ratified 1958 US Atomic Energy Act.[53] Irrespective of the outcome of these discussions, the DRPC recommended that 'as a matter of urgency' a command guidance version of Bloodhound should be developed to carry a nuclear warhead.[54]

DEFENCE AGAINST BALLISTIC MISSILES

The determination with which the DRPC advocated the development of a command guidance version of Bloodhound was strongly influenced by Britain's decision, taken in the mid-1950s, to embark on an antiballistic missile (ABM) programme. Prior to this decision, early studies in the ABM field had been confined to a small team working at RAE's Aerodynamics Department, with research conducted on techniques rather than development. The new four-year programme, approved by the DRPC in October 1953, concentrated on two specific areas: the improvement of detection techniques capable of tracking ballistic missiles, and the development of active measures leading ultimately to the interception and destruction of the warhead. In early 1954, design contracts for an ABM early-warning radar and an ABM missile system were placed with Marconi and English Electric. These two investigations later formed the basis of AST 1135, which called for the development of a defensive system capable of detecting ballistic missiles at ranges of up to 5000 nautical miles (nm).[55] The programme was to be completed by 1957 and incorporated into the planning arrangements for stage 2 of the UK air defence system.

The Soviets' production of a lightweight megaton nuclear warhead combined with a ballistic missile resulted in an acceleration of the programme. The exact specifications for an ABM system, however, were hampered by a lack of knowledge and sufficient technical detail on the current status of the Soviet missile programme. These details were required to assess the immediacy of the threat, establish performance characteristics, and anticipate future developments. Moreover, although it was known that the Soviets were actively engaged in rocket development, no reliable evidence was available on the current status of the programme.[56] To obtain this information, intelligence organisations in both Britain and the USA were directed to prepare a combined estimate of the Soviet ballistic missile capability.

The broad details of the Soviet missile programme were already known to the British, derived in part from informal discussions held between Sir Stuart Mitchell, the Controller of Guided Weapons and Electronics, and Alan Dulles, the Director of the CIA. These contacts formed the basis of the first combined intelligence estimate on the probable development of the Soviet guided weapons programme, which was produced in November 1954.[57] To consolidate these exchanges, a tripartite conference on ballistic missile defence was held in London between 18 and 20 January 1956, attended by representatives from the USA, Britain and Canada. The conference was informed that the majority of the intelligence now available on Soviet missile development had been obtained from German technicians repatriated from the Soviet Union during the period 1945–50. During this period, the Soviets had initiated two development projects: the R.10, based largely on the German V-2 but with an increased range, and the R.14, a design study to produce a ballistic missile capable of delivering a 3000 kg warhead to a range of 3000 km. The development programme for the R-10 was expected to be completed by 1960 and deployed operationally by 1962.[58]

These assessments formed the basis of a briefing paper presented to the new Minister of Defence, Duncan Sandys, in January 1957. The paper detailed work on three separate Soviet rocket programmes with ranges of 150, 350 and 650 nm. It further revealed that the 650 nm missile (SS-3) had been test-fired 60 times and possessed an accuracy of 3 nm. The deployment schedule for the missile was expected to be late 1957 if fitted with a kiloton warhead, and 1959–60 for a megaton warhead.[59] On 4 October 1957, the Soviets launched *Sputnik 1*. The event was greeted with apprehension in the West, as it demonstrated that the USSR could now strike at intercontinental distances with weapons of mass destruction.[60]

In 1959, to counter developments in the Soviet missile programme, the British government approved an ambitious five-year ABM research programme. This envisaged the development of a passive defence project in two years, leading to the deployment of an active defence system by the end of the five-year period. Although passive techniques were considered attainable, the design of an active defence system was regarded as more problematic, requiring an extensive research and development programme estimated at £2.5 million.[61] Moreover, to undertake such a programme a command guidance version of Bloodhound was regarded as essential, a requirement forcibly conveyed to the Defence Committee by the Chief of the Air Staff, Sir Dermot Boyle: 'command guidance nuclear-headed Bloodhound is certainly required, it is the only foreseeable lead-in to an active ballistic missile defence'.[62] In response to these demands, Bristols were invited to commence development studies into a command guidance version of Bloodhound to be used for ABM defence.

The initial proposal put forward by Bristols for the operation of an ABM system employing the Bloodhound Mk 3 missile was as follows:

> Guidance: Although all missiles can carry terminal homing, in the ABM role only the command link set will be used. The Yellow River tracking radar is the only existing set capable of providing the required angular accuracy. The target tracker, the command link and the ground computer are the new elements in the system. Instructions from the ground computer will be fed to the missile by pulse position modulation of the missile tracking radar. In the command guidance system proposed, the missile is controlled by the ground computer which derives missile acceleration demands from the missile and target positions obtained from the tracking radars.[63]

To support the ABM test programme, research was conducted at RAE on various missile configurations using the Leopard multistage hypersonic test vehicle. Initial studies were based on the trajectory characteristics of a basic threat. This corresponded to an IRBM re-entering the earth's atmosphere at a steep angle with a velocity of 17,000 ft/sec. These studies resulted in the development of Project 29, a modified Bloodhound with larger fins and improved ram jets (Fig. 4.4).[64] To supplement this work, further investigations were carried out using a two-stage missile known as Project 36, designed for a maximum speed of Mach 9.0 and employing a cone/cylinder/flare layout; test firings of Project 36 were scheduled to begin at Woomera in July 1961.[65] Preliminary studies were also conducted on three-stage systems using motors and design data obtained from the Rocket Propulsion Establishment at Westcott. Further research was also undertaken at RRE into the development of an advanced discrimination radar to be sited on the island of Barbuda in the Caribbean.[66]

To coordinate these various programmes and to facilitate the exchange of information with the USA, the Tripartite Technical Cooperation Program (TTCP) was established. Eventually including the four nations of Britain, America, Canada and Australia, the TTCP fostered a number of collaborative ventures in the ABM field.[67] These joint programmes, conducted under the names of Gaslight, Dazzle and Sparta, were designed to investigate re-entry phenomena and were carried out using the facilities at Woomera, the ballistic range at CARDE in Canada, and the high-Mach-number experimental facility at Fort Halstead.[68] Britain's decision to participate in

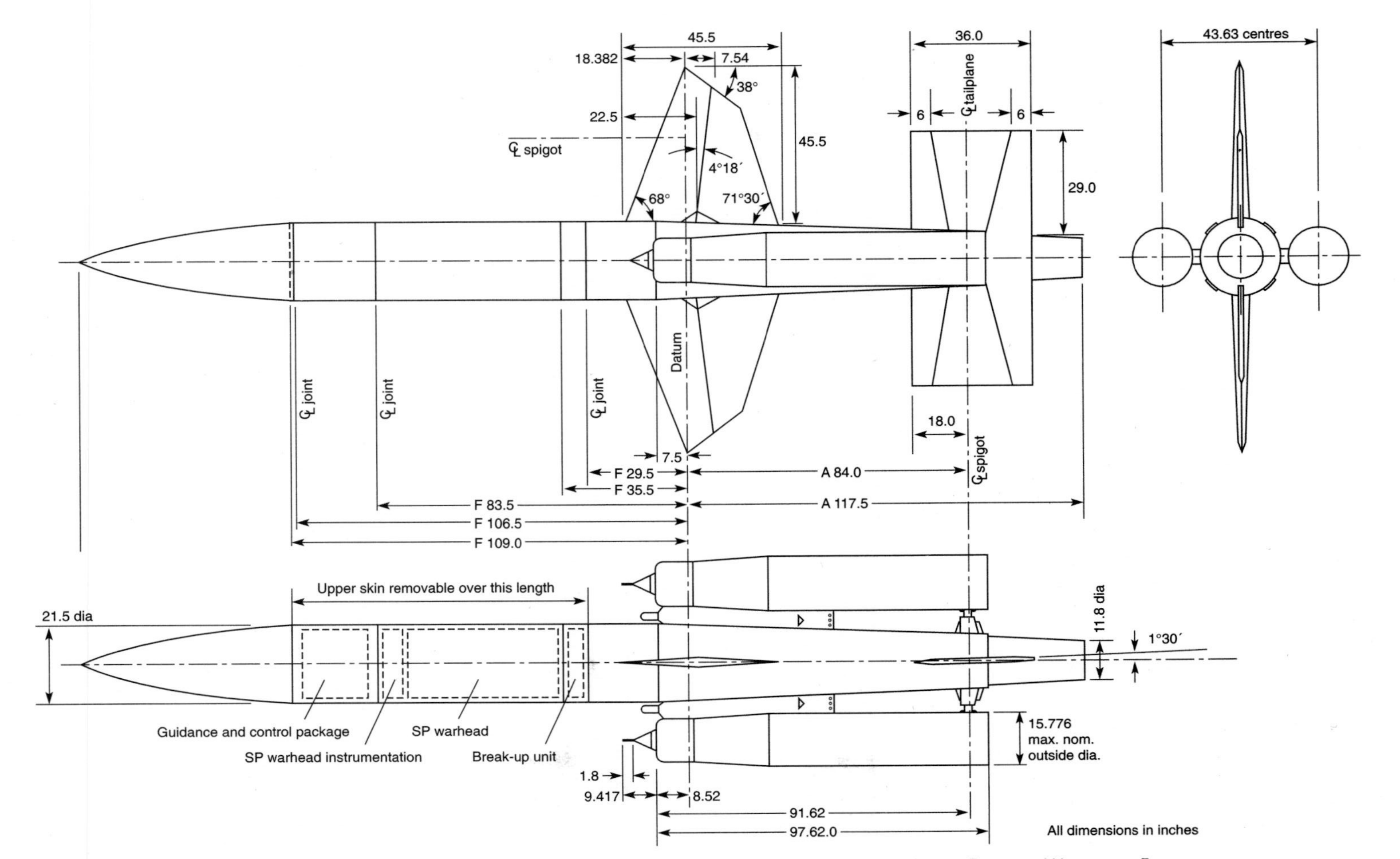

Figure 4.4: Bristol Aircraft Project 29, Red Duster Series 2 Mk 2 (Bristol Aircraft Ltd, Guided Weapons Engineering Department, redrawn)

the joint project carried certain obligations, including the requirement to supply to the USA details concerning the full extent of the British ABM programme. This condition was nevertheless accepted on the grounds that:

> We do not believe that there is any particularly outstanding area of work in this field in which we could claim to be particularly adept. We should, therefore, be prepared to consider undertaking any specific investigations which the US might wish under this interdependence scheme. We think it unlikely that they will leave any really important aspect to us. We should, therefore, leave the development of EW [early warning] equipment to the US and obtain the few such equipments we would require from the US. This would give us a quicker EW capability. As regards missile research, we should work on the problems associated with high speed capability.[69]

Despite collaboration with the USA, however, progress remained slow. To explore the policy options, the government invited Dr W.H. Penley of RRE to prepare a report detailing the current state of research in the ABM field and make recommendations. Completed in 1962, the paper's conclusions were less than sanguine: 'an adequate active defence is unlikely to emerge in practical form in less than 10 years.'[70] The Penley Report effectively marked the end of the British ABM programme.

Despite its abrupt termination, the programme was not without its successes. First, as funding was primarily aimed at maintaining a link with America, eventual British participation in the US Ballistic Missile Early Warning System (BMEWS) was a significant achievement.[71] Second, the wealth of research material and design data generated by the research establishments did not go to waste, but was applied to many other British guided missile systems, such as Red Top, Sea Cat, Sea Dart, Sea Wolf, Javelin, Martel, ALARM, and the most successful of all, in terms of overseas sales, Rapier.

RAPIER

Originally conceived in 1959 to combat the threat posed by supersonic low-to-medium-altitude targets, the Rapier air defence system almost never materialised. Designed to meet a highly ambitious RAF requirement for an all-weather, day or night automated antiaircraft weapon, the project was eventually cancelled in 1962 but was continued as a private venture by BAC, leading eventually to a less sophisticated fair-weather system known as ET 316.[72] To be used in conjunction with the US blind-fire system Mauler, the dual system offered all-round air cover for British forces serving overseas. Following the cancellation of Mauler in 1965, the service requirement for low-level air defence was reassessed. As a result of the review, the Weapons Development Committee (WDC) authorised the development of ET 316 and a 12-month project study for a blind-fire capability (Rapier DN181), a decision later endorsed by the Cabinet in March 1968.[73] During this early period the role of the defence establishments proved vital. Indeed, speaking in 1969, the then director of the Royal Signals and Radar Establishment Dr E. V. Glazier contended that both the initial proposal and continued interest in Rapier originated exclusively from within RSRE, and that without the support of the establishment the project would ultimately have foundered.[74]This view was later endorsed by Sir Solly

Zuckerman, the MoD's Chief Scientific Adviser, who expressed his admiration (and astonishment) for 'how the development team working on Rapier was maintained after the project was officially cancelled.'[75]

In its original version, first deployed in 1974, the Rapier system comprised three main elements: a launcher capable of firing four missiles; an automatic surveillance radar with built-in IFF (identification friend or foe); and an optical tracker to direct the missile to the target.[76] The system operated in the following manner. First, the horizon was continually scanned by the surveillance radar (housed in a central dome on the launcher), with all targets subjected to automatic interrogation. Second, if no friendly IFF response was detected both the optical tracker and the missile launcher were automatically aligned on to the target. Finally, on receiving a signal from the tracking computer that the target was within range, a missile was fired by the operator. As this was fitted with rear-mounted flares any deviation of the missile from the target line of sight was monitored and new instructions were sent to the missile by a command transmitter. The process continued until the missile reached the point of aim. The target was destroyed either by direct impact or by the detonation of a small warhead. An automatic command to line of sight (ACLOS) system, Rapier's main advantage was its mobility, ease of operation and low manpower commitment, requiring a detachment of only five men to deploy and operate the entire system (Fig. 4.5). Disadvantages were also apparent. First, in common with all early command systems, the requirement to follow the missile until impact before engaging a further target restricted its engagement rate. Second, in the absence of an all-weather capability, the system could only be used in daylight operations.[78]

The lack of an all-weather capability was a major concern to the MoD, who strongly contended that a blind-fire capability 'would make a very big difference to sales prospects for Rapier since the all-weather capability vastly increased the attractiveness of the missile system.'[79] These arguments proved persuasive, and in May 1971 final authority was given by the DEPC for the development and production of the DN 181 Blindfire radar. Developed in conjunction by RSRE and Marconi Command and Control Systems, the DN 181 Blindfire employed a monopulse radar operating in the millimetric frequency band, producing a narrow beamwidth provided by an aperture of 1.5 metres.[80]

Introduced into limited Service use in 1977 and described by the Commons Defence Committee as 'an outstandingly effective system', Rapier was eventually deployed at all RAF stations in West Germany, with a further two units protecting airfields in Scotland.[81] In 1981, Rapier received further endorsement when a Memorandum of Understanding worth over £140 million was signed with the US Air Force for the protection of American forces stationed in Britain.[82] Eventually deployed by the armed forces of 11 nations, Rapier export sales totalled £400 million (1976 prices), with BAC being awarded two Queen's Awards for Industry in recognition.[83] A less successful development in terms of export sales was the production of the tracked Rapier. Originally conceived to meet the requirements of the Iranian armed forces, the contract was cancelled after the downfall of the Shah in 1979. Despite this setback, the BAe Tracked Rapier (based on a modified M548 chassis)

FIGURE 4.5: RAPIER AIR DEFENCE SYSTEM DURING LAUNCH (MATRA BAE DYNAMICS)

was eventually adopted by the British Army of the Rhine (BAOR), which took delivery of 72 systems in 1985.[84]

The main research establishment responsible for the development of Rapier was RSRE, with one of the three divisions within its Guided Weapons Department exclusively engaged on managing the Rapier project. Designated the project R&D Authority, RSRE was responsible for coordinating the work undertaken by other government establishments, including the Royal Armament Research and Development Establishment at Fort Halstead and Langhurst, and the Fighting Vehicles Research and Development Establishment at Chobham.[85] Further establishments involved in Rapier's development included RAE, for the provision of test facilities at both Aberporth and Woomera, and the Army Personnel Research Establishment (Farnborough) which pioneered the development of a training programme specifically designed for the Rapier system.[86] In both its optical and blind-fire variants the prime industrial contractor for Rapier was British Aerospace (formerly BAC), with major subcontractors including Decca Radar (responsible for the surveillance radar), Marconi Command and Control Systems (formerly GEC Marconi and responsible for the Blindfire tracking radar and missile safety and arming system), Cossor (IFF) and Thorn EMI (proximity fuse and telemetry).[87]

In contrast to most other guided weapons projects developed in the UK, Rapier was one of the few systems ever to be used in active combat. Deployed during the 1982 Falklands conflict, the system was described by the MoD as having 'had a major impact on the Campaign'.[88] Although it was deployed only in its optical mode, thanks to correct intelligence that the Argentine Air Force had only a limited night-flying capability Rapier proved a major deterrent against low-level attack. Roughly one-third of the missiles that successfully engaged a target scored a direct hit.[89] Indeed, according to the Commanding Officer of the 12th Air Defence Regiment, Royal Artillery, given the adverse operating conditions, the performance of the Rapier system 'must be considered a remarkable achievement, the best low-level defence system in the world'.[90] Despite such commendations, a number of operational difficulties were also revealed. One particular problem was the generator, which proved noisy, difficult to maintain and incapable of performing on a continuous 24-hour basis. Accepting these criticisms, the MoD announced an immediate improvement package, in tandem with a long-term evolutionary development programme designed to enhance Rapier's operational capability and enable deployment to continue to the end of the century.[91]

The first fruits of this development programme were soon realised with the deployment of Rapier Field Standard B and B2 in the mid-1980s. In both these models the emphasis was on improving reliability, increasing the engagement rate and providing enhanced ECM protection. The development of Field Standard C (FSC), however, proved more problematic. Designed for multiple target acquisition, FSC was 'essentially a new weapons system'.[92] Indeed, although full development work was initiated in 1983, lack of project definition and cost overruns led to significant delays. Given these problems, the project was relaunched in 1986, with development costs increased by 65 per cent.[93] As in previous Rapier programmes, the research establishments 'provided technical advice and guidance throughout ...[and] were closely involved with both industry and the project office'.[94] In practice, this was achieved by senior personnel from RSRE attending quarterly technical reviews at which BAe submitted a comprehensive report on all aspects of the FSC's development programme. One particular area in which the involvement of the research establishments proved vital was the provision and evaluation of over 8000 hours of test flights, in which 300 targets were tracked under a wide range of operational conditions. The work undertaken at RSRE on digital filtering techniques and travelling wave tube (TWT) technology also proved invaluable. Both techniques formed an integral aspect of the new 'Dagger' surveillance and Blindfire radar systems employed on Rapier FSC.

In 1991 BAe announced the successful completion of the FSC development programme, and released detailed specifications on the performance of the now-renamed Rapier 2000. Possessing similarities to earlier models, the main improvements to the system included a higher rate of fire, greater operational flexibility, new and more effective missiles, extensive built-in test capability, enhanced reliability and multiple target acquisition. Consisting of three major elements (launcher, surveillance radar and tracker), each of which is towed by a Leyland-DAF 4-tonne truck carrying 15 missiles, the Rapier 2000 can engage targets through 360° to a range of 8 km at an altitude of 3 km.[95] Designed to carry eight missiles (as opposed to four in earlier models), the launcher is equipped to fire the new Mk 2 missile, of which there are two versions: the Mk 2A, which retains the semiarmour-piercing warhead of the Mk 1 but with increased range owing to a new rocket motor developed by Royal Ordnance; and the Mk 2B, which has a dual-purpose semiarmour-piercing fragmentation warhead triggered by an active infrared proximity fuse. This was developed by Thorn EMI under a £10 million contract awarded in 1986, and was designed to engage small airborne targets such as Cruise missiles and remotely piloted vehicles.[96] Scheduled for full Service deployment in the late 1990s, Rapier 2000 is expected to provide low-level air defence for British forces well into the next century, and to place British industry in a favourable position to exploit the future needs of the air defence market.[97]

AN ASSESSMENT

To assess both the role and the function of the research establishments in the development of guided weapons, a suitable benchmark is required against which the numerous activities undertaken by them can be measured. This is a difficult task. One approach would be to employ the management techniques of today. However, to evaluate the activities of the establishments over 50 years by using terms such as 'mission statement', 'value-added' and 'performance indicators' appears largely inappropriate. Indeed, adopting such a methodology would be akin to criticising the building techniques of the pharaohs for failing to pay due care and attention to safety at work. Given these problems, how are we to proceed?

One useful way is to assess the performance of the establishments against their own designated functions, namely:

1. To perform basic research likely to be of assistance to the British defence industry.
2. The provision of expert advice to the government, allowing the evaluation of work undertaken by industry under contract and the effective monitoring of extramural research work.
3. To contribute to work undertaken by industry, particularly in the development stage of projects where no civil application is possible.
4. To provide and maintain large capital equipment for common use by industry, including firing ranges, wind tunnels and testing facilities.[98]

To evaluate the overall effectiveness of the research establishments by reference to these criteria, a number of important qualifications need to be addressed. First, there is the danger of not comparing like with like and placing equal importance on each of these functions. This tendency must be resisted. In both scope and importance these four activities are qualitatively and quantitatively different. This aspect is illustrated by reference to the provision of basic research and test sites. In the former, insight and imagination are essential, whereas in the latter managerial and maintenance skills are the primary requirement. Second, the more narrowly defined the objective, the greater the confidence in attaining the required outcome. Consequently, the more opaque the requirement, the greater the difficulty in establishing the validity of the output. Finally, these criteria are static and fail to take account of the dynamic nature of the development process.

A prime example is the provision by the establishments of missile ranges and testing equipment. Comprising the facilities at Larkhill, Aberporth, West Freugh and Woomera, the British guided weapons programme was widely regarded as possessing some of the best facilities in the world.[99] Indeed, when viewed in isolation, this outcome is a notable achievement. Nevertheless, despite accepting that test facilities constitute a necessary element in the guided weapons programme, their provision *per se* cannot be regarded as representing the central core of the establishments' work. In April 1992, in recognition of this position, all test sites were transferred to the Directorate General of Test and Evaluation (DGT&E).

In contrast, whereas the provision of test sites may be peripheral to the overall activity of the research establishments, basic research is central. The standard of basic research conducted by the establishments, however, is open to question. On the one hand it can be argued that advances pioneered at RAE, particularly in the use of test vehicles, proved invaluable in the early development process; on the other hand there is an equally valid argument (and one consistently expressed by industry) that research undertaken within the establishments was 'too academic', and consequently of marginal use to production.[100] In defending their position, the establishments have conceded that although such criticism 'may have been justified in the past', the practice adopted since the 1970s has been largely dictated by industrial need, so that 'as soon as research revealed the emergence of a new technique, it was the establishments role to bring industry into future planning'.[101] A mitigating factor cited by the establishments in providing basic research (and one peculiar to guided weapons) was the large number of establishments engaged in the programme. Indeed, a suitable structure to manage these activities efficiently has never been fully devised, with proposals ranging from the complete autonomy of individual sites to the creation of a centralised establishment encompassing the entire guided weapons field. The decision taken in the mid-1970s to transfer the development of Navy, Army and Air Force projects to the Sea, Land, and Air Controllerates was primarily designed to balance these conflicting demands. In contrast, the creation of the DRA in the 1990s indicates that the centralisation of basic research was considered (at least by the Conservative government of the day) to represent the most suitable structure for managing complex technological development.

An equally important function undertaken by the establishments is the provision of independent expert advice to government departments. Moreover, in matters relating to the defence of the state sound scientific, technical and financial judgement is vital. The future of many defence projects, particularly at the feasibility and project definition stages, was often determined on advice submitted by establishments. The requirement to provide expert advice was a role for which the establishments were well suited, being able 'to draw on unrivalled experience of a whole range of past and current guided weapons projects and on a fund of information derived from the broad spectrum of intra and extra mural research'.[102] This function was less relevant in the design and development phase, where the initiative (and risk) was considered to lie with the contractor. In providing expert advice the role of the establishments remained essentially advisory, with responsibility for project decisions ultimately vested with project managers at ministry level. Given this position, any attempt to assess in isolation the efficacy of the establishments in providing expert advice is problematic: political, economic and strategic factors are equally important.

A further area in which the establishments were expected to play a full role was involvement with industry, especially in the development stage of projects where no civil application was possible. In judging the effectiveness of the establishments in performing this function, the main problem is to determine where development ends and production begins. Two examples illustrate this division. The first was the complete in-house development in the 1950s by the Admiralty Signals and Radar Establishment of Sea Slug's radar guidance system, a development that 'saved' Armstrong Whitworth £40 million.[103] The second relates to the Vigilant antitank weapon, which was conceived and developed entirely by Vickers at the company's own expense. Between these two extremes is Rapier, developed partially as a private venture but benefiting considerably from work undertaken at the establishments. The development contribution is also difficult to quantify when projects are abruptly cancelled: the ABM programme is a prime example. Despite these difficulties, the engineering techniques pioneered by British scientists compare favourably with developments in the USA, with British expertise in the design of rocket motors, re-entry vehicles and silo construction reportedly forming a significant element in the US ICBM programme.[104]

A further concern expressed by industry relates to the length of the development schedule for modern weapons, and the tendency for establishment scientists 'to work too high on the curve of diminishing return'.[105] For industry, the preferred solution was to introduce guided weapons into Service use as soon as possible, rather than awaiting development of the perfect system. In this way development could proceed incrementally, removing the risks inherent in the development of complete new systems. This criticism has largely been accepted by the Ministry of Defence, with the result that the progressive development of weapons systems, as witnessed in the production of Rapier, is now an accepted aspect of the process.

To optimise this procedure, effective scrutiny of each development phase is essential. The role played by the establishments in judging the cost-effectiveness of the work undertaken in industry is therefore of considerable importance to both the taxpayer and the Services alike. It is therefore worrying that in the case of guided weapons there is strong *prima facie* evidence to suggest that this was the least successful role undertaken by the establishments. The final development costs of Sea Slug, Thunderbird and Firestreak, for example, were respectively 30, 10 and 6 times the original estimates.[106] Explaining this outcome, the establishments contended that in the early stages of the programme 'there was no experience of what was involved of turning the principles into reliable weapons; and to a large extent the delays and related increases in cost were a reflection of the inherent impossibility of making reliable estimates at the outset, or during the earlier stages of this kind of project, of the amount of testing that would be required'.[107] Despite not wishing to diminish the validity of these factors, the repetition of similar problems in Rapier FSC gives cause for concern: it indicates that earlier lessons were either ignored or not acted upon with sufficient vigour. The adoption of new management procedures in the 1980s is designed to address these concerns and ensure that 'technical risk will be better contained and eliminated'.[108] The episode nevertheless reinforces criticisms made by Sir Frank Cooper, a former Permanent Secretary at the MoD, that too many British defence projects are bedevilled by 'the over-optimism (one almost tends to say arrogance) of the activists in research and development and the continuing gross under-estimation of difficulties, timing and cost and the inability to deal with industrial pressure'.[109]

Given these factors, the contribution of the research establishments to the development of guided weapons in the UK is mixed. This outcome is not wholly the responsibility of the establishments. Indeed, starved of resources by successive governments and subjected to a series of inconclusive strategic defence reviews, the standard and scope of work conducted within the establishments is a considerable achievement. Perhaps, above all else, the main contribution of the establishments was in providing continuity: continuity of thought processes; continuity of knowledge and understanding of the ever-evolving threat; continuity of weapons concepts and systems to take advantage of advances in technology; and continuity of capability and effort at the most basic level. In short, the research establishments provided the institutional memory vital to the success of the British defence industry.

1. *Defence Research*, Second Report from the Select Committee on Science and Technology, HC 213, Session 1968–69 (HMSO, 27 March 1969).

2. *Defence Research* (n. 1 above).
3. DO(50)96, PRO CAB 131/9 (18 November 1950).
4. The Guided Weapons Department, RAE, was established in 1947 replacing the Controlled Weapons Department. In 1962, the Guided Weapons and Armament Departments were combined to form the RAE's Weapons Department which in October 1974 became the Air Armaments, Guided Weapons and Systems Assessment Division, RAE.
5. For further details, see S. Twigge, *The Early Development of Guided Weapons in the United Kingdom 1940–1960* (Harwood, 1993): 117–20.
6. *Government Organisation for Defence Procurement and Civil Aerospace,* Cmnd 4641 (HMSO, April 1971).
7. *Guided Weapons,* Seventh Report from the Expenditure Committee together with part of the Minutes of the Evidence Taken before the Defence and External Affairs Sub-Committee, HC 597, Session 1975–76, Vol. 1, app. B (HMSO, July 1976).
8. H.W. Pout, 'The evolution of guided weapons', *Journal of the Royal Aeronautical Society,* 73 (June 1969): 547–52.
9. See for example, Sir Alwyn Crow, 'The rocket as a weapon of war in the British forces', Thirty-Fourth Thomas Hawksley Lecture, *Proceedings of the Institute of Mechanical Engineers,* 158 (January–December 1948): 15–21.
10. *Future Developments in Weapons and Methods of War,* COS(45)402(0), PRO CAB 80/94 (16 June 1945).
11. Slessor to VCAS, PRO AIR 2/12027 (16 July 1945).
12. TWC (46) 15, Revise, PRO DEFE 2/1252 (6 July 1946).
13. 135th Meeting, COS (46), PRO CAB 79/51 (3 September 1946).
14. *Guided Weapons Organisation 1946–47,* PRO AVIA 54/1438 (7 October 1946).
15. DO (49) 62, Annex 1 (5 August 1949) PRO DEFE 9/9.
16. 77th Meeting, Confidential Annex to COS (49), PRO DEFE 4/22 (25 May 1949).
17. *Defence Research and Development Priorities,* DO (50) 52, PRO CAB 131/9 (5 July 1950).
18. The 1950 Global Strategy Paper, DO (50) 45, was recently published in *Korea: June 1950 – April 1951, Documents on British Policy Overseas,* edited by H.J. Yasamee and K.A. Hamilton, 2 series, Vol. IV (HMSO, 1991): 411–31.
19. DO (50) 95, PRO CAB 131/9 (18 November 1950).
20. 2nd Meeting, DRP (52), PRO DEFE 7/266 (22 January 1952).
21. *Brief for Minister's Visit to USA, Weapons Projects,* PRO AVIA 54/1907 (June 1954).
22. *Statement on Defence, 1954,* Cmd 9075 (HMSO, 1954).
23. Figure 4.1 gives a comparison of the expenditure undertaken by industry and the research establishments on guided weapons for the period 1946–58, from PRO AVIA 54/1907.
24. Interview between Sir Robert Cockburn and Mark Venables, 3 September 1984, cited in M. Venables, *The Place of Air Power in Post-War British Defence Planning, and its Influence on the Genesis and Development of the Theory of Nuclear Deterrence, 1945–52* (Ph.D. thesis, King's College, London, 1985).
25. Note by CAS, *Deployment of Surface-to-Air Guided Weapons in Defence of the United Kingdom,* PRO AIR 20/10060 (n.d.).
26. For details of Thunderbird and Sea Slug, see G.R. Jefferson, 'The development of Thunderbird', *Journal of the Royal Aeronautical Society* 67 (June 1963); P. Hampton, 'Surface to Air Guided Weapons (Naval)', *Journal of the Royal Aeronautical Society,* 78 (April 1974).
27. T.L. Smith, 'RAE guided weapon test vehicles in the 1950s', *Journal of the Royal Aeronautical Society,* 69 (February 1965): 101.
28. J. Gough, *Watching the Skies: a history of ground radar for the air defence of the United Kingdom by the Royal Air Force from 1946 to 1975* (HMSO, 1993): 104.
29. D.J. Farrar, 'The Bloodhound', *Journal of the Royal Aeronautical Society,* 63 (January 1959).
30. *Guided Missile Trials—Firing Results,* DRP/P(60)32, PRO DEFE 7/1338 (5 April 1960).
31. For further details, see A.G. Biggs, A.R. Cawthorne, 'Bloodhound missile evaluation', *Journal of the Royal Aeronautical Society,* 66 (September 1962).
32. DRP/P(58)82, PRO AIR 20/10805 (15 October 1958).
33. For details of the air defence debate, see PRO AIR 41/86, C. James, *Defence Policy and the Royal Air Force 1956–1963* (Ministry of Defence, Air Historical Branch, 1984): chap. 6.
34. *The Bloodhound Mark 2 System,* PRO DEFE 7/1338 (5 December 1960).
35. *Project Management,* RRE, Malvern, PRO AVIA 7/3734 (9 August 1962).
36. *Deployment of Bloodhound II,* AUS(A)/7978, PRO DEFE 7/1338 (2 May 1962).
37. Hill to Tuck, PRO DEFE 7/1126 (22 November 1960).
38. 7th Meeting, D(60), PRO CAB 131/23 (25 July 1960). For further details, see F. Cain, 'Missiles and mistrust: US Intelligence responses to British and Australian missile research', *Intelligence and National Security,* 3, no. 4 (October 1988).

39. *RRE Half Yearly Progress Report*, PRO AVIA 20/2047 (July 1966).
40. *House of Commons Debate*, 190, cols. 256–60 (1 May 1991).
41. J. Neufield, *Ballistic Missiles in the United States Air Force 1945–1960* (Office of Air Force History, United States Air Force, 1990): 56–65.
42. AD(WP2)(54)18, PRO DEFE 8/48 (23 July 1954).
43. Air Council, *Conclusions of Meeting*, PRO AIR 20/10060 (21 February 1957).
44. Annex to AC(57)35, PRO AIR 20/10060 (14 May 1957).
45. *Proposals for Future SAGW Development*, PRO AVIA 65/906 (May 1957).
46. In addition to its defensive potential, the R-1 effect also had implications for Britain's offensive strategic nuclear forces. In short, if the Soviets were also aware of the R-1 effect, Britain's megaton warheads would require neutron shielding to guarantee nuclear detonation–a development which would divert resources, delay the introduction of the weapon and decrease the effective range due to the additional weight required by the resultant modifications. Uncertain whether an immune warhead could be developed, the Air Council recommended that the present development programme should be continued with further emphasis placed on the shielding and immunity of nuclear warheads. Note by DCAS, the Research and Development Programme, PRO AIR 20/10060 (n.d.).
47. Note by CAS, PRO AIR 20/10060 (June 1957).
48. Annex to AC(57)35 (n. 44 above), extract from JIC(57)7.
49. *Nuclear Warheads in SAGW*, record of meeting, PRO AVIA 65/906 (January 1958). In a command guidance system, trajectory control signals are transmitted to the missile from the ground. This information enables the missile to calculate its own position and that of the target and to calculate the most suitable co-ordinates for collision. The merits of the system are that most of the complex computational equipment is located on the ground, is less prone to damage, and can readily be updated. The system also possesses high accuracy against targets travelling at high altitude and speed. The main deficiency was that first generation systems were only capable of guiding one missile at a time. Consequently, subsequent engagements of the target could only be undertaken after the explosion of the first missile. For further details of command guidance techniques, see *Jane's Land Based Air Defence* (Jane's Information Group, 1998).
50. *SAGWs with Nuclear Warhead*, PRO AIR 20/10805 (3 April 1959). The decision was also influenced by the development schedules and performance characteristics exhibited by each type of system. CW guidance systems, for example, could be deployed by the late 1950s but were vulnerable to ECM and provided only limited defence against powered bombs. In contrast, command guidance systems did not possess such limitations but were not expected to be fully operational until the mid 1960s.
51. *Defence: outline of future policy*, Cmnd 124, par. 59 (HMSO, 1957).
52. Note by the Secretary of the Cabinet, *Fissile Material for Nuclear Weapons*, D(57)14, PRO CAB 131/18 (27 July 1957).
53. Memorandum for the Secretary of Defense, *Further Cooperation with the United Kingdom and Canada in Military Applications of Atomic Energy*, US National Archives, RG 218, CCS 350.05 (3–16–48) Section 12, JCS 2220/140 (6 June 1958). For background, see Ian Clark, *Nuclear Diplomacy and the Special Relationship: Britain's Deterrent and America 1957–1962* (OUP, 1994): 77–106.
54. DRP/P(59)35, PRO AIR 20/10805 (3 April 1959).
55. DWR(D) to CSAR, DERA Archive (Farnborough, 14 August 1954).
56. DRP/P(53)50, PRO DEFE 7/698 (13 November 1953).
57. JIC/2473/55, PRO DEFE 7/698,(13 September 1955).
58. DBM/M(56)1, PRO DEFE 7/698, (8 February 1956). The R.10 was deployed operationally as the SS-3 (NATO designation Shyster) and the R.14 entered service with the Strategic Rocket Forces as the SS-5 (NATO designation Skean).
59. *Brundrett to Sandys*, PRO DEFE 7/1756 (18 January 1957).
60. For a discussion of intelligence estimates concerning Soviet capabilities, see *Espionage, Security and Intelligence in Britain, 1945–1970*, edited by R. Aldrich, Documents in Contemporary History, pt II (Manchester University Press, 1998); L. Freedman, *US Intelligence and the Soviet Strategic Threat* (Macmillan, 1986): chap. 4; F. Kaplan, *The Wizards of Armageddon* (Stanford University Press, 1983): chaps 10, 19; J. Prados, *The Soviet Estimate* (Princeton University Press, 1986); see also S. Koch, *CIA Cold War Records: selected estimates on the Soviet Union 1950–1959* (Washington: CIA Center for the Study of Intelligence, 1993); S. Twigge and L. Scott, *Fail Deadly? Britain and the Command and Control of Nuclear Weapons 1945–1964* (Nuclear History Program Report, 1997): 331–46.
61. *Briefs for visit of CAS to US and Canada*, PRO AIR 8/2324 (26 October 1959).
62. D(59)23, PRO CAB 131/22, (8 July 1959).
63. Bristol Aircraft Limited, Guided Weapons Engineering Department (BGWED), *The Use of Red Duster for Phase 1 ABM Defence*, ABM Study Note No. 7 (25 September 1957).

64. BGWED, ABM Study Notes Nos 23, 27, 41, 42 (June–December 1958).
65. BGWED, ABM Study Note No. 94, XTV 20 Firing Programme (14 December 1959). A cone-cylinder flare layout refers to the design of the fully assembled rocket, i.e. conic nose section, cylindrical main body stage and flared tail-end.
66. BGWED, *A Review of the Aerodynamic and Propulsion Work to Date on ABM Assessment and Test Vehicles*, ABM Study Note No. 101 (8 February 1960).
67. J. Gough, *Watching the Skies* (n. 28 above): 201.
68. For further details, see R. Dommett, *Ballistic Missile Defence*, Lecture to the Guided Flight Group of the Royal Aeronautical Society (DRA, April 1994).
69. *Interdependence on ABM Project*, PRO AVIA 65/1564 (14 August 1958). I would like to thank Dr Susanna Schrafstetter for bringing this document to my attention.
70. C. James, *Defence Policy and the Royal Air Force* (n. 33 above): 144.
71. Further collaboration with the US on ABM defence was conducted as part of the SDI program. In October 1994, a pre-feasibility study for the design of an ABM system was awarded to BAe, research for the project being undertaken at DRA Farnborough. For details, see *House of Commons Debate*, 271, col. 723 (16 February 1996).
72. The original work on the Rapier system can be traced back to studies undertaken by English Electric in 1959 into a short range ground to air weapon called PT 428. Following the absorption of English Electric into BAC, work continued on a simpler version of the project known as ET 316, later renamed Rapier. *RRE Half Yearly Progress Report*, PRO AVIA 26/2045 (July 1965).
73. Seventh Report from the Expenditure Committee (n. 7 above), Annex D to Appendix H.
74. *Defence Research* (n. 1 above), Evidence, Q 1259.
75. S. Zuckerman, *Monkeys, Men and Missiles: an autobiography 1946–88* (Collins, 1988): 211.
76. Air Vice-Marshal Pocock, 'The introduction of Rapier into the Royal Air Force', *Aeroplane* (December 1974): 22–7.
77. S.C. Dunn, 'Rapier', *Journal of the Royal Aeronautical Society*, 74 (October 1970): 847–52.
78. Lt.-Colonel A.J.A. Brett, 'Air Defence of the Field Army', *Journal of the Royal United Services Institute for Defence Studies*, 118, no. 1 (March 1973).
79. Seventh Report from the Expenditure Committee (n. 7 above), par. 19.
80. RSRE, *1985 Research Review* (HMSO, 1985): 3.
81. First Report from the Defence Committee, *Allied Forces in Germany*, HC 93, Session 1981–82, par. 28 (HMSO, 1982).
82. *House of Commons Debate* 999, col. 388 (25 February 1981).
83. *The Financial Times* (15 May 1979).
84. *Statement on Defence Estimates, 1985*, Cmnd 9430, par. 424 (HMSO, 1985); *Jane's Defence Weekly* (29 September 1990): 587.
85. G.E. King, 'Project Management Techniques of Rapier', *Journal of the Royal Aeronautical Society*, 74 (January 1970): 75–81.
86. The need to monitor human factors emerged after it was discovered that although operators could track targets perfectly under simulated conditions, their accuracy deteriorated markedly when live missiles were fired. Evidence that stress levels rose dramatically during live firing was later confirmed by the outcome of tests conducted by APRE which monitored the chemical composition of the operators' urine.
87. During its early development under the Ministry of Technology, the management structure was somewhat different. For although BAC was still the major contractor it was not the prime contractor, with Imperial Metal Industries (responsible for the rocket motor), GEC Portsmouth (responsible for the safety and arming unit) and EMI Hayes (responsible for the transmitter valve) under the direct control of the Ministry.
88. *The Falklands Campaign: the lessons*, Cmnd 8758, par. 231 (HMSO, 1982).
89. Lt.-Colonel M.C. Bowden, 'After the war was over', *Journal of the Royal Artillery* (March 1983).
90. Lt.-Colonel M.C. Bowden, 'After the war was over' (n. 89 above).
91. *Implementing the Lessons of the Falklands Campaign*, HC 345, Session 1986–87, par. 207 (HMSO, 1987).
92. Seventh Report from the Defence Committee, *The Procurement of Rapier Field Standard C*, HC 273, Session 1989–90, par. 4 (HMSO, May 1990).
93. *Jane's Defence Weekly* (23 June 1990): 1233.
94. Seventh Report from the Defence Committee, *Further Memorandum Submitted by the Ministry of Defence*, question 4 (22 February 1990).
95. For detailed specifications, see *Jane's Defence Weekly* (7 September 1991): 423–6.
96. *Jane's Defence Weekly* (15 June 1991): 1002.
97. BAe Dynamics is currently investigating the design and development of various medium range SAGW systems and is participating in a number of joint ventures in conjunction with DRA on the NATO requirement for Short Range and Very Short Range Air Defence (SHORAD/VSHORAD) in the twenty-first century.

98. *Defence Research* (n. 1 above), par. 188–95.
99. See for example, P. Morton, *Fire Across the Desert: Woomera and the Anglo-Australian Joint Project 1946–1980* (Australian Government Publishing Service, 1989).
100. The view that the research establishments placed theory before practice has been prevalent since the inception of the guided weapons programme. Indeed, in 1947 it was strongly argued that no progress would be made in the development of guided weapons until the programme was removed from the 'basic research outlook' of Sir Ben Lockspeiser, the Chief Scientific Advisor in the Ministry of Supply, 'and placed under virile leadership'. *Carroll to Tizard*, PRO PRO DEFE 9/9 (17 June 1947).
101. *Defence Research* (n. 1 above), par. 62.
102. Seventh Report from the Expenditure Committee (n. 7 above), par. 10.
103. Second Report from the Committee of Public Accounts, *1959–1960*, HC 256, Session 1959–60 (HMSO, July 1960), par. 3052.
104. *Memorandum from Wing Commander S.H. Bonser, Director of Guided Weapons (R&D) Ministry of Supply*, PRO AVIA 54/2128 (31 May 1956).
105. Seventh Report from the Expenditure Committee (n. 7 above), par. 48.
106. For further details, see S. Twigge, *The Early Development of Guided Weapons* (n. 5 above): 220–33.
107. *Civil Appropriation Accounts, 1958–59*, HC 53, Session 1959–60, par. 27 (HMSO, December 1959).
108. Seventh Report from the Defence Committee (n. 92 above), par. 33.
109. Sir Frank Cooper, preface to C. James, *Defence Policy and the Royal Air Force* (n. 33 above): xi.

Chapter 5

Armoured Fighting Vehicles

Richard Ogorkiewicz

INTRODUCTION

The history of armoured fighting vehicles is primarily that of tanks, which played a major role in the Second World War and which were deployed on a large scale throughout the Cold War. During that period of confrontation between the Soviet Union and the Western nations the number of tanks fielded by the Soviet Army and its Warsaw Pact allies reached almost 60,000. To counter them NATO forces built up their own tank strengths to about 25,000 and, by the numbers in which they were operated by the opposing armies, tanks clearly demonstrated their importance.

As the most important and demanding of the armoured fighting vehicles, tanks attracted most of the research effort and their requirements drove the majority of the technological developments in the field of armoured vehicles. The following account is therefore very largely confined to tanks, although there are many other types of armoured fighting vehicles, ranging from armoured scout cars to armoured infantry fighting vehicles.

HISTORICAL BACKGROUND

Until the outbreak of the Second World War in 1939 the UK occupied a leading position on the tank scene. To start with, it enjoyed the prestige of being the first country to develop and successfully use tanks during the First World War. Subsequently its standing was enhanced by the lead taken in the development of new and more effective ways of employing tanks, and in promoting new designs for them. However, this lead was lost during the Second World War, in spite of the very considerable amount of effort devoted to the development and production of tanks. In fact, during the first five years of that conflict the UK produced more tanks than did Germany.[1] Yet in spite of this, in the latter part of the war British armoured formations were equipped largely with US-built tanks because of the inadequate armament and unreliability of British tanks.

The shortcomings of British tanks were due to a large extent to the doctrine adopted by the War Office during the 1930s. This divided them into 'infantry' and 'cruiser' tanks, and thereby led to a dispersion of the available development effort, and forces in the field, between two overspecialised categories,[2] one intended to support the infantry at slow speed and the other for a more mobile role akin to that performed previously by horse cavalry. Moreover, the emphasis placed on armour protection in the case of the infantry tanks, and on mobility in that of the cruisers, obscured the even greater importance of tank armament, to the detriment of British tank gun development.

The dispersal of effort was aggravated by the vacillating requirements of the War Office which, together with the preoccupation with the other, less important, characteristics of tanks, led to no fewer than 12 different tanks being designed and built in the space of six years, all armed with the same 40 mm gun. The resulting situation was made worse by the fact that the experience available at the time in the UK in the design and development of tanks was extremely limited, being confined to two small engineering teams, one of which was the tank design department at Woolwich Arsenal and the other at the experimental works of Vickers-Armstrongs at Chertsey. To cope with the number of designs demanded by the War Office several other companies were brought in on the eve of and during the war. But they came from the motor and railway industries and had no previous experience of tanks, which was bound to have an adverse effect on the characteristics and performance of British tanks.

The position did not begin to improve until the end of the war, with the demise of the pernicious division into infantry and cruiser tanks and the advent of the Centurion tank. The latter was the outcome of the first design undertaken during the war by the Department of Tank Design (DTD) of the Ministry of Supply, which made use of its expertise and wartime experience to such good effect that the Centurion not only passed very successfully the troop trials of its prototypes in 1945, but was still in service, albeit modified, in several parts of the world 50 years later.

GENEALOGY OF RESEARCH AND DEVELOPMENT ESTABLISHMENTS

DTD was a continuation of the tank design department set up at Woolwich Arsenal in 1921, and it became known as such when it was located in 1942 at the new establishment built on Chobham Common, near Chertsey. After the war it was renamed the Fighting Vehicles Design Department, and in 1948 the Fighting Vehicles Design Establishment (FVDE), which in 1952 was amalgamated with the Fighting Vehicles Proving Establishment to form the Fighting Vehicles Research and Development Establishment (FVRDE).

FVRDE continued under the Ministry of Supply until that was abolished in 1960. It then came under the War Office and, in 1964, under the newly formed Ministry of Defence. In 1970 it was amalgamated with the Military Engineering Experimental Establishment (MEXE) to form the Military Vehicles and Engineering Establishment (MVEE), which in 1984 was merged with the Royal Armament Research and Development Establishment, becoming RARDE Chertsey. On the creation of the Defence Research Agency in 1991 it became DRA Chertsey, and in 1997 the Land Systems Sector of the Defence Evaluation and Research Agency (DERA).

Until 1986 FVRDE and its successors were responsible not only for research, but also for the initial design of tanks in response to the requirements of the War Office or the Ministry of Defence, although the greater part of detailed design and the construction of prototypes were carried out under contract by industry. This involved Vickers-Armstrongs in Newcastle and, at first, also Leyland Motors or, in the case of light tanks, Alvis, and after 1973 the Royal Ordnance Factory in Leeds. However, as a result of the government's policy to devolve the responsibility for the design and development, as well as the production, of armoured fighting vehicles to industry, these activities were taken away from RARDE Chertsey and consequently its functions became confined to research and the provision of advice to the Ministry of Defence.

RARDE, with which MVEE was merged, grew out of the Armament Research Department (ARD) and Armament Design Department (ADD), which were transferred in 1942 from the Woolwich Arsenal to Fort Halstead, a mobilisation site built in the 1890s as part of the defences of London. In 1955 the two departments were amalgamated to form the Armament Research and Development Establishment (ARDE), which in 1962 received the Royal appellation. It became RARDE Fort Halstead when the merger with MVEE took place and then in 1991 DRA Fort Halstead, which was no longer responsible for the design of tank guns and other weapons but became confined to a research and advisory role similar to that of the Land Systems Sector of DERA, operating as its Weapons Systems Sector.

TANK GUNS

Work on tank guns at Fort Halstead did much to restore the reputation of the UK in the tank field by its contribution to the development of the Centurion and in particular by providing it with progressively more effective main armaments.

The prototypes of Centurion, and its early Mk 1 and Mk 2 production versions, were still armed with the wartime 76.2 mm 17-pounder gun. But the Mk 3, which entered production in 1948, was armed with a newly designed and much more powerful 83.8 mm gun, or 20-pounder, as it was called under the archaic system of gun designation still used at the time by the War Office (Fig. 5.1).

The 20-pounder was designed by ADD with what was then the standard APCBC (armour-piercing, capped, ballistically capped) tank gun ammunition. Its general characteristics were similar to those of the German 8.8 cm KwK 43 L/71 gun, which was introduced towards the end of the Second World War on the Tiger II heavy tank, and was probably the most effective tank gun of that conflict.[3] However, the 20-pounder was superior to the German gun in terms of its armour-piercing performance because it was provided not only with APCBC but also with APDS (armour-piercing, discarding-sabot) ammunition. Projectiles of this type incorporated a very hard, high-density, subcalibre penetrator of tungsten carbide which concentrated the kinetic energy of the projectiles on a smaller area of the target, and thereby penetrated much more armour than conventional full-calibre armour-piercing projectiles.

Ammunition with tungsten carbide penetrators was actually introduced in German tank guns during the war and, among others, was produced for the 88 mm gun of Tiger II. However, in the case of the German, and then similar Soviet and

FIGURE 5.1: CENTURION TANK ARMED WITH THE 83.8 MM 20-POUNDER GUN WHICH FIRED TUNGSTEN-CARBIDE-CORED APDS (ARMOUR PIERCING DISCARDING SABOT) PROJECTILES WITH A MUZZLE VELOCITY OF 1465 M/S—HIGHER THAN THAT OF ANY OTHER TANK GUN WHEN IT WAS INTRODUCED IN THE LATE 1940S (© CROWN COPYRIGHT, DERA)

US tank gun ammunition, the penetrator was carried in a full-calibre projectile body which travelled with it all the way to the target, and thus caused a rapid loss of velocity and armour penetration with range. As a result this type of ammunition, which came to be called armour-piercing, composite, rigid, or APCR, was only significantly better than the more conventional APCBC at short ranges. On the other hand, there was no such loss of performance with range with APDS ammunition, because the pot-like sabots in which its penetrators were carried were discarded at the muzzle, so that the penetrators travelled to their targets without them and their aerodynamic drag was consequently considerably lower than that suffered by APCR projectiles.

The successful development of APDS ammunition was a very major achievement on the part of ADD and ARD. It was initiated at Fort Halstead by L. Permutter, who came to the UK from Belgium at the beginning of the Second World War. The first experimental firings took place in June 1941, using a 25 mm antitank gun provided by the Free French Forces.[4] It was then produced for the 57 mm 6-pounder guns of Churchill infantry tanks, to be used by them in the Normandy campaign of 1944. Towards the end of the war APDS was also produced for 17-pounder tank guns, but far greater penetration was achieved with the 20-pounder of the Centurion. Its APDS was fired with a muzzle velocity of 1465 m/sec, which was higher than that of any other type of tank gun ammunition produced until then, and not only greatly increased armour penetration but also resulted in a flatter projectile trajectory and therefore a greater probability of hitting the target. As a result, the 20-pounder became an exceptionally effective tank gun.

Armed with the 20-pounder, Centurion 3 performed with credit when it was first used in 1951 in the Korean War, and the three British armoured divisions which were deployed in Germany during the 1950s and were re-equipped with the Centurions formed the most effective counter to the Soviet armour massed in what was then East Germany. Moreover, the capabilities of the Centurions led to their use by no fewer than 16 other countries, which acquired them during the 1950s either by direct purchase, like Sweden and Switzerland, or by delivery under US military aid programmes, like the Netherlands and Denmark, and, later, by procurement from third parties, like Israel. In fact, more than half of the total production of 4423 tanks was exported.

The development of the 20-pounder and its APDS ammunition was followed by the even more successful evolution from it of the 105 mm L7 tank gun. The original experimental version of this gun was actually made in 1954 by boring out a 20-pounder barrel from 83.8 to 105 mm. It proved capable of still greater armour penetration, not only because of its larger calibre but also because its greater expansion ratio allowed more of the propellant gas energy to be imparted to the projectiles.

The armour-piercing capabilities of the 105 mm gun were further increased by the progressive development of its APDS ammunition. The main feature of this development was a gradual move away from penetrators made of tungsten carbide, which for all its other advantages was relatively brittle, causing penetrators to shatter on well-sloped armour. In their place came tungsten alloys, which were less hard but had even greater density and, what is more, were tougher. The ultimate result of this was the 105 mm L52 APDS projectile, developed during the early 1960s, which could penetrate 120 mm of steel armour inclined at 60° at ranges of up to 1830 m, or almost twice what an APCBC projectile fired from the same gun would have penetrated.

On the strength of this ARDE proposed that the Centurions be rearmed with the 105 mm L7 gun, and in 1959 this began to be implemented. In 1958 the 105 mm L7 was also adopted with some modifications by the US Army, in preference to two competing US tank guns, and it became the main armament of the US battle tank of the 1960s and 1970s, the M60. In addition, the 105 mm L7 was also adopted, with or without modifications, for several other tanks, including the German Leopard 1, the Swedish S-tank, the Swiss Pz.61, the Indian Vijayanta, the Japanese Type 74, the Israeli Merkava and even the original version of the US M1, designed in the mid-1970s, as well as the South Korean K1, designed still later, and the Chinese Type 80. Altogether, about 35,000 different tanks have been armed with the 105 mm L7 or its derivatives, making it the most widely used tank gun since the Second World War, although not as numerous as the 100 mm D-1OT mounted in about 90,000 Soviet T-54 and T-55 tanks and their Chinese, Type 59 copies.

In the meantime the threat of war with the Soviet Union, which became manifest by 1950, led to the view that even the then newly produced 20-pounder would not be powerful enough to defeat the contemporary Soviet IS-3 heavy tanks. To do this a 120 mm gun was considered necessary, but the development of such a gun had not even started in the UK. In consequence, a decision was taken to adopt the only gun of this calibre that was available which was a US 120 mm gun developed from an antiaircraft gun, originally for the US M34 experimental heavy tank designed in 1945. However, ARDE developed for the British 120 mm L1 version of this gun two new and very different types of ammunition from those used at the time in US and other tank guns.

One of these was a larger-calibre version of the APDS ammunition already so successfully used with the 20-pounder. The other was of a novel type called high-explosive squash head, or HESH, whose projectiles consisted of thin-walled shells filled with a plastic high explosive which squashed on impact against the surface of the target before being detonated. This type of ammunition was devised in the UK during the Second World War by Vickers-Armstrongs for use against concrete fortifications, but was abandoned before the end of the war. However, it was then realised that HESH was effective not only against concrete and similar structures, but also against armour, because, by exploding in close contact with the outer surface, it caused the spalling of lethal metal scabs from the armour's inner face. As a result, HESH was adopted as a complement to APDS for the US-based 120 mm L1, and subsequently also for the 105 mm L7 and later British tank guns. It was also taken up in the USA under the name high-explosive plastic (HEP), and in Germany when both countries adopted the 105 mm L7, but only as a tertiary type of ammunition, and neither country continued to use it in later types of guns. In the UK, on the other hand, HESH was considered so effective that it was used as the principal type of ammunition, not only with the 76 mm gun mounted in the Scorpion light tank as well as the Saladin armoured car, but also of the 183 mm L4 tank gun developed by ARDE during the mid-1950s.[5]

The 183 mm L4 is believed to be the largest-calibre tank gun ever built. Even the Soviet Union, which was given to gigantism, does not appear to have built a tank with a gun of more than 152 mm, and although in 1952 the US Army contemplated the design of a heavy tank with a 175 mm gun, it never built one with a gun of more than 155 mm. The 183 mm L4 was to arm the Heavy-Gun Tank No. 2, or FV 215, which was intended to defeat the expected successor of the Soviet IS-3 heavy tank. Design studies started in 1950, but its development was abandoned in 1957, when it had progressed no further than a full-size mock-up. However, the 183 mm gun was test-fired in 1955 from a modified Centurion chassis. Its rather odd calibre was merely a metric conversion of 7.2 inches, the calibre of the gun/howitzer that formed the equipment of the heavy regiments of the Royal Artillery during the Second World War and which was later used for trials of large HESH projectiles.

The 120 mm L1 gun, which established the use of HESH, was mounted in a development of a tank designated FV 201. This was designed by FVDE as the postwar Universal Tank, but it was never adopted as such, and the 20-pounder originally intended for it was mounted in the Centurion which, in effect, took over its role. Instead, FV 201 was developed into FV 214, which was to be Heavy-Gun Tank No.1 and was given the name Conqueror.

The development of FV 214 as a heavy-gun tank and the redesignation of the Centurion as a medium-gun tank followed a tripartite American, British and Canadian agreement, reached in 1948, that there should be three categories of tanks: light gun, medium gun and heavy gun. This represented a departure from the previous British Army concept of a universal tank but, mercifully, did not constitute a return to the discredited doctrine of infantry and cruiser tanks, which did not differ from each other so far as their guns were concerned. However, the agreement did not hold for very long. In 1957 the Fourth Tripartite Armour Conference, held in Quebec, recommended that medium- and heavy-gun tanks be replaced by a single main battle tank (MBT). By then the Conqueror heavy-gun tank had gone into production, with the first being delivered in 1955. However, only about 180 were built, and soon after the last entered service it lost much of its purpose because the armour-piercing capabilities of its 120 mm gun were approached and eventually matched by those of the 105 mm L7, with which the more versatile Centurions were being rearmed.

The concept of the MBT was first embodied in what was originally envisaged as Medium-Gun Tank No. 2 and eventually became the Chieftain. Studies leading to it started at FVDE as early as 1951, and at first considered the use of a 105 mm gun of US origin. However, as its design began to crystallise in 1956, it incorporated a new British 120 mm L11 gun (Fig. 5.2), which differed radically from all earlier tank guns in using two-piece ammunition with separate, bagged propellant charges.[6]

FIGURE 5.2: CHIEFTAIN MAIN BATTLE TANK ARMED WITH A 120 MM GUN WHICH WAS THE MOST POWERFULLY ARMED AS WELL AS THE MOST HEAVILY ARMOURED TANK AVAILABLE TO NATO FORCES DURING THE 1970S (© CROWN COPYRIGHT, DERA)

Separated ammunition had been used before, with Conqueror's 120 mm L1 and other large-calibre tank guns, as one-piece rounds would have been too heavy in their case to manhandle. However, it still used brass cartridge cases, which were heavy. By comparison, bagged charges were much lighter, took up less space and, being fully combustible, eliminated the problem of noxious fumes given off within tanks by spent cartridge cases. Moreover, trials carried out by FVRDE showed that bagged charges were less of a fire hazard when a tank's armour was perforated than were the traditional metal-cased charges.[7]

Bagged charges were, however, incompatible with the standard sliding-block type of tank gun breech. Heavy artillery and naval guns used them, but with a screw-type breech which was too slow for tanks. The problem was resolved by ARDE, which adopted a sliding-block breech which, instead of relying as usual on brass cartridge cases for obturation, used an elastic steel sealing or obturating ring, first employed in 1944 in a German 150 mm howitzer.[8] The 120 mm L11 gun could penetrate more armour with its APDS projectiles than could the contemporary 105 mm L7 and its derivatives, which made the Chieftain the most potent tank available to NATO forces in Central Europe in the late 1960s and early 1970s.

During this period more guns were developed by RARDE, but with less success. The first was the 110 mm EXP-7, which was conceived to exploit the widespread use of the 105 mm L7 and its derivatives by being designed so that it could directly replace them when the expected demand for it arose. However, this demand did not materialise. Moreover, EXP-7 had serious problems with the extraction of its brass cartridge cases, so nothing came of it except for the development of another 110 mm gun, the EXP-14, which was the first British tank gun to use semicombustible cartridge cases with a metallic stub case.

EXP-14 fired APDS with a higher than ever muzzle velocity of 1578 m/sec and, in spite of its slightly smaller calibre, could penetrate more armour than the 120 mm L11. Its performance made MVEE favour it for future tanks, and in 1974 it was put forward as the main armament of the FMBT, or future main battle tank, which the UK was by then trying to develop in collaboration with the Federal Republic of Germany.[9] However, in 1975 a trilateral evaluation was carried out by the UK, the Federal Republic and the USA in an attempt to find a common armament for the future tanks of the three countries. As a result of this the UK discarded the 110 mm gun in favour of a new 120 mm M7. However, the M7, like the EXP-14, used separated ammunition, and this made it incompatible with the XM-1 tank being developed by the US Army. To suit the XM-1, RARDE was made hurriedly to develop yet another gun, the M13A, which fired one-piece ammunition. This was entered in the second round of the trilateral competitive trials, which took place in 1977 in the USA at the Aberdeen Proving Ground, and which resulted in the adoption by the US Army of the Rheinmetall 120 mm smooth-bore gun. But the M13A came a close second to the German gun, which was all the more creditable in view of its hurried development and the fact that RARDE had only recently taken up what had become the most important type of tank gun ammunition, in the trilateral competition and in general, namely APFSDS, or armour-piercing, fin-stabilised, discarding-sabot.

APFSDS

The importance of APFSDS ammunition was due to its long arrow-like, fin-stabilised penetrators, which concentrated the kinetic energy of projectiles on a smaller area of the target and thereby achieved considerably greater penetration than other armour-piercing projectiles. Recognising its potential, the US Army started working on this type of ammunition in the 1950s, but it was seduced away from smooth-bore guns firing APFSDS by the more futuristic appeal of guided missile launchers: their development was taken up in the mid-1960s in Germany. The Soviet Army had also started developing smooth-bore guns firing APFSDS in the late 1950s, and in 1961 it decided to adopt one for the T62 tank in the form of the 115 mm U5-TS. However, as late as 1973 the official UK view was that APFSDS was inferior to APDS because it was considered to have greater dispersion, leading to a lower probability of hitting targets, and a more rapid loss of velocity, and therefore of armour penetration, with range.[10]

However, by then it was clear even to outside observers that '...fin-stabilised, hard-core projectiles represent, potentially, as much of an advance on APDS projectiles as the latter did on the original armour piercing ammunition ...'.[11] Unfortunately, the great success achieved with APDS made its developers at Fort Halstead reluctant to admit that another type of ammunition could be even better. By the time they did, APFSDS was already established elsewhere, and RARDE lost the world leadership it had enjoyed as the result of its development of the 20-pounder and the 105 mm L7 guns and their APDS ammunition.

The place of the 105 mm L7 as the international standard was taken over by the Rheinmetall 120 mm smooth-bore gun, which was widely adopted or copied outside the former Communist bloc. In fact, the UK was the only Western country not to develop a gun which was similar to the Rheinmetall Rh 120. Instead, it remained wedded to rifled guns, which were not interoperable with all the other 120 mm tank guns. An often-quoted justification of this continued use of rifled guns, when the advent of APFSDS made rifling redundant, was that they were necessary to fire HESH, which the British Army preferred to high-explosive or multipurpose rounds as secondary tank gun ammunition. In fact, HESH could be fired from smooth-bore guns, albeit at the cost of some reduction in its explosive content because of the weight of the fins that would be required.

The 120 mm M13A, as well as the M7, was considered for the new MBT-80 tank, work on which started after the 1976 collapse of the Anglo-German FMBT programme, but neither gun was adopted. Instead, it was proposed to use an evolutionary development of the Chieftain's L11 gun and then a new EXP-28M1 gun. However, the requirement for the MBT-80 was withdrawn and the direction of gun development changed yet again. This time it was aimed at a new 120 mm rifled gun, which would be used to rearm the existing Chieftain tanks and to arm their development, Challenger. Because of this the gun became known as CHARM, an abbreviation of Chieftain/Challenger Armament.

The design of this gun, designated the 120 mm L30, was constrained by the requirement that it should be retrofittable in Chieftains, but ironically, before it could be produced, a decision was taken by the Ministry of Defence to withdraw Chieftains from service. Later the same thing happened with Challenger 1, which in the meantime was armed with the same 120 mm L11 as the Chieftains. The L30 was finally only fitted in Challenger 2, where it was free of the constraints imposed by the older tanks and, in particular, by their turrets. Nevertheless, the use in its design of a monobloc autofrettaged barrel made of electroslag refined steel enabled the L30 to have a chamber pressure 25 per cent higher than that of the L11. Like the EXP-28M1, it also incorporated a novel type of split-block breech mechanism, which ARDE had already patented in 1959. This combined the rapid action of the sliding-block type of breech with the Crossley pad obturation used with the screw-type breeches of artillery guns, and consequently made possible the continued use of fully combustible propellant charges while being more robust than the breech of the L11 gun with its steel obturating ring.[12]

However, after designing the L30 and further studies, RARDE fell in with the world trend and accepted a return to the use of single sliding-block breech, and again took up combustible propellant charges with stub cases. It also finally agreed to abandon the rifling of gun barrels in favour of smooth bores. All this happened in the context of the development of the Future Tank Main Armament (FTMA), a 140 mm gun which evolved from studies started in 1982. Similar studies were taking place at the time in other countries, in particular in the USA, where tanks with 145 mm guns were already being designed. This led to an agreement in 1988 between the UK, USA, France and Germany about a future interoperable 140 mm smooth-bore gun, which then governed the characteristics of FTMA, or 140 mm EXP-38. Prototypes of it were built and fired from a Centurion chassis, but by 1995 the British as well as other armies had lost interest in 140 mm calibre tank guns and decided to continue using 120 mm guns, whose performance was considered capable of further improvement.

UNCONVENTIONAL GUNS

In addition to all the work on conventional solid-propellant guns, the ARD, RARDE and then DRA worked on unconventional guns intended for tanks. The earliest of these were liquid-propellant guns, which were seriously considered in the 1950s as a candidate for what became the Chieftain tank, after the US 105 mm and before the British 120 mm L11 gun was developed.

Work in the UK on liquid-propellant (LP) guns came close on the heels of their development in the USA, which started in the late 1940s following the lead taken in Germany during the Second World War in the application of liquid propellants to rockets. It started in 1952 and led, in the first instance, to the construction in 1953 of an experimental LP version of the 83.8 mm 20-pounder. The experimental gun had a regenerative injection system and used a hypergolic bipropellant consisting of hydrazine fuel and red fuming nitric acid oxidiser, which inevitably caused severe corrosion. This, together with a number of other problems, resulted in a redirection of research effort to bulk-loaded 76.2 mm guns using monopropellants. However, these failed to offer any significant advantage over solid-propellant guns and work in the UK on LP guns was terminated in 1957. A similar fate overtook the work on LP guns in the USA, where it was virtually abandoned around 1960.

There was little further interest in LP guns until the early 1970s, when there was a revival in the USA as a result of the development of new monopropellants. These were pioneered by the US Navy for torpedoes and were based on hydroxyl ammonium nitrate (HAN), with water as a solvent and diluent. HAN-based monopropellants were much more attractive than the earlier liquid propellants because of their low toxicity and low flammability and they were also regarded as potentially less vulnerable than solid propellants.[13]

Attempts to exploit the new monopropellants began in the USA in the mid-1970s and led to a revival of interest in LP guns at RARDE in 1981. Work on them resulted in the setting up in 1987 of a Research Technology Demonstrator Programme, aimed at exploring the use of LP guns as tank armaments,[14] but before the investigations had progressed very far interest began to shift from tank to artillery applications, and in mid-1991 all work at Fort Halstead on LP guns was curtailed, and wound up completely in 1995. Much the same thing happened in the USA, where studies of the application of LP guns to tanks also began in 1987, only to be abandoned in 1990. The development of a 155 mm howitzer, which had started in 1986, did better as it resulted in the construction of experimental guns, but this project was eventually terminated in 1996, although LP guns were considered to offer much greater advantages as artillery than as tank weapons. However, even then their potential advantages were outweighed by the complexity of their regenerative injection systems and by their combustion problems.

Even before the revival of its interest in LP guns, RARDE also began in 1980 to study electromagnetic (EM) guns. The concept of using such guns can be traced back to the turn of the century, but it was only in the late 1970s that they began to be regarded as a practical proposition. This was due very largely to the experiments carried out around 1970 at the Australian National University, where an EM launcher was used to accelerate small, 3 g pellets to 6000 m/sec, which was much higher than the velocity achieved with any solid- or liquid-propellant gun. These and other experiments prompted the US Defense Advanced Research Projects Agency (DARPA) to fund the development of EM launchers, and in 1983 one of them accelerated a realistic 317 g projectile to 4200 m/sec.

Studies and experimental results such as this led RARDE, as well as its US counterparts, to overoptimistic views of what could be done, and in particular about the possibility of mounting EM guns on armoured vehicles. Thus, in 1987, RARDE proposed the construction of a technology demonstrator consisting of an EM gun on a Challenger tank chassis, but it was soon realised that this was not yet a practical proposition and RARDE adopted a more measured approach to the development of EM guns.

One feature of this approach was an agreement between RARDE and DARPA to construct an EM launcher facility, jointly funded by UK and US governments, at Kirkcudbright in Scotland, which would complement the facilities already built in the USA. The facility was commissioned in 1993 and became the only one in the whole of the Western world where full-size projectiles accelerated by EM launchers could be tested at ranges of more than a few metres and against realistic armoured targets.

As well as EM launchers, the Kirkcudbright facility can also be used to test electrothermal–chemical (ETC) guns, which represent a more immediate proposition than EM guns. They do so because only a part of the launch energy of their projectiles is electrical, as the rest can come from the chemical reaction of a solid or liquid propellant, and they do not require electrical equipment as large as that of EM guns. However, having already taken up work on EM guns, RARDE did not take as readily to ETC guns when they began to be developed in the USA in the 1980s, and where they were already being proposed, albeit prematurely, in 1989 for use in M1 tanks. But in 1991 DRA took up the study of ETC guns in earnest, although with its work directed at artillery rather than tank applications.

DESIGN OF TANKS

The development of tank guns has inevitably been interrelated with that of the tanks themselves, which are their mobile protected platforms and which increase their battlefield effectiveness by making them more likely to survive as well as more mobile.

The degree of survivability of tanks has been very largely a matter of the thickness of their armour, which has increased steadily over the years in pursuit of immunity from opposing tanks and antitank guns. However, this also led to increases in the weight of tanks, to which there is a limit. This was reached in the early 1950s by the Conqueror heavy-gun tank, developed from the FV 201 universal tank designed by FVDE immediately after the Second World War, which weighed 66 tonnes, only two tonnes less than the heaviest tank ever used, which was the German Tiger II of 1944–45.

To avoid such weights and yet retain a high level of armour protection, FVRDE decided in the mid-1950s, when it began work on the design of the FV 4201 or Chieftain, to depart from tradition. This involved a novel hull configuration in which the driver adopted a supine position instead of sitting upright. This significantly reduced the height of the hull and therefore of the frontal area of the whole tank, which is always covered by the heaviest armour and accounts for much of the tank's weight.

This novel design was adopted by FVRDE after extensive functional and physiological tests of alternative driving positions, carried out between 1955 and 1957. It was preceded by the construction in 1956 of an experimental FV 4202 tank based on Centurion components, which had a lower hull as the result of incorporating a reclining driver's position. With a lower hull the weight of the Chieftain was kept down to 55 tonnes, but the level of its armour protection was greater than that of the Conqueror. For example, the front glacis plate of its hull was 120 mm thick and, as it was well sloped, at 72° from the vertical, this implied a horizontal shot-line thickness of 388 mm of armour, compared to the 258 mm of the Conqueror; and yet the Chieftain was 11 tonnes lighter. In fact, the Chieftain was better armoured over its front than any other tank available to the NATO forces until 1980. The effectiveness of its design is attested by the areal density of its frontal armour, amounting to 3 tonnes per square metre, which has not been surpassed to any great extent even by much heavier tanks built later.

In spite of the novel features incorporated in its design the Chieftain still had the appearance of a conventional turreted tank, but even before a prototype of it was built in 1959 FVRDE began to work on other, radically new designs.

One of the most radical was for a tank very appropriately called Contentious.[15] Work on it started in 1956 in response to the perception that the advent of what was called at the time 'nuclear warfare' placed greater emphasis than ever before on mobility. This led to the concept of a vehicle considerably smaller and lighter than contemporary tanks, so much so that it could be air-transportable or even parachutable. In fact, it was expected to weigh between 20 and 30 tonnes and yet enjoy a high level of protection against nuclear as well as conventional threats. To this end the crew of Contentious was reduced to two, and it dispensed with a turret, its gun being mounted in the hull. The gun mounting provided a limited traverse of ±20° but no elevation, which was achieved by tilting the whole hull by means of an adjustable suspension. Other features of Contentious included a unique photoelectric gun control system and a pulsed-light ranging system.

To test all these novel features an experimental vehicle was built using components of an earlier Comet tank, but the concepts embodied in it did not advance beyond trials, which continued into the mid-1960s.

Another major departure from convention pursued by FVRDE consisted of the design of tanks with guns which were not mounted in turrets but rather externally, on pedestals. Such tanks were considered as early as 1951, in the initial phase of the studies that led to the Chieftain, as a way of keeping down the weight of tanks. None of the original designs was adopted, but they were ahead of other designs for tanks with externally mounted guns devised in the USA and elsewhere.

In the 1960s FVRDE returned to the concept of a tank with an externally mounted gun as part of its studies for a successor to the Chieftain. In 1966 these came to include the design for a lighter MBT with an externally mounted gun which would weigh about 35 tonnes, and which was considered to have a number of distinct advantages over more conventional designs.[16] In fact, tanks with guns mounted externally on pedestals had the advantage of a considerably smaller frontal area, particularly in defensive, hull-down positions, thereby offering a smaller target to enemy weapons. They also had the advantage of concentrating the crew in the hull, where they could be better protected. However, they required a highly reliable automatic loading mechanism because manual loading was impossible, and they made it difficult to protect their guns against ballistic attack.

As part of its studies, in 1968 FVRDE built an experimental vehicle with an externally mounted 83.8 mm gun based on a Comet tank chassis. This came to be called COMRES 75, after the research programme that made possible its construc-

FIGURE 5.3: COMRES 75 EXPERIMENTAL VEHICLE OF 1968, THE FIRST TANK EVER TO BE BUILT WITH AN EXTERNALLY MOUNTED GUN (© CROWN COPYRIGHT, DERA)

tion, and was the first tank with an externally mounted gun to be built anywhere (Fig. 5.3). The concept of externally mounted guns was then carried into the initial stages of the Anglo-German FMBT programme. A design for a medium-weight tank based on this was in fact put forward as one of three UK contributions to that programme. However, after an Anglo-German technical symposium in May 1972, it was abandoned in favour of other concepts that were considered more promising, although the concept was sufficiently attractive to be pursued further elsewhere, particularly in Sweden into the early 1980s and in the USA into the early 1990s.

Of the designs that were pursued further under the Anglo-German FMBT programme the most promising was considered to be a turretless tank with a semifixed gun. This was one of the other UK contributions and it was agreed that it should be the main object of MVEE studies. The third UK contribution, which consisted of a turreted design, was retained as a reserve concept.[17]

The emergence of the turretless tank as the main object of studies was due very largely to the development in Sweden of the turretless S-tank.[18] This was the first and still is the only tank to be built with a fixed gun mounting, and offered a number of advantages over earlier turretless vehicles as well as turreted tanks. Some of these were foreshadowed by the Contentious experimental tank mentioned earlier, but the design of the S-tank was undoubtedly original and attracted a great deal of attention when it appeared in the early 1960s. In consequence, two S-tanks were brought over for trials at the Royal Armoured Corps Centre at Bovington in 1968, and these led to the conclusion that it represented a viable concept, although it was recognised that it suffered from the major disadvantage of not being able to fire effectively while on the move.

Interest in turretless tanks resulted in their being included in the second stage of the studies on 'AFVs for the 1980s', on which FVRDE embarked in 1969, and these led to the turretless UK contribution to the FMBT programme. However, the FVRDE—or by then the MVEE—concept of a turretless tank differed in several respects from the S-tank. In particular, the gun mounting was not fixed but allowed the gun to move in elevation from –10° to +20° and ±2° independently of the hull. This meant that the tank did not require an adjustable suspension, and that the accuracy with which the gun was aimed in azimuth did not depend entirely on the tank's steering system. The turretless tank proposed by MVEE was also much heavier, its estimated weight being 54 tonnes, compared to the 39 tonnes of the S-tank. It would still not be able to fire effectively on the move, but its design allocated more weight to its armour than any of the other concepts.

Notwithstanding the differences between the MVEE design and the S-tank, the British Army remained concerned about the inability of turretless tanks to fire on the move, and in 1971 demanded a re-examination of the S-tank concept. As part of this, it borrowed 10 S-tanks from Sweden for tactical evaluation trials, which were carried out in Germany by one of the British tank regiments stationed there. The outcome of the trials, called Exercise Dawdle, was a recommendation against the development of a turretless tank for the British Army.

Nevertheless, an experimental vehicle representing MVEE's turretless tank, except for the gun mounting, was built in 1973–74. In deference to the Anglo-German programme of which it formed a part, it was called CTR, or Casemate Test Rig, after the name given by the Germans to their turretless designs. Studies of it continued until 1976, when the concept of a turretless tank with a semifixed gun was abandoned with the end of the Anglo-German programme.

The decision against further development of the turretless tank with a semifixed gun was not confined to the MVEE concept of it, but applied also to the German concept of this type of vehicle, which formed the principal German contribution to the FMBT programme. Like the MVEE concept, the German casemate tanks were inspired by the S-tank but differed from both in having not one but two guns, so that they could fire salvoes of two rounds to increase the probability of hitting targets, although hardly in a very efficient way. They also differed from the British concept, which emphasised armour protection. Moreover, they were meant to overcome the inability of turretless vehicles to fire on the move by firing when the guns were in coincidence with the gunner's independently stabilised line of sight while they zig-zagged or snaked instead of moving in a straight line. However, this was only practicable in very favourable, flat terrain, and for all their sophistication the German casemate tanks did not prove acceptable to their potential users any more than MVEE's simpler semifixed gun tank concept.

After the concept of tanks with semifixed guns was abandoned with the termination of the FMBT programme in 1976, both the UK and the Federal Republic of Germany concentrated once again on the separate development of conventional, turreted tanks. Further development of such tanks was greatly encouraged in the UK, and subsequently in other countries, by the possibility of considerably increasing their armour protection by means of a new and much more effective type of armour, devised at FVRDE since the mid-1960s.

DEVELOPMENT OF ARMOUR

Until the 1950s the development of armour protection for tanks consisted primarily of progressive increases in the thickness of steel. In the UK this process reached its limit with the Conqueror and Chieftain, as further increases in thickness would have resulted in unacceptable increases in vehicle weight. As it was, the thickness of the Chieftain's armour was considered to provide adequate protection against the kinetic energy armour-piercing projectiles of contemporary enemy tanks, which constituted the principal threat.

More sophisticated forms of protection against armour-piercing projectiles were considered but were not adopted. In particular, in the aftermath of the Second World War armour consisting of three spaced layers was considered effective against the contemporary APCBC projectiles, with the first layer stripping their caps, the second breaking up the body of the shots, and the third absorbing the remaining energy of the fragments.[19] However, spaced armour was only considered seriously against the new threats posed by HESH ammunition, and by shaped—or, as they were originally called in the UK—hollow-charge weapons.

Of the two, HESH was at first considered to be the more effective and to pose a more serious threat, although it was a British development and no foreign army was known to be using it. However, it must have been assumed that HESH would be regarded elsewhere as highly as it was in the UK, and that the enemies would come to use it. In fact, the UK view of HESH was almost unique. This is well illustrated by the fact that the first antitank guided missile to be adopted by the British Army in 1962, the Malkara, had a HESH warhead, whereas all other antitank missiles produced around the world have had shaped charge warheads.

The view taken of HESH in the UK led to the development of new types of armour which would be more effective against it than solid steel. These took the form either of ribs welded on to the outside of the main armour, or of single spaced 'burster' plates mounted in front of it: the former were incorporated in the design of Contentious, and the latter were tested, among others, on a Conqueror.[20] Both types of armour prevented HESH shells from exploding in intimate contact with the main armour, and thereby reduced their effectiveness. However, neither was adopted and the threat posed by HESH was eclipsed in the 1960s by that of shaped charge weapons, and especially of antitank guided missiles.

Shaped charge weapons had come into use during the Second World War, but in the immediate postwar period they consisted mainly of handheld infantry weapons, which were marginally effective against the frontal armour of the latest tanks and, even if they perforated it, their behind-the-armour effects were limited. In the circumstances the only measure against them that was deemed necessary and practicable was the adoption on the Centurion of thin, additional side plates or suspension skirts. Plates of this kind were originally fitted during the war to German tanks as additional protection against Soviet 14.5 mm antitank rifles, but they were subsequently retained for protection against antitank rocket or grenade launchers, which led to their being called 'anti-bazooka' plates, after the name of a musical instrument bestowed on two widely used US weapons of this kind. The side plates detonated shaped charges at a distance from the tanks' hulls and, as the optimum standoff distance of the early shaped charges was relatively short, this reduced their penetration. In consequence, suspension skirts improved the protection of the Centurion, and then of the Chieftain, against side attack.

The threat of shaped charge weapons became much more serious with the development of antitank guided missiles which had much larger warheads and, therefore, considerably greater penetration capability. The first of these was the SS-10, developed in France from the basis of earlier German work, which went into production for the French Army in 1952. At the same time the US Army ordered a number of SS-10 for trials, and in 1957 adopted it as its first antitank guided missile. By then the SS-10 had also been ordered by Israel, Sweden and Germany, and in 1956 a second and more powerful missile, the SS-11, became operational with the French Army.

The SS-10, and even more the SS-11, could perforate the armour of any contemporary tank and, as the thickness of conventional, monolithic steel armour could not be increased much further, this created a need for some other form of protection against shaped charges which would be more effective in relation to its weight. As part of its studies leading to the Chieftain, FVRDE responded to this need in 1954 by producing designs for tanks with arrays of three or four relatively thin but fairly widely spaced sloping plates in front of the main armour.[21] Arrays of such plates separated by air gaps caused ablation—or what was later called foreshortening—of shaped charge jets, and by so doing reduced their penetration.

FVRDE's designs of multiplate arrays represented a remarkably early and original attempt to counter the emerging threat of large shaped charges. However, none of them was adopted, because of uncertainties about their effectiveness.[22] Little more was then done in the UK about shaped charges for several years. The SS10, as well as other missiles, was tested, only to lead in 1958 to doubts about the lethality of shaped charge warheads of its size.[23] Doubts about the effectiveness of shaped charge weapons were also reflected in the attitude of ARDE, which saw no requirement for them, having so successfully developed APDS ammunition and having also developed HESH.[24] An interesting illustration of the contemporary situation is provided by the fact that when Vickers-Armstrongs (Aircraft) in 1956 began, as a private venture, to develop the Vigilant, which was to become the British Army's first antitank guided missile with a shaped charge warhead, this was originally designed not in the UK but by the Swiss company CML.

What little was done during the latter part of the 1950s about the development of protection against shaped charges consisted of investigations by FVRDE into the use of fuel tanks as part of it. These stemmed from the recognition several years earlier that the penetration of shaped charge jets was essentially hydrodynamic, which meant that its depth was inversely proportional to the square root of the density of the target material. In consequence, low-density materials and liquids could be more effective than steel in resisting shaped charge jets in relation to their weight.

To exploit this, fuel tanks were incorporated in the design of the Contentious as a major element of its protection, and tests subsequently carried out showed that they could withstand the impact of large-calibre kinetic energy projectiles as well as shaped charge jets.[25] However, by the time diesel fuel tanks had been developed as protective devices, better ways of resisting shaped charges were being devised at FVRDE and the idea of using them was abandoned. But it was sound, and fuel tanks were adopted during the 1960s as part of the frontal protection of the Swedish S-tank and of the US–German MBT-70, and then of the Israeli Merkava.

Before fuel tanks began to be considered as part of tank protection and FVRDE produced the designs of the original multiplate arrays, firing trials showed that glass was considerably more effective in resisting shaped charge jets than its density would indicate.[26] However, no attempt was made by FVRDE to exploit the properties of glass or similar materials, in contrast to what was done in the USA, where silicious armour began to be incorporated into the design of tanks in 1953.[27] However, although silica-cored armour was at first considered much more effective against shaped charges than solid steel, it was never adopted because of production and cost problems associated with it, and because by 1961 the requirements of the US Army were met without it by the M60E1 tank.[28]

The armour of the next US tank, the MBT-70, designed between 1964 and 1966 in collaboration with Germany, advanced beyond solid steel plates and castings, but only to a system of two plates separated by an air gap. A similar arrangement, although without the benefit of the outer layer being made of high-performance high-hardness steel, was tried by FVRDE and found to be effective against kinetic energy armour-piercing projectiles as well as HESH, but not to offer greater protection against shaped charges than solid plates of the same total thickness.[29] There remained therefore a need for some other kind of armour which would give tanks greater immunity from the threat of antitank guided missiles with shaped charge warheads.

The need was finally and very successfully met as the result of a research programme initiated at FVRDE in 1963, when the threat posed by antitank guided missiles was clearly recognised. The programme was largely of an empirical nature and was directed by Dr G.N. Harvey, then Assistant Director of Research at FVRDE, who is generally credited for its successful outcome but who carried it out in collaboration with J.P. Downey, who was responsible for its extensive series of firing trials.

The research programme began to bear fruit in 1964, and by the following year had resulted in a new form of armour which was more than twice as effective against shaped charges as rolled homogeneous armour of the same weight, and at least as effective as the latter against kinetic energy armour-piercing projectiles. The precise nature of the armour devised by Harvey and Downey is at the time of writing still

FIGURE 5.4: FV 4211 TEST VEHICLE BUILT IN 1971 USING CHIEFTAIN TANK AUTOMOTIVE COMPONENTS TO DEMONSTRATE FOR THE FIRST TIME THAT THE USE OF CHOBHAM ARMOUR WAS PRACTICAL (© CROWN COPYRIGHT, DERA)

a closely guarded secret, but it is obviously a form of spaced armour which incorporates non-metallic materials as well as steel.

The new armour came to be called Chobham armour, after the location of FVRDE, and in 1968 work began on applying it to tanks. By 1970 a viable armour system had been developed, and in February of that year a decision was taken to build an experimental tank based on Chieftain 3 components, which would incorporate it.[30] The test vehicle was built at FVRDE in the remarkably short space of 13 months, and became FV 4211, the first—if only experimental—tank ever built with the new type of armour (Fig. 5.4).

In addition to having Chobham armour, and because of its effectiveness, FV 4211 was also the first battle tank to have a hull of aluminium armour, which helped to keep down its weight. Its design demonstrated that Chobham armour could be successfully used on tanks, and might have been expected to lead to the development of a new, Chobham-armoured tank for the British Army. But the UK was at the time embarking on the joint FMBT programme with Germany and, as a result, decisions concerning a new tank for the British Army were postponed until this politically driven programme was terminated. By the time a decision was taken Chobham armour had been adopted by the US Army and, in consequence, it was the latter and not the British Army that was the first to put a Chobham-armoured tank into production and service.

The adoption of Chobham armour by the US Army followed from the close military and political links between the UK and the USA. These resulted in the Chief

Scientific Advisor to the UK government, Sir Solly Zuckerman, briefing the US Director of Defense Research and Development about the development of the new armour by FVRDE in December 1964, when it had only just produced its first successful results. This was followed by technical presentations to those involved with tank armour in the USA, but their reaction was remarkably unenthusiastic. Ostensibly, they considered the new armour to be too bulky to be of practical use, but there were other considerations that influenced their negative attitude, apart from the common reluctance to admit that others could do better. The principal one was that in 1965 the US Army had just began to work on the new US–German MBT-70, which was to be armed with a 152 mm gun-launcher as the result of a recommendation of the US Army ARCOVE study, accepted in 1957, that future tanks should be armed with guided missiles. Moreover, the 152 mm shaped charge warhead of the missiles to be fired by MBT-70 was considered capable of defeating any tank armour, and an admission that there was now a much more effective form of it would have undermined the whole of the contemporary US tank development policy.

Admittedly, the 152 mm gun-launcher was being stretched for MBT-70, from its original XM81 form into the longer-barrelled XM150, which could fire high-velocity kinetic energy projectiles as well as guided missiles. But this was being done very largely under German influence. Like FVRDE and RARDE, the German tank developers regarded high-velocity guns as the most effective form of tank armament —so much so, that when the US Army abandoned work on its very promising 120 mm smooth-bore Delta gun, which fired APFSDS projectiles, they took it up in 1965 as a potential alternative to the 152 mm gun-launcher. From this came the 120 mm Rheinmetall gun which, ironically, the US Army later adopted for its next tank.

In the meantime some work was done on Chobham armour at the US Army Ballistic Research Laboratory (BRL), but it had no impact on US tank development until 1972. By then MBT-70 had been abandoned and the US Army was beginning to consider the design of another new tank. To define its characteristics a task force was formed whose head happened to visit FVRDE in the spring of 1972, and only then learnt about Chobham armour and its effectiveness, in spite of BRL's having known about it for several years.[31] As a result of this visit the US Army took up Chobham armour in all seriousness, which led to its being incorporated in the design of the new US XM-1 tank, after the signing at the end of 1972 of an Anglo-American memorandum of understanding which covered the transfer of the relevant technology.

Chobham armour was actually incorporated into the design of the XM-1 after another visit to FVRDE in July 1973, this time by representatives of Chrysler and General Motors, who were competing for the XM-1 development contract. They were shown the FV 4211, and after the visit both companies modified the designs they had already produced to accommodate Chobham armour instead of the simple, two-layer spaced armour and the arrays of several plates separated by air gaps which they had previously considered. This led to each company building a prototype, which was delivered by the beginning of 1976 and, after the selection of the Chrysler design, to the production of it as the M1 tank with what was by then an American version of Chobham armour. Deliveries of the M1 from production commenced in 1980, and eventually totalled 8937 tanks. In addition, more than 1000 sets of the American version of Chobham armour have been produced for the South Korean K1 or Type 88 tank.

Before a tank with Chobham armour was produced for the British Army it might have been produced not only for the US but also for the Iranian Army. This possibility arose out of the large purchases of British tanks by Iran, which ordered 707 Chieftains in 1971 and then took delivery of 187 more. These orders were followed by an Iranian requirement for more tanks, but with improved automotive characteristics. In response, FVRDE offered a new version of the Chieftain with a more powerful and more modern power pack, as well as other improvements, and then another version which not only had all this but also a hydropneumatic suspension, a new turret and, what is more, Chobham armour. So far as its armour was concerned, the design of the second version was based on FV 4211, but its hull was made of steel instead of aluminium.

The offer was accepted, and in 1974 Iran ordered 125 of the first and 1225 of the second version, which came to be called Shir 1 and 2, respectively. This constituted the biggest-ever export order for tanks, and held the prospect of considerable economic benefit to the UK, but it created a bizarre situation because it meant that the UK government was agreeing to the export of tanks with a new type of armour that the British Army was not going to have for several years, and which the Ministry of Defence still considered highly secret 20 years later, to a country whose political stability was questionable and whose regime had been overthrown once already by elements hostile to the UK. The sale agreement also created ill-feeling in the USA, where it was regarded as a betrayal of Allied secrets, in spite of the fact that the contemporary Iranian regime was supported by the US government, which had no qualms about selling it the latest fighter aircraft as well as M60 tanks.

As it was, no Shir 2 with Chobham armour was ever delivered to Iran, because its development was overtaken by the overthrow of the Shah's government, after which, in March 1979, the new Iranian regime cancelled the whole order. However, by then five prototypes of Shir 2, or 4030/3 as it was designated in the UK, had been built, and these were subsequently used for the development of the first tank with Chobham armour for the British Army.

Originally this was meant to be the MBT-80, on which MVEE began work in 1976 after the collapse of the Anglo-German FMBT programme. MBT-80 was to be a relatively heavy, conventional turreted tank, which represented a further extension of the experience gained with FV 4211. However, the consequences of Iranian revolution faced the Royal Ordnance Factory in Leeds with the prospect of having no tanks to produce for several years. To prevent this happening, and to preserve employment in the UK tank industry, the Ministry of Defence decided in July 1980 to abandon the development of MBT-80, which would not be ready for production for some time, and to adopt instead a derivative of 4030/3, which became the

Challenger. A number of changes had already been made to the design of 4030/3 in anticipation of this, and in 1981 the Royal Ordnance Factory received an order for 237 Challengers, the first of which was delivered in December 1982. Thus, the British Army received its first tank with Chobham armour. Eventually 420 of the original Challengers were built, the last in July 1990.

TANK ENGINES

Compared with what it achieved in the field of tank armour, FVRDE was less successful or fortunate in that of the automotive components of tanks. Unlike armour, these components were not in general developed by FVRDE itself, but by means of research and development contracts placed with industrial companies. To some extent, therefore, it was not responsible for everything that happened, but even if some of the shortcomings were due to the failings, or misjudgements, of the British industry, the decisions concerning it taken by FVRDE (and before it by FVDE) were not always as judicious as they might have been. This was particularly true when they allowed themselves to be seduced by overoptimistic assessments of their capability to support the development of unconventional or novel engines, none of which lived up to its promise.

During and immediately after the Second World War British tanks were powered by adaptations of engines designed for other purposes. The best of these was the Rolls-Royce Merlin gasoline aeroengine produced for the Hurricane and Spitfire fighters of the Royal Air Force which, in a derated form, in the latter part of the war became available for tanks as the Meteor engine. After the war the Meteor was used to power Centurion tanks and, when more power was required by FV 201 and then the Conqueror, its output was raised from 650 to 810 horsepower. This was done by following the example set by the final version of the German Maybach tank engine, and involved the replacement of carburettors by gasoline injection, which was also adopted at the time for US and French tank engines.

In the meantime, as resources and time became available after the war, it became possible to develop a new engine specifically for tanks. Given the general trend of development the obvious candidate was a diesel engine. The Russians had already used on a large scale a four-stroke V12 diesel which was to prove effective for 50 more years, and whose advantages were well known at the time.[32] Instead, FVDE decided to promote the development of something entirely different and new, namely a gas turbine tank engine.

This was prompted by the discovery in the aftermath of the Second World War that in Germany, which had already led in the development and employment of gas-turbine-driven jet aircraft, studies had been pursued of the design of a gas turbine of about 1000 hp for tanks.[33] As a result of this, within seven months of the end of the war FVDE awarded a contract to C. N. Parsons & Company to investigate the design of a 1000 hp gas turbine for tanks. This led to the conclusion in 1948 that such an engine should be built.[34] When it was completed it only developed 600 hp, but nevertheless in 1954 it was installed in a Conqueror heavy-tank chassis and became the world's first gas turbine tank engine.[35]

A second gas turbine engine was to develop 910 hp, but its fuel consumption proved to be unacceptably high, as might have been expected.[36] In consequence, FVRDE very sensibly abandoned the further development of gas turbines which, at best, was premature. In retrospect it is difficult to understand why it was undertaken so readily, except for the contemporary euphoria surrounding gas turbines as a result of the world lead in aircraft development that the UK enjoyed at the time.

FVRDE considered the possible use of gas turbines in the 1960s, when one of its reports rashly predicted that by 1980 they would be more attractive than diesels.[37] Fortunately, nothing followed, except for tests in 1969 of a 140 hp Rover automotive gas turbine. However, similar views arrived at in the USA at the time led to the development of the AGT-1500 gas turbine and its subsequent adoption for the M1 tank, which resulted in its consuming, in the 1980s, twice as much fuel as a comparable diesel-engined tank.

A similar handicap might have been inflicted on the MBT-80 as a result of pressure exercised in 1978 by the US industry for the use in it of the AGT-1500. MVEE actually considered this as one of three engine options, but a diesel was selected instead and the development of MBT-80 itself was abandoned, as already mentioned.

In the meantime, in the mid-1950s FVRDE embarked on the development of a tank diesel engine which had been delayed not only by the misguided attempt to develop a gas turbine engine for tanks, but also by a US-led policy that military vehicles should use gasoline engines, on the grounds that gasoline was more readily available than diesel fuel. In fact, it was only in 1958 that the US Army finally agreed to the use of diesel fuel, although its Ordnance Department had already initiated work in 1954 on the conversion of the standard US AV-1790 gasoline tank engine into a diesel.

Nevertheless, questions about the availability of the different fuels persisted and, to please everybody, NATO in 1957 adopted a policy that fighting vehicles should be powered by multifuel engines capable of running on fuels ranging from light diesel oil to 80-octane gasoline. Conventional diesel engines could be adapted to meet this requirement, particularly if they had indirect injection systems with swirl or precombustion chambers, like those which were being developed at the time for French, Swiss and German tanks. However, FVRDE decided that the requirement for operation on a wide range of fuels could be met even better by an unconventional, opposed-piston two-stroke diesel, and in 1957 recommended such an engine for the Chieftain tank, which was then being developed.[38]

The recommendation was based on investigations that went back to 1952 and about two years' running on a range of fuels of an existing opposed-piston two-stroke diesel, the 105 hp TS3 truck diesel produced by the Rootes Group. A much larger engine was required for tanks, and the order for it went in 1958 to Leyland Motors. They had no experience of opposed-piston two-stroke diesels but based their design on the Junkers Jumo diesel produced for aircraft in Germany during the 1930s, and had their first L.60 engine running in 1959.

The opposed-piston configuration of the Leyland L.60 resulted in combustion taking place between pairs of hot piston crowns, which was very advantageous from the point of view of operating on a wide range of fuels. Unfortunately, highly rated engines of this kind are liable to suffer from thermal stress problems in their complicated cylinder liners, and from the high thermal loading of their pistons, which were overlooked or underrated by their proponents. In consequence, the L.60 was for many years afflicted with piston and cylinder liner as well as other problems, and although it was designed to produce 700 hp, when Chieftain tanks powered by it entered user trials in 1962 it could only develop 550 hp. An intensive development programme was launched in 1968 to correct its shortcomings, and eventually raised the output of the L.60 to 720 hp, but its problems marred the reputation of the Chieftain throughout its service life.

However, the problems encountered with the L.60 did not discourage FVRDE from becoming involved in the development of yet another unconventional engine. This followed a study of future engines that the Motor Car Division of Rolls-Royce was commissioned to carry out in 1962, and which concluded that the most promising engine was a Wankel-type rotary engine then beginning to be developed in Germany. On the strength of this conclusion FVRDE decided in 1964 to investigate this type of engine through Rolls-Royce, who were funded to acquire a licence from NSU Wankel and who subsequently began to develop a novel, two-stage diesel version of the rotary engine. By 1968 this produced the first experimental rotary diesel, which was followed by the construction of two more experimental units and the design and manufacture, in 1970, of a larger, 350 hp 2R6 engine.[39]

The Rolls-Royce rotary diesel was thought to offer the advantage of being more efficient than the gas turbines and less bulky than conventional, piston-type diesels. It was therefore considered by MVEE as a potential tank engine during the 1969–72 period and was included as such in some of the contemporary design studies. However, its development ran into problems inherent in its configuration, and in 1974 a conclusion was reached at MVEE that no further support should be given to it.[40] A similar negative verdict was passed on the original, single-stage spark-ignition version of the Wankel rotary engine by the world's major motor manufacturers, who explored its possible use in the early 1970s, following an initial wave of enthusiasm for it.

Once work on the rotary engine was abandoned, development effort was concentrated on conventional four-stroke diesels, as it should have been all along. An engine study carried out at FVRDE in 1969 had already come out in favour of a V12 four-stroke diesel as a tank engine, and from about 1972 onwards the use of such an engine, and of a related V16 diesel, was envisaged in tank design studies.[41] Neither engine existed at the time, but they were being designed at Rolls-Royce as part of a family of V-configuration diesels based on the experience with an earlier, six-cylinder inline Eagle truck engine.

The design of this family of engines was initiated in the late 1960s by the Diesel Engine Division of Rolls-Royce as a private venture, while its Motor Car Division was still promoting rotary engines. The first of them, which was conceived not as a military but as a commercial engine, was a V8, which was expected to deliver 400–750 hp. A prototype was built in 1973, but no order was placed for a V12 until 1974, and then only in response to the Iranian demand for tanks with better automotive performance than that of the Chieftains—which led MVEE to develop them into 4030/2.

The first V12 engine was built and running at its full rated output of 1200 hp in the remarkably short space of 18 months, and by the end of 1976 it was already powering the first prototype of Shir 1, or 4030/2. However, while production was getting under way the Shah was overthrown and orders for Shir 1, as well as Shir 2, were cancelled. In 1979 a modified version of 4030/2 was sold to Jordan as the Khalid battle tank in even greater numbers than that of the Shir 1 order. In fact, the Jordanian order amounted to 274 tanks, the first of which was delivered in 1981 and became the first British-built tank to go into service since the early part of the Second World War with a sensible, thermally efficient, four-stroke diesel.

The same V12 diesel had also been ordered for Shir 2, or 4030/3, and when this tank was developed further and produced as the Challenger the engine began to enter into service with the British Army in 1983.

TRANSMISSIONS AND SUSPENSIONS

In contrast to their willingness to support the development of unconventional or novel engines, FVDE and then FVRDE were more conservative when it came to tank transmissions. In fact, not only the Centurion, but also the FV 201 and the Conqueror, were still fitted with the Merritt–Brown transmission originally designed at the beginning of the Second World War for the Churchill infantry tank, which continued to be produced by the David Brown company. This was robust but based on a rather crude, manually operated crash-type gearbox which was not easy to use. However, it was the first to incorporate a very effective, triple-differential steering system devised by Dr H.E. Merritt, who was the Superintendent of the Tank Design Branch at the Woolwich Arsenal.

After the war the triple-differential type of steering system was adopted for other transmissions which were developed at the time in the USA as well as the UK. The US transmission was the CD-850 developed by the Allison Division of General Motors, which began to be produced in 1949. It incorporated a hydrokinetic torque converter and, so far as forward motion was concerned, already only required the driver to select one of two speed ranges. Its British counterparts were simpler but still required tank drivers to change gear, although the crash-type gears were replaced by epicyclic trains.

The first of the new British transmissions was the Merritt–Wilson TN 10, which was designed for the FV 301 light tank conceived in 1947. It consisted, in essence, of the Merritt triple-differential steering mechanism grafted on to a Wilson-type epicyclic gearbox. This combination was originally proposed by G.V. Cleare, who was responsible for transmissions at FVDE and who considered it 'ideal'.[42]

The TN 10 transmission was actually designed and built by Self-Changing Gears, a company founded by Major W.G. Wilson, who was the leading designer in the UK of tank transmissions from the middle of the First World War to the beginning of the Second, and whose epicyclic gearboxes were used successfully by the British motor industry. Its pedigree was therefore very promising, but only three TN 10 transmissions were built, as the development of FV 301 was abandoned in 1953 after the construction of one prototype. However, when the Chieftain began to be developed in the mid-1950s TN 10 served as the basis of the TN 12 transmission adopted for it. The same basic design was also adopted, in a scaled TN 15 form, for the Scorpion light tank, or Combat Vehicle Reconnaissance (Tracked) as it was officially called, developed during the 1960s.

However, by then the design of the Merritt–Wilson transmission began to fall behind the general trend towards fully automatic transmissions with hydrokinetic torque converters and double-differential steering systems with infinitely variable hydrostatic drives. This trend started with the Allison CD-500 transmission, which began to be produced in 1951 for the US M41 light tank, and the transmission developed in Switzerland in the late 1950s for the Swiss Pz.61 battle tank. The former incorporated a torque converter with a lock-up clutch to minimise its relative inefficiency, as well as a double-differential steering system, albeit still mechanically controlled; the latter did not incorporate a torque converter but already had a hydrostatically driven double-differential steering system.

By 1960 the Renk company in Germany had developed the HSWL 123 automatic transmission for the contemporary *Jagdpanzer*, which had both a torque converter and a hydrostatically driven double-differential steering system. These were also incorporated in the mid-1960s in the HSWL 354 transmission developed by Renk for the US German MBT-70, and subsequently adopted for the Leopard 2 tank. But FVRDE did not begin to take an interest in this type of transmission until 1967, when it produced a study of one, designated TN 16X.[43] The actual design of a transmission of this kind did not begin until 1973, when MVEE embarked on the design of the TN 17, intended for a 30-tonne mechanised infantry combat vehicle.[44] In part, all this reflected the contemporary negative attitude of the British motor industry towards automatic transmissions with torque converters, and MVEE did not fall in with the world trend until 1975, when it adopted for the 4030/2 tank the TN 37 transmission designed and made under contract by the David Brown company.

The conservative attitude of FVDE and FVRDE towards transmissions was echoed in their approach to tank suspensions. In fact, from the point of view of ride dynamics their practice was even regressive at first, because it involved abandoning the Christie-type independent suspensions of the wartime Cromwell and Comet tanks in favour of the Horstmann-type bogie suspension, consisting of pairs of road wheels connected by coil springs. An earlier version of the Horstmann suspension was used successfully on British light tanks of the 1930s, but its performance could not compare with that of independent suspensions. Nevertheless, it was adopted not only for the Centurion and the postwar Conqueror, but later also for the Chieftain, and it was still being used in the Khalid in the 1980s.

However, although the ride the Horstmann-type suspensions of the Centurion and Chieftain offered over rough ground was inferior to that of the Cromwell and Comet, they did have the merit of not taking up any of the critical hull width, which the

Christie suspensions did. Moreover, being externally mounted, they could be repaired more easily if damaged than the independent suspensions with torsion bar springs, which were generally adopted following the example set before and during the Second World War by Swedish, German and Soviet tanks.

A torsion bar suspension was tried in the prototype of the FV 301 light tank and was adopted by FVRDE for the FV 432 armoured personnel carrier, but so far as tanks were concerned it was only adopted later for the Scorpion light tank. When MVEE eventually decided to abandon the use of Horstmann-type suspensions it leapfrogged independent suspensions with torsion bars or other mechanical springs, and in 1974 adopted a hydropneumatic suspension for the 4030/3. This suspension was the outcome of the progressive development of hydropneumatic suspensions pursued by FVRDE since 1967, when it installed one in TV 15000, an experimental tracked vehicle which was a forerunner of the Scorpion family of light armoured vehicles. This was followed by another hydropneumatic suspension for the 30-tonne mechanised infantry combat vehicle mentioned earlier, and later by a third, which was adopted for the 4030/3 tank. Tests of this suspension began in 1976 and it went into production in the Challenger tank in 1982.[45]

Challenger was not the first tank to be produced with a hydropneumatic suspension: the Swedish S-tank preceded it in 1966, and the Japanese Type 74 in 1975, but its suspension represented a considerable advance on those of the earlier tanks and placed the Challenger in the forefront of tank development in this respect.

SOFT-GROUND OPERATION

Suspension characteristics, together with their power-to-weight ratio, play a crucial role in the ability of tanks to move at speed over rough ground. If the ground is soft their performance depends primarily on the interaction between the tracks and the soil. An understanding of this interaction is therefore an essential ingredient of tank design, and has been the subject of extensive studies, particularly during the 1950s and 1960s, at FVRDE and elsewhere. Prior to that an important contribution to the subject was made by E.W.E. Micklethwaite, working at the School of Tank Technology which at the time was colocated with the Ministry of Supply's Department of Tank Design on Chobham Common.[46] This consisted, in essence, of the application of Coulomb's equation for the shear strength of soils, and made it possible to predict the maximum possible tractive effort a tracked vehicle could develop in a given soil. What remained to be done was to find a way of predicting the sinkage of tank tracks in soft soils, and the work consequently absorbed in soil deformation.

A considerable amount of effort was devoted to this subject at FVDE and FVRDE. Some of the incentive for it was provided by the contemporary interest in large-wheeled vehicles as a more mobile alternative to tanks, which involved, among others, the development of non-pneumatic but resilient wheels with automatically retractable spuds.[47] No wheel of this kind was adopted, but an experimental six-wheeled vehicle with large pneumatic tyres was built in 1958 as a possible basis for an alternative to the tracked Contentious, and was originally called Contentious II, although it was later renamed TV 1000, by which name it became known.

To make it more competitive with tracked vehicles by eliminating the space taken up by conventional wheeled vehicle steering, TV 1000 was skid-steered like tracked vehicles. It was the first large military wheeled vehicle to be steered in this way and, so far, there has been only one other: the French AMX 10 RC six-wheeled armoured reconnaissance vehicle developed during the 1970s.

In addition to the more fundamental studies of soil–vehicle mechanics, work also continued at FVRDE on the more practical problem of finding a better general comparator of the ability of tracked vehicles to operate on soft soils than the time-honoured concept of nominal ground pressure (NGP). This implied that the pressure exerted on the ground by the tracks was uniform, whereas in practice it varied, with peaks occurring under the centres of the road wheels. This was recognised even before the Second World War in the Tank Design Department at Woolwich Arsenal, and it was the peaks of the ground pressure, and not its average value represented by NGP, that governed sinkage into the ground. But no satisfactory method of taking these variations in ground pressure into account was found until D. Rowland, working at MVEE, put forward the concept of mean maximum pressure (MMP)—i.e. the average of the peaks of the ground pressure under the tracks—and derived a relatively simple equation to calculate it.[48]

Comparisons of their MMP values explained differences in the relative soft soil performance of tanks which could not be accounted for by their NGP, particularly when their running gear was significantly different. This and its better physical basis has led to MMP being accepted as superior to NGP as a means of assessing the relative soft-soil performance capabilities of tanks.

CONCLUSION

In retrospect, research into the technology of tanks and their development in the UK during the 30-odd years that followed the end of the Second World War resulted in several notable successes. This was particularly true in the field of armour protection and, for a number of years, of tank guns. But efforts devoted to other aspects or components of tanks, such as their engines, were not, in several instances, correspondingly successful.

1. M.M. Postan, *British War Production* (HMSO, 1952); R. Shilling, *History of German Tank Development*, Combined Intelligence Objective Sub-Committee (SHAEF, 1945); R.M. Ogorkiewicz, *Design and Development of Fighting Vehicles* (Macdonald, 1968).
2. R.M. Ogorkiewicz, *Technology of Tanks*, Vol. 1 (Jane's Information Group, 1991).
3. *Movement and Fire Power*, Royal Armoured Corps Tank Museum (Bovington Camp, 1965).
4. L. Permutter and S.W. Coppock, *First Report on the Sabot Projectile*, ARD Terminal Ballistics Report No. 1/44 (1944).
5. *FV 215 Heavy Gun Tank No. 2*, Report PC 41 (FVRDE, May 1958).
6. T.L.H. Butterfield, *Development of FV 4201—Chieftain*, unpublished typescript (FVRDE, ca. 1962).
7. *Relative Fire Risk of 20 pr QF Cartridges and Bag Charges*, Report AR 157 (FVRDE, November 1955).
8. *Illustrated Record of German Army Equipment 1939–1945*, Vol. II (Artillery, War Office, 1948).
9. *AFVs for the 1980s*, Preliminary Study, Stage 2, Final Report, Report No. 71098 (MVEE, February 1972).
10. Deputy Master General of the Ordnance, letter to R.M. Ogorkiewicz, Ministry of Defence (17 April 1973).
11. R.M. Ogorkiewicz, 'The next generation of battle tanks', *International Defense Review* 6, no. 6 (December 1973).
12. J.B. Coleman, 'Tank gun research since 1940', unpublished typescript (DRA Fort Halstead, June 1994).

13. D.C.A. Izod *et al.*, *Liquid Propellant Gun Technology*, Memo 3/84 (RARDE, July 1985).
14. *Liquid Propellant Gun Systems Industry Seminar* (RARDE, April 1989).
15. *Contentious: a possible future main battle tank*, Report AR 180 (FVRDE, October 1959).
16. *AFVs for the 1980s*, Preliminary Study, Stage 1 (FVRDE, November 1969).
17. L.C. Monger, *Future Main Battle Tanks—UK/FRG Feasibility Studies, UK Concept Assessment Report*, unpublished report (MVEE, September 1976).
18. R.M. Ogorkiewicz, 'S-Tank', in *Modern Battle Tanks*, edited by D. Crow (Profile Publications, 1978).
19. J.W. Thomas, *Spaced Armour Assemblies Depending for Their Effectiveness on the Break Up of Shot*, Ministry of Supply, Permanent Records of Research and Development, Monograph No. 5.034 (June 1947).
20. *The Use of HESH as a Substitute for HE Shell in AFVs*, Report TR-1 (FVRDE, September 1956).
21. Photographs 29550/53 and 54 and 299550/40 (FVRDE, January 1955).
22. L. Monger in *Chieftain*, edited by G. Forty (I. Allen, 1979).
23. *The Anti-Tank Lethality of Hollow Charge Warheads Including the French SS-10*, Report TR-15 (FVRDE, April 1958).
24. R.F. Eather, N.Griffiths, *Some Historical Aspects of the Development of Shaped Charges*, Report 2/84 (RARDE, March 1984).
25. *Report on Frontal Protection for an AFV by an Armoured Fuel Tank*, Report BR 182 (FVRDE, February 1965).
26. *Armour Trials*, Report AT 343 (FVRDE, July 1953).
27. R.P. Hunnicutt, 'Abrams', *A History of the American Main Battle Tank*, Vol. 2 (Presidio Press, 1990).
28. R.P. Hunnicutt, 'Patton', *A History of the American Main Battle Tank*, Vol. 1 (Presidio Press, 1984).
29. *Spaced Armour for Glacis of Centurion*, FVRDE Memo. TR 4/61.
30. *AFVs for the 1980s*, Preliminary Study, Stage 2, Supplementary Report, Report 71098 (MVEE, February 1972).
31. O. Kelly, *King of the Killing Zone* (Norton, 1989).
32. J.D. Barnes and D.M. Pearce, *Report on Russian C. I. Tank Engine Type V2 from T-34 Cruiser Tank* (School of Tank Technology, Chertsey, May 1944).
33. R.H. Bright, 'The development of gas turbine power plants for traction purposes in Germany', *Proceedings of the Institute of Mechanical Engineers*, 157 (1947): 375–86.
34. *Gas Turbine for Heavy AFVs*, Report PP 140 (FVDE, November 1948).
35. *Development Tests of a Gas Turbine Power Plant in a Heavy Tracked Test Vehicle*, Report PP 198 (FVRDE, May 1956).
36. R.M. Ogorkiewicz, 'Gas turbines for tanks?', *ARMOR*, 61, no. 5 (September–October 1952): 16–19.
37. J.W. Fitchie, *Gas Turbines for Military Vehicles*, Report BR 198 (FVRDE, December 1967).
38. *Development History of the L. 60 Engine for Chieftain Battle Tank*, VT-1 Division Note 4/88 (RARDE Chertsey, July 1988).
39. F. Feller, 'The two-stage rotary engine—a new concept in diesel power', *Proceedings of the Institute of Mechanical Engineers*, 185 (1970–71): 139–58.
40. P.S. Notley, 'The Rolls-Royce rotary diesel', Report 74012 (MVEE, March 1974).
41. *Analysis of Engine Design Studies for Future Main Battle Tank*, Report PP 237 (FVRDE, December 1969).
42. G.V. Cleare, 'Tank transmission development, part II', *Journal of Automotive Engineering*, 4, no. 6 (December 1973): 23–30.
43. *Design Proposals for a Mechanical Transmission Nominally Rated at 1000 HP*, Report TN 164 (FVRDE, November 1967).
44. M.D. Cawley, A.L. Morris, *The Design of the TN 17 Transmission*, Report 72032 (MVEE, October 1973).
45. P.J. Bosomworth, N.A. Wort, *Development of Hydrogas Suspension for a Main Battle Tank*, Report 77021 (MVEE, September 1977).
46. E.W.E. Micklethwaite, *Soil Mechanics in Relation to Fighting Vehicles* (School of Tank Technology, Chertsey, 1944).
47. F.L. Uffelman, 'The performance of rigid cylindrical wheels on clay soil', *Proceedings of the 1st International Conference on the Mechanics of Soil-Vehicle Systems* (Edizione Minerva Tecnica, 1961): 111–30.
48. D. Rowland, *Tracked Vehicle Ground Pressure*, Report 72031 (MVEE, July 1972); D. Rowland, 'Tracked vehicle ground pressure and its effect on soft ground performance', *4th International Conference of the International Society for Terrain-Vehicle Systems* (Stockholm, 1972): 353–84.

Chapter 6

Aircraft Carriers and Submarines: Naval R&D in Britain in the Mid-Cold War

Tom Wright

INTRODUCTION

In 1945 there can have been few people in Britain who did not consider that one of the pillars of Britain's greatness was its standing as a maritime nation, a view bolstered by heroic tales of seafaring, derring-do and rousing choruses of 'Rule Britannia'. The fundamental force that had converted an obscure island perched off the north coast of Europe into a global imperial power, was maritime might. This took the form of the world's largest merchant fleet and a Navy whose job it was to protect trade, project imperial authority and defend the shores of Britain when threatened by foreign enemies.

In 1995, stripped of empire and merchant fleet and an ambivalent member of the European Community, with mass global travel undertaken by air, the notion of Britain as a nation looking to the sea for its 'greatness' and salvation could at best be regarded as being wildly sentimental. However, no matter what the fall in popular interest or knowledge of matters maritime, Britain's island status and its strategic position as a staging post for the projected two-power-bloc confrontation on the north German plains maintained the role of the Navy as a crucial element of Western defence strategy, even if in its trade defence role the ships it might be defending were more likely to be flying the flag of Liberia or Cyprus rather than the Red Ensign.

The above is put forward to emphasise that, although naval matters lost the high degree of visibility they had enjoyed in the public imagination at the start of the Cold War, they lost none of their importance to Britain as a sovereign state throughout the period. Although the advent of intercontinental nuclear-tipped ballistic missiles and the contemplation of MAD[1] changed the terms—and certainly the image—of the Cold War, the demands of naval warfare were present in any conflict involving less than near-instant total annihilation. Although naval warfare had lost none of the importance it had had in a non-MAD world, it had to cope with the burgeoning exotica of the war's baroque armoury. Indeed, with the advent of the nuclear-powered Polaris submarine, naval warfare acquired a new and central role in the West's strategic posture.

In what follows the aim is to attempt to investigate, and to contribute to, an understanding of the relationship of the Royal Navy's strategic task in the Cold War to the technology that was produced by its own R&D efforts. To do this the scope of the work is directed towards the activities of its research establishments, the types of issues they confronted, and the sort of products with which they were concerned. This therefore excludes what might be thought of as the political economics of naval procurement and the operational use to which equipment was put, except in so far as it was fed back into the R&D knowledge process. In doing this the aim is to provide a sense of what was involved in both an engineering and a scientific sense in naval R&D, while at the same time taking into account the wider social contexts involved, both bureaucratic and individual.

To meet these aims this chapter provides a brief overview of how the organisation of the naval R&D establishments changed over the duration of the Cold War, thus defining the background necessary for the examination of a number of case studies. These focus on the period 1955–65, principally because it coincides with the coming of the nuclear submarine in its various forms. The latter date is significant in Britain's Cold War history, as it marks the official transference of responsibility for the nation's nuclear deterrence from the airborne V-bomber to the Polaris submarine.

BRITAIN'S ROLE AS A NAVAL POWER DURING THE COLD WAR

The end of the Second World War found Britain near bankrupt, seeking to reassume its imperial role as head of a maritime empire held together by trade. Although the Admiralty looked to resume business as before, the inherently unstable nature of the enterprise was early recognised with the granting of independence to India in 1946, an event overseen by the Viceroy of India, (later) Lord Louis Mountbatten, who was to play a pivotal role in the shaping of the Navy during the Cold War, initially as First Sea Lord (the Navy's most senior officer), from 1955 to 1959 and then as Chief of Defence Staff from 1959 to 1965.

Apart from the operational requirement to supply naval forces to meet unanticipated needs such as the Suez campaign, the Korean War, the Cod War, the Falklands War and numerous counterinsurgency and policing operations, the general thrust of naval policy following the end of the Second World War was one of retreat from empire and the redefinition of a strategic role in a world increasingly influenced by nuclear weapons, even if the retreat was not completed until 1975. As with the other Services, naval attention at any given time was focused upon a continuous succession of defence reviews, driven in the main by a weak home economy. The one enduring political goal of the Navy in these circumstances was to retain a blue-water globally interventive capacity, rather than be reduced to providing an extended form of coastal defence in a missile-dominated war in Europe.[2]

Central to all these concerns was the identification of the Soviet Union as the threat to Europe, from the North Cape in Norway via the north German plains to the Mediterranean, as being the most likely battleground for a third world war. The founding of NATO in 1949, with its elaborate plans for trans-Atlantic reinforcement of men and materials at the outbreak of war, brought with it not only the operational need for strategic and tactical collaboration with other nations, most noticeably the USA, but also for technical and scientific cooperation, the latter requirements specifically being mentioned in its Charter. Given Britain's bitter memories of the Battle of the Atlantic in the Second World War, and its own geographical siting, the protection of the North Atlantic sea lanes within the NATO alliance became a major role. However, it must be recognised that the Royal Navy saw this as only a part of its mission within NATO and did not wish to become typecast so that it could only fight one sort of war. That said, the North Atlantic and the ability to launch an attack against the Soviet Union from it in terms of aircraft strikes or amphibious warfare remained an enduring concern.

The importance of air power had been amply demonstrated by the Battle of the Atlantic, a point not lost on the Soviets, who systematically built up their land-based naval aviation, an initiative matched by a massive expansion of their Navy and, in particular, their submarine fleet. The Soviet Navy had played a minimal role in the Second World War, and its massive expansion from a coastal defence force subordinated to the Army, into a truly global power projector, is one of the most remarkable facets of the Cold War, as has been its subsequent catastrophic decline.[3]

Thus from Britain's perspective the threats to the North Atlantic were seen to reside in ever-increasing Soviet naval power, particularly in terms of naval aviation and submarines. Much time and effort over the years of the Cold War were invested by the Admiralty in attempts to persuade the politicians to undertake a massively expensive aircraft-carrier-building programme, which would deal with both and provide the added benefits of flexibility. The adjunct to the need for carriers was a requirement for aircraft, particularly early-warning aircraft, and radars capable of locating Soviet aircraft well out to sea before they could fire their missiles. Given the reluctance of politicians to divert scarce national resources to new carrier construction, the pressure on the Navy was to make do with what they already had in terms of existing carriers, while facing an increasingly demanding environment and adversary.

Until the coming of the nuclear submarine in the 1960s, the submarine threat to the North Atlantic was similar to that of the Second World War, only more developed. The Soviet operational submarine, of which the Zulu class of 1952 was a fair representative, was diesel propelled with an improved snorkel for running under water, being essentially a close-attack weapon depending upon torpedoes and stealth, deeply influenced by the German U-boat type XXI. Defence against such vessels was seen as an extension of the Second World War tactics principally undertaken by aircraft and surface ships fitted with radar, sonar and depth charges. With the introduction of the nuclear submarine USS *Nautilus* in 1955, it became clear that submarine warfare was going to change: *Nautilus* was capable of doing everything the conventional submarine could do, but could do it faster and remain under water almost indefinitely.

In October 1957 *Nautilus* took part in 'Rum Tub', a Royal Navy antisubmarine convoy defence exercise conducted in the Atlantic south-east of Ireland. Detailed analysis of the results revealed:

> The overall impression is that the available anti-submarine defences are not capable of providing protection to any surface force against a high speed, deep diving true submarine.
> The outstanding feature of the submarine operations was NAUTILUS's ability to close an escorted force from any direction, hold or break off contact at will.[4]

As befitted his station as Commander in Chief, Mountbatten's concerns were somewhat broader:

> It is clear that the visit by USS *Nautilus* has precipitated the need for a decision that only an immediate and fundamental revision of our priorities will save the Royal Navy from delegation to a minor role in the future of naval warfare.[5]

Mountbatten's answer to this dilemma was the building of Britain's first nuclear submarine, *Dreadnought*. The process of change wrought upon the Royal Navy by the advent of *Nautilus* was further enhanced by the appearance of the large US nuclear-deterrent Polaris submarines in the early 1960s.

It was at this time that the role of the submarine was beginning to be extended, from being a vessel mainly concerned with the business of sinking surface ships to also becoming a hunter and destroyer of other submarines, a role accelerated by the arrival of the missile-carrying submarine the Soviets had been developing continuously since the end of the Second World War. This change, itself a development of late Second World War American Pacific experience, had begun to be recognised in Britain by 1951 when, in discussing a series of trials to test the hunter concept that had been conducted using conventional submarines, it was noted that: 'The primary operational function of British submarines in the time of war is the interception and destruction of enemy submarines'.[6] This conclusion was confirmed by Rum Tub some six years later: 'It can be argued that the only effective counter to nuclear-powered submarines is a force of similar submarines employed in the A/S role and as escorts.[7]

The hunter/killer concept implicitly placed a high emphasis on the ability of the hunter to seek out the presence of the prey and either deal with it directly or simply act as a listening platform to observe and track the potential enemy without itself being spotted.

The building and design of *Dreadnought*, launched in 1960, represents this major shift in thinking, a decision rapidly reinforced and developed by the building of the various classes represented by the first all-British design of hunter/killer, *Valiant*, of 1963, followed by two boats of the Churchill class of 1970 and thereafter the Swiftsure class of 1973. The first British Polaris submarine, *Resolution*, was commissioned in 1967. The arrival of *Resolution* completed the transformation of the Royal Navy from a force whose operational focus was essentially that of the Second World War, to one whose primary weapon was at the heart of Cold War strategy.

NAVAL SCIENCE AND COLD WAR R&D

Of the three Services the Royal Navy could arguably claim to have had the longest history of sustained, organised scientific research aimed at solving particular issues relevant to its functioning. In the period of relative global peace from the end of the Napoleonic Wars to the First World War, with its accompanying transformation from the wooden walls of *Victory* to the steel hull and steam turbine of the *Dreadnought* (1906), the Navy saw a steady increase in its investment in scientific research. These efforts spanned a range of enterprises, from the gentlemanly voyages of scientific exploration, such as those of the *Beagle* and the *Challenger*, to the establishment of specialised institutions dealing with ship-related matters, such as the Admiralty Compass Laboratory of 1832 and the Admiralty Experimental Works (AEW) of 1870, founded to undertake the scientific design of ships' hulls using models and towing tanks.[8]

The First World War brought not only an increase in the size of, and commitment to, scientific research, but also moves towards improving its organisation to make it more responsive to the Admiralty's needs in an increasingly complex world. This was accomplished through various transformations of the Admiralty's Board of Invention and Research of 1916, into the Department of Scientific Research and Experiment (SRE) of 1920, which in turn led to the founding in 1921 of the Admiralty Research Laboratory (ARL) at Teddington.

In 1946 the SRE was reorganised into the Royal Naval Scientific Service (RNSS), which was composed of four directorates reporting to the Third Sea Lord. These four directorates—Operational Research, Research Programmes and Planning, Aeronautical and Engineering Research, and Physical Research—interacted with a range of facilities, some of which were under their direct administrative control and some of which reported to other directorates within the Admiralty. The professional staff who manned these establishments were not only members of the RNSS, but also came from the Royal Naval Engineering Service (RNES) and the Royal Corps of Naval Constructors (RCNC), the organisational grouping that designed and oversaw the construction of ships and ran the dockyards. These staff were civilians and, as civil servants, belonged to a large number of other civilians who provided backup and logistical support to the Navy in all its many guises, including the provision of the interface between the Navy and industrial suppliers of equipment and services.

It is important to appreciate, however, that naval R&D was not limited exclusively to Admiralty-controlled establishments. Thus in the case where R&D was associated with naval aircraft and their fittings, the Royal Aircraft Establishment (RAE) at Farnborough, and latterly its offshoot RAE (Bedford), through their naval aircraft sections, were the principal authorities.

In any given research establishment it was quite usual to find members of these nominally different groups working alongside each other on particular projects, which would also frequently have attached to them serving naval officers with operational expertise. In addition, different research establishments frequently shared the same site. Of particular relevance to the present considerations was the role of ARL at Teddington, near London, whose main purpose was to be on the more 'scientific' end of the R&D business, whereas laboratories such as AEW were deemed more 'practical' and concerned with developing operational equipment. In reality, the differences between 'pure' and 'applied' science were not as great as the above

TABLE 6.1: THE NUMBER AND LOCATION OF ADMIRALTY EXPERIMENTAL ESTABLISHMENTS—SEPTEMBER 1948 (TAKEN PRINCIPALLY FROM ADM 268/56 WITH AMENDMENTS AS SHOWN FROM A VARIETY OF OTHER SOURCES)

Directorates responsible for establishments
Director of Underwater Weapons—DUW
Director of Naval Construction—DNC
Engineer in Chief—E. in C.
Director of Physical Research—DPR
Director of Aeronautical and Engineering Research—DAER
Director of Electrical Engineering—DEE
Director of Naval Ordance—DNO
Director of Compass Department—DCD

R&D establishments maintained primarily as centres of R&D for particular types of equipment

1. Underwater Detection Establishment (UDE): underwater detection, DUW, Portland, Dorset
2. Admiralty Mining Establishment (AME): mining, mine sweeping and antisubmarine weapons, DUW, Havant,West Leigh, Leigh Park, Fareham and Edinburgh
3. Torpedo Experimental Establishment (TEE): torpedo development, DUW, Greenock
4. Underwater Launching Establishment (ULE): underwater launching, DUW, West Howe, Bournemouth
5. Admiralty Experimental Works (AEW): propellers and ship form, DNC, Haslar
6. Naval Construction Research Establishment (NCRE): effect of underwater explosions, DNC, Rosyth
7. Admiralty Fuel Experimental Station (AFES): boilers and fuel economy, E. in C., Gosport
8. Admiralty Engineering Laboratory, Engineering (AEL, Eng.): propulsive machinery, E. in C., West Drayton
9. Admiralty Research Laboratory (ARL): general research, DPR, Teddington
10. Services Electronics Research Laboratory (SERL): inter-Service electronic valves, DPR, Baldock, Herts
11. Admiralty Materials Laboratory (AML): materials research, DAER, Holton Heath, Dorset
12. Admiralty Signal & Radar Establishment (ASRE): signals radio and radar, DRE, Haslemere, Witley, Waterlooville
13. Admiralty Engineering Laboratory, Elec (AEL, Elec.): low-frequency electrical problems, DEE, West Drayton
14. Admiralty Gunnery Establishment (AGE): fire control and stabilisation, DNO, Teddington
15. Admiralty Compass Observatory (ACO): compasses, DCD, Slough
16. Royal Naval Physiological Laboratory (RNPL): physiological research, MDG, Alverstoke

Establishments maintained either for giving local services or as a trials service

12. Admiralty Hydro-Ballistics Research Establishment (AHBRE): underwater ballistics, DAER, Coalport and Glen Fruin
13. Admiralty Signal Research Establishment Extension, Tantallon: radio, radar and infrared, DRE, Tantallon (North Berwick)
14. The Frythe, Welwyn: hydrogen peroxide engine for torpedoes, DAER, Welwyn, Herts
15. Loch Goil: underwater noise range, DPR, Loch Goil
16. Admiralty Craft Experimental Establishment (ACXE, Hornet): small craft testing, DCAM, HMS *Hornet*, Gosport

17. Services Valve Life Test Establishment (SVLTE): valve life testing, DPR, Liss, Hants
18. HMS *Excellent* (Gunnery): operational naval establishment with a small group of scientific staff engaged in trials work.
19. HMS *Vernon* (mine warfare): as for *Excellent*
20. HMS *Lochinvar* (submarine warfare): as for *Excellent*

Establishments engaged partly in giving service to the departments and partly on development work

21. Chemical Department Portsmouth: chemical advice, DAER, Portsmouth
22. Central Metallurgical Laboratory (CML): metallurgical, DAER, Emsworth
23. Bragg Laboratory: metallurgical, DAER, Sheffield
24. Naval Ordnance Inspection Laboratory (NOIL): propellant testing, DAER, Caerwent Mon

---

Note: The following establishments were also being operated in 1948 but are not listed in the basic PRO reference:
Admiralty Development Establishment (ADE), Barrow[109]
Admiralty Vickers Gear Research Association (AVGRA), Barrow
Yarrow Admiralty Research Department (YARD)

may suggest, and in areas such as the control characteristics of underwater bodies the various groups at ARL and AEW knew each other well, and tended to divide the work between them, depending upon other priorities.

During the Cold War there were a number of reorganisations of R&D effort and, to give some idea of the initial state of affairs, Table 6.1 provides a list of establishments as they existed in 1948 and which remained intact until the first regrouping in 1960. In 1972 the Procurement Executive of the Ministry of Defence was formed, following the Rayner Review, resulting in the further amalgamation. This process was carried further with the founding in 1990 of the Sea Systems Sector of the Defence Research Agency (DRA), leading to another reduction in the number of establishments.

To give a sense of some of these changes, and at the same time provide the background to events to be discussed below, it is instructive to consider the example of the Admiralty Underwater Weapons Establishment (AUWE), situated at Portland Naval Base, Dorset. AUWE is not listed in Table 6.1 as it came into existence in the 1960 reorganisation and disappeared as a site and set of buildings in 1993, its function, if not its material, being taken over by DRA (Winfrith).

AUWE was formed by the merging of four establishments listed in Table 6.1. The first was Her Majesty's Underwater Detection Establishment (HMUDE, or more usually UDE), which had been at Portland since 1927, apart from the war years. It dealt principally with the means to detect vessels and weapons which operated underwater, using water as the detecting and energy-transmission medium. HMUDE was joined by the Underwater Countermeasures and Weapons Establishment, which transferred from Havant in 1959; the Underwater Launching Establishment, which transferred from Bournemouth in 1959; and the Torpedo Experimental Establishment, which transferred from Greenock, also in 1959. The titles of these establishments define their concerns well, and the functional definition of AUWE reflects their functional amalgamation, namely:

The AUWE is the R&D establishment charged with the design, development, ship fitting and trials to prototype stage of underwater weapon systems and the provision of full production information. This covers:

- Equipment for the detection, location and classification of underwater objects such as submarines, torpedoes, mines etc.
- Underwater weapons for the destruction of targets on and below the surface to the water.
- Fire control for such weapons.
- Countermeasures to enemy underwater weapons such as torpedoes and mines.
- Instructional equipment for the above.
- Equipment for communication, identification and navigation involving the exploitation of underwater physical phenomena.

Manufacture in quantity for service is not carried out but full production drawings etc. are supplied to manufacturers.[9]

To undertake these responsibilities AUWE possessed the following trials facilities in the Portland Area: sound range, Balaclava Bay; floating laboratories, Portland Harbour; torpedo range at Bincleaves; acoustic laboratory at Bincleaves; swell recorders in Weymouth Bay and Chapman's Pool; minesweeping evaluation range off Chesil Beach; and a general trials range at Portland Bill. Remote from Portland the establishment also operated Weston Super Mare aircraft minedropping range; Horsea Island explosive and towing trials range; Culdrose torpedo trials unit; Predannack Down antisubmarine mortar range; Alexandria (Dumbartonshire) RN Torpedo Factory; HMS *Vernon* (Portsmouth) mining tank; Coulport weapon net; two frigates, four minesweepers, five trials vessels at Portland, 10 small vessels at outstations and one helicopter.

To service these requirements, 93 naval staff and 1741 civilians were employed, of which 462 belonged to the professional scientific grades and 346 were craftsmen. The rest worked in administration, the drawing office and other support services. The total cost of the establishment in 1962/63 was £4,142,580, of which salaries and wages represented some £1,604,230. AUWE's role was principally centred upon R&D associated with producing new equipment, the process being formally triggered by the issuance of a staff target by the Staff Division of the Admiralty which expressed the broad function and performance required. Theoretically, AUWE would then undertake feasibility studies, from which the staff requirement was generated; this could then lead into a development period involving design and experiment to prototype stage, the device then being passed to industry with all the necessary plans and other documents for production. In addition, a great deal of extensive trials work was carried out to assess and improve existing equipment and, as with AEW, there was a close relationship with ARL over matters of common interest, such as torpedoes and acoustics. In 1984 AUWE changed its title to the less specific Admiralty Experimental Establishment (Portland), and in 1993 was shut down and a much reduced unit transferred to Winfrith in Dorset as DRA (Winfrith).

For the 1960s period under review, Admiralty R&D was coordinated and planned by the Directorate of Research Programmes and Planning (DRPP), which produced an annual plan under a variety of general headings, such as 'Thermal Mechanical and Chemical Engineering'; 'Radiological Phenomena and Defence'; 'Human Factors' etc., and laid out in general terms what would be looked at by a range of establishments under these headings, together with which Directorate would be responsible. The numbers and grades of establishment staff were listed, together with the costs associated with using any outside research organisations, such as the British Iron and Steel Research Association. Thus for the 1962/63 plan[10] under Category A, which dealt with 'Shape structure and propulsion of ships, submarines and underwater weapons', the Director of Naval Construction (DNC) was responsible for running the programme coordinating the efforts of the Naval Construction Research Establishment (NCRE), AEW and ARL.[11] The areas to be investigated were propeller design, pump-jet propulsion, boundary layers, water entry of projectiles, stability and control of pump-jet-propelled bodies, behaviour of ships in waves, submarine design (effects of changes in form and behaviour of high-speed submarines) and manoeuvrability of surface ships, including hydrofoils and aerodynamics of ships' structures. Some of these areas, such as pump-jet propulsion, had relatively large numbers of people working on them (12 scientific and experimental officers) and others were minimally or part-time manned. The total number of staff involved in the whole 'Ships' category was 55, with £39,300 being allotted for extramural research.

One of the enduring issues faced by the research establishments was shortage of staff. Thus, in considering the results of Rum Tub, one reviewer could opine:

> It is regrettable that so often the brake upon the development of important projects is the shortage of R&D staff and facilities. It is hoped the steps taken by NATO countries to pool their resources and knowledge will lead to better equipment.[12]

The extent to which NATO, as an organisation, provided a means of pooling resources appears to be have been very patchy. For instance, in the area of oceanography, which became crucially important with the advent of the Polaris submarines in the early 1960s, various avenues of international cooperation were assessed by the DPR when the Military Oceanographic Group was set up at ARL.[13] In considering the various NATO groups involved it was noted that the NATO Science Committee's Subcommittee of Oceanography did not deal with military oceanography, and the NATO ASW Centre at La Spezia (Italy) was accorded respectful if rather lukewarm praise. What did emerge was that CANUKUS's (Canadian, UK, US) subgroup G seemed to provide the best means for sustained joint action. Such sentiments did not exclude other groupings with strong mutual interests, such as the US/UK/ Norwegian Data Exchange Agreement on Underwater Research of 1962–63,[14] which was basically concerned with the oceanographical properties of the waters off Norway, and techniques of signal processing for the siting and operation of seabed passive arrays to detect Soviet submarines as they left their northern bases.

REPRESENTATIVE EXAMPLES OF COLD WAR NAVAL R&D

In seeking to find representative examples of naval R&D as exemplars of technical innovation developed in response to strategic needs, two areas emerge from the considerations discussed above. The first is the significance of the non-appearance of a carrier programme, and the second is the overnight obsolescence of ASW in the face of the arrival of *Nautilus.* On the basis of these observations, the case histories below are chosen to illuminate two distinct aspects of naval R&D. The first deals with the technological implications of takeoff and landing on those aircraft carriers that were in commission, and the second addresses the issues of stealth and detection in (mainly) nuclear submarines.

Takeoff and Landing

In considering the continual striving of the Naval Staff to build new aircraft carriers, and in the light of their failure to do so (the euphemistically entitled 'through deck cruisers' fitted with Harrier VSTOL aircraft dating from 1972, had a different and uncertain genesis),[15] it is reasonable to ask whether the increasingly obsolete carriers they did have managed to cope. The answer is that they did, more or less, and among a number of reasons that can be adduced for this success are three key inventions, the steam catapult, the angled flight deck and the mirror landing system.

The primary cause for the introduction of these devices was the realisation that became apparent towards the end of the Second World War that heavy, fast jets taking off or landing from a carrier designed to handle comparatively light and slow piston-driven aircraft were going to present long-term problems of carrier design and operation. In what follows, for the sake of brevity, only the steam catapult and the angled flight deck will be discussed.

Prejet catapults for launching aircraft used the power of hydraulically driven rams to transmit their force to the aircraft by means of wire ropes run through a complex system of pulleys and sheaves, these methods having been slowly developed since the 1920s. In 1938 C.C. Mitchell, who worked for the engineering firm of MacTaggart, Scott & Co, manufacturers of catapults, filed a private patent[16] to overcome the problems of transmission of the higher forces to aircraft and circumvent the cumbersome nature of the aircraft fixing. In essence, he proposed to propel a piston along a tube, fitted into the flight deck, which had cut into it a longitudinal opening, the slot, opening on to the deck. The aircraft was to be fixed to the piston by a fin protruding from the slotted tube, and motive power for propelling the piston was to be provided by an unspecified fluid. The idea was an old one, having been used in the nineteenth century for atmospheric railway propulsion, whose ultimate failure resided in an inability to seal the moving piston and thus stop the loss of pressure across it. The problem of sealing was addressed at some length in the patent.

The Admiralty did not take up the idea, probably because there was no need at the time for such a potentially powerful device. In 1944 Mitchell, now working in

FIGURE 6.1: BLACKBURN BUCCANEER TAKING OFF WITH THE ASSISTANCE OF A STEAM CATAPULT (FLEET AIR ARM MUSEUM)

the Engineer-in-Chief's Office at the Admiralty, went to France to view the V-1 rocket-launching sites overrun by the Allies, and brought back to the armaments testing range at Shoeburyness a launching ramp for tests and experiments. The V-1 catapult employed a similar slotted tube/piston arrangement to that proposed by Mitchell, but used hydrogen peroxide to produce steam as the propelling agent.[17] (Fig. 6.1) In September 1947 a 12-ft prototype power cylinder for the steam catapult gear was ordered by the Engineer-in-Chief (E. in C.)[18] from the Edinburgh firm of Brown Bros, and a trial was carried out at their factory in June 1948 on a prototype by J. Lewis of the Admiralty Engineering Laboratory (West Drayton). The trial was conducted to assess the stresses encountered and to provide data for further design work. The method used was unable to provide peak stresses in areas of concentration, and in August 1949 further trials were carried out by J. Paffet of the Naval Construction Research Establishment (NCRE) using photoelastic models.[19] The device went through an extensive, if leisurely, development programme, the first full-sized prototype being fitted to *Perseus* in 1951, with some 1500 trials being conducted to assess its effectiveness, including a visit to the USA for the purposes of demonstration. By August 1953 the E. in C. was able to note that, on the basis of the *Perseus* trials, an order for a further 18 catapults, designated Type BS 4, was placed with Brown Bros, and that a number of improvements to increase performance had been introduced, resulting in a steam saving of some 40 per cent, with reduced loads on the structure.[20] These trials were so successful that they led directly to foreign sales, including five sets plus manufacturing rights to the US Navy, the intention being to develop the device further in conjunction with the USN via the Mutual Weapons Development Programme,[21] using the Royal Aeronautical Establishment (Bedford) as the developing establishment.

In February 1954 a team from the Admiralty, including C.C. Mitchell, who was employed by Brown Bros and the USN as a consultant, visited the USA to view the catapult trials of the first US vessel fitted with a steam catapult, the USS *Hancock*. In addition to discussing the operating experience of both the British BS4 and the American C11 variant derived from it, further developments were also mooted.[22] By 1957 a special unit had been set up at RAE (Bedford), complete with a raised deck 6 ft off the ground and special steam facilities, to develop the steam catapult further.[23]

In 1944 the RAE (Farnborough) was considering the general issue of improving the performance of high-performance aircraft and, among the options that were put forward, was the suggestion to remove the undercarriage.[24] In November 1944 the Director of RAE was informed that the Controller of Naval Research's (CNR) interest was such that:

> The practicability of operating high speed aircraft from Carriers, without undercarriages, has been discussed with CNR who does not see any insuperable objection and would like this application considered.
>
> Will you therefore please make an appreciation of the advantages to be gained by this means in Naval Aircraft and a review of any difficulties that you would foresee in their operation.
>
> This should also be considered in the review you are making of the practicability of operating jet propulsion aircraft from Carriers.[25]

The scheme that emerged from these considerations was the proposal that the undercarriageless aircraft should be launched from the flight deck of the carrier by means of a cradle propelled by a powerful catapult (presumably the concurrent experiments at Shoeburyness were in mind here). The problem of landing the aircraft on an unsteady flight deck then became the issue to be resolved, and it was proposed that a flexible deck (basically a thick rubber coating over a portion of the normal flight deck) should be tested to see whether belly landing on such a device was a realistic proposition.

The Farnborough scientist Lewis Boddington was put in charge of these experiments and, by January 1949, 65 live landings had been carried out on flexible decks, 44 on a simulated deck at Farnborough and 21 at sea.[26] What became clear was that, although it was entirely possible to land an undercarriageless plane in this way, no space remained for deck parking on the carrier, with the consequent loss of operational flexibility. To overcome this problem Boddington suggested building another deck on top of the conventional flight deck, using it for landing and the deck below for taking off.[27] The advantage of such an arrangement would be the ability to conduct takeoff and landing operations at the same time.

Experiments with the flexible deck idea continued; in July 1952 representatives of the USN Bureau of Aeronautics visited Farnborough to view the project and in January 1953 Boddington visited the USA to advise upon their own flexible deck trials. However, in August 1951 Captain Dennis Cambell, Assistant Controller Naval Research, in discussing the problems of the flexible/double-deck concept with Boddington, and just prior to an internal conference on the matter, suggested that if the second deck were brought to the level of the original deck but slanted at about 8° to it, with the end protruding over the port side, the flexible decking could be mounted on this skewed deck. Thus aircraft could land on the skewed deck and the takeoff problem could be solved by using the conventional foredeck (the steam catapult that was now being concurrently tested meant that this deck was not as long as it once needed to be). Both participants recognised this discussion as significant, and the skew-deck arrangement was immediately sketched out (literally on the back of an envelope).[28] In late 1951 Boddington decided that, in view of the geometry of the arrangement, a jet with an undercarriage could land on the extra length the slanted deck provided, with suitably repositioned arrester wires. He then ditched the original concept of the flexible deck, with the added benefit that the triangular deck area between the angled deck and the starboard side could act as an excellent deck park, a major concern in all the previous discussions.[29]

A model was made of *Illustrious* to demonstrate the viability of the geometry, and in January 1952 a draft staff target was issued for an 'angled flight deck'. By February 1952 *Triumph* had her deck marked out with the scheme, and non-landing trials were conducted that showed there were no problems in terms of approach and flying technique.[30] In April 1952 the Director of Naval Air Warfare opined:

> the angled flight deck opens up new possibilities in the field of operating modern aircraft from carriers; it is of interest to note that the USN are already carrying out full scale trials of the British scheme and have plans to incorporate it and the steam catapult in the *James V. Forrestal*.[31]

The full-scale trials referred to above were being conducted on the USS *Antietam*, which was being refitted at the time and was structurally altered to become the world's first angled-flight-deck carrier. When the *Antietam* visited Britain in June 1953, the School of Aircraft Handling conducted an exhaustive series of takeoffs and landings and produced an enthusiastic report on the concept.[32]

Flexible-deck trials continued until 1955, by which time the concept had gone through five different stages of testing and development, involving some 575 live landings (Fig. 6.2). Although technically feasible, the original requirement clearly no longer existed.

Stealth and Detection

Nautilus was built as a testbed to evaluate the concept and practicability of nuclear propulsion, and in doing this it achieved outstanding success. However, the Rum Tub exercises demonstrated that she made a great deal of noise and was thus easily detectable by the escorts, even if they could not localise and track her owing to her high underwater speed of 20 knots.

In fact, by the time Rum Tub took place it had already been decided to proceed with building a British nuclear submarine, and by the end of 1956 an initial staff requirement had been formulated.[33] While Rum Tub was under way a full staff

FIGURE 6.2: HMS *ALBION* SHOWING SEA HAWKS AND GANNETS AND ANGLED FLIGHT DECK (FLEET AIR ARM MUSEUM)

requirement for the vessel was issued[34] which emphasised its hunter/killer and nuclear testbed roles. The history of the transition of *Dreadnought* from an all-British-technology nuclear submarine to an American/ British hybrid has been dealt with elsewhere,[35] but in its latter form a revised staff requirement was issued in September 1959 which specified a maximum submerged speed in excess of 25 knots, and a maximum submerged depth of not less than 750 ft.[36] The result was essentially a British bow section grafted on to an American after section which contained the all-important reactor as well as the propulsion unit.

As the purpose of *Nautilus* was to test nuclear propulsion with the minimum of delay, it was essentially a conventional twin-screw submarine fitted with an experimental nuclear reactor. In its design no account was taken of findings emerging from a parallel programme of hydrodynamic research associated with the teardrop-shaped USS *Albacore.* The *Albacore* programme, which was stimulated by the results of the German advanced U-boat programme, of which the Type XXI was an example,[37] underlined what had long been known: that the shape of the vessel had a major impact upon various types of flow regimes experienced over the surface of the hull, which in turn created noise from turbulence, vortex shedding and wake formation.

These flow effects were increased and exacerbated when the submarine increased its speed and manoeuvred under water. Indeed, one of the few sightings of *Nautilus* during Rum Tub was by an aircraft flying at 2000 ft, which reported that the submerged vessel was clearly visible, with large turbulent plumes streaming from her after end and conning tower.[38] Thus the transformation of the submarine from a vehicle that operated principally on or near the surface at low speed, to one that operated in three dimensions at high speed, brought to the fore a set of hydrodynamic design problems akin to those faced by aircraft designers, with the additional challenge caused by the creation of unwanted noise.

Although *Nautilus* clearly had flow noise problems the principal source of the noise detected during Rum Tub was her two propellers. This noise, called cavitation, had been known about—and partially solved in a rough and ready way—since the days of Parson's experiments with the famous *Turbinia* of the late 1890s. It was caused by the creation of very low pressure, usually (but not always) over the back of the propeller blades, which allowed dissolved air in the water to emerge and form bubbles or cavities. These cavities then collapsed, giving off noise and significantly reducing the efficiency of the propeller. Such effects were influenced by the shape of the vessel's afterbody, which defined the flow of water into the face of the propeller disc, as well as the propeller's diameter, blade cross-section, number of blades, speed of rotation, and the power fed into it by the engines.

The other main source of radiated noise was the vessel's internal machinery, as pumps moved liquids around the engine room, internal combustion engines provided main power or generated electricity, or gearboxes transmitted power in shafting systems. All these noise problems were exacerbated in the nuclear submarine by the high powers that had to be generated and transmitted by machinery, and which then had to be absorbed by the water to develop the speeds required. For a non-nuclear submarine there was the additional noise created when running under water with the snorkel in operation, piercing the surface. In general, the higher the speed the more noise created, and vice versa.

During the Second World War and just after it was believed that the majority of the sound radiated by a submarine was in the high audiofrequency region of about 10 kHz. By the early 1950s it had become clear that a lot more noise was being radiated, but at much lower frequencies.

Although ARL, AEW, AEL and the NCRE were the principal research establishments working on the problems of noise, the overall direction and coordination of their efforts was handled by a special panel of experts known as the R&D Panel on Noise Reduction. This was chaired by the Director of Physical Research and met for most the period covered above. In addition to members representing research and operational groups interested in the matters of noise reduction, representatives from the US Naval Attache and the US Office of Naval Research in London were also members, as was a representative of the Canadian Joint Staff. In this sense the Panel was very much an example of the close cooperation in these matters that existed at all levels between what became known as the ABC countries.

The principal means used to detect submarines under water was sonar (sound navigation and ranging), the American abbreviation for what the Royal Navy had called asdic[39] from the early days of the First World War, and which they continued to so call until 1960. Active sonar transmitted an acoustic signal whose reflection from an object was processed to give direction, range and speed. Passive sonar received sounds generated by the submerged body which, when processed, provided bearing information only, range being measured by triangulation. In the case of passive sonar the design of the receiving transducer differed somewhat from the active transducer and was called a hydrophone, but the basic physics of both types was essentially similar.

Until the advent of the hunter/killer concept and the coming of *Dreadnought*, Britain's interests in sonar development had been concentrated on active sonar, as ASW was seen almost exclusively as a matter for surface ships escorting convoys. Since surface ships under way created a lot of noise, which would swamp any sensitive passive sonar mounted on the hull, and since their radiated noise would advertise their presence to an enemy submarine anyway, the emphasis on British sonar design had always been on active systems which would locate the submarine as far away as possible, so that it could be dealt with before it came within torpedo range. On the other hand, the German U-boat experience of the Second World War favoured passive sets, as their operational speeds were low and any active sonar would betray the U-boat's presence; this emphasis on passive listening reinforced the concomitant desire to quieten the U-boats as much as possible.

Although a high underwater speed could, in the right circumstances, overcome knowledge of a submarine's presence, it clearly made sense to reduce the amount of radiated noise any submarine could make to a minimum, particularly with the appearance in the mid-1950s of seabed installations of passive hydrophone arrays, and the ever-present possibility that some unknown passive set might be listening. The cumulative effects of the different noise sources were such as to create a unique acoustic signature for each vessel, which opened up the possibility of being able to identify targets by their signatures.

The importance of propeller cavitation as a source of noise, rather than a mechanical inefficiency, was made apparent in the Second World War with the invention by the Germans of the homing torpedo. To counteract this, HMUDE undertook a series of experiments in 1941 with sponge-covered propellers designed to release the pressure of the collapsing vapour cavities and which gave encouraging results.[40] However, by 1951 experiments were under way with the 'Agouti' masking system, whereby the blades were provided with air ducts running near the leading edge in which air holes released air over the blades.[41]

Masking or camouflage systems had also been tried by the Germans in the Second World War when, in 1944, Meyer and Oberst collaborated on project 'Alberich', whose aim was not to stop generated noise radiating outwards, but to absorb or minimise the reflection of any active sonar energy impinging upon it. After the war Meyer and Oberst were debriefed and in 1948 their work on 'Alberich' was translated and circulated to interested parties by the Director of Physical Research.[42]

Meyer and Oberst had undertaken theoretical, laboratory and full-sized trials to develop a coating that could be applied to U-boat hulls which would reduce the reflectivity of the hull to impinging sonar pulses. Their experiments demonstrated that the parameters of concern were the thickness of the hull plating and whether air or water was behind it. On this basis they developed a two-ply rubber sheet 4 mm thick, with the inner ply perforated by 2- and 5-mm holes which determined the resonant nature of the system. Although the theory proved useful, the material was developed in an empirical fashion involving the testing of some tens of thousands of different materials. Trials were carried out with a full-sized U-boat in the Skagerrak late in the war which seemed to show the efficacy of the system, with a claimed reduction in reflectivity of 20 per cent. As this was accompanied by a reduction in detection range of 60 per cent, these claims were treated sceptically by the report's translators. By 1951, ARL had started work on its own acoustic camouflage system using rubber filled with heavy particles mounted in a steel mesh.[43]

This work was part of a wider programme started at ARL in 1951 into the whole issue of noise suppression, which focused on a range of issues that included a mathematical investigation into the basic bubble-collapsing mechanism that lay at the heart of cavitation. Work was also undertaken into generating an adequate theory of marine propellers to assess the effect of such characteristics as hollow blade profiles, suction points, rake, skew, shrouds and ducts.[44] By 1958 it was possible to design a propeller that appeared to be non-cavitating, but when one was fitted to the new Porpoise class of submarines it was found that it exhibited the disturbing propensity to strongly radiate noise at discrete frequencies, and to 'sing'. This problem initiated a series of experiments at NCRE and AEW which demonstrated that the likely cause was the rhythmic shedding of vortices from the tips of the propeller, which caused the blades to resonate. As the blades were made out of manganese bronze, which had low damping characteristics, they vibrated at frequencies related to their resonance frequencies, thus creating the 'singing' effect.[45] This problem was solved by increasing the internal damping of the blades by cutting grooves in their surfaces, filling them with a compound of lead and Araldite, and by cutting the trailing edge to a 25° wedge shape.[46]

One of the contributory elements influencing a propeller's propensity to cavitate more generally was the complex inflow of the water to the propeller, caused by the shape of the vessel's afterbody, with the water meeting the blade at different angles and velocities over the face of the propeller's disk. This highly disturbed flow meant that there were time variations in thrust experienced by each blade as it revolved through different regions. These thrust variations were related to the number of blades and the speed of rotation, the effect being known as blade rate noise. By late 1963 ARL had developed their propeller theory to the point where they had produced designs for *Valiant*'s propeller that incorporated skew as a central feature (blades with high sweepback, increasing towards their tips), to optimise performance to take account of this effect; these skew designs were tested and verified at AEW using physical scale models.[47] Meanwhile, a research programme initiated to investigate an alternative to the propeller, the pump jet, was nearing fruition.

The difference between a propeller and a pump jet was that the propeller worked in open water (albeit behind the afterbody), whereas the pump-jet propeller or rotor blades rotated within a highly confined duct or annular shroud. With correct shaping of the annular cross-section and the variation of its internal diameter over its length, the flow of water in the duct could be made to accelerate or decelerate at will. The duct was supported and mounted on the hull by a combination of vanes and stators, the stators being shaped to provide the best conditions for the water entering and/or leaving the rotor. In other words, 'swirl' could be imparted to the water entering the rotor so that when it came out the wake would be straight, or else an after-stator could be used to take out the swirl (so imparting extra forward propulsive force) as it left the rotor; or there could be some combination of these two techniques. In addition, the fine clearance between the rotor tip and the duct was such as to allow high loadings to be incurred over smaller lengths of blades, as the shroud stopped the formation of tip vortices.

From the above it will be seen that the design of a pump jet was a complex combination of siting on the afterbody, overall dimensions in terms of diameters and lengths, hydrodynamic profiles of the encasing shroud, rotor, stator(s) and their blades. In addition to the reduced cavitation offered by such a device the speed of rotation of the rotor was lower, the diameter smaller and the propulsive efficiency higher than for an equivalent conventional propeller. Against this, the device was much more complex and heavier than a traditional propeller, with the extra weight being placed right at the stern of the vessel, which created extra strength problems. In addition, the reversing characteristics of the device were not good.

Pump jets in the above form made their appearance in Britain in 1951, following a tour by a team from ARL to the US to visit various hydrodynamic laboratories concerned with underwater ballistics and torpedo technology. In the team's report[48] a review of new methods of torpedo propulsion being investigated by the USA was included, with a description of the system being proposed for the new American

Mk 41 torpedo:

> The scheme proposed for propulsion of the Mark 41 torpedo involves the use of a single propeller operating in a duct with fixed vanes. The flow over the afterbody enters the duct, passes through 11 fixed vanes, then though the 12 rotating propeller blades and finally leaves the duct through 13 fixed vanes. Clearance between the prop blades and duct is small. This arrangement is intended to give improved efficiency and reduced noise as compared with conventional arrangements (contra rotating props).[49]

As a result of this visit it was decided in May 1951 that a programme of research should be undertaken at ARL into the various methods seen in the USA, as part of the larger noise reduction and propulsion programme being undertaken. In 1952 A. Mitchell investigated 'Some Possibilities for Ship, Submarine and Torpedo Propulsion',[50] concentrating on the pump jet. The conclusion from this theoretical study was that a substantial improvement in maximum quiet speed and cavitation performance could be achieved for surface ships, submarines and torpedoes if the pump jet was adopted. The advantage of the pump jet over other novel propulsion systems was that the power plant required was the same as for a normal propeller, with the added advantage that the rotor in the pump jet would probably turn more slowly.[51]

The pump-jet idea does not seem to have been afforded any special priority, as it was not until 1959 that the theory was sufficiently developed to enable the device to be compared with experimental results to gauge its adequacy for design. There seems little doubt that this 'increased interest in pump jets'[52] was due to the appearance of the *Nautilus*, and the requirement to reduce noise on the nuclear submarines that were to follow the one-off *Dreadnought*.

The pump-jet experiments undertaken at ARL used the 31-in diameter water tunnel with a 17-blade rotor and 15-blade after-stator incorporated into a body of revolution of well known hydrodynamic characteristics amenable to computation. The theory developed was based upon aerodynamic compressor theory and, in addition to measuring the power characteristics of the device, a great deal of attention was directed towards predicting the onset and characteristics of cavitation. The experiments produced good agreement with the theory and made the point that, unlike a normal propeller, it was impossible to consider the pump jet in terms of open-water characteristics, as it was such an integral part of the body/hull that it was difficult to compare it with the standard methods used for propeller data.

By mid-September 1961 the development programme had moved towards considering the effect of the pump jet, not only on forward propulsion but also on the steering, or directional stability, characteristics of the hull.[53] Because of the close relationship between pump jet and hull, the flow characteristics were such that for even a small change in axisymmetric flow conditions the duct generated asymmetric lift, which was transferred to the hull as a destabilising force. Further, more detailed studies were initiated using a series of ducts that were systematically varied to find the effects and to test for the best position of the pump on the afterbody.

At the end of 1962 the Naval Staff agreed that the results of the pump-jet programme were sufficiently promising to be taken further, and a series of full-scale experiments was arranged with a pump jet fitted to a Mark 8 torpedo, which was tested at the Arrochar torpedo range.[54] These trials produced speeds of 41 knots with no blade-modulated or cavitation noise. The Naval Staff had also agreed to fit a pump jet to a full-sized coastal mine sweeper (CMS) as part of the programme to build a new class of very quiet mine countermeasures vessel (MCMV).[55] It was also decided to fit a pump jet to *Excalibur*, one of the two E-class hydrogen peroxide submarines laid down in 1951, completed in 1956 and used as high-speed craft for ASW training. The effects of fitting a pump jet on the directional stability and control of *Excalibur* had already been investigated at AEW and, although a destabilising effect was discovered, it was easily overcome by adjusting the hydroplane balance.[56] In the event *Excalibur* was not available, and in October 1963 it was suggested that the USA be approached to collaborate on a full-sized development programme. The proposal was rejected on the grounds that such an approach would involve a delay to the programme of several years.[57] However, the CMS pump-jet programme was proceeded with and tenders were sought in 1963.

Until May 1965 pump-jet design had been dealt with almost exclusively by ARL, but by this time the design procedure was sufficiently established that AEW began to collaborate with ARL over the pump-jet designs for the new nuclear submarine SSN04, and that to be tested on the CMS.[58] These trials were conducted in 1965 on board the *Shoulton*, where a Dowty 12-in diameter water jet was mounted on a specially designed cradle slung over the starboard side and an extensive series of experiments undertaken into the slow speed and noise characteristics of the device.

The first full-sized submarine pump-jet tests were carried out on the nuclear hunter/killer *Churchill* in the late 1960s (*Churchill* becoming operational in 1970), the intention being to fit a high-rotational-speed installation initially and to replace it with a low-speed unit two years later. Following on from this it was decided that a pump jet should be fitted to the 'improved Valiant class' as an integral part of its design, and ARL and AEW collaborated on both this and the MCMV project.[59] From 1973 and the advent of the Swiftsure class the pump jet became a standard feature of British nuclear submarine propulsion, and it was found that, by careful matching of the pump jet to the hull, their mutual integration allowed a much fuller (blunter) stern to be fitted than would otherwise have been the case, thereby providing more space aft and better structural support for the pump jet itself.[60] (Fig. 6.3)

Running parallel to the various propeller and pump-jet programmes were similar programmes concerned with acoustic isolation of the engine and auxiliaries from the hull by means of resilient mountings. When the German U-boat that was later to become HMS *Graph* was captured in 1943, it was found that the auxiliary machinery was mounted on resilient rubber mountings, designed by Meyer, with a subsequent reduction in the vibrations transmitted to the hull.[61]

The techniques of acoustic isolation, using a variety of measures such as resilient mountings, flexible couplings and acoustic insulation, were taken up and developed in the postwar years principally by the Naval Construction Research Establishment (NCRE), Dunfermline.[62] The results of these programmes appeared in the long-running and overlapping Porpoise and Oberon classes of conventional submarines,

FIGURE 6.3: ROLLOUT OF HMS *VANGUARD*, A BRITISH NUCLEAR STRATEGIC MISSILE SUBMARINE WHICH CARRIES PUMP-JET PROPULSION. DETAILS OF THE PUMP JET ARE STILL CLASSIFIED. (© VSEL)

commenced in 1958, which were:

> the result of an extensive quietening programme and are significantly quieter at a given speed either on electric motors or main engines than any known class in Western or Soviet Navies.[63]

Given close attention to detail and the use of acoustic isolation devices, the radiated noise of the Porpoise class when snorting was reduced to a staggering 3 per cent of what had previously been the norm.[64] In 1961, E.J. Risness of ARL considered that:

> At present nuclear submarines are considerably more noisy than the Porpoise class, and to reduce their noise to a comparable value with the reactor running may well be very difficult.[65]

However he also noted:

> In principle there is the possibility of a great deal of further noise reduction in submarines ... There remains the problem of flow induced noise which is already dominant at high speeds and will become progressively more important as other sources of noise are eliminated.
>
> In practice, however, further improvements may turn out to be quite difficult and slow to achieve. Now that the big reductions have been made in the two most important single sources of noise, the main engine and the propellers, further reductions involve, for example relatively small improvements on a large number of auxiliary machines.[66]

The original all-British nuclear submarine design called for a quiet speed of 10 knots at periscope depth, and to achieve this the main turbines, turbine gearing and electric drive generators were all mounted resiliently on a 'raft', which was in turn resiliently mounted on the hull. The American machinery design incorporated into the final *Dreadnought* design was unable to accommodate these features, so that the rafting was abandoned and the maximum silent speed for the revised *Dreadnought* was reduced by over 50 per cent, to 4.5 knots.[67]

The original idea of mounting machinery on a raft had a respectable pedigree, having first been demonstrated in 1944, when noise trials had been carried out on a midget submarine, the XT5. The main motor and the reduction gearing had been mounted on a subframe, which was in turn mounted on 10 special ARL resilient mountings, in addition to a flexible rubber sandwich coupling linking the engine to the propeller shaft. Tests were carried out with the subframe both bolted down and 'floating', together with changes in the type of flexible shaft coupling and oils in the gearbox. Engine noise reductions in the region of 50 per cent were claimed for the technique.[68]

By 1956 a theory of vibration reduction through 'compounding' had been developed by J. Snowden of Imperial College, whereby the noise source was mounted on resilient mountings on an intermediate beamlike mass, which was in turn resiliently mounted on a solid foundation. By the judicious use of the various masses and resiliences the system could be 'tuned' to reduce the amount of vibration transmitted to the foundation.[69] Two proposals were put forward to take advantage of this idea: the first was to have two concentric hulls flexibly connected, with the machinery flexibly mounted in the inner hull. The second was to mount the engine flexibly on a secondary mass, which in turn was flexibly mounted on the hull.[70] It was this proposal that had been incorporated into the original all-British design of nuclear submarine, with the secondary mass being called the raft.

In the *Valiant* of 1963 the more-or-less original pre-*Dreadnought* design was reverted to and rafting reintroduced, a considerable feat considering the sizes and weights involved and the need for all non-rafted machinery and piping interacting with the rafted equipment to be flexibly joined. The dynamics of the rafting were such that, at normal operating speeds, such as convoy protection, the raft 'floated', whereas at high speed it was locked in position, with the result that *Valiant* made a great deal of noise. The later Swiftsure class of 1971 went one step further and mounted the condenser, as well as the rest of the propulsion machinery, on the raft.

Although the Rum Tub exercises demonstrated conclusively that the Royal Navy's standard sonar equipment was incapable of tracking such a high-speed target as a nuclear submarine, the finding came as no surprise. In June 1956 a preliminary tactical study had been undertaken at HMUDE on a simulator which simulated an attack on a convoy by a nuclear submarine capable of a maximum speed of 30 knots. The escort vessels were equipped with the latest type of sonar currently at the prototype stage of development, the 'searchlight'-style Type 177. The conclusion of the study was clear: the Type 177, which represented the development of Second World War experience, was simply inadequate for dealing with a nuclear-powered submarine. The study further showed that, owing to the speeds involved, there was an increased need for early warning of the presence of such a submarine, which in turn required long-range sonars. Given the standing of the Type 177, this in turn required a totally new system to be developed.[71]

Although the detection of submarines by sonar remained the standard technique throughout the Cold War, there were always continuing concerns about investigating other means. By 1951 research was under way at ARL on non-acoustic methods, such as improving magnetic loop detection (magnetic anomaly detection, MAD), which had been developed by the USA, and analysing the wake of a submarine using infrared methods to detect heat emissions.[72] When the *Nautilus* took part in Rum Tub a series of experiments was mounted using an aircraft and the British submarine *Undaunted*, to see if her presence could be detected by sensing any associated external radioactivity. Calculations suggested that a stationary nuclear submarine might produce a detectable field. A trial was set up to see if the *Nautilus*'s wake contained any gamma flux which could be detected under water by *Undaunted*, and which could be detected from the air by means of an aircraft-mounted ionisation detector. In the event *Undaunted* could not find *Nautilus*, and although the aircraft made contact it too lost her. The general conclusion from the fragmentary data that were collected was that no gamma or ionisation effects were detectable.[73]

By 1963 various other non-sonar methods of detecting submerged submarines had been or were being investigated. These included the optical detection of surface waves of the order of 1 inch in amplitude which, theory suggested, would be generated by a submarine at 300 ft moving at 25 knots, and also the production of a trailing slick generated by the reduction of surface tension.[74] These investigations were complemented by range of other methods, such as underwater optical radar, a laser device, but 'None shows sufficient promise at present to be likely to supersede the more conventional methods'.[75]

Where success had been achieved was in the field of passive sonar detection using fixed underwater hydrophones. After the war the Americans discovered that a submarine radiated a lot more noise at lower frequencies than had previously been suspected and that these lower frequencies, in the region of 40–100 Hz, propagated much further than the few miles of the 10 kHz that had previously been thought the limit. In the early 1950s means to exploit these discoveries were sought; in the USA this led to the 'Jezebel' programme, and in Britain the 'Corsair' experiments. Corsair used the idea, developed by radioastronomers to detect radio sources in space, of using correlation techiques, which involved multiplying together signals received from two or more spaced points, over a wide frequency band, and averaging this product over a relatively long timescale. Initial sea trials using a pair of hydrophones were conducted off the coasts of Cornwall and Northern Ireland in 1952. The results suggested that a detection range in the order of 100 miles was conceivable, and eventually in 1955 an array 1200 feet in length, consisting of hydrophones spliced into a cable, was laid 40 miles off the northern tip of the Shetlands. This was then followed by a 375-ft array laid 70 miles off the Shetlands which could be electronically 'steered' to form 40 beams covering a 180° arc from west through north to east. The results of these efforts did not bear out the early optimism, and in December 1957 a further trial, 'Thermostat IV', showed that a snorting submarine could be detected at 70 miles. However, interference from fishing vessels was extremely heavy and the results could only be interpreted with difficulty;[76] from this date the array was closed down on a care and maintenance basis, and used for propagation and other studies.

The American Jezebel project, instead of using signal correlation techniques, used the narrowband spectrum analyser developed by Bell Telephone Laboratories for speech analysis, as its data analyser. The application detected snorting submarines at a range of 100 miles or more, and development of this low-frequency spectrum analyser (LOFAR) technique also provided a means whereby the unique acoustic signature of a ship, both submerged and on the surface, could be recorded, stored and analysed. The long-range passive detection performance of the LOFAR technique led directly to the establishment of the well known American sound surveillance systems (SOSUS) in the Pacific and Atlantic Oceans. These used arrays of 40 hydrophones with 40 separate beams covering an arc of 120°.[77]

One of the reasons for closing down the Corsair experimental array off the Shetlands was that, with the appearance of the Porpoise class of submarines, the whole advantage gained by Corsair techniques was cancelled overnight, and detection ranges were once again back to a few miles, or possibly tens of miles.[78]

With conventional submarines having demonstrated the power of standard noise reduction techniques, and the appearance of the fast but noisy *Nautilus*, it is perhaps not surprising that Mountbatten could note, at the end of 1957, that:

> the urgent requirement now is for a complete re-appraisal of our anti-submarine equipment with a view to modernisation of existing and provision of new equipment to deal with true submersibles. This requirement comes at a time when our R&D effort over the last 10 years to deal with conventional submarines was just reaching fruition.[79]

Their Lordships were, however, 'well aware of the advantages of the nuclear submarine in the antisubmarine role' and they attached 'great importance to get[ting] the *Dreadnought* to sea as soon as possible', the 'development of long range active asdic ... being given a high priority so that suitable equipment will be available for fitting in *Dreadnought*'.[80]

At the heart of the problems of the searchlight type of sonar employed by the defending forces in Rum Tub was a set of basic physical issues that made themselves apparent in the new circumstances. Given an active sonar, and the speed of sonar radiation in water being about 5000 ft/sec, it would take about 9 seconds to send a signal and receive it back from a target 4 nautical miles (nm) away. For a beam of angular width 12°, which for other physical reasons would be normal, the single beam could then be turned through a 360° sweep in about 4.5 minutes, so that the target's change in position would be about 0.6 nm if the submarine was doing 8 nm/h (knots). However for a target at 25 nm distance the echo would be returned in 1 minute and it would take 30 minutes to sweep through 360°, receiving in that time only one echo per beam width. If the target was a nuclear submarine doing 25 knots it would have changed its position by 12.5 nm in this time.

For these reasons the concept of using multiple- rather than single-beam formation was introduced on the prototype Type 177, which mounted its transducers in the form of an array on a flat plate. This plate was mechanically rotated about the vertical axis, in an acoustically transparent dome protruding from the bottom of the escort. The rotating transducer array of the Type 177 measured 4 feet square, weighed 2 tons and produced four beams, each of 10° width covering a 40° sector.[81] However, the performance of the Type 177 during Rum Tub fulfilled the simulator's predictions and was judged 'disappointing'.[82]

In view of the physically limiting nature of the searchlight type of sonar considerable research had been undertaken after the Second World War, particularly in the USA, into developing the notion of a steerable beam capable of being moved in both the horizontal and the vertical planes simultaneously, but without using moving parts. The idea had been tried as far back as the late 1930s at ARL, but had been abandoned because of its complexity.[83] Essentially what was required was to introduce phase differences or time delays between the transmission/reception of separate transducer/hydrophone elements in the array. This ploy had the effect of moving the line of maximum intensity of the beam off to one side. By then altering the phase differences over time, the beam could be made to rotate about a vertical axis and, depending upon the shape and size of the array, in a combination of vertical and horizontal directions as well. Given the non-mechanical nature of the switching and timing electronics required, the beam could be steered or made to scan very quickly. A relatively simple system using a linear array of passive transducers had been used during the war by the Germans, and by 1957 the Americans had developed their BQR series of passive arrays in cylindrical—and thence spherical—geometry, the former being fitted to *Nautilus*.[84]

The team assembled at HMUDE in April 1957 to start the design of the new sonar system for *Dreadnought* had behind them not only a very thorough institutional experience and infrastructure of sonar devices and their operation, but also access to new ideas and research data. These data had been generated by HMUDE,

ARL and, increasingly, the USA, with whom collaboration had always been very close. Apart from informal personal contacts between the British and American communities, developed through many visits and joint meetings, a particularly useful source of information for the British was the classified American *Journal of Underwater Acoustics*,[85] which reflected the involvement of the American academic community in its very large sonar development programmes. This use of academics was not reflected to any great extent in the British organisation, which relied almost entirely upon official establishment staff. The initial design study for *Dreadnought*'s sonar was completed just after Rum Tub, by a team at HMUDE composed of G. Horton, S. Mason and R. Barker.[86] In recognition of the complexity of what was being proposed, compared to the Type 177, the equipment was given the designation Type 2001.

From 1941 onwards, and throughout the whole of the Cold War, large oceanographical research programmes were initiated by the USA and Britain to improve the acoustic understanding of the oceans. In Britain this was organised through the ARL, the Hydrographic Department of the Navy, and what later became the government-funded civilian organisation the National Environmental Research Council (NERC). This work involved the elucidation of the detailed physics of the reasons for attenuation (loss of strength) of the sound in sea water, and in particular losses caused by the reflection of the signal off small suspended particles, called reverberation. Late in 1956 the Americans revealed to the 2001 team their ideas on the physical possibilities of bouncing sonar beams off the sea bed and using the technique to create a bottom-bouncing sonar.[87] This was followed by an exchange of papers which deeply influenced the design of the 2001.[88] From 1958, the US Navy's Underwater Sound Laboratory (USNUSL) began experiments with its bottom-reflected active sonar system.[89]

Ever since the early days of the Second World War it had been known that the temperature gradient of sea water did not necessarily fall off in a linear fashion, from warm at the surface to cold at the bottom, and that under the right conditions of location, time of year, local weather and so on, the vertical temperature and salinity structure of the sea could vary considerably. These spatial variations meant that sound waves in sea water could be refracted so that they did not necessarily travel in straight lines, and were also capable of being reflected off both the surface and the bottom. Under the right refractive conditions the sound could be trapped in a convergence layer which acted as a form of wave guide, transmitting the sound over very much larger distances and focusing on annular convergence zones. These various refractive conditions could also create 'dead' areas in which sound could not penetrate, making a submarine difficult to detect in them. The idea behind bottom bounce was that, by reflecting the sonar wave off the sea bed, these dead areas could be entered and reflections from the target could be detected via a return bounce path. Unlike radar, the environment in which sonar had to work had a massive effect on the propagation of the signal.

The Type 2001 was designed from the beginning to give continuous sonar cover over an arc of 240°, that is, 120° either side of the centre line, and to be both active and passive at the same time. For the active mode, and depending upon *Dreadnought's* speed and the aspect of the target, the range varied from a maximum of 25 miles at 5 knots to 5 miles at 20 knots. The decrease in range with increase in speed was due to the self-noise effect generated by hydrodynamic flow. In deep water it was expected that the focusing effects of temperature and density on sonar waves to produce convergence zones would enable the set to provide detection in the first zone of 30 miles quite frequently; under poor conditions it was the intention to use bottom bounce.[90] For passive operation in good conditions, with *Dreadnought* moving at up to 5 knots, the detection range was estimated to be 30 nm for a snorting submarine target moving at 8 knots; this was reduced to 17 nm and 6 nm when *Dreadnought's* speed increased to 10 knots and 20 knots, respectively.

The equipment was to be designed such that there would be a long-range passive channel for initial detection and crude bearing information, with a short-range passive channel being used for increased bearing accuracy. The active channel used Doppler techniques to measure the target velocity in 4-knot increments up to 30 knots. To cover the 240° spread 24 beams, each of 10° angular width, were to be formed, all steered electronically. To provide the active bottom-bounce capability, means were incorporated to also depress the beam 40° below the horizontal. To meet these requirements the transducer array was to be fixed and 'wrapped' around the bow of *Dreadnought* (hence the term 'conformal array'), sloping backwards at an angle of 20° from the vertical for the purposes of streamlining, and for mounting on the upper part of the bow above the torpedo tubes (Fig. 6.4). This unconventional arrangement seems to have caused objections from the naval architects, the disagreement being decided personally by Lord Mountbatten.[91] The height of the array was to be 6 ft and the total length of the horseshoe arrangement 40 ft, comprising 218 separate elements and a total weight of between 20 and 30 tons.[92]

The other major need created by the new design was for massive processing and switching power, which standard valve technology simply could not provide in the space and environment available. This led to the risky decision to incorporate transistors right from the start of the project: initially germanium, which was later replaced by silicon as silicon transistors with the right characteristics became available.

Because of the novelty of the design, the team appears to have been more or less left to get on with the task and, given the complexity of the design and its many subcomponents, work proceeded in parallel on many of the items. Progress was swift: having started in April 1957 with a clean sheet of paper, a half equipment was completed to start beam pattern trials in November 1959. This was a considerable feat which involved the design and building of 14 cabinets of electronics, a large transmit/receive switch and a large photographic display, all from scratch, although using standard components wherever possible. In addition, the HMUDE trials vessel had to be modified to take the highly complex array made up of 800 transducers, and a complex trials programme organised.

Following the trials and the essential confirmation of the system and its parameters, detailed design and manufacture were developed over a further two years, in

FIGURE 6.4: HMS *DREADNOUGHT:* TYPE 2001 TRANSDUCER ARRAY, VIEW LOOKING AFT ON BOW, FORWARD END (ROYAL NAVY SUBMARINE MUSEUM, © VSEL)

conjunction with the Cottage Laboratories of Plessey, the defence contractor. As a measure of the jump in complexity involved in the system's design, the Type 177 used 450 valves, whereas the *Dreadnought*'s used 16,000 transistors housed in some 28 cabinets.

During the development of the 2001 a series of trials had been conducted using the bottom-bounce mode where active pulses of various lengths had been used, the intention being to determine the gross propagation loss as a function of range. The results from this work seemed to suggest that if the very considerable losses encountered were to be offset, then considerably more sophisticated signal processing was required.[93] To tackle this issue and provide information for future bottom-bounce sonars, it was decided to conduct a series of trials with the 2001 in the bottom-bounce mode using broadband signals and correlator-type processing, once *Dreadnought* had been completed. These trials were conducted by teams from AUWE and ARL and took place in the Eastern Atlantic from 1963 to 1965, using an experimental version of the 2001 fitted to the frigate *Verulam.* In March 1965 a series of joint trials was conducted with the US Navy Underwater Sound Laboratory in the western Atlantic to compare the results of British and American measurements of bottom-bounce losses under comparable conditions. The conclusions from these studies were that for bottom bounce to be truly effective, larger transducer arrays were necessary.[94]

There is little doubt that the 2001 was a massive step forward in sonar technology and set the standard for the new generation of sonars designed to deal with nuclear submarines. However, its very sensitivity was such that it experienced major problems with self-noise, which affected its performance accordingly and produced false echoes. To find the origins of this effect, an extensive monitoring and assessment exercise was undertaken which suggested that, in addition to the flow noise characteristics of the bow, the false echoes were being generated by the steam condenser system. Here the noise was transmitted to the sea by the main seawater cooling system, which was then picked up by the bow-mounted transducers, having been reflected off the sea surface.[95]

Sonar, Secrets and Spying

In considering the range of items discussed above, in the fields of both aircraft carrier and submarine technologies, it is important to recognise that the ultimate aim of the exercise was to be in a position of superiority with respect to the Soviet Union should the Cold War become a hot one. Similarly, the Soviet Union was not an idle bystander and, as is well known, made strenuous efforts across the whole field of Western military technology to find out about developments, using both overt and covert means.

The role of espionage in the Cold War is well recognised, although its actual usefulness is a matter of considerable debate. Britain, and in particular the Admiralty, was plagued by a number of sensational cases over the years, of which the harbinger was the Portland Spy Ring.

Harry Houghton was an ex-Master at Arms (the Chief Petty Officer responsible for discipline on board ship) who had seen extensive action in the Second World War. On leaving the Navy at the end of 1945 he had been employed as a clerk, first in a naval training establishment and then in Portsmouth Naval Dockyard.[96] In March 1951 he applied for posting abroad and was appointed clerk to the Naval Attaché at the British Embassy in Warsaw. Given his position and privileged diplomatic access to Western goods of all kinds, he was soon heavily and profitably engaged in black-market deals and amorous pursuits.[97] It was not long before the

Polish secret service became aware of these activities and ensnared him in the time-honoured fashion via girlfriends and sex orgies.

Houghton's heavy drinking was certainly common knowledge in the Embassy, and it is likely that his black-market activities were also known about. On the basis of his unsatisfactory record he returned to Britain at the end of 1952, after only one tour of two years. In November 1952 he was posted to HMUDE as a clerk in the non-secret personnel department, and in January 1957 transferred as clerk to the Port Auxiliary Repair Unit, the unit responsible for the upkeep and maintenance of the large number of small, auxiliary and trials vessels required by both the naval base and HUMDE. In early 1960 he became responsible for the acceptance, distribution and filing of all papers and correspondence that passed through the unit.

With Houghton's appointment to Portland the Polish secret service maintained contact and either threatened or persuaded him to start providing information, which he willingly did for payment.[98] Initially he appears to have supplied easily accessible information of low value relating to such things as the nearby firing ranges and their times of use. At some time control of Houghton passed from the Poles to the KGB, who provided him with a miniature camera for photographing any documents that passed his way.

According to Houghton's own later account, he soon found himself engaged in a range of activities, from receiving spies landed in nearby coves by submarine, meeting controllers in Dublin and Austria, to breaking into HMUDE at night with his controller and photographing documents. It is difficult to know how true all this is, but one thing is certain: in his position as clerk to the Auxiliary Repair Unit he was well placed to know what in general was going on.[99] For instance, he was able to tell his controller that HMUDE's trials vessel *Decibel* was going to be fitted out for some new trials, although it is clear that Houghton had no sense of what the name *Decibel* stood for, or what the vessel did.

While these events were unfolding, Houghton's marriage deteriorated and he started going out with Ethel Gee, a clerk who had worked in the stores section at HMUDE. Gee worked in this section until 1955, when she transferred to the Drawing Office Records Section. At some point control of Houghton was passed to Gordon Lonsdale, a KGB agent later revealed to be Konan Trofimovich Moldi, who had assumed the identity of Lonsdale, a long-dead Finnish Canadian.

Lonsdale wanted more detailed technical information than Houghton was able to provide, and was aware of both the affair with Ethel Gee and her position in the Drawing Records Office. He must have been a very persuasive talker, since at a meeting with Gee he was able to convince her, with Houghton's support, that he was working for the USA, and that, by passing information over to him via Houghton from her key position in the plans section, she would be assisting an old ally. Houghton backed Lonsdale up with the story and, naive as it may seem, Gee became the conduit for funnelling information directly from the heart of HMUDE to the KGB.

The date when Lonsdale took control of Houghton and thence recruited Gee is open to some doubt: Houghton claimed it was July 1960, but he could well have been trying to shield Gee. However, regardless of the date, as the most important project being handled by HMUDE at this time was the Type 2001 and its fitting to *Dreadnought*, and as the whole project was entering its final stages, this meant that all the theoretical, experimental, trials and engineering aspects of the 2001, from its very inception to its near completion, were available for inspection. Given the tight timescales involved for the development of the 2001, a large part of the design work existed only in the form of drawings, and all of them had to go through the Records Centre. How systematic the plundering was appears to be unknown. Neither Houghton nor Gee had the technical background to understand anything but the most general of information, but it is clear that a lot of material was removed and photographed, and once the classification system was understood it would have been simple for Lonsdale to direct them to follow sequences.

By July 1960 MI5 had become aware of leaks emanating from HMUDE, thanks to information supplied by a Polish defector, and it did not take them long to get on the trail of Houghton, Gee and Lonsdale. Several exchanges and meetings were observed over a period of nearly 6 months which led, via Lonsdale, to Peter and Helen Kroger, antiquarian booksellers living in a bungalow in Ruislip, a London suburb. The Krogers turned out to be Morris and Lena Cohen, who had been involved in the Fuchs and Rosenburg ring in the USA, having disappeared in 1951. All the usual paraphernalia of high-powered radio transmitters, one-time pads and so on, was found at the bungalow when it was raided in January 1961. It was later revealed by Peter Wright of *Spycatcher* fame that Lonsdale's own biweekly radio transmissions to Moscow had been monitored and decoded for some time, and that Houghton's code name was Shah.[100]

The Portland Spy Ring, as they became known, were convicted in March 1961, with Houghton and Gee receiving 15 years each, the Krogers 20 years and Lonsdale 25. In the way of the Cold War, Lonsdale was released after three years in exchange for Greville Wynne, the Krogers were exchanged after eight years for Gerald Brooke, and Houghton and Gee were released after 10 years and then married.

The trial took place just after that most celebrated of Cold War events, the U2 incident, which led to the scrapping of the May 1960 Paris summit by Khrushchev, with the subsequently highly publicised Moscow trial of the pilot, Gary Powers. There seems little doubt that the Portland trial was viewed as a counterbalance to these events, at least in Britain.

Although much touted as a triumph for British Intelligence, the Admiralty had been informed of Houghton's activities over a period of nearly four years prior to his apprehension. In June 1956 the Director of Naval Intelligence (DNI) was informed via a Naval Welfare Officer that 'It is understood that Mrs Houghton alleged that her husband was divulging secret information to people who ought not to get it'.[101] Even the local probation officer was told that 'her husband was in touch with Communist agents in London and was passing information to them'.[102]

The effect of the discovery of the Portland Spy Ring upon personnel at HMUDE—or AUWE as it was by the time of the trials—seems, not surprisingly, to have been one of deep communal betrayal and shock over their activities.[103] Certainly the affair had all the elements of a Cold War drama—farce, love, betrayal of country, betrayal of love, trusting naiveté, clandestine meetings in the 'Bunch of Grapes' in South Kensington, etc. At this stage it is impossible to assess the impact of the information Houghton and Gee purloined upon Soviet sonar design. Given that the Type 2001 was the major concern of AUWE at the time, that it was designed from the outset to hunt other submarines, it is reasonable to assume that it was a very big prize. In an assessment of the Soviet submarine threat to be faced in the 1970s, compiled for the Director of Under Surface Weapons in 1963, it was considered that:

> The Soviets are behind the West in the development of long range active and passive sonars but it is probable that by 1970 they will have sets of similar performance in use today by the West such as the 2001 and 186.[104]

As the above was written with the full knowledge of the effects of the Portland Spy Ring, it seems likely that any information obtained by them would not have made an operational impact for about seven years, by which time NATO development would have moved on.

NAVAL R&D IN THE MID COLD WAR—SOME COMMENTS

It is no surprise that, on the basis of the case studies presented, the dominant influence on British Naval R&D in this phase of the Cold War was the very close relationship that existed with the USA and the mutual advantage that seems to have accrued to both parties. From the British perspective there can be little doubt that access to new physical knowledge generated by the Americans through large research programmes was a key element that helped it maintain a technological respectability. This is certainly so in the case of the Type 2001, and was neatly summed up in a paper presented to the Director of Physical Research in 1963 when considering the need for the establishment of a Military Oceanographic Group at ARL. It was thought:

> very fortunate that through subgoup G machinery [we have] reasonably full access to American knowledge and thought [on oceanograpy], without such access progress would be very slow.[105]

Although there was a NATO component, as in the data exchange projects with the Norwegians, the main axis was undoubtedly with the USA. It seems to have suited the USA very well to have had a relatively independent informed ally to be used as a sounding board and consultant. The numerous study tours and the formal relationships engendered through membership of joint committees, both within the NATO structure and outside it, seemed to have worked to everybody's advantage. Even though the USA had almost unlimited economic ability to do what it wanted to regardless of anybody else, the technological possibilities were so great that it could not do everything. Thus in the matter of the pump jet, although its origins lay in the USA, it was Britain that converted it into a realistic proposition, a process in which the USA declined to be involved. The circle was squared with the device being re-exported back to the USA at a later period of the Cold War.

In considering the transatlantic R&D interaction via the studies, there is the distinct impression that, because of economic stringency, the British teams involved in two-way exchanges of progress and information (as certainly happened with the angled flight deck and 2001) tended to be much smaller and less specialised than their American counterparts. This seems to have had ramifications in terms of management, communications and response times, and perhaps indicates national differences in style, a topic worthy of further investigation.

In terms of the case studies themselves, it is clear that a great deal of inventive capacity and ingenuity was brought to bear on a range of problems, from very physical but scientifically straightforward problems concerned with the spatial layout of flight decks, to highly theoretical and complex issues centring on the physics of acoustic transmission and electronic processing. Although individuals had a major impact upon outcomes in the projects discussed, it is also clear that the institutions where they worked provided strong technical infrastructure and a great deal of tacit knowledge about what was possible and what was not.

A number of general points emerge from considering Table 6.1, the list of naval R&D establishments in the period from the end of the Second World War to the first reorganisation in 1960. First, the large number and variety of establishments involved (27), and second, the number of establishments required to conduct trials of one sort or another (9). This latter point illustrates well the requirement of the R&D process in general to have the facilities and organisation to enable the jump from a laboratory scale of investigation to something resembling reality. This reality is frequently rather stylised, in the sense that it is a halfway house wherein the full-sized artefact can be tested in a reasonably controlled, semirealistic operating environment in which a number of factors are present over and above those isolated in the laboratory. Having noted this, the scope of trials and their complexity varies enormously. What might be taken as a given in a laboratory setting can present major problems when put to the full-sized test. One has only to read the trials reports of *Undaunted*, blundering around trying to find and intercept the wake of *Nautilus* so that it could analyse it for traces of residual radiation using specially constructed instrumentation, to appreciate the difficulties involved. Similarly, the extent of, need for and investment required for full-sized testing is well illustrated by the listing of the static and moving facilities required by the AUWE. In general, it may be said that the function of trials, as a validating process akin to the role of the experiment in laboratory science, has received little attention in the literature, and is clearly worthy of deeper study.[106]

In considering the activities of the Intelligence authorities in the matter of the Portland Spy Ring, and their seeming incompetence surrounding Harry Houghton's double life, it is worth reading the memoirs of Markus Wolf, the former Head of the East German Foreign Intelligence Service, to realise that farce was an endemic part of Cold War espionage, and not confined to any one side.[107] In keeping with the age of James Bond, a film was made following the conviction of the Portland

Ring that provided a rather flattering view of the abilities of MI5 and Special Branch. On release from prison Houghton got his own back with a lurid biography entitled *Operation Portland*, but the enduring image is to be found not within the book but on its dust jacket. Here a photograph of the intrepid spies shows Harry and Ethel standing beside their little Renault 'Dauphine' on a hot summer's day, against a background of sandy beach, sea and white cliffs, the very picture of Darby and Joan respectability enjoying an afternoon's outing at the seaside. Such benign images are not encountered in speaking to surviving Portland employees, where the betrayal is still keenly felt and has left deep impressions. It has been reported that the US Navy sought to break off British/US intelligence exchanges in 1965, and used the Houghton case as an example of the poor security that was said to be accorded to US secrets.[108] This is doubly ironic in view of the later history of Soviet spies in the US Navy, and the fact that this was the very argument used by Lonsdale to persuade Gee to work for him.

In attempting, finally, to highlight the attitudinal and organisational changes that affected naval science over the course of the Cold War, much can be learned from the terminology used. In the period up to the 1960s reorganisation, the outlook was very much concerned with the specific: 'Director of Physical Research', 'Underwater Weapons Launching Establishment', asdic, torpedo etc. By the 1990s the language had became much more general: 'Sea Systems Sector', DRA (Winfrith), with generic terms such as 'sensor', 'integration' and 'platform' becoming the lingua franca. This process no doubt reflects the impact of 'systems' thinking upon both the material and organisational aspects of naval R&D, with its emphasis on the notions of totality and holism at the expense of the singular and particular. The need to deal with growing technical complexity appears to have been addressed linguistically and practically by introducing new layers of descriptive categories, a process reflected by the increasingly important role of computation and simulation. In this sense the strongly typed—and service defined—'naval science' of the 1940s and 1950s was transformed into the less typed and more amorphous 'defence research' of the 1990s.

1. Mutually Assured Destruction, not Magnetic Anomaly Detector.
2. The best review of Britain's naval policy during the Cold War is to be found in E.J. Grove, *Vanguard to Trident: British naval policy since World War II* (United States Naval Institute Press, 1987).
3. For an early 1980s view of the growth of the Soviet Navy which relates the development of ship types to strategy see B. Ranft, G. Till, *The Sea in Soviet Strategy* (Macmillan, 1983), which also provides a full bibliography for entry into a very extensive field.
4. *Analysis of Operation Rum Tub*, PRO ADM 1/26756, enclosure 1, par. 12 and 13 (August 1958).
5. *Memo from Commander in Chief*, PRO ADM 1/26756 (17 December 1957).
6. *Submarine Versus Submarine Attack Reports*, PRO ADM 1/25252 (21 April 1951).
7. *Analysis of Operation Rum Tub* (n. 4 above), par. 8.
8. T. Wright, *Ship Hydrodynamics 1710–1880* (Ph.D. thesis, CNAA, Science Museum, 1983).
9. *An Introduction to AUWE Portland 3rd ed.*, PRO ADM 302/147 (1963).
10. *Admiralty Research Plan 1962/63*, PRO ADM 282–18.
11. In fact the NCRE and AEW organisationally reported to the DNC not DPR.
12. Memo, PRO ADM 1/26756 (May 1958).
13. *Formation of Military Oceanographic Group at ARL*, PRO ADM 1/29000 (1963).

14. *US/UK/Norwegian Data Exchange Agreement on Underwater Research, 1962–63*, PRO ADM 1/29117 (1962).
15. E.J. Grove, *Vanguard to Trident* (n. 2 above).
16. Patent Specification 478, 427, *Improvements in and relating to Devices for Accelerating Aircraft for Launching Purposes* (accepted 18 January 1938).
17. D. Hargreaves, 'Aircraft Catapult Development', *Journal of the Royal Naval Scientific Service* 12 (1957): 139–43.
18. *Stress Analysis of Power Cylinder for Steam Catapult Gear*, PRO ADM 227/779 (October 1955).
19. *Photoelastic Stress Analysis of Section of Steam Catapult Cylinder and Cover*, PRO ADM 280/32 (August 1949).
20. *Steam Catapult Development*, PRO ADM 1/27359, file 1, par. 3 (October 1955).
21. PRO ADM 1/27359, file 2, par. 1.
22. *Report of Visit to U.S.A. June-July 1954 Of Admiralty Team To Discuss Catapults And Arresting Gears and Witness Trials In U.S.S. Hancock*, PRO ADM 265/80 (August 1954).
23. D. Hargreaves, 'Aircraft Catapult Development' (n. 17 above).
24. *Chronicle of the Development of Undercarriageless Aircraft...*, Science Museum, technical file T/1993–2546.
25. *Chronicle of the Development of Undercarriageless Aircraft* (n. 24 above).
26. *Flexible Deck*, PRO ADM 1/21735 (January 1949).
27. PRO ADM 1/21735, sec. 7.
28. P. Beaver, *The British Aircraft Carrier*, 2nd ed. (P. Stephens, 1984).
29. *Angled Flight Deck*, RAE Tech. Note NA 247, app. 1, PRO ADM 1/22421 (February 1952).
30. *Staff target for an 'Angled Flight Deck' Carrier*, PRO PRO ADM 1/22421 (3 January 1952).
31. *Minute of Director of Naval Air Warfare*, PRO ADM 1/22421 (1952).
32. *Report on Characteristics of USS ANTIETAM*, PRO ADM 1/24536 (16 September 1953).
33. *Staff Requirement of the First Nuclear Submarine*, PRO ADM 1/27372 (19 December 1956).
34. *HM Submarine Dreadnought Staff Requirements*, PRO ADM 1/26843 (7 October 1957).
35. E.J. Grove, *Vanguard to Trident* (n. 2 above): 230–3.
36. *Revised Staff Requirement for SSN01 HMS DREADNOUGHT*, PRO ADM 1/27568 (September 1959).
37. G.E. Weir, *Forged in War: the naval–industrial complex and American submarine construction 1940–1961* (Naval Historical Center, 1993).
38. *Nuclear Submarine Detection*, PRO ADC 204/1752 (October 1957).
39. W. Hackmann, *Seek and Strike: sonar and anti-submarine warfare and the Royal Navy 1914–54* (HMSO, 1984): xxv. The origin, and indeed the meaning, of 'asdic' is uncertain; it most likely stood for 'pertaining to the Anti-Submarine Division' or 'Anti-Submarine Division-*ics*'.
40. *Ship Noise Investigations, Past and Present*, PRO ADM 259/26 (December 1951).
41. *Ship Noise Investigations Past and Present* (n. 40).
42. *Submarine Acoustic Camouflage . . .*, PRO ADM 285/10 (June 1948).
43. *Basic Research for the Navy: Appendix to Report to CNRSS*, PRO ADM 204/977 (May 1951): 10.
44. *Basic Research for the Navy* (n. 43 above).
45. *Vibration Response of PORPOISE Propellers*, PRO ADM 280/519 (September 1963).
46. *R&D Panel on Noise Reduction: Group Report the the Main Panel*, PRO ADM 285/35 (September 1963).
47. *R&D Panel on Noise Reduction, Group 1*, PRO ADM 285/37 (October 1963).
48. *Report on a Visit to Hydrodynamic Laboratories in the US 1950*, PRO ADM 204/744 (1950).
49. *Report on a Visit to Hydrodynamic Laboratories in the US 1950* (n. 48 above): 15.
50. *Some Possibilities for Ship, Submarine and Torpedo Propulsion*, PRO ADM 204/959 (February 1957).
51. *Some Possibilities for Ship, Submarine and Torpedo Propulsion* (n. 50 above): 10.
52. *The Propulsion and Cavitation Performance of a Pumpjet Propeller Test Vehicle*, PRO ADM 204/2304, introduction.
53. *Measurement of Force on the Duct of a Pumpjet Propelled Torpedo Model*, PRO ADM 204/1738 (September 1961).
54. *R&D Panel on Noise Reduction*, 33rd Meeting, PRO ADM 285/21 (27 December 1962).
55. *Trial of Jet Propulsion in a Minehunter*, PRO ADM 1/29212 (12 March 1964).
56. *R&D Panel on Noise Reduction, Group 1*, PRO ADM 285/35 (September 1962).
57. *R&D Panel on Noise Reduction, Group 1*, PRO ADM 285/37 (October 1963).
58. *Ship Department Annual Research and Development Progress Report*, PRO ADM 281/129 (May 1965).
59. *Ship Department Annual Research and Development Progress Report* (n. 58 above): 3.
60. N. Friedman, *Submarine Design and Development* (Conway, 1984): 88–9.
61. *The Use of Simple and Compound Mounting Systems For the Reduction of Structure Borne Vibrations*, PRO ADM 204/1312 (1956), quotes report ARL/R1/95.03/D (November 1943).
62. *Vibration Isolating Mountings . . .*, PRO ADM 280/210 (September 1957).

63. *Submarine Detection . . . The Present State of the Art and Future Trends*, PRO ADM 302/223, sec. 116 (August 1963): 18.
64. *Some thoughts on the recent history, present status and future prospects of passive acoustic submarine detection*, PRO ADM 204/2547: 3 (May 1961).
65. *Some thoughts on the recent history*, (n. 64 above): 18.
66. *Some thoughts on the recent history*, (n. 64 above): 18.
67. *Revised Staff Requirements For SSNOI H.M.S. Dreadnought*, PRO ADM 1/27568 (September 1959).
68. *Noise Trials on XT5 fitted with Flexibly Mounted Machinery*, PRO ADM 204/1248 (1944).
69. *The Use of Simple and Compound Mounting Systems for the Reduction of Structure Borne Vibrations*, PRO ADM 204/1312 (November 1956).
70. *An Appreciation of the Use of Palliatives in Submarines for the Reduction of Underwater Machinery Noise*, PRO ADM 285/50 (1955).
71. *Trade Convoy Protection Against Nuclear Submarines*, PRO ADM 259/526 (June 1956).
72. *Basic Research for the Navy* (n.43 above): 8.
73. *Nuclear Submarine Detection*, PRO ADM 204/1752 (October 1957).
74. *Optical Detection of Submarines*, PRO ADM 204/2432 (April 1963).
75. *Submarine Detection . . . The Present State of the Art and Future Trends*, PRO ADM 302/223 (August 1963).
76. *Some Thoughts on . . . Passive Acoustic Submarine Detection*, PRO ADM 204/2547 (May 1961): 2–3.
77. *Some thoughts on the recent history* (n. 64 above): 4.
78. *Some thoughts on the recent history* (n. 64 above): 3.
79. Memo, PRO ADM 1/26756 (17 December 1957).
80. Memo, PRO ADM 1/26756 (May 1958).
81. W. Hackmann, *Seek and Strike* (n. 39 above): 349.
82. *Analysis of operation Rum Tub*, PRO ADM 1/26756, sec. 4 (1 December 1957).
83. W. Hackmann, *Seek and Strike* (n. 39 above): 150.
84. W. Hackmann, *Seek and Strike* (n. 39 above): 353.
85. S.D. Mason, private communication (31 January 1997): 1.
86. *Design Study for Asdic Type 2001*, PRO ADM 259/254 (October 1957).
87. S.D. Mason (n. 85 above): 2.
88. S.D. Mason (n. 85 above): 2.
89. *Presentation to AUWE . . . on the Physics of Bottom Bounce*, PRO ADM 302/276 (1965).
90. *Design Study for Asdic Type 2001* (n. 86 above): app. C.
91. S.D. Mason, private communication and interview (28 January 1997).
92. *Design Study for Asdic Type 2001* (n. 86 above).
93. *Some Factors in Bottom Bounce Sonar*, PRO ADM 302/282 (1966).
94. *Some Factors in Bottom Bounce Sonar* (n. 93 above), par. 4.
95. AUWE Publication 10945 (September 1964): 3.
96. *Trial Details H. Houghton*, PRO ADM 1/27369 (1961).
97. C. Pincher, *Traitors* (Sidgwick and Jackson, 1987): 91.
98. H. Houghton, *Operation Portland* (Hart Davis, 1972). In his autobiography Houghton made no bones about his willingness to cooperate, with money rather than ideology being the reason.
99. H. Houghton, *Operation Portland* (n. 98 above). Houghton, for a variety of reasons which were probably mainly financial, seems to have wanted to capitalise upon his notoriety, but as bizarre as some of the events he describes appear some of them do have the ring of authenticity to them.
100. P. Wright, *Spycatcher* (Viking, 1987): 132.
101. *Trial Details H. Houghton* (n. 96 above).
102. *Trial Details H. Houghton* (n. 96 above).
103. S.D. Mason, Interview (28 January 1997).
104. 'App. 1. Performance of Submarines and Weapons', *Submarine Threat in the 1970s in Limited War*, sec. 15, PRO ADM 1/28884 (6 January 1964).
105. *Formation of Military Oceanographic Group at ARL* (n. 13 above): 4.
106. E.W. Constant, *The Origins of the Turbojet Revolution* (John Hopkins University Press, 1980), comes nearest to considering the matter in his notion of 'testability' in which he points out that 'testing involves construction of complex test rigs that are themselves major technological achievements': 21.
107. M. Wolf, *Memoirs of a Spymaster* (Pimlico, 1997).
108. P. Wright, *Spycatcher* (n. 100 above): 269.
109. Although the impact of Rum Tub was to underline to the Royal Navy the need for Britain to develop her own nuclear submarine force, the potential of the fast submarine had been recognised by the Admiralty in the immediate post-war period, stimulated by what had been discovered of the U-boat developments. ADE was

set up in 1946 and run as a department of Vickers Armstrong, to develop the Walther hydrogen peroxide system of submarine propulsion. Due to the highly complex technical and financial difficulties experienced by the programme its first fruit, the hydrogen peroxide submarine EXPLORER was not completed until the end of 1956 and its second vessel EXCALIBUR the beginning of 1958, both vessels neatly bracketing Rum Tub. However from their inception, with their high underwater speeds of 20 knots, they were seen as experimental in nature with their operational role of being defined as high speed underwater targets for ASW training.

Chapter 7

Thermal Radiation and Its Applications

ERNEST PUTLEY

INTRODUCTION

Seeing in the dark has long been a dream of military strategists. In the twentieth century it became a real possibility and a challenge to science, as it seemed that the thermal radiation emitted by bodies even at night could be detected. Thermal radiation is the electromagnetic radiation emitted by warm bodies by virtue of their temperature. It has been studied from classical antiquity, but it is only since the beginning of this century that its characteristics have been fully understood. It generally includes waves whose frequency is less than that of red light—typically described as infrared. Because its intensity and spectrum are determined by the body's temperature, an obvious application is to the measurement of temperature. This was the first, and is still one of the most important. Moreover, because the glow of warm bodies could give away their presence even if not visible to the naked eye, detectors that could pick up infrared would enable the owner to 'see' in the dark. Military applications were first suggested during the First World War, and were studied by all the major combatants.

In 1915, the Oxford physicist Dr F.A. Lindemann (later Lord Cherwell) proposed several ways of exploiting thermal radiation.[1] These were all examined, but at the time none proved feasible. It was not until near the end of the Second World War that new technology made it possible for Lindemann's suggestions to be followed. Today this development is almost complete. Major contributions to it have been made by those laboratories now incorporated within DERA.

Among the applications suggested by Lindemann was the detection of aeroplanes or ships by the thermal radiation they emitted. Another was covert signalling carried out using a simple thermal source such as a large bonfire, with suitable screening to cut off the visible. Although Lindemann did not suggest any form of imaging system, infrared can be detected on a photographic plate and this method of detection had long been suggested as an extension of the use of the photographic plate in visible spectroscopy. During the interwar years these suggestions were followed up at the Admiralty, where a simple but very effective communications system was devised which, with a suitable screened lamp in conjunction with a photoemissive image converter (the RG tube) formed an active viewing system good enough to be used as a night-driving aid.[2] (Fig. 7.1)

(a)

(b)

FIGURE 7.1: NIGHT DRIVING EQUIPMENT, 1944. (A) JEEP FITTED WITH SCREENED HEADLIGHTS AND VIEWING BINOCULARS USING RG IMAGE-CONVERTER TUBES. (B) VEHICLE AS SEEN WITH IT. (© CROWN COPYRIGHT, DERA)

The development of systems using the radiation emitted by the targets proceeded much more slowly because no suitable detectors had so far been discovered. Towards the end of the Second World War, however, the Germans developed a detector made from lead sulphide.[3] The electrical conductivity of a crystal of this compound can rise when infrared light falls upon it. For the first time, engineers had a detector that could detect long-enough wavelengths for it to be able to detect the radiation from hot bodies such as aircraft engines or ships' funnels, and which was sufficiently fast and robust for military use. The war ended before the Germans were able to complete the development of this device, but the work was taken up by the Allies. The British then made major contributions at the Admiralty Research Laboratory (ARL, later SERL, Baldock) and TRE. These constituted the first contributions from organisations that would later constitute DERA, to the development of modern infrared (IR) technology.

The availability of suitable detectors is probably the major consideration in designing a thermal imager. An obvious goal was to design the IR equivalent of the vertebrate eye. This consists of an optical system that focuses an image on to a screen—the retina—which is a two-dimensional array of about a million detectors. The signals from these detectors are combined in an image-processing system in the brain to produce the picture we 'see'. Examination of a modern equipment therefore reveals that it will employ a sophisticated optical system comparable with that used in an optical instrument in the visible. It will use a detector responding to the far infrared, but in many cases approaching the eye in complexity, and to exploit this it will have signal-processing electronics to analyse the information received. Finally it will either have a display unit to present the results or further electronics to respond to the input—setting off a fuse, reporting an intruder, or perhaps just talking to him. None of these components was available in 1915 when Lindemann first proposed using thermal radiation, nor even in 1945 when the lead sulphide cell was discovered. Since then, however, significant contributions to their development have stemmed from the establishments that would constitute DERA.

The progress of research from 1945 until today falls into a number of phases. First, the lead salt era, in which the development of PbS (lead sulphide) and its close relatives was completed and their likely applications explored. This was followed by a period of solid-state research, leading to a fuller understanding of infrared photoconductivity and the discovery of CMT (cadmium mercury telluride, the 10 μm detector). Next there was a kind of interregnum when research was concentrated on far infrared for thermonuclear diagnostics. Then, 'seeing in the dark', when the relative advantages of low light and infrared technologies were critically examined. By the end of this phase the merits of thermal radiation were clearly established. The final phase is one of large-scale exploitation in which both the most advanced and the very simplest infrared systems are playing a part. Research is not quite finished. The microwave thermal imagers are still on their way, and improved, cheaper versions of the older systems are coming along and will lead to greater exploitation of this technology.

ORGANISATION

Lindemann's suggestions were followed up during First World War by Professor Callendar at Imperial College, and by the Cambridge Instrument Company. The Armed Services themselves, however, apparently did nothing until after the war. Then the Admiralty Signal School began to develop near-infrared signalling systems using signal lamps or searchlights fitted with infrared filters as sources. For detectors they first used ultraviolet (UV)-activated phosphors, and later Case thallium sulphide photoconductive detectors. Systems of this type were used to a limited extent in the Second World War by light craft off enemy coasts to provide secure communication.

During the 1920s and 1930s some study was made of thermal radiation both at the Admiralty Research Laboratory and the Royal Aircraft Establishment (and also the National Physical Laboratory) to investigate the possibility of detecting ships or aircraft by their thermal radiation. This work was supported by the Tizard Committee (Committee for the Scientific Study of Air Defence) but given low priority after the rapid progress of radar research.

In 1938 it was decided that the Admiralty (at the Admiralty Research Laboratory—ARL) would undertake this research on behalf of the three Services, which they continued to do throughout the Second World War. During this time little progress was made with the development of thermal radiation systems, mainly because no adequate detectors were available, but also because the development of radar reduced the need for alternative systems.

Although ARL was responsible for the research, the individual Services were responsible for the applications. Thus ASE continued with their work on naval signalling while ADRDE, on behalf of the Army, considered antiaircraft applications and the AORG studied night-driving aids for Army transport and tanks. Both RAE and TRE were involved in various aircraft applications, mainly with identification and communication.

This organisation worked very well until near the end of the war. Then, under pressure to detect submarines equpped with Schnorkels (see below) from the air, the management at the Ministry of Aircraft Production's Telecommunications Research Establishment decided to follow up the German lead on lead sulphide and embark on infrared research. This decision was at first bitterly contested by the Admiralty, but their objections were overcome by the argument that, as the Admiralty would not let anyone else work on their ships, it was not unreasonable for the Air Ministry to adopt a similar policy with respect to aircraft. In practice, a good working relationship soon developed between the Telecommunications Research Establishment (TRE) and ARL in which TRE took the lead in developing the specialised electronic equipment now required. Both ARL and TRE took an active part in the development of the new photoconductive detectors. Attempts were made to avoid too much duplication of effort, but the difficulty was that the results depended very much on the skill or know-how of the individual research worker. Thus a very flexible administration was essential to achieve the optimum results. Overall, across departments, the number of people actively working in the area lay between 10 and 20.

After the end of the war, the Admiralty set up the Services Electronic Research Laboratory (SERL) to continue on a tri-Service basis the work of valve development, which it had been doing from before the war. The infrared group from the Admiralty Research Laboratory was transferred to SERL, where it continued to work in loose collaboration with TRE (later the Radar Research (Royal Radar) Establishment) until their merger to form RSRE in 1976. At the end of the war TRE was reorganised, with the bulk of the research effort concentrated in the new Physics Department headed by R.A. Smith, with F.E. Jones in charge of experimental physics, G.G. MacFarlane theoretical and A.M. Uttley electronics. In addition to infrared research, Uttley's work on optical homing (initially for stellar navigation) and on the computer TREAC contributed to the evolution of the thermal systems. The Royal Aircraft Establishment (RAE) at Farnborough was not involved in infrared research at the end of the war, but when F.E. Jones left TRE to become a Deputy Director at Farnborough in 1953 he set up an infrared group to concentrate on countermeasures in the belief that this could be done more effectively in an organisation not directly concerned with measures.

At this time the Signals Research and Development Establishment (SRDE) was responsible for Army requirements, including night-vision aids. From their studies of the IR situation at that time they concluded that, in the immediate future at least, the Army's requirements could be satisfied by the then rapidly developing low-light/near-IR systems,[4] a conclusion that could not be faulted until the rapid development of thermal imaging in the 1970s.

THE LEAD SALTS DISCOVERY

Until the end of the Second World War the British made no serious attempt to exploit infrared thermal radiation because there were no suitable detectors. Although infrared had been detectable in the laboratory for two centuries, the apparatus was large, cumbersome and slow. Two events near the end of the war restored interest in the longer-wave thermal radiation. The first was the capture by the Allies of German detectors made of lead sulphide.[5] This solid-state detector would transform the technology of transducing infrared rays into electrical signals. The second factor was the introduction by the Germans of the Schnorkel, which enabled submarines to cruise under water using their air-breathing diesel engines, rather than relying on batteries for necessarily short submersions. This made the submarine virtually undetectable by radar and, had it been available earlier, could have prevented Allied successes in the Battle of the Atlantic.

As a last hope, the Admiralty asked for infrared trials on the detection of Schnorkels or exhaust emissions from U-boats.[6] This work continued for several years and was to make an important contribution to the development of thermal imaging, but a much more interesting German equipment was captured at the end of the war. This was the Kielgerät airborne aircraft detection system. The only complete installation in a JU88 was found by Air Commodore Chisolm[7] and flown to England, where it became the starting point for our development of IR-guided missiles.

After a preliminary examination it was agreed to fly the JU88 to Defford for detailed flight trials of the equipment. Unfortunately, the aircraft was wrecked in a flying accident before this could take place. Luckily the apparatus was salvaged and fitted in a Tudor. Flight tests then confirmed the German claims. Originally it had used a PbS cell cooled with solid carbon dioxide, but by this time more sensitive lead telluride (PbTe) detectors cooled with liquid nitrogen were being made at TRE. Even better performance was now obtained. In the original Kielgerät, a visual display indicated the bearing of the target. It was now decided to combine the German optical system with a lock-follow scanner taken from an aircraft interception system. This produced an infrared sensor that automatically followed its target. At this time the development of guided missiles had just started (see Chapter 3). These experiments showed the way to design an infrared-based guidance system.[8] The first successful IR homing missile was the first version of Firestreak, produced in 1953–54. It was also possible to devise an IR fuse, thus ensuring that a direct hit was not necessary for the missile to be effective.

THE FIRST THERMAL IMAGERS

From 1945 the Admiralty and TRE carried out a series of trials to determine the effectiveness of infrared in detecting ships at sea using an airborne radiometer. The long-term objective was to examine the possibility of using IR to detect submarines.

The postwar Admiralty project also developed from German wartime research. The German firm E. Leybold's Nachfolger[9] was the prototype from which many thermal imagers, both British and American, were later developed. This contained a double optical system in which an infrared detector was scanned across the scene while at the same time a spot of light was scanned in a corresponding pattern across a film. The output from the detector was used to modulate the light spot, so that when the detector picked up infrared emission, the light scanning the film became more intense. As a result, a visible picture corresponding to the thermal scene was built up. At the end of 1944, when the PL3 version of this equipment was built, there were no photodetectors sensitive enough to pick up the radiation, with waves 10 μm or so in length, which is emitted by bodies at room temperature. The Germans, however, did have fast bolometers with which they were able to obtain a simple picture of a person.

In 1946 ASE Haslemere decided to build a similar apparatus to establish whether thermal images of outdoor scenes could be produced, and to see if these would be of value for the detection and identification of objects at night. They immediately obtained encouraging results and arranged for a better version to be built by GEC, which was used for the first detailed study of thermal imaging by ARL from 1948.[10] The scanning optics was a modified German coast-watching equipment and the detector was a thermistor bolometer designed by BTH based on German research. The imager had to use a very slow scan rate, taking about 20 minutes to form a picture. The purpose of it was to explore the nature of thermal pictures, for at that time it was not clear whether thermal imaging had a military use. The conclusions

FIGURE 7.2: ADMIRALTY THERMAL IMAGER. THE MIRROR FOCUSED A POINT IN THE SCENE ON TO A BOLOMETER DETECTOR. THIS WAS SCANNED BOTH HORIZONTALLY AND VERTICALLY, TAKING ABOUT HALF AN HOUR TO FORM A PICTURE. (© CROWN COPYRIGHT, DERA)

reached from these trials depended upon who was viewing the pictures. To those with the vision to see the future advance of infrared technology they were very promising, but to those who wanted operational equipment now they at best were too far in the future (see Seeing in the Dark) (Fig. 7.2).

Another experiment, code-named 'Yellow Duckling', showed that the main difficulty was in interpreting the returns from the sea before the days of data processing. When the equipment was flown over land meaningful signals were soon revealed, indicating that with a properly designed apparatus a maplike image of the land should be obtained. This was the first suggestion of the infrared linescan imaging system which by 1980 had become an indispensable reconnaissance tool.[11] In the first experiments a PbTe detector was used, later to be replaced by Cu-doped Ge (copper-doped germanium). InSb (indium antimonide) and CMT (cadmium mercury telluride) detectors were used in the later versions.

THEORETICAL STUDIES

The discovery of the PbS cell led to a re-examination of the potential of the thermal spectrum. Some work on the ultimate sensitivity of thermal detectors had already been undertaken at Oxford, in response to a request from the Admiralty that they should follow up the suggestion originally made by Watson-Watt, that the very long-wave region (up to the microwave region) should be studied. This work was continued by W.B. Lewis at TRE, whose results form the basis on which current theories of the noise-limited sensitivity of infrared detectors are based.[12]

In addition to this basic theoretical work, applied work on the possibility of detecting submarines, on applications to guided missiles, and on the general value of infrared, was carried out at an early stage as a guide for the developing experimental programme.

When the PbS detector was discovered knowledge of solid-state physics was rudimentary. The major task confronting the developers of improved detectors was to find a material sensitive to the 10 μm waves emitted by bodies at room temperature. The obvious first step (which the Germans had begun) was to examine the related compounds PbSe and PbTe (the selenium and tellurium salts of lead). This soon revealed that a worthwhile extension of spectral response was achievable, but not enough to reach 10 μm.

One thing was very clear: there appeared to be no simple relation between the properties of the material of the detectors, comprising thin films prepared by chemical deposition or by evaporation, and the then-known bulk properties of the lead salts. Detailed measurements of all relevant optical and electrical properties of the films, including electrical conductivity, thermoelectric power and Hall effect, were undertaken, but there seemed to be fundamental differences between the behaviour of PbS in the form of a film and in its purest available bulk form of natural crystals of galena (or bleiglanz, as it is sometimes called).[13] These results suggested that the critical feature of the films lay in the intergranular contacts between particles in the layers, which must account for the much greater resistivity of the films than of the bulk material.

Elaborate theories of intergranular contacts were developed to account for photoconductivity. These were derived from the well known rectification effects observed in the purest samples of galena, which had led to one of the first successful PbS detectors (the German Bleiglanznetz cell, in which a fine platinum mesh was pressed against the polished face of a galena crystal). These theoretical studies anticipated the Bell Laboratories' discovery of the p–n junction, key to the development of the transistor, but they did not advance the search for a 10 μm detector.

In 1948 Sir George MacFarlane (then head of the TRE Theoretical Physics Division) reviewed the situation and concluded that the only way forward was to resort to fundamental research on the infrared sensitive semiconductors. At that time similar work had already begun in the USA aimed at the microwave detector materials—silicon and germanium—which had proved so valuable in the development during the war of centimetric radar. TRE scientists were, of course, very interested in these, but at that time were not undertaking any fundamental research.

The infrared research programme started at TRE included a critical review of photoconductivity, and of the statistical mechanics of semiconductors and one of the first attempts to calculate the band structure of PbS using the new digital computer developed at TRE (TREAC).[14] In parallel with these theoretical studies a small team was set up to prepare and study single crystals of the lead salts.[15] At that time the technological importance of single-crystal material was not widely recognised. At Bell Laboratories in the United States proponents had to overcome the opposition of W. Shockley, later known as the pioneer of the transistor! It is curious to reflect how today's emphasis on materials technology developed from that initial effort to produce 10 μm detectors.

SOLID-STATE PHYSICS

The lead salt detectors had demonstrated the potential of infrared, but they had several serious limitations.[16] The most fundamental was that their spectral response did not extend far enough to cover the 10 μm room-temperature radiation band. Second, they were very difficult to manufacture, consisting of thin polycrystalline chemically unstable layers produced by chemical precipitation or vacuum evaporation. Finally their mode of operation was not understood, giving no guidance on how to seek for a 10 μm detector. A research programme began by growing single crystals of lead telluride and studying their basic electrical properties. This was, it is believed, the first attempt to apply single-crystal technology to the study of electronic devices.

The study of single crystals established the intrinsic conductivity of lead salts.[17] Today it may seem obvious that one way to extend the spectral response was to discover materials with smaller energy gaps, but in 1951 the concept of intrinsic conductivity was far from being generally accepted. A search was therefore begun for a suitable material. By this time a considerable effort in Germany and elsewhere was being made to find new transistor materials. This was considered necessary because at that time the energy gap in Ge (0.76 eV) was considered too small for stable operation, and methods for preparing sufficiently pure silicon had not been discovered. One result of this search was the discovery by Welker of indium antimonide (InSb).[18] This has an energy gap roughly the same as PbTe, but InSb detectors could be made using single-crystal material. The InSb devices were easier to make and more convenient to use than PbTe, and soon were replacing it for use in the 3–5 μm band. InSb found a use not only in infrared linescan, but also in thermal imagers for medical research. Developed jointly by St Martin's Hospital, Bath, Smith's Instruments and TRE (but not on the official programme),[19] it took about 10 minutes to form an image of the patient but provided a permanent record on a form of photographic paper. InSb remains in use today.

The search for a 10 μm detector discovered several possible materials. Almost 50 were investigated before one with adequate chemical and physical stability and with the correct energy gap was discovered.[20] This was the mixed crystal $Hg_xCd_{1-x}Te$ (a mixed crystal of mercury and cadmium tellurides), whose fate was to dominate this story.[21] At least one other possible material—$Pb_xSn_{1-x}Te$ (LTT, a mixed crystal of lead and tin tellurides)—was discovered and later received limited development, but CMT was found to be superior and LTT was dropped.

When it was first discovered the detection of 10 μm radiation by single-crystal filaments of CMT was demonstrated, but the development of detectors was not undertaken. MacFarlane in fact had the work on CMT stopped in 1957. This may seem a strange decision now, as it did to some of the team at the time. However, considering all the circumstances, I now think it was correct. At the time when CMT was discovered, InSb detectors were beginning their development as 3–5 μm detectors to replace PbTe for guided-missile projects and for some of the early research on thermal imaging. There was then a pressing operational requirement for these devices, and MacFarlane believed that we did not have enough effort to support both the InSb and the CMT developments.[22] InSb proved more difficult to develop than expected, and our later experience with CMT showed that this was even more difficult, thereby justifying MacFarlane's decision.

Another reason for suspending work on CMT was the US development at this time of another type of long-wave photodetector, the impurity ionisation detector.[23] This used a wide bandgap material such as Si or Ge containing carefully chosen impurities which are ionised by the far infrared. In this way radiation longer than 100 μm can be detected and it is fairly easy to prepare material for 10 μm. Detectors of this type were made at RRE for Yellow Duckling[24] and for the first infrared linescan systems.

There were, however, several disadvantages in using the impurity-doped detectors. First, because the concentration of the active impurities must always be small compared with the density of lattice atoms, the absorption will be less and therefore a thicker piece of material will be required. For the same reason, to avoid saturation of the active centres by ambient thermal radiation impurity detectors must be cooled to a lower temperature.[25] Thus, whereas InSb or CMT do not need cooling below 77 K (boiling liquid nitrogen) the corresponding doped Ge detectors required cooling to 20 K (boiling liquid hydrogen), which raised practical difficulties. Nevertheless, the doped Ge detectors proved valuable as research tools and were used in some of the first manufactured linescan equipments.

CRYOGENICS

It has long been known that the performance of infrared detectors improves by cooling below normal ambient temperature, but the reasons do not seem always to have been fully understood. There are two basic reasons. First, a detector will be immersed in the radiation field characteristic of its surroundings, but in a thermal system it will be trying to detect radiation from (usually) a hotter body. Clearly, the less radiation received from its surroundings the greater its sensitivity (in terms of signal/noise). For this reason the Germans cooled the PbS detectors in their early equipments with dry ice—frozen carbon dioxide—and most high-performance detectors are now cooled with liquid air. This process cannot be continued indefinitely because to do its job the detector must see more of the outside world than just its target, and the radiation from this small element of the ambient background will become the dominant source of noise when the detctor's temperature falls below a certain value. This argument applies to terrestrial scenes, but in outer space the situation may be somewhat different.

Second, low temperatures may be needed because certain detectors employ effects observed only at very low temperatures. Thus, as already mentioned, impurity photoconductors require lower temperatures than intrinsic ones. Another example (which was considered for Yellow Duckling but abandoned in favour of Cu-doped Ge) is the superconducting bolometer which has to be cooled to the transition temperature.[26] The most sensitive bolometers used by the infrared astronomers operate below 1 K.

Provision of the necessary cryogenic facilities for use in Service equipment posed a logistics problem. Today liquid hydrogen and helium are produced industrially and can be purchased, whereas in the 1960s TRE had to build its own H_2 and He liquefiers. In the laboratory liquid nitrogen could just be poured into the cells, although automatic apparatus using the Leidenfrost principle was designed for this. For use in operational equipment Parkinson at TRE designed a miniature liquefier for nitrogen based on the Joule–Thomson effect (whereby rapidly expanding gases cool), which fitted into the cell blank and which, when fed with compressed air, rapidly produced enough liquid to keep the detector cooled for the time of flight of an infrared homing missile. It could also be operated continuously when required for an Army thermal imager such as Shortie (see below).

More elaborate equipment was required for detectors operating below the liquid air range. In the Yellow Duckling experiments it was planned to use a niobium nitride superconducting bolometer, which would have required cooling with liquid hydrogen. This detector was replaced by a Cu-doped Ge photoconductive detector, but it still required liquid hydrogen. Again, a miniature Joule–Thomson liquefier was designed, but it had to be a two-stage device because hydrogen has to be cooled with liquid air before the Joule–Thomson expansion.[27]

Some extrinsic Ge detectors require cooling with liquid helium (4 K). They were used in early linescan systems. The elements were mounted in large stainless steel dewars which could hold enough liquid helium for a test flight. Detectors used for CTE diagnostics (see below) were cooled below the helium λ-point, but this was in a laboratory installation.

These systems were quite impractical for operational equipment. The first step was to develop detectors operating at higher temperatures. Before the development of CMT the best choice for the 8–13 μm band was mercury-doped germanium, which could operate at 35 K. This temperature could be achieved using a refrigerating engine, such as the Stirling. These machines were very difficult to design and construct, and their working life was rather short, but they eliminated the need for supplies of special refrigerants.[28]

With the steadily increasing numbers of elements in detector arrays the sensitivity per element can be reduced and thus the maximum operating temperature can rise. With hundreds of thousands of elements, operation at normal ambient temperatures without loss of performance is possible. However, there may be a need for temperature stabilisation because some detectors are very temperature sensitive, and the large integrated circuit arrays may need stabilisation. A very satisfactory way of meeting this requirement is to use a Peltier temperature control module, which can cool the arrays to below 273 K (0°C) or maintain a convenient temperature.[29]

Much of the work of developing these later systems has been carried out by the manufacturers, either as DERA contractors or as private ventures. However, work on all the systems described here originated at DERA from the time of TRE onwards. Close contact with the manufacturers has always been maintained, to our mutual benefit.

ELECTRONICS

In 1915, when Lindemann persuaded the Cambridge Instrument Company to try to detect radiation from aircraft in flight, the most sensitive detector available was the Boys Radiomicrometer, which was essentially a very sensitive type of moving-coil galvanometer.[30] This was not a very practical instrument and not sufficiently sensitive to detect background noise.

It was not until the late 1920s that galvanometer amplifiers were introduced capable of detecting thermal noise.[31] These, with thermopiles, were then the most sensitive radiation detectors, but they were strictly laboratory instruments. Thus in spectroscopy a galvanometer was the standard type of measuring instrument used with very slow thermopiles. In 1937 R.V. Jones built an airborne infrared aircraft detector using a thermopile, a three-stage RC amplifier and a vibration galvanometer tuned to 40 Hz.[32]

For military use a stable high-gain thermionic amplifier was required. The first step was to consider replacing the galvanometer with a DC amplifier, but these were notoriously difficult to stabilise. The trick was to convert to AC. In the first instance this could be done by modulating the input to the amplifier, first by a mechanical switch and later by a vibrating condenser. A good example was the General Motors chopper amplifier, which was used in our atmospheric transmission measurements.[33]

This solution did not eliminate the problem of drift in the detector itself, but this could be overcome by modulating the radiation incident upon the detector, usually by a rotating shutter. The modulation frequency had to be set to suit the type of detector being used. With thermopiles or Golay cells it had to be less than 10 Hz, but with photoconductive detectors it could be up to 1000 Hz.

TRE undertook the design of amplifiers. These required more careful design than standard audio or video amplifiers because they had to work down to noise level. Carefully designed input transformers tuned to 5 and 800 Hz were used for thermopiles and photodetectors, respectively.[34] When the use of a superconducting bolometer was proposed, an amplifier with a liquid-helium-cooled input transformer was designed.[35] A major advance was the use of phase-sensitive rectification. A reference signal was generated from a small lamp, which was modulated by the input chopper and enabled postdetector integration to be used.

With the introduction of the transistor, discrete component solid-state amplifiers replaced thermionic valve circuits. The early transistors were much noisier than the best valves, so that for some time a triode input stage had to be used when the lowest noise was required.[36] With the introduction of low-noise field-effect transistors (FETs) which could be optimised to match the output circuits of the detectors the older valve circuits were finally superseded. When these FETs were being developed it was found that AERE Harwell had similar requirements to ours for their solid-state nuclear radiation detectors. In the 1970s, therefore, through the agency of CVD (Components, Valves and Devices), we set up with AERE a joint development programme at Texas Instruments, Bedford.

The introduction of integrated circuits saw a big reduction in the size of the detector amplifiers, to the extent that it became common practice to incorporate the input stages within the detector's encapsulation. This led not only to a valuable reduction in size, but also to an improvement in performance by eliminating the troublesome interconnecting leads at the detector output.[37]

With the development of large detector arrays (more than 1000 elements) for use in thermal imagers, the need for corresponding large-scale integrated (LSI) circuits arose.[38] This raised economic problems because the high cost of developing these circuits could only be sustained if there were a large-scale requirement for thermal imagers. At first, the scale of the requirement was not very clear because operational experience of the capability of the imagers and other thermal devices was required. This definitely retarded the development of the staring array systems. However, after the successes of the thermal imagers in the Gulf War and other recent events,[39] the value of thermal systems has been clearly established and the wisdom of adequate investment in their development finally accepted.

NUCLEAR INTERLUDE

When the InSb development had been completed the military authorities saw no further need for infrared research and in 1956 proposed that the team be disbanded. This suggestion was not as perverse as it seems now because at that time low light vision techniques were advancing more rapidly than thermal imaging; it was nonetheless wrong. Furthermore, it was not put into effect. This was the time of the first experiments at Harwell on a controlled-fusion reactor (Zeta), which revealed the need for better controlled thermonuclear energy diagnostics. One area felt worth exploring was the use of submillimetre or millimetre radiation. The development of these techniques was a substantial research project which Atomic Energy Research Establishment at Harwell was unable to tackle, but which was ideal for the RRE infrared and millimetre-wave teams. Design studies started in 1957 and the measurements on Zeta in 1959.

Spectrometers and other instruments for analysing wavelengths below 1 mm were successfully designed and constructed at Malvern, but the limiting factor was again a suitable detector. Until this time the only types of detector for this part of the spectrum were the slow, non-selective detectors such as thermopiles, bolometers or Golay cells, which were unable to follow the rapid transient response of the thermonuclear machines. Research at RRE on the properties of InSb at very low temperatures (down to 1 K), initially carried out to assess the purity of the material used for the intrinsic detectors, led to the discovery of a new type of photoconductive effect—the free carrier, or electronic bolometer effect—which was most sensitive at wavelengths longer than 1 mm and was sufficiently fast ($t \approx 100$ ns) for time-resolved spectroscopy of the CTE discharge experiments.[40]

After several years' work this project was terminated by the handing over to UKAEA Culham of several sets of submillimetre diagnostic equipment. The basic techniques developed are still used on machines such as the European experimental thermonuclear reactor JET, but modern developments in instrumentation have led to considerable improvements in the ease of use of this technique.

SPECTROSCOPY ATMOSPHERIC PROPAGATION

Although the general topic of spectroscopy was not of concern, that of atmospheric propagation was crucial in the endeavour to see through fog or even clear air. The effects of humidity and altitude on transmission bands and absorption regions need to be understood. In short-range communications one needed to understand absorption to avoid false alarms and to communicate without fear of eavesdropping.

In 1945 it had long been known that for most of the spectral region between the visible and the microwaves (wavelength >1 mm) the atmosphere absorbed thermal radiation, although one or two transmission windows were known to exist. Although in some circumstances long-wavelength rays seemed to carry further than visible light, transmission was known to depend on humidity and, despite much popular belief to the contrary, dense fog was effectively opaque. That, by contrast, performance was not seriously impeded by battlefield smoke, was also becoming known.

It was therefore clearly important that the spectroscopic characteristics of the atmosphere (the real atmosphere, not a collection of gases in a long glass tube) be accurately known before setting up expensive research programmes which might fail if set in the wrong waveband. The interest was not just in regions of maximum transmission: for certain tasks, such as covert short-range signalling or minimising false alarms in proximity fuses, the use of an absorption band is desirable.

To satisfy this need a joint experiment was set up by ARL and TRE, located at Tantallon Castle in Scotland, where a transmission path crossing the sea could be set up. The results obtained have been accepted as a definitive measure of the atmospheric characteristics ever since.[41] These measurements were made at sea level. At high altitudes, as with airborne equipment, the behaviour will be modified owing to the dependence of absorption upon pressure. To investigate the effect of this on system performance a simple high-speed spectrometer was designed and flown at operational altitudes. These results were particularly useful for designing infrared homing missiles.[42]

Later measurements extended the Tantallon results to cover the whole spectrum to microwaves.[43] These confirmed that beyond about 20 μm the atmosphere was highly absorbing. There are several absorption bands in the millimetre region, but beyond about 1.5 cm the atmosphere becomes perfectly transparent. In fog the problem is scattering by the water vapour droplets. This becomes less as the wavelength increases. Thus at 10 μm it is less than in the visible, but still too great to be able to see through it. At wavelengths of about 1 mm the scattering is sufficiently small that it becomes possible to consider viewing systems that can penetrate fog.

OPTICAL MATERIALS

In the visible there is a wide range of materials to use in constructing optical instruments, but until fairly recently the corresponding range for infrared applications has been very restricted. Considerable efforts had to be made to remedy this. The DRA laboratories have made several contributions to this development.

A range of materials was developed for infrared spectroscopy, most suitable only for laboratory use. Infrared detectors needed transparent windows, as ordinary glass does not transmit much beyond 1 μm. Two materials with suitable optical properties are rock salt and diamond: rock salt is soluble in water, and diamond is not normally economic, so the first requirement was for window materials both for the detectors and for infrared domes for the missiles. Synthetic sapphire and periclase were amongst the first cell window materials. DRA was responsible for developing a range of glassy compounds for infrared domes.[44] The first to be used (on Firestreak) was arsenic trisulphide. Much later, when requirements for transistors had led to large single crystals of silicon and germanium becoming available, they proved important not only for windows but also for manufacturing high-quality lenses and other optical components required for the infrared sytems. One problem with germanium is that it reflects a large proportion of the incident radiation. This can be cured by evaporating refractive layers (blooming), with a layer of diamond acting both as refractive material and as a hard material coating. A vapour-deposition technique has been developed at RSRE and is now widely used.[45] This has proved one one of the key techniques in the development of the advanced thermal imagers.

SEEING IN THE DARK

In the 1950s, although the requirement for improved night vision equipment was accepted, it was not obvious that this could be met by a thermal imager. Real-time thermal imagers had not yet been designed, and the quality of the best thermal pictures was rather poor. There had, however, been considerable progess in developing low-light vision systems. These were a logical development of the active near-infrared night-vision systems developed and widely used by the British during the war. Shortly thereafter, developments in photomultipliers had increased the sensitivity of these systems to the point where it became possible to dispense with

FIGURE 7.3: IMAGE INTENSIFIER RIFLE SIGHT. THIS AMPLIFIES THE RESIDUAL LIGHT, ENABLING THE RIFLEMAN TO SEE HIS TARGET. TOP RIGHT PICTURE: TARGET AS SEEN WITH SIGHT. (© CROWN COPYRIGHT, DERA)

the near-infrared source and use the residual daylight present on all but the darkest nights.[46] The pictures obtained (Fig. 7.3) had the advantages that, being very similar to visible pictures, they were much easier to interpret than most of the infrared pictures then available, and they lent themselves to binocular vision systems—important for night driving or flying, and a facility not yet available with the latest thermal imagers.

At this time there was considerable interest in developing low-altitude—terrain-following—flying techniques for the advanced strike aircraft TSR-2, intended for the RAF in the 1960s, and similar aircraft. This required much new instrumentation, and one problem was photographic reconnaissance. This had always been a hazardous task for low-flying aircraft, but the new flying techniques meant that the aircraft would go too low and too fast to enable the normal type of camera to be used. Instead, it was proposed to use a form of flying spot system in which a line transverse to the direction of flight would be scanned by the camera and, as the aircraft flew, successive lines would be combined to form a complete picture of the terrain. This optical linescan system was found to give useful results in daytime, but by 1965, when TSR-2 was cancelled, its use at night was not practicable. The usual photographic flares were not suitable, and the use of a scanned light spot raised both logistic and security problems. The solution finally adopted was to revive in improved form the infrared scanning system used in Yellow Duckling. This, now called infrared linescan, proved so satisfactory that, as well as night-time use, it took over the daytime role planned for optical linescan.

MORE ON DETECTORS

British interest in thermal imaging was revived in the late 1960s by US success. Although the British team had stopped research on CMT, following such work which had continued in France and the USA came under the general terms of reference 'do what you like but get some results'. By combining relatively small multielement arrays with scanning optics the US Air Force was able to use forward-looking imagers in its aircraft, giving real-time pictures which were usable over a greater range of atmospheric conditions than the low-light vision systems. This development coincided with another US advance in detector technology. Urgent military requirement for better FLIR (forward-looking infrared) imagers had arisen from Vietnam, and Minneapolis-Honeywell made the startling claim that they had succeeded in making CMT detectors with adequate performance at the relatively high temperature of 77 K (–196°C). The first FLIR used InSb 3–5 mm detectors. To use the more sensitive 8–13 μm band they would have had to use mercury-doped germanium detectors cooled by a refrigerating engine to 35 K. Because they had persisted with research on CMT, Minneapolis-Honeywell were now able to reveal that they had succeeded in manufacturing liquid-nitrogen-cooled 8–13 μm CMT detectors.

At first the team at RRE was not certain whether to believe the claims of their US allies, but Putley judged from the very guarded response he received when he visited them that their claims were well founded. The RRE team then resumed work on CMT, which it had dropped seven or eight years earlier. The task was now to try and catch up with American detector and thermal imaging technology. For once the team succeeded. The first steps were taken by Mullard, a British subsidiary of Philips, but later on Malvern made major advances, culminating in the invention of SPRITE (signal processing in the element) by C.T. Elliott at RSRE (1981) and its successful manufacture by Mullard in Southampton.[47]

SHORTIE

Work on real-time thermal imagers began with RRE designing a series of experimental models, which were manufactured by Barr & Stroud.[48] The detectors were small arrays of, first, InSb and later, CMT, manufactured by Mullard, Southampton. The earlier imagers used mirror optics, but later germanium lenses were used. Shortly after this programme began the Army issued a specification for a night-viewing equipment which would become their standard. This specification had been drawn up with a low-light system in mind, but RRE maintained that they could design a thermal imager that would meet the specification. After some discussion it was agreed that a thermal imager could be submitted for trial.[49] This was done, again with Barr & Stroud manufacturing it, and when the trials took place this thermal imager easily outperformed the low-light entries. This was when the Northern Ireland troubles were at their height, and the Army immediately seized the imager

FIGURE 7.4: THERMAL IMAGER SHORTIE. A GERMANIUM LENS FOCUSED THE SCENE ON TO A CMT DETECTOR COOLED WITH LIQUID NITROGEN. A RAPID OPTICAL SCANNING SYSTEM BUILT UP A REAL-TIME PICTURE. THIS WAS THE FIRST TO BE USED OPERATIONALLY BY THE BRITISH ARMY, *c*.1972. (© CROWN COPYRIGHT, DERA)

(codenamed Shortie) and, despite its weight, took it on foot patrols. It was also used in helicopters. Although it was then considered very secret it could be seen on several TV news films sticking out of the helicopter's window (Fig. 7.4).

When the early thermal imagers were compared with the low-light systems it was found that the thermal imagers were much better at detecting targets, but the latter were better at recognition. An attempt was made to combine the advantages of both by attaching a non-image-forming thermal detector (a 'thermal pointer') to a low-light system to direct it to likely targets.[50] Considerable progress was made with this scheme using a TGS pyroelectric detector in the thermal pointer, but it was rendered obsolete by the rapid improvement of the thermal imagers with the later introduction of SPRITE. The combined system can be thought of as a precursor of modern aircraft installations, in which the pictures from a forward-looking imager are combined with those from a pair of low-light night goggles. The thermal pointer development was not entirely wasted, because the experience gained in developing operational pyroelectric devices contributed to the development of battlefield sensors which are now part of standard Army kit, and which in their turn paved the way for the massive civil developments in this field.[51]

The success of Shortie brought demands for thermal imagers for many specific tasks, such as night observation devices for tanks and other vehicles. The combination of a thermal imager with a carbon-dioxide laser rangefinder proved very effective. This growth of interest created a further problem: the design effort required to produce all these models was becoming too much, and their cost was becoming excessive. Detector arrays with up to 100 elements were being called for. The solution to the first problem was to copy the American idea of breaking the imager system into a series of subunits—'common modules'—which could be combined in various ways.[52] The problem with the detectors was that the arrays had to be built up from individual elements because CMT could not then be grown in large enough plates to enable a monolithic array to be fabricated, again making the cost excessive. The solution to this not only reduced the cost but improved the performance. This was the invention of signal processing in the element (SPRITE), in which each element was equivalent to ten or more simple detectors.[53] The introduction of this innovation into the common modules gave them the edge over other designs, making them the most cost-effective imagers available anywhere.

The Falklands conflict saw the first major British use of these new imaging and sensing systems,[54] but they were not fully deployed until the Gulf War. Several military commentators have described how they enabled our Forces to operate at night as well as in daytime.[55] The equipments in use ranged from the most comprehensive surveillance/rangefinding battletank installations to handheld viewers used by long-range patrols.

The exploitation of this new technology was not confined to military applications: there has been a continuous, if rather weak and informal, link with medical research. An advanced imager for medical use was designed, financed by the Rank Prize Fund, which is interested in supporting new applications of electronics to medicine and which had marked the personal contributions of several members of the DRA staff with individual prizes.[56]

BELT AND BRACES—PYROS

When work on CMT resumed at Malvern, Sir George MacFarlane (then Director, RRE) was worried that unforeseen difficulties might prevent the succesful production of CMT. He had earlier been involved in the development of PbTe and InSb detectors, and was very aware of the difficulties that could arise. He believed in the importance of developing infrared technology. As an insurance against the failure of the CMT development, he had a study of alternative methods of detection started at RRE. This resulted in the development of pyroelectric detectors[57] and the pyroelectric vidicon.[58]

An alternative to the use of digitally integrated elements was to design an infrared equivalent of the camera tubes used in television. These mostly use a layer of a high-resistance photoconductor. When this is exposed to light a charge pattern is built up on it corresponding to the changes in intensity in the scene. This charge pattern is read out by a scanning electron beam and used to generate the picture on the display. In addition to being sensitive to the incident radiation, the layer has to have a sufficiently high resistance that the charge pattern will not decay before the scanning beam can read it out. In the visible this is no problem, but infrared-sensitive photoconductors at room temperature are of much lower resistivity and so cannot be used. Attempts were made (not by DERA) to make camera tubes containing helium-cooled detectors, which were hardly compatible with thermionic electron tubes! An alternative approach was to replace the photoconductive layer with an infrared-sensitive bolometric layer or, as in the absorption edge image converter,[59] by a material whose optical properties were very temperature sensitive and were changed by the heating pattern produced by the radiation from the scene. These devices were not of inadequate sensitivity but their picture resolution was seriously degraded by thermal diffusion of the heating pattern from the scene during the beam scan time.

The solution to these problems came with the development of pyroelectrics. These materials are of high resistivity and, because they are AC coupled, are not affected by ambient temperature drifts, unlike most bolometric devices. Even so, to achieve the highest spatial resolution the sensing layer had to be constructed to minimise thermal spread. The pyroelectric vidicon has made possible the widespread use of thermal imaging. Its simplicity and robustness has enabled fire brigades worldwide to work in smoke and hazardous surroundings in a way that had not been possible before. It is likely that, in the near future, the rapid progress of microelectronics will replace it with an all-solid-state imager.

In the event the Malvern CMT programme was a brilliant success. This led to attempts to shut down the pyroelectric development, but it was soon realised that pyros were suitable for applications for which CMT could not be used. Two examples are battlefield surveillance devices (Fig. 7.5), which led directly to the widely used domestic intruder alarm (Fig. 7.6)—easily the largest application of any infrared device—and the fire brigade smoke viewing camera, which is widely used by the emergency services and which is carried in large numbers by both the Royal Navy and the US Navy.[60]

The battlefield surveillance systems usually include several types of sensor, of which the pyroelectric infrared is one. This reduces the risk of false alarms and enables a simple type of signature analysis to be used to identify the type of intruder. These systems have proved effective in guarding perimeters, even extended frontiers.[61] It has recently been suggested that they could be employed with advantage in combination with smart protective devices, which could replace the dangerous landmines we all hope will be banned.[62]

SEEING THROUGH FOG

Fog remains a natural obstacle which we have still not mastered. Before the limitations of atmospheric transmission were properly understood, many hoped that

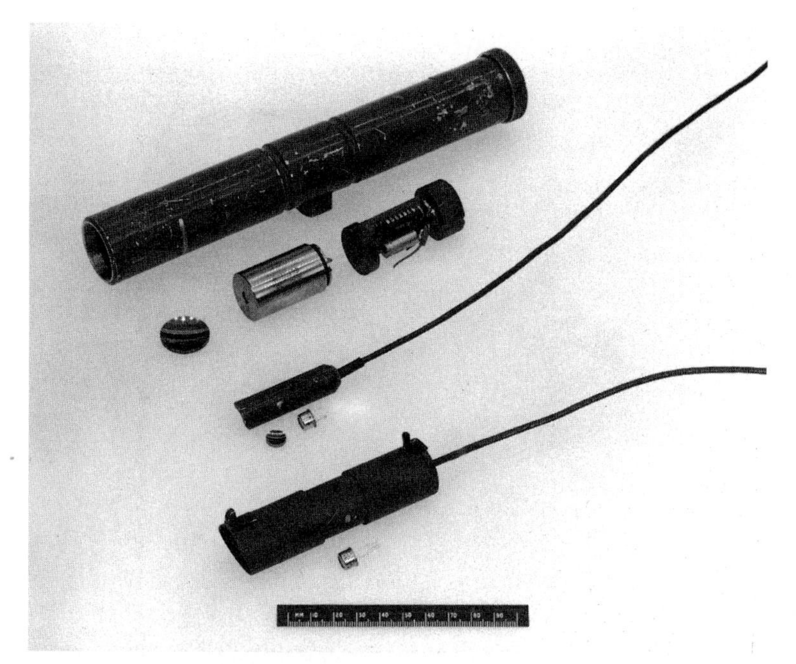

Figure 7.5: Infrared intruder alarms using pyroelectric detectors and germanium lenses to detect intruders crossing the line of sight. The prototype (top left) used a triglycine sulphate detector and was successful in police surveillance operations. The centre and lower models use pyroceramic detectors and are used with the Racal Classic battlefield surveillance kit. (© Crown copyright, DERA)

Figure 7.6: House fitted with a commercial pyroelectric intruder alarm. These are now widely used to control lighting at night or to activate burglar alarms. (© Richard Putley)

infrared would provide a means of overcoming this hazard. Because he apparently did not understand the difference between fog and smoke, Logie Baird, the inventor of television, believed his Noctovision (a television system working in the near infrared) would solve the problem.[63] As we now know, only long-wave radiation (wavelength greater than 100 μm) is transmitted without serious scattering, but these long wavelengths are strongly absorbed by atmospheric gases.

Experiments with active millimetre-wave systems produced crude pictures which could be difficult to interpret because the number of picture points was severely limited by geometrical factors.[64] Only recently, by means of the latest signal-processing techniques, has this problem been solved, enabling passive millimetre-wave thermal imagers to be designed.[65] At present the technique is only adequate for static scenes, but with the expected progress with large-scale integration this limitation should be overcome in the near future. We will then at last have a fog-vision system capable of general use by all forms of transport.

Normal smoke obscures the visible but does not significantly scatter the 10 μm radiation. Hence the employment of the infrared fire service camera pioneered by RSRE. With fog the situation is much more difficult. It is true that the infrared is not scattered as much as the visible, but this is only sufficient to give a useful improvement in light haze, not in dense fog: to penetrate dense fog one must use a much longer wavelength. The difficulty now arises that the clear atmospheric absorption is too high until the millimetre-wave region is reached. We then encounter further problems of technique. Recent progress in millimetre-wave development has gone a long way to solve the receiver problem. It is important to note that the spectral distribution curves are not symmetrical, the intensity falling off much more slowly on the long-wave side of the peak. Thus it has for some time been possible to construct sensitive microwave radiometers.

AN INFRARED EYE

The key to eliminating the optical scanner would obviously be the fabrication of a large-enough two-dimensional array (say more than 10,000 elements). In fact, an even bigger problem was what to do with the 10,000 connections emerging from this array. Although the task of making the detector array was not trivial, it was the design of the microelectronics, complete with interconnections, which proved to be the greatest hurdle.

Whereas the first attempts to use large detector arrays failed owing to the lack of the follow-on microelectronics, this barrier can now be overcome. It is necessary to look more closely at the detector array itself. The obvious advantage of using a large array is that it removes the need for a complex optical scanner. A less obvious advantage is that for most applications, as pointed out above, it will not require cryogenics.

There is now a greater choice of materials for these large arrays because each element will have the whole frame time to make its contribution to the picture. Not only could CMT of suitable composition be used, but also a pyroelectric, as in the pyroelectric vidicon. Three types of structure must be considered: first, an all-silicon

FIGURE 7.7: FIREMAN AT MALVERN FIRE STATION WITH A HELMET-MOUNTED THERMAL IMAGER FOR SEEING THROUGH SMOKE OR LOCATING CASUALTIES IN THE DARK (© CROWN COPYRIGHT, DERA)

one using some form of extrinsic detector; second, using an established detector material and constructing the LSI element from it; and third, making the sensing array from a convenient material and bonding this to the silicon circuit. The choice will depend on factors such as ease of manufacture and reliability in use.[66]

An example of things to come is the second-generation firecrew camera (Fig. 7.7). This uses a pyroelectric array with 10,000 elements and can be attached to a helmet to keep the operator's hands free.[67] An imager of this type has many other applications. In large-scale production it should become relatively cheap, making thermal imaging available for applications where up to now it has been thought too expensive. Experimental arrays with more than 100,000 elements are now in use.

COLLABORATION WITH INDUSTRY

Unlike most branches of military electronics there was no equivalent civil activity. Therefore, as thermal technology developed it was necessary to support in various ways the industrial development that would be required to move from the laboratory prototype phase to industrial production and Service use. An exception to this was television technology. Here the the Second World War near-infrared work rested heavily on early television development, and this applied to the early postwar low-light development and to pyroelectric imaging, where both the camera tubes and the cameras themselves drew heavily on television experience.

The development of an industrial base began as soon as ARL succeeded in repeating the German successes with PbS. BTH Rugby were the first British manufacturers of infrared photocells, producing versions of the ARL cells both for room-temperature and cooled (dry ice) operation. They also produced thermistor bolometers based on German designs, which at that time were the best available detectors for experiments in thermal imaging.[68]

As research proceeded, other firms were brought in. Mullard, first at Redhill and later at Southampton, were given contracts to develop the TRE PbS, PbSe and PbTe group of detectors; Barr & Stroud took on the development of infrared optical materials, starting with arsenic trisulphide glass, and infrared optical instruments. One of their early jobs was to produce infrared double-prism monochromators based on a German design taken over by TRE. Later they were to build a range of experimental thermal imagers, including Shortie. In these thermal imagers they pioneered the use of germanium refractive optics in place of the mirror systems originally used.

During the late 1960s (when the establishments were in the Ministry of Technology) considerable effort was made to find new industrial applications for infrared. Senior staff from RRE, headed by the Director (then Dr E.V. Glazier), visited firms and their staff were invited to RRE, both on an individual basis and to conferences. Trials of the industrial use of our thermal imagers were conducted on promising applications. Particularly encouraging results were obtained at paper manufacturers.[69] With the abolition of MinTech this effort fizzled out. At the time the Malvern team members were very disappointed and puzzled that industry did not appear to take advantage of their efforts. The only company that seemed to profit from their help was Land Pyrometers, who were already manufacturing measuring instruments for heavy industry and who were able to apply our recent advances. The exchange was not one way: Malvern also benefited from their industrial experience. With the benefit of hindsight it is clear, however, that in general the step from research apparatus to the production line was too great. Not until the intermediate stages of development were completed would research conducted for specialised military applications be valuable in industry.

Some time later, when the pyroelectric vidicon was beginning to show promise, ICI approached Malvern for help to explore its use in the maintenance of large chemical plants. The results were extremely encouraging, enabling the downtime for essential maintenance to be considerably reduced.[70] Unfortunately this effort did not become widely known because ICI did not want to help their competitors.

The introduction of InSb brought an expansion of Mullard's activities and introduced Plessey, Caswell, to the scene. By this time infrared homing missiles were coming into production and their detector requirements (PbS PbSe and InSb) were met by Mullard. The cryogenic equipment they required was manufactured by Hymatic,[71] Redditch, who also produced Leidenfrost equipment. When interest in the 10 μm region revived, Mullard began making extrinsic Ge detectors and the helium cryostats they required. These were used mainly for experimental studies. No British 10 μm equipment went into service before the introduction of the CMT detector, but Mullard were given contracts for its development and manufacture, including the SPRITE.

When research on pyroelectrics began, Mullard, Redhill, and Plessey, Caswell, were both asked to develop pyroelectric materials. When promising results were obtained, Mullard, Southampton, were given a contract for detectors, but Plessey began developing them as a private venture. English Electric developed the pyroelectric vidicon with help from Imperial College on the triglycine sulphate target material. The most important development after the introduction of CMT was the common module programme. The Rank Laboratories became the major contractor, and have successfully manufactured these high-performance imaging units.

The pyroelectric progamme has seen not only the large-scale manufacture of cameras by EEV, but also the production of battlefield surveillance systems by Plessey, Towcester and Racal. This must be seen against the background of a commercial mass market for intruder alarms, which is the first example of a defence-independent industrial development in the thermal radiation area. In defence development the construction by GEC of large-scale imaging array systems, based on research at Malvern, has led to the production (in the USA) of the latest firecrew smoke vision imager, and may possibly lead to a similar non-military expansion of thermal imaging.

A notable feature from the beginning of the industrial exploitation of thermal radiation was the close collaboration between industry and the military research establishments. The establishments tended to concentrate on the more basic initial phases of development, and industry on the final production phases, but there was never a rigid demarcation line. Some notable advances were initiated by industry,[72] and it was not unknown for the establishments to step in to help solve manufacturing problems, or even to set up small production lines to meet urgent user requirements[73]—as had proved so effective in the wartime introduction of radar. The moral seems to be to avoid too rigid an organisation but to retain administrative flexibility.

INTERNATIONAL COOPERATION

There has been close Anglo-American cooperation from early in the Second World War. In 1940 the British mission to Washington, led by Henry Tizard, initiated the close scientific liaison between the governments that was to make an outstanding contribution to victory. In the case of radar this entailed the well-known transfer of the British-invented cavity magnetron to the USA. In the case of infrared collaboration has consisted primarily of the exchange of information. After the end of the war the USA set up an organisation for exchanges on military science with its principal allies. It was divided into subgroups dealing with specific subjects, that dealing with infrared being Sub Group J. This held regular conferences exchanging classified material and organised working parties to tackle specific questions such as the standardisation of procedures for characterising detector performance.[74] With the growth in importance of the infrared field the Americans found it necessary to set up organisations for the internal exchange of information, such as the Infrared Information Symposium (IRIS). Although these were primarily national organisations, they did from time to time invite international participation. This served a very useful purpose in exchanging current research results. Some of the most useful reviews of British progress are to be found in the IRIS Proceedings.[75]

Close collaboration was also maintained between the European countries, not limited to NATO or EEC members. Several European conferences were held at Malvern, at which not only were the various goverments represented but also the leading manufacturers. These meetings were also classified, enabling a much wider exchange than possible at a normal scientific conference.[76] These meetings served a valuable purpose while infrared technology was becoming established as a recognised discipline, but now that it has become an accepted field of applied science the proliferation of normal scientific conferences fulfils this requirement.

Relations with the Soviet Union were closer than might have been expected during the Cold War. This came about because first, study of the published literature revealed that, prior to 1950, Soviet semiconductor research was of a very high standard and its study proved valuable when we were setting up our early research programme. We came to appreciate the quality of the Soviet work well before the launch of the Sputnik (1957), which brought more general recognition of Soviet research.

The growth of research into controlled thermonuclear energy led to direct contact between UK and Soviet research teams via the Vienna agreement. In particular, in our case there was a direct exchange of information on the InSb submillimetre detector. This was our discovery, but the Soviets were the first to explain the nature of the effect. We had exchange visits between our laboratories and we exchanged samples of InSb.

When the Soviets built the first Tokamak they invited a team from UKAEA Culham to visit them to verify their results. The equipment taken to the USSR for this purpose included the InSb submillimetre detector designed and manufactured in Malvern for the UKAEA.

THE STATE OF THE ART

Today we have a mature industry exploiting thermal radiation, which has contributed to our successes in two recent wars and in numerous minor incidents and policing activities. In 1945, at the beginning of the solid-state era, we did not know whether we could develop the technology required, nor, if we could, whether target and atmospheric conditions would yield sufficiently worthwhile results to justify this development. Research at the DERA establishments has made a major contribution to answering these questions in the affirmative. Unlike many areas of military electronics, there was no civil development to fall back on. Instead, the reverse situation has obtained: the developments necessary for the military applications have made various civil applications economic which otherwise would not have been viable.

The developments in the technology of thermal radiation originally inspired by military requirements are already affecting items of daily use. Thermal imagers are now widely used by the emergency services, as well as in industrial and medical applications. Simple intruder burglar alarms, first developed as electronic sentries for Army use, now adorn many suburban housefronts.[77] The most widely used infrared system must be the simple communicators used with most TV sets and as intercouplings in some data-processing systems. These are the direct descendants of Lindemann's signalling system, although they do not come within our terms of reference because they use LED sources, not thermal radiation. If (or rather when) a real-time fog viewing system is finally developed this could create another mass market for thermal technology, possibly the largest so far.

One yardstick for assessing the relevance of the applied research projects undertaken by DERA laboratories is to note the Queen's Awards for Technology.[78] These may indicate a measure of the quality of the research, but they are also an indication of its relevance because only those projects that have achieved significant success in the marketplace are eligible for consideration. Since 1979 13 awards have been received. Four of them were for projects within the scope of this chapter, and another four were for electronic materials research, an area derived directly from the early emphasis on materials research arising from our basic studies in infrared devices. Thus over half of the Queen's Awards received by DERA establishments can be traced back to the original programme of infrared research. This indicates the far-reaching consequences of this project, and this is probably why it has been the source of so much civil spinoff.[79]

Today's industry is now a multibillion-pound operation. The steps bringing this about seem to be as follows:

1. Basic research in the fundamentals and applications of infrared. Despite being the first to demonstrate infrared missile homing and making major contributions to the first effective (but not real-time) thermal imagers, we did not succeed in exploiting these advances. So everyone uses Sidewinder and not Red Top, and the first operational thermal imaging equipment used by the RAF was American using a Hg-doped Ge detector, not InSb or CMT.
2. Early successes with real-time thermal imagers established their superiority over low-light systems, despite the usefulness of the latter for certain purposes. Unfortunately, the British were not yet able to compete with the American common module programme.
3. The Malvern team now made two strides forward. The invention of the SPRITE CMT detector gave our common module programme a lead over all earlier thermal imaging systems, and the introduction of pyroelectric devices enabled thermal imaging to be used in situations where the use of cryogenics was impracticable or uneconomic.

Before the introduction of the SPRITE, Malvern's common modules used a 4 × 12 (48-element) matrix CMT array. With a single-element SPRITE an experimental serial-scanned imager was demonstrated in 1980; however, an eight-element SPRITE was then introduced which was equivalent to the earlier 48-element matrix but which needed only about one-tenth of the signal-processing electronics and gave a significant reduction in spatial noise. Subsequently 16-element SPRITES have been developed which give a further improvement in the high-performance range of common modules (TICMsII). A two-waveband version to enable the two main infrared windows to be compared simultaneously has also been developed.

The initial success of the pyroelectric imagers was achieved using a type of vidicon camera tube which (we must admit) was a rather clumsy throwback to the vacuum tube era. Two-dimensional arrays using specially developed pyroelectric ceramics (sometimes called ferroelectric bolometers because they need biasing to maintain stable operation near their Curie point) have now been developed. The first to be manufactured has 100 × 100 elements and is used in the latest fire brigade imager. It is sufficiently compact and light that it can be mounted on a helmet, leaving the hands free, but its performance is superior to the pyro vidicon. Larger pyro arrays of 256 × 128 and 384 × 288 elements have been produced since 1994, so we can expect further improved uncooled thermal imagers in the near future.

As the soldiers in Desert Storm have reported, the common module thermal imagers turned the night on the battlefield into day, with consequences Clausewitz never imagined. The impact of the pyroelectrics was more subtle. They first came into prominence in the Falklands Campaign—by their absence. After the disasters to UK ships the Navy discovered that the hitherto-neglected pyro imagers were just what was needed for emergency rescue and damage control. Luckily, an RSRE/Home Office-approved fire service camera had just been designed by EEV, but they had not sold any. The Royal Navy took all the stock of about 50, and the US Navy placed substantial orders. Thereafter it became a must for all up-to-date emergency services throughout the world. It has even been featured on the popular BBC series 'Casualty' at least twice.

Two types of TICM are now in use worldwide. The Class I models are lightweight handheld equipments, whereas the Class II types are for vehicle mounting. The early Class Is were used with great success in the Falklands, but subsequent developments have reduced their weight from about 12 kg to around 3 kg. Since 1984 about 3000 have been manufactured, and about 1000 of these have been exported to all parts of the world. The more sensitive Class II TICMs have been fitted to the main battle tanks, the Rapier and other Army and Navy missile systems, and in the Tornado and other aircraft. Their impact on the Gulf conflict in 1991 was, in the words of the troops, to turn night into day. Over 3500 of these systems have been sold worldwide.

Now a new generation of thermal imagers is beginning to appear, all solid-state staring cameras. They will require focal plane arrays of up to a million elements. The most successful one so far uses a 100 × 100 pyroelectric array coupled to a corresponding integrated circuit via a solder bump array joining each detector element to its corresponding readout circuit, and is already in production. However, it is too early to say whether this will be the ultimate technology. It now seems possible, as in the visible, to combine both the detector and signal processing arrays in a monolithic device, and the development of both photoconductive and pyroelectric arrays is being studied.

The research starting in 1945 had two outstanding achievements. First, in the common modules the British produced the most effective military thermal imagers so far devised. Second, the family of pyroelectric devices stemming from the introduction of this technique has produced the nearest to a mass consumer market of any infrared application so far envisaged. Looking to the future, although it seems likely more smart devices will emerge,[80] the main factor setting the pace of future developments will be progress in large-scale integration. This will determine how widely we will be able to avail ourselves of our ability to see in the dark, through smoke, and, ultimately, through fog.

1. Lindemann Advisory Committee for Aeronautics, *A Suggestion for the Detection of Aircraft by Long-Wave Radiation*, Report T510 (April 1915). This suggestion followed from the discovery that all bodies emit electromagntic radiation (thermal radiation) characteristic of their absolute temperature. For a body at room temperature this radiation peaks at a wavelength of about 10 μm. As the temperature increases, the intensity of the radiation increass and the peak moves to shorter wavelengths. Thus the radiation from the sun (surface temperature about 1000K), giving us our visible light, peaks at about 0.5 μm wavelength.
2. *Records of Admiralty Wartime Research and Development*, Part IR, ADMIRALTY (1950); T.H. Pratt, 'An infra-red image-converter tube', *Journal of Scientific Instruments*, 24 (1947): 312–4; T.H. Pratt, 'The infra-red image converter tube', *Electronic Engineering*, 20 (1948): 274–8.
3. The German progress in IR, especially the invention of the PbS detector, which led to renewed British effort on thermal radiation, is detailed in Air Scientific Intelligence Report No. 251 23/1/45. This report includes a spectral response curve for a German PbS cell and a description and photograph of the WPG 15 (see notes 6 and 13 below). T.F. Johns, *German Photoconducting Cells for the Detection of Infra-Red Radiation*, BIOS Final Report, No. 2, Item No. 9 (Stationary Office, n.d.); see also T.F. Johns, 'German photo-cells for the infra red', *Journal of the Television Society*, 4 (1946): 280–1; A. Elliott, 'Recent advances in photo-cells for the infra-red', in *Electronics and Their Application in Industry and Research*, edited by B. Lovell (Pilot Press, 1947): 97–133; P. Nolan and E.R. Rickin, *German Infrared Equipment in the Kiel Area CIOS*, Item No. 1, File No. XXX-3 (n.d. c1946).
4. A.M.J. Mitchell, *Mechanical Scanning Systems for the use with Infra Red Detectors*, SRDE Tech. Memo. No. Res. 189 (August 1956).
5. T.F. Johns, *German Photoconducting Cells* (n. 3 above); T.F. Johns, 'German photo-cells for the infra red' (n. 3 above); A. Elliott, 'Recent advances in photo-cells for the infra-red' (n. 3 above); P. Nolan and E.R. Rickin, *German Infrared Equipment* (n. 3 above).
6. S.E. Toulmin, *The Detection by Radio Methods of Temperature Radiation from U-Boat Exhaust Gases*, TRE Report L/M42/SET/MGNH (1 February 1945); E. Lee and H.A. Gebbie, *Detection of Schnorkel Exhausts by Non-Selective Thermal Receivers* ARL Reports ARL/R2/E570 and ARL/R1/E570; R. Stirling, A.S. Baxter and H.A. Gebbie, *Detection of Schnorkel Exhausts by Lead Sulphide Cells*, ARL Report ARL/R3/E570 (31 August 1945); H.A. Gebbie, R. Stirling and J. Beattie, *Detection of Schnorkel Exhausts by Thermal Radiation*, ARL Report ARL/R4/E570 (11 October 1945); R.A. Smith, 'Radar detection for schnorkel', *TRE Journal*, 8 (1946): 92–111. According to ARL (see reference 11 in ARL/R5/E320, 1945) the Germans had claimed that their IR coast watching equipment WPG 15 (see Reference 11), could detect a schnorkelling U-boat at 1 to 2 km range. This equipment had a 150cm collecting mirror and a 15 or 30 mm PbS cell. The Germans claimed this was the most sensitive of all their coastwatching equipments, although the Donaugerät's performance was comparable at shorter ranges. The Donaugerät's optical system was used in the first ARL Thermal Imager (see References 5 and 10 in ARL/R5/E320, 1945).
7. Roderick Chisolm, *Cover of Darkness* (Chatto and Windus, 1953).
8. B. Milnes, *Air-Air Infra Red Detection Trials (Blue Lagoon)*, TRE Memo. 280 (June 1950); B. Milnes, *The RRE Mk 3b Homing Eye*, TRE Tech. Note 203 (October 1953); L. Manns, *A Balanced Concentric Chopper for Blue*

Sapphire, TRE Tech. Note 60 (April 1950); J.C. Smith, *The Servo Response of the RRE Mk 3b Homing Eye*, TRE Memo. 812 (Oct. 1953); S. Jones, *The Two-Colour Optical Chopper*, TRE Memo. 676 (December 1952); P.R. Franks, 'Some aspects of the Firestreak Guidance System', *Proceedings of the IRIS*, 4 (1959): 49–55; G.H. Hough, 'Proximity-fuze for Firestreak', *Proceedings of the IRIS*, 4 (1959): 75–85.

9. Leybolds Nachfolger, Hans Brucker, (4 December 1944) *A Preliminary Short Description of the Thermal Picture Equipment 'Potsdam'*, ARL File EO45/3.
10. *Thermal Pictures*, extract from ASE R&D Prog. Rep. ASE/6M/2 (1947); W. Rogulski and R.P. Deans, *Thermal Picture Forming Equipment*, ARL Report ARL/R1/E360 (17 February 1950); W. Rogulski, 'An assessment of the military value of thermal pictures', ARL Report ARL/R2/E360 (16 July 1951).
11. R. Gill, *Field Trials of Yellow Duckling Equipment*, RRE Tech. Note 541 (December 1954); R. Gill, *Further Trials of Yellow Duckling*, RRE Tech. Note 606 (December 1956); R. Gill, *A Simple Thermal Mapping Scanner*, RRE Tech. Note 600 (March 1957); R. Gill, *Measurement of Thermal Radiation from Ground Targets*, RRE Tech. Note 627 (December 1957); R. Gill, *Measurement of the Infra-Red Radiation from Military Vehicles Using a Cooled PbTe Detector*, RRE Memo 1324 (April 1957); *Iris Conference on Thermal Mapping USA Visit June 1959*, Memo. 1681 (March 1960); V. Roberts, *A Proposed Apparatus for Experimental Observation on Thermal Radiation from Submarine Wakes*, TRE Memo. 529 (November 1951); V. Roberts, *Estimation of the Performance of Yellow Duckling Using a Lead Telluride Cell*, Memo. 691 (February 1953); R. Gill, *Thermal Mapping in the UK*, *Proceedings of the IRIS*, 4 (1959): 357–62; R. Gill, 'Optical Linescan', *Proceedings of the IRIS*, 4 (1959): 352–6; J.S.C. Wheeler, *Development of IR Linescan*, Paper read at Sub-group J Meeting (Malvern, 21 June 1965); J.W. Sabey, *Assessment of RRE IR Linescan Maritime Results*, Paper read at Sub-group J Meeting (Malvern, 21 June 1965).
12. W.B. Lewis, *The Ultimate Limits of Infra Red Technique*, TRE Report L/M50/WBL (2 January 1945); J.G. Daunt, *The Ultimate Sensitivity of Thermal Detectors*, Clarendon Reports to SRE, CL Misc. 52, 54, 55 (1945) and 61 (with R. Kompfner) (1946); W.B. Lewis, 'Fluctuations in streams of thermal radiation', *Proceedings of the Phyical Society*, 59 (1947): 34–40.
13. G. Starkiewicz, L. Sosnowski and O. Simpson, *Report on Lead Sulphide Cells*, ARL Report ARL/R8/E320 (1947); L. Sosnowski, J. Starkiewicz and O. Simpson, *Photoconductive and Photovoltaic Effects in Lead Sulphide*, ARL Report ARL/R9/E320 (1947); L. Sosnowski, J. Starkiewicz and O. Simpson, 'A theory of photoeffects in lead sulphide', *Nature*, 159 (1947): 818; J. Starkiewicz, L. Sosnowski and O. Simpson, 'Photoconductive and photovoltaic effects in high-resistance semi-conducting films', *Nature*, 158 (1946): 28.
14. G. G. Macfarlane, *The Scientific Approach to the Study of Photo-Conductive Infra-Red Detectors*, TRE Maths Group Memo. 61/GGM (1948); D.W. Sciama, *Some Notes on Photoconductivity with a Summary of the Literature*, TRE Memo. 43 (December 1948); T.S. Moss, *Photoconductivity in the Elements* (Butterworths 1952); P.M. Woodward, 'Quantum statistics as applied to semiconductors', *TRE Journal*, 23 (1950): 52–81; D.G. Bell, D.M. Hum, L. Pincherle, D.W. Sciama and P.M. Woodward, 'The electronic band structure of PbS', *Proceedings of the Royal Society of London*, A217 (1953): 71–91.
15. W.D. Lawson, 'A method of growing single crystals of lead telluride and lead selenide', *Journal of Applied Physics*, 22 (1951): 1444–7; W.D. Lawson, 'Some Methods of Growing Single Crystals', *TRE Journal*, 29 (January 1952): 60–89; W.D. Lawson and S. Nielsen, *Preparation of Single Crystals* (Butterworths, 1958); E.H. Putley, 'The Hall coefficient electrical conductivity and magnetoresistance effect of lead sulphide, selenide and telluride', *Proceedings of the Physical Society*, B68 (1955): 22–34; E.H. Putley, 'Thermoelectric and galvanomagnetic effects in lead selenide and telluride', *Proceedings of the Physical Society* B68 (1955): 35–42; E.H. Putley, *TRE Journal 'The Hall Effect'*, 29 (January 1952): 1–59; E.H. Putley, *The Hall Effect and Related Phenomena* (Butterworths, 1960).
16. T.S. Moss, 'Lead telluride and lead selenide infra-red detectors', *Research*, 6 (1953): 258–64. See also C.J. Milner and B.N. Watts, 'Lead sulphide photocell', *Research*, 5 (1952): 267–73; B.N. Watts, 'Increased sensitivity of infrared photoconductive receivers', *Proceedings of the Physical Society*, A62 (1949): 456; A.S. Young, 'Photoconductive detectors for infra-red spectroscopy', *Journal of Scientific Instruments*, 32 (1955): 142–3.
17. R.A. Smith, 'Electrical and optical properties of certain sulphides, selenides and tellurides', in *Semiconducting Materials*, edited by H.K. Henisch (Butterworths, 1951): 198–207; R.A. Smith, 'Infra-red photo-conductors', *Philosophical Magazine*, suppl. 2 (1953): 321–69; R.A. Smith, 'The electronic and optical properties of the lead sulphide group of semiconductors', *Physica*, 20 (1954): 910–29; E.H. Putley, 'Lead sulphide selenide and telluride', in *Materials Used in Semiconductor Devices*, edited by C.A. Hogarth (Interscience, 1965); R.A. Smith, F.E. Jones and R.P. Chasmar, *The Detection and Measurement of Infra-Red Radiation*, 1st edn (Oxford University Press, 1957), 2nd edn (Clarendon, 1968). A comparison of both editions reveals the rapid progress of infrared technology in the decade between the publications.
18. H. Welker, 'On new semiconducting compounds', *Zeitschrift für Naturforschung* 7a (1952): 744–9; D.G. Avery, D.W. Goodwin, W.D. Lawson and T.S. Moss, 'Optical and photo-electrical properties of indium antimonide', *Proceedings of the Physical Society*, B67 (1954): 761; C. Hilsum and I.M. Ross, 'An infrared photocell based on the photoelectromagnetic effect in indium antimonide', *Nature*, 179 (1957): 146; D.G. Avery, D.W. Goodwin

and A.E. Rennie, 'New infra-red detectors using indium antimonide', *Journal of Scientific Instruments*, 34 (1957): 394–5; D.G. Avery, D. W. Goodwin and A. E. Rennie, 'Photoconductive and P-N junction detectors using InSb', *Proceedings of the IRIS*, 3 (1958): 88–93; D.W. Goodwin, 'InSb detector development at the Royal Radar Establishment', *Proceedings of the IRIS*, 5 (1960): 125–33; K.F. Hulme, 'Indium antimonide', in *Materials Used in Semiconductor Devices*, edited by C.A. Hogarth (Interscience, 1965).

19. K. Lloyd Williams, C. Maxwell Cade and D.W. Goodwin, 'The electronic heat-camera in medical research', *J. Brit. IRE*, 25 (1963): 241–50;. C. Maxwell Cade and Boris V. Barlow, 'Thermography: imaging in the far infrared', *Contemporary Physics*, 11 (1970): 589–610.
20. A.F. Gibson and T.S. Moss, 'The photoconductivity of bismuth sulphide and bismuth telluride', *Proceedings of the Physical Society*, A63 (1950): 176–7; J.G.N. Braithwaite, *Photoconductivity in Certain Semiconductors*, TRE Memo. No. 207 (January 1950); L. Pincherle and J.M. Radcliffe, 'Semiconducting intermetallic compounds', *Advances in Physics*, 5 (1956): 271–322; W.D. Lawson, S. Nielsen, E.H. Putley and V. Roberts, 'The preparation, electrical and optical properties of Mg_2Sn', *Journal of Electronics Series 1*, 1 (1955): 203–11.
21. W.D. Lawson, S. Nielsen, E. H. Putley and A. S. Young, 'Preparation and properties of HgTe and mixed crystals of HgTe-CdTe', *Journal of Physics and Chemistry of Solids*, 9 (1959): 325–9.
22. Sir George MacFarlane, private communication.
23. E. Burstein, J.J. Oberly, J.W. Davisson and B.W. Henvis, 'The optical properties of donor and acceptor impurities in silicon', *Physical Review*, 82 (1951): 764.
24. D.A.H. Brown, *Impurity Photoconduction and Majority Carrier Lifetime in Copper Doped Germanium*, RRE Memo. No. 1261, 1956, *Copper Doped Germanium as an Infra Red Detector*, RRE Memo. 1300 (January 1957); D.A.H. Brown, 'Majority carrier lifetime in copper doped germanium at 20°K', *Journal of Electronics and Control*, 4 (1958): 341–9.
25. E.H. Putley, 'Far infra-red photoconductivity', *Physica Status Solidii*, 6 (1964): 571–614.
26. L.M. Roberts and S.J. Fray, *Niobium Nitride Superconducting Bolometers*, RRE Memo. 1127 (April 1955); E.H. Putley, 'Thermal detectors', in *Infrared Detectors*, edited by R.J. Keyes (Springer-Verlag, 1977).
27. D.H. Parkinson, *A Miniature Air Liquifier Suitable for Cooling Infra-Red Photo-Conductive Cells*, RRE Memo. 1209 (January 1956); D.A.H. Brown, *A Miniature Hydrogen Liquifier and Cryostat for Airborne Use*, RRE Memo. 1276 (January 1957).
28. D.N. Campbell, 'Cryogenic cooling of infrared detectors', *IEE Conference Publication*, 204 (1981): 48–54; A.K. de Jonge, 'Small split Stirling coolers for IR detectors', *IEE Conference Publication*, 204 (1981): 55–9.
29. E.H. Putley, *Miniature Refrigerators*, RRE Memo. 1283 (December 1956); H.J. Goldsmid and R.W. Douglas, 'The use of semi-conductors in thermoelectric refrigeration', *British Journal of Applied Physics*, 5 (1954): 386–90; R. Marlow, R.J. Buist and J.L. Nelson, 'System aspects of thermoelectric coolers for hand held thermal viewers', *IEE Conference Publication*, 204 (1981): 60–64.
30. C.C. Mason, *Detection of Aircraft by Long-Wave Radiation*, Advisory Committee for Aeronautics, Report T510a (June 1915).
31. W.J.H. Moll and H.C. Burger, 'The thermo-relay for recording small rotations', *Philosophical Magazine*, 50 (1925): 624–6.
32. R.V. Jones, 'Infrared detection in British Air Defence 1935–38', *Infrared Physics* 1 (1961): 153–62.
33. M.D. Liston, C.E. Quinn, W.E. Sargeant and G. G. Scott, 'A contact modulated amplifier to replace sensitive suspension galvanometers', *Review of Scientific Instruments*, 17 (1946): 194–8.
34. D.A.H. Brown, 'A tuned low-frequency amplifier for use with thermocouples', *Journal of Scientific Instruments*, 26 (1949): 194–7; D.A.H. Brown, *A Homodyne (Coherent Detector) Equipment for use with Photoconductive Cells at Frequencies 100–800 c/s*, TRE Memo. 384 (December 1950).
35. P.J. Baxandall, 'The design of a very-low-noise 800c/s selective amplifier and input transformers', *RRE Journal*, 37 (October 1955): 1–41.
36. J.H. Ludlow, *A Low Noise Preamplifier for Liquid-Helium-Cooled Infra-Red Detectors*, RRE Memo. No. 2220 (August 1965).
37. Plessey/Mullard, Manufacturers' Data Sheets.
38. Plessey, *Infrared Activities at the Allen Clark Research Centre*, NATO AC243, Panel IV, RSG2 Report (27 February 1975).
39. Not only military uses (see notes 55 and 56) but also civil. From the gas explosion wrecking a block of flats in London in January 1985 to the unexpected floods in Worcestershire Easter 1998 thermal imagers used by the fire brigade and the police have helped reduce casualties and their value is widely appreciated.
40. E.H. Putley, 'Electrical conduction in n-type InSb between 2 K and 300 K', *Proceedings of the Physical Society*, 73 (1959): 280–90; E.H. Putley, 'Impurity photoconductivity in n-type InSb', *Proceedings of the Physical Society*, 76 (1960): 802–5; E.H. Putley, *Development of Infrared Techniques for Plasma Diagnostics*, RRE Memo. No. 2215 (August 1965) and papers cited therein; M.F. Kimmitt, 'Far infrared measurements on thermonuclear plasmas', *RRE Journal*, 52 (April 1965): 113–40; M.F. Kimmitt, A.C. Prior and V. Roberts, 'Far-infrared

techniques', in *Plasma Diagnostic Techniques*, edited by R.H. Huddlestone and S. L. Leonard (Academic Press, 1965).

41. A. Elliott and G. G. MacNeice, 'Transmission of the atmosphere in the 1–5 micron region', *Nature*, 161 (1948): 516; H.A. Gebbie, W.R. Harding, C. Hilsum, A.W. Pryce and V. Roberts, *Atmospheric Transmission in the 1–14 Micron Region*, TRE/T2129 and ARL/R4/E600 (December 1949); H.A. Gebbie, W.R. Harding, C. Hilsum, A.W. Pryce and V. Roberts, 'Atmospheric transmission in the 1–14 micron region', *Proceedings of the Royal Society of London*, A206 (1951): 87–107.
42. D.A.H. Brown and V. Roberts, 'A simple high-speed spectrometer for the infra-red region', *Journal of Scientific Instruments*, 30 (1953): 5–8; D.A.H. Brown and V. Roberts, *A Simple High-Speed Spectrometer for the Infra-Red*, TRE Tech. Note No. 147 (January 1952); V. Roberts, *Preliminary Report on Airborne Spectroscopic Measurements Connected with Green Garland*, TRE Tech. Note No. 187 (April 1953); F.E. Jones and V. Roberts, 'Solar infra-red spectroscopy from high-altitude aircraft', *Proceedings of the Royal Society of London*, A236 (1956): 171–5; J.T. Houghton, T.S. Moss and J.P. Chamberlain, 'An airborne infra-red solar spectrometer', *Journal of Scientific Instruments*, 35 (1958): 329–33; T.S. Moss, 'Atmospheric transmission measurements using an airborne grating spectrometer', *Proceedings of the IRIS*, 4 (1959): 213–20.
43. C.B. Farmer and F.J. Kay, 'A study of the solar spectrum from 7 to 400 microns', *Applied Optics*, 4 (1965): 1051–68; P.J. Berry, C.B. Farmer and J. Swensson, 'Atmospheric transmission measurement in the 4.3 µm CO_2 band at 5200 m altitude', *Applied Optics*, 4 (1965): 1045–50; C.B. Farmer and S.J. Todd, 'Absolute solar spectra 3.5–5.5 microns 1 experimental spectra for altitude range 0–15 km', *Applied Optics*, 3 (1964): 453–8.
44. S. Nielsen and J.A. Savage, *RRE Newsletter and Research Review*, 3 (1964): 17 ff.; J.A. Savage and S. Nielsen, 'Chalcogenide glasses transmitting in the infrared between 1 and 20 µm—a state of the art review', *Infrared Physics*, 5 (1965): 195–204; J.A. Savage and S. Nielsen, 'Preparation of glasses transmitting in the infra-red between 8 & 15 microns', *Physics and Chemistry of Glasses*, 5 (1964): 82–6; J.A. Savage and S. Nielsen, 'The effect of oxygen on the infra-red transmission of Ge-As-Se glasses', *Physics and Chemistry of Glasses*, 6 (1965): 90–4; J.A. Savage and S. Nielsen, 'The infra-red transmission of telluride glasses', *Physics and Chemistry of Glasses*, 7 (1966): 56–67; M.O. Lidwell, 'Developments in high resolution thermal optics', *Optica Acta*, 22 (1974): 317–26. J.A. Savage, 'New infrared window and dome technologies', *Journal of Defence Science*, 4 (1999): 81–7.
45. G.W. Green and A. Lettington, unpublished note (1981); A.H. Lettington and G.J. Ball, *Multilayer Antireflection Coatings for Germanium Optics Using Diamond-Like Carbon Layers*, RSRE Memo. No. 3298 (May 1982); A.H. Lettington and G.J. Ball, *The Protection of Front Surfaced Aluminium Mirrors with Diamond-like Carbon Coatings for use in the Infrared*, RSRE Memo No. 3295 (January 1981).
46. P. Schagen, 'Electronic aids to night vision', *Philosophical Transactions of the Royal Society*, A269 (1971): 233–69.
47. C.T. Elliott, 'New detector for thermal imaging systems', *Electronics Letters*, 17 (1981): 312–3; C.T. Elliott, D. Day and D.J. Wilson, 'An integrating detector for serial scan thermal imaging', *Infrared Physics*, 22 (1982): 31–42; A.B. Dean, P.N.J. Dennis, C.T. Elliott, D. Hibbert and J.T.M. Wotherspoon, 'The serial addition of sprite infrared detectors', *Infrared Physics*, 28 (1988): 271–8.
48. W.D. Lawson, 'Thermal imaging', in *Electronic Imaging*, edited by T.P. McLean and P. Shagen (Academic Press, 1979): 325–64; W.D. Lawson and J.W. Sabey, 'Infrared techniques', in *Recent Techniques in Non-Destructive Testing*, edited R.S. Sharpe (Academic Press, 1970): 443–79.
49. D. Griffiths and J.W. Sabey, *A Proposed Thermal Imaging Equipment for Army Night Observation*, RRE Memo. 2575 (September 1970); Maj. K.W.E. Ferguson, W.D. Lawson, D. Griffiths and N.J. Smith, *Proposed Thermal Imaging Equipments for a Night Observation Device*, RRE Memo. 2682 (1971); K.F.N. Savage, L.G. Wykes, L.C. Charman and M.J. Massey Beresford, *Feasibility/Project Study for Night Observation Device to GSR 3365*, SRDE Rpt. 70052 (November 1970).
50. D. Pardy and P.J. Wright, *Location of a Military Target by Thermal Radiation*, SRDE Rpt. 1148 (November 1965); P.J. Wright, *Design for a Hand Held Thermal Pointer*, SRDE Memo. Q10/67 (September 1967) (re-issued as SRDE Rpt 69001, January 1969); T.J. Davies, *Application of Hand Held Thermal Pointer to Infra-Red Passive, and Daylight Sights*, SRDE Memo. Q4/69 (April 1969); D. Pardy, *Thermal Pointer for Chieftain—A Preliminary Study*, SRDE Memo. Q6/69 (August 1969).
51. IRIS (Infra-Red Intruder System manufactured Marconi-Elliot Systems Ltd) was the first Army system. This used a beam from an IR LED to detect intruders at up to 200m but was difficult to set up and use. The first thermal intruder detector was built by Mullard using a TGS thermal pointer detector (see *Applications of IR Detectors* [Mullard, 1971]). This aroused considerable interest at RRE, resulting in the design of a thermal replacement for IRIS and later the manufacture by Plessey of an improved version designed for use in conjunction with the seismic (Geophone) intruder alarms also being developed at RRE and manufactured by RACAL. An off-shoot of this was the introduction by Plessey of the first domestic thermal burgular alarm, employing their ceramic detectors, cheaper and more reliable than TGS.
52. D.B. Webb, 'Thermal imaging via cooled detectors', *Radio and Electronics Engineer*, 52 (1982): 17–30; R.D.

Hoyle, 'The Class I Thermal Imaging Common Module—an update', *IEE Conference Publication*, 263 (1986): 163–9; S.P. Braim and G.M. Cuthbertson, 'The dual waveband imaging radiometer', *IEE Conference Publication*, 263 (1986): 174–9; S.P. Braim, 'Technique for the analysis of data from an imaging infrared radiometer', *Infrared Physics*, 28 (1988): 255–61.

53. C.T. Elliott, 'New detector for thermal imaging systems' (n. 47 above); C.T. Elliott, D. Day and D.J. Wilson, 'An integrating detector for serial scan thermal imaging' (n. 47 above); A.B. Dean *et al.*, 'The serial addition of sprite infrared detectors' (n. 47 above).
54. Max Hastings and Simon Jenkins, *The Battle for the Falklands* (Pan, 1983): 345.
55. Gen. Sir Peter de la Billière, *Storm Command* (Harper Collins, 1992): 286–7; David Shukman, *The Sorcerer's Challenge* (Hodder & Stoughton, 1995): 135–43.
56. Rank Prizes were received for the discovery of CMT, the invention of the SPRITE and the development of pyroelectric vidicon.
57. E.H. Putley, 'The pyroelectric detector', *Semiconductors and Semimetals*, 5 (1970): 259–85; E.H. Putley, 'The pyroelectric detector—an update', *Semiconductors and Semimetals*, 12 (1977): 441–9; S.G. Porter, 'Pyroelectric detector Arrays', *IEE Conference Publication*, 263 (1986): 30–2.
58. Putley, R. Watton, W.M. Wreathall and S.D. Savage, 'Thermal imaging with pyroelectric television tubes', *Advances in Electronics and Electron Physics*, 33A (1972): 285–92; R.G.F. Taylor and H.A.H. Boot, 'Pyroelectric image tubes', *Contemporary Physics*, 14 (1973): 55–87; R. Watton, 'Infrared television: thermal imaging with the pyroelectric vidicon', *Physics in Technology*, 11 (1980): 62–6; D. Burgess, *Medical Thermography with a Pyroelectric Vidicon Camera*, RSRE Memo. No. 3365 (March 1981); D. Burgess, 'Pyroelectric infrared sensors', in *Recent Advances in Medical Thermology*, edited by E.F.J. Ring and B. Phillips (Plenum Press, 1982): 241–50.
59. C. Hilsum, 'A survey of passive imaging systems, with particular reference to absorption edge image tube', *SERL Journal*, 43 (1957): 2.2–2.8; C. Hilsum and G. Harding, 'The theory of thermal imaging, and its application to the absorption-edge image-tube', *Infrared Physics*, 1 (1961): 67–93.
60. H. Hughes, J. Owen and J. Sabey, *Infrared Smoke Penetration Trials at the Fire Service Technical College, Moreton*, RRE Memo. No. 2512 (1–2 August 1968); R. Watton, 'Pyroelectric materials: operation and performance in thermal imaging camera tubes and detector arrays', *Ferroelectrics*, 10 (1976): 91–8; R. Watton, D. Burgess and P. Nelson, 'The thermal behaviour of reticulated targets in the pyroelectric vidicon', *Infrared Physics*, 19 (1979): 683–8.
61. The RACAL CLASSIC system is an example, with acoustic and seismic sensors as well as thermal.
62. Gen. Sir Hugh Beach, Letter, *The Times* (9 September 1997).
63. T. McArthur and P. Waddell, *Vision Warrior* (The Orkney Press, 1990): 165–6.
64. L.A. Cram and S.C. Woolcock, 'Development of model radar systems between 30 and 900 GHz', *Radio and Electronics Engineering*, 49 (1979): 381–8.
65. R. Appleby, D. Gleed and A. Lettington, 'The military applications of passive millimetre-wave imaging', *Journal of Defence Science*, 2 (1997): 227–32.
66. DERA research has concentrated mainly on hybrid structures in which an array of detectors is bonded to a silicon circuit array. Experimental CMT arrays of 768 × 16 and 128 × 128 detectors have been fabricated while pyroelectric arrays of 100 × 100, 128 × 256 and 384 × 288 elements have been built. The 100 × 100 pyro array is in production and is used in the new Fire Service camera. In this array the pyro elements are bonded to their silicon circuits via an array of solder dots. This apparently crude approach has proved very successful, but a method of depositing the pyroelectric array directly on the silicon is already giving encouraging results. This technique will be almost essential for the much larger arrays under development. Similar techniques are being devised for the photoconductive arrays, but their development is not as far advanced as the pyro ceramics.
67. 'Uncoded detector research', *Imaging 96* (DERA, 1996).
68. C. J. Milner and B. N. Watts, 'Lead sulphide photocell', *Research*, 5 (1952): 267–73; B.N. Watts, 'Increased sentistivity of infrared photoconductive receivers', *Proceedings of the Physical Society*, A62 (1949): 456.
69. T.P. McLean (Chairman), *Report of RRE Working Party on Civil Applications of Infrared Techniques* (31 March 1967).
70. N.P.H. McDonnel, *Thermal Imaging with Pyroelectric Vidicons* (ICI Petrochemicals Division, 1975).
71. S.W. Stephens, 'Miniature infrared cooling systems', *IERE Conference Proceedings*, 22 (1971): 173–85; D. Marsden, 'A Stirling cycle cryogenic cooler for infra-red detectors and systems', *IEE Conference Proceedings*, 263 (1986): 128–32.
72. Innovation by industry. One of the best examples is the evolution of the Intruder Alarm. This began as a laboratory demonstrator at Mullard, using the TGS detector they had developed for the Thermal Pointer. This led directly to the RRE development of battlefield surveillance devices. Plessey, who were developing pyroelectric ceramics for RRE, realised that they could make very simple detectors with them

which immediately made the domestic intruder alarm a practical possibility. So much so that the business was soon taken over by the cheap mass producers, leaving Mullard and Plessey to retain the military and space markets.

73. Fairly large scale in-house sample manufacture was often necessary as part of the development process, and these samples were often donated to help other researchers and also to obtain user experience, but equipment was sometimes sold. One example, the InSb submm detector, complete in cryostat before Mullard had begun manufacture. A better example was the diffraction gratings used in the submm spectrometer. To attain the precision required in their manufacture a special ruling engine was designed and built by the RRE Engineering Department. Other spectroscopists wanted these gratings but it would not have been economic for a manufacturer to build another ruling engine and therefore RRE was able to sell them. With the development of new spectroscopic techniques this requirement ceased and the engine was eventually dismantled.
74. One example of Sub-group J's activity was that in 1965 it met in Malvern. The main item on the agenda was a series of papers on British military IR research, reviewing IR Linescan, detectors, cryogenics, optical materials, atmospheric transmission and defence against missiles.
75. The IRIS Proceedings cover mainly classified American developments, but they do include several useful accounts of British work including papers on specific topics, some of which have been cited above and general reviews, for example: D.G. Avery, L. Manns and R. Gill, 'Military IR activities in the UK', *Proceedings of the IRIS*, 3 (1958): 103–7; V. Roberts and R.A. Smith, 'IR research at RRE', *Proceedings of the IRIS*, 5 (1960): 135–41.
76. European Symposia on Military Infrared were held in Malvern in 1967, 1969, 1971 and 1974. Each was opened by a keynote review given by a prominent scientist. The contributions by Sir George MacFarlane (1969) and Dr F.E. Jones (1974) describe the early development of IR within the MOD. The meetings also included working displays of equipment.
77. Plastic frogs fitted with sensors warning them to croak when approached have been on sale in local garden centres (see Mrs Harriet Lear, Letter, *The Times*, 6 October 1997). A recent XTRAS novelty catalogue described a toilet night light activated by IR indicating whether the seat was up or down! A more serious application pioneered by the Japanese is the talking vending machine which welcomes approaching customers and tells them how to work it.
78. The Queen's Awards for Technological Achievement can only be granted for innovations which can demonstrate a commercial success. This would appear to rule out DERA from receiving them, but it is accepted that many successful products start life in the research laboratories before the manufacturer comes in. This makes DERA, in collaboration with the manufacturers, eligible for awards. Awards are particularly valuable to the DERA Laboratories because they demonstrate the economic value of research which at first sight to many seems completely useless. Queen's Awards were received for items coverd in this chapter: Pyroelectric Detectors (with Plessey), 1982, CMT Detectors (with Mullard), 1983, and the Pyroelectric Vidicon (with English Electric Valve Co.), 1987. Of the further ten Awards received by DERA, five were for projects derived from the materials research originating from the early IR research.
79. E.H. Putley, 'Infrared spin-off', *Physics in Technology*, 17 (1986): 32–7; D.H. Parkinson, 'Defence research and civil spin-off', *Physics in Technology*, 18 (1987): 244–9; B.R. Holeman, *The Impact of RSRE on the UK Economy*, RSRE Memo. No. 3867 (August 1985). See also Green Paper CM 3861 March 1998, *Defence Diversification: Getting the Most out of Defence Technology*: 2, 5.
80. T. Ashley, I.C. Carmichael, C.T. Elliot, N.T. Gordon, A.D. Johnson and G.J. Price, 'Negative luminescence techniques for military applications', *Journal of Defence Science*, 1 (1996): 30–36.

CHAPTER 8

Open Systems in a Closed World: Ground and Airborne Radar in the UK, 1945–90

JON AGAR AND JEFF HUGHES

INTRODUCTION

The development of radar since 1939 has had an enormous impact on warfare. From air traffic control and navigation systems to guided missiles and weather forecasting, the harnessing of electromagnetic radiation for military purposes has fundamentally shaped the defensive thinking and the offensive capability of the Armed Forces. Radar has also played a central part in shaping and holding together the complex web of military, industrial and academic institutions that characterises postwar technoscience. Through the diaspora of radar scientists and engineers who returned to and reoriented civil science after the war, those who remained in government employment and continued the development of electronic systems for military purposes, and those who helped create the new electronics industries servicing both civil and military sectors, radar technology and the skills that produced it became pervasive elements of the postwar sociotechnical ensemble, part of the reason why the academic–military–industrial complex existed, and simultaneously the 'glue' that held it together.[1]

Yet precisely because of this pervasiveness it is difficult to write a 'history of radar development'. Radar technology does not exist in and of itself: it is always part of some larger system, be it the Chain Home reporting system established during the Second World War and entrenched today as air traffic control and early warning systems, the guidance system for a ballistic weapon or a naval navigation system. At the same time, however, it is another characteristic feature of radar that particular assemblies or subsystems might well have multiple uses. Even during the war, for example, it became clear that with suitable modifications a radar set designed for airborne interception might well have application (and often better application) in a shipboard context. A particular piece of radar technology could thus be deployed and made useful in various ways which depended on complex negotiations between producers (scientists and engineers at the research establishments) and operational users. It is this *systemic yet flexible* aspect of radar that makes it difficult to abstract the development of radar *per se* from the development of the wider networks in which it is designed to play a part.[2]

In the postwar context, too, radar subsystems displayed this sense of flexibility in use—of having the capacity to be embedded in different systems which gave them different uses and different meanings. Now, however, with a radically different institutional framework and a very different set of relationships between the producers and the users of radar technology, the development of these component technologies was subject to different demands and an entirely different institutional philosophy of innovation. The end of the wartime 'crash' programmes allowed the radar and electronics establishments to devote considerable resources to 'basic' research aimed at developing and exploiting new materials and techniques, as well as to the extension of existing methods and approaches. The relative emphasis placed on fundamental and developmental research in the establishments varied considerably with changing political and military imperatives, of course, and in what follows we shall see numerous examples of the fluidity of research and development programmes as establishments responded to policy shifts and to unexpectedly sudden military requirements. In such a context the quantity of speculative fundamental research actually carried out far exceeded the number of successful applications that reached the stage of production and operational deployment. Indeed, much of the postwar work of the establishments could be characterised as incremental improvement of existing technology and its reconfiguration into new subsystems and systems (the wartime Rebecca beacon-finding system, for example, was still in use in its Mk 12 version in the 1970s).

Rather than a single, unified history of radar development, then, we are confronted with the task of accounting simultaneously for the development of a diverse and complex array of components, techniques, subsystems and the larger systems—guided missiles, aircraft navigation and weaponry systems, civil and military air traffic control and so on—in which they were destined to feature. We are also faced with a complex story in which basic research, the applied development of new techniques and the incremental improvement of existing techniques coexisted in a constantly changing political and military situation. In this chapter, we attempt to map significant features of the technical, institutional, political and strategic geography within which radar was developed in postwar Britain. Confining ourselves mostly to ground and airborne radar systems (naval systems are covered elsewhere in this volume), we attempt to show how the work of the research establishments responsible for radar development—principally the Telecommunications Research Establishment (TRE) and the Radar Research and Development Establishment (RRDE) at Malvern (which merged in 1953 to form the Radar Research Establishment (RRE)), the Signals Research and Development Establishment (SRDE) at Christchurch and the Joint Services Electronics Research Laboratory (SERL) at Baldock—took shape in response to wider political and military developments.[3]

In addition to the notion of open, adaptable systems and subsystems, we have found useful Paul Edwards' analysis of the 'closed world' constituted by the electronic technologies and ideological discourses of the Cold War. Going beyond recent 'revisionist' historiography, which continues to emphasise the role of nuclear weapons as constituting the superpower arms race, Edwards explores the widely distributed electronic technologies that made the nuclear world possible. He characterises the closed world as 'a dome of global technological oversight ... within which every event was interpreted as part of a titanic struggle between the superpowers'. Its key themes, he argues, were 'global surveillance and control through high-technology military power', and the cyborg-like integration of human and machine in 'weapons, systems and strategies whose human and machine components could function as a seamless web, even on the global scales and in the vastly compressed timescale of global nuclear war'. In this regime, computers—and, we argue, the radars and other electronic devices which were linked to and through them—'made the closed world work simultaneously as technology, as political system, and as ideological mirage'.[4]

Again, based primarily on the American experience of the Cold War, Edwards' analysis raises a number of interesting questions about the British case, particularly about the ways in which a nation like the UK situated itself technologically, as well as politically and ideologically, within the closed world largely constituted by the superpowers. Edwards emphasises the shift towards the centralisation of global information and control systems during the height of the Cold War, a shift with significant implications for the UK air defence system and, consequently, for the work of the research and development establishments. The decline of Empire, Britain's engagement with NATO in the 1950s and increasing European integration in the 1960s and 1970s underpinned a profound shift in the work and role of the radar establishments as imperial and postimperial Britain sought to chart a role for herself in the Cold War. Our main conclusions centre on the relations between research establishments and service users, the trend towards integration of systems, but also their surprising vulnerability—'open', then, in several senses of the word.

A WORLD OF THEIR OWN? THE RADAR ESTABLISHMENTS AND THEIR WORK, 1945–49

When the war ended in September 1945, the scientists, engineers and technicians populating the many government research establishments which had sprung up in Britain during the war faced a stark choice. For some, the obvious move was to return to the universities whence they had come in 1939, and resume their teaching and research, putting the war—if not always the military—behind them. For others, the natural choice was to return to industry, exploiting the close links that had been forged between the research establishments and industrial concerns under the constraints of war. Perhaps the least obvious (and most difficult) choice for those facing demobilisation in 1945 and 1946 was to remain in government service, hoping that the research establishments would continue to prosper in an uncertain postwar world.

At the Telecommunications Research Establishment (TRE) a very particular institutional style of innovation had developed during the war, in which very close communication between entrepreneurial innovating scientists and potential users was a key part of the process by which new technologies came into being and were put into active service. Typically, informal contacts between Service staff (whose visits to the research establishments were promoted by events such as A.P. Rowe's celebrated 'Sunday Soviets' at TRE) and scientists would lead to two-way discussion of operational problems and suggested solutions, and of new devices and possible applications.[5] Confluences of interest could be negotiated in this informal setting, and scientists 'selling' a new idea would usually identify closely with the needs of a particular user or group of users, who would become their 'product champion'. Authority to produce a 'mock up' of proposed new equipment was easily obtained, and only successful and very promising results would usually be drawn to the attention of higher levels of management. Only after many prototypes and trials, when an innovation reached the stage of engineering for production, would the management of a development project become formalised. Crucially, TRE would continue to be involved with applications as they were introduced into operational service, an important factor in the success of TRE in developing new systems to help counter the Nazi threat.[6]

A good example of the TRE style can be found in the development of the blind-bombing system 'Oboe'. Frank Jones, who was appointed leader of the 'Oboe' development team by TRE Chief Superintendent A.P. Rowe in 1941, was a fervent advocate of electrical engineer A.H. Reeves' original idea, and typifies TRE practice:

> Frank was no back-room boy; he got out and about advancing Oboe, the cause in which he fervently believed. He was to be found at HQ 8 Group, at Squadrons talking to the pilots, at the Air Ministry briefing Sir Robert Renwick, who influenced Churchill to back Oboe, at Bomber Command and at 60 Group, who built and operated the oboe stations.[7]

Similarly, Bernard Lovell's promotion of the H_2S navigation system put a new project on the books and created new opportunities for technical development, operational use and professional advancement within the TRE hierarchy.[8]

Responsive and efficient, the informality of innovation at TRE worked admirably well during the war. The close liaison between providers and users essential to the success of TRE all but ceased with the end of the war, however, when the researchers in the establishments were not required to react quickly to unfolding events but to work systematically within a long-term research programme. This change of approach fostered a very different philosophy of innovation, which must have been quite alien to those who had come to scientific maturity during TRE's war years. The change was made all the more difficult by the fact that the long-term research programme was itself the product of a complex series of negotiations as establishment management and senior staff sought to create a new role for themselves in an uncertain world. Project development now became vulnerable to the vicissitudes of financial, as well as operational planning, with the result that the views of a much wider range of governmental agencies and other establishments had to be taken into account in the planning and design of new technologies.

In a memorandum written in May 1945, W.B. Lewis, deputy to A.P. Rowe, called for the establishment to devote more attention to basic research, including peacetime applications and electronics for industry and the atomic energy programme. Within weeks, Stafford Cripps, Minister of Aircraft Production, visited TRE to thank the scientists for their wartime efforts and to announce that the government would continue to support further work on radar, including an expansion of research into shorter wavelengths and civil applications of radar technology. It seemed as if Lewis would get his way. In October, Lewis succeeded Rowe as Chief Superintendent. By this time, however, the wartime coalition government had been replaced by a new Labour administration, and with this change came a new climate of uncertainty for TRE.

In a new memo of October 1945, Lewis argued again for continued support for TRE and for diversification of the establishment's activities. In the changed political climate, however, his pleas met with a cooler reception. J.C. Wilmot, Joint Minister of Supply/Aircraft Production (the two ministries were amalgamated in April 1946), wrote that TRE 'ought to have clear practical objectives' since 'a Government department doing applied science without any real objective tends to drift'.[9] In the first instance, however, it was the staff of TRE who drifted away from Malvern to jobs elsewhere: in March 1946 Lewis told Henry Tizard that the scientific staff at TRE had been cut from its wartime peak of 880 to 660, and that he had been ordered to make a further reduction of 230–250.[10] Lewis' problems with what A.P. Rowe called the 'decay of research' at TRE were compounded by the need to move TRE out of Malvern School to a nearby ex-Admiralty site, HMS *Duke*, and by the fact that his staff were divided between the two sites for the period of his superintendency (this state of affairs only ended in 1948, when the move was completed). Organisationally, the link with RRDE (as the Air Defence Research and Development Establishment, ADRDE, had been renamed in 1944) was also a delicate problem to be negotiated. Nevertheless, Lewis was able to push through a reorganisation of the internal structure of TRE, involving the creation of three departments: Radar, Physics and Engineering. His successor, W.J. Richards, formerly Head of the Physics and Instrumentation Department at RAE Farnborough, thus inherited an establishment firmly geared towards research as well as the development of existing technologies.

Among the established projects on the books at TRE in 1945 were air-to-air and air-to-ground guided weapons (AAGW, AGGW), high-power search-and-track air interception (AI) equipment, collision and cloud warning radar, meteorology and wind-speed measurement, telemetry, and infrared and ultrasonic techniques. The philosophy of the establishment moved away from the 'crash' programmes that had characterised its wartime activities (the 'second and third best "tomorrow", on which we have lived for years') and towards long-term work, particularly on miniaturisation, tropicalisation and 'better designs than crash programmes have allowed us to make'.[11] Much of the development work was devoted to the improvement of H_2S, with a numerical bomb computer (NBC) under design for use with H_2S Mk IVA, and a Mk IX version planned for installation in the E3/45 jet bomber. The development of robust instrumentation for use in the tropics was important too, as attested by the formation of TRE (Asia) in 1946. And it was towards ease of maintenance in far-flung areas of the empire that TRE officially adopted the principle of 'sub-unit, rather than component replacement' in 1945.[12] Clearly, TRE staff still had strong ideas about adapting their work to developing operational requirements.

That the establishment had learned from the wartime experience of design organisation was clear: the E3/45 was flagged as 'the first aircraft to be designed as a complete fighting weapon, the radar being an integral part of the weapon instead of an accessory added after the aircraft design is complete', so that 'consultation between TRE and the aircraft designers is ... taking place at the initial stages of the design'.[13] The Orange Putter tail warning system was also under development with an eye to the E3/45. With projects such as air interception radar Mk IX, and the issue of specifications by all three Services for an airborne interception control radar with a range of 200–250 miles and comprehensive beacon, identification and communication facilities, the importance of airborne systems in the immediate postwar period is clearly indicated.[14] Ground radar, on the other hand, had relatively low priority at immediate postwar TRE, with just a handful of staff completing work on Type 70 radars (particularly for use in the Far Eastern theatre) and thinking about the development of civil aviation radar (for which the Blind Landing Experimental Unit was established in the summer of 1945). Together with some work on telemetry and other electronic systems for the Long Shot and Journey's End missiles (which formed the basis for the Blue Boar guided bomb, eventually cancelled in the mid-1950s) and the commencement of work on a new continuous-wave homing radar for use in an antiaircraft projectile, the TRE programme by the summer of 1946 was beginning to diversify significantly. Indeed, with the TRE staff working on radio requirements for aviation reduced to 27 per cent of its peak wartime strength, and an expansion of civil aviation, there was likely to be a 'marked change in the capacity now available for the satisfaction of purely military requirements'.[15]

At the nearby RRDE, similar conditions prevailed. Here, as at TRE, the main projects on the books in late 1945 were continuations of the wartime development programmes, chiefly in coastal defence (both in the detection of shipping and in control of naval defence artillery) and in air defence, especially gun-laying radar for the control of AA guns, searchlight radar and searchlights, and early-warning and 'putting on' (tactical control) radars. There was also some work on a surface-to-air guided weapon (SAGW) codenamed LOPGAP (liquid oxygen, petrol, guided air projectile), but this was transferred to RAE in 1947. By 1946, the coastal defence work and searchlight work had been abandoned, and the researchers transferred to work on gun-laying radar, while the early warning radar work was upgraded. Yet, as at TRE, there was a sense of 'drift' about the programme at RRDE, with little strong sense of direction coming either from users or from central authority.[16]

It was partially to overcome this sense of inertia that in November 1946 the Research Branch of Fighter Command (FC) issued a memorandum on future requirements for ground radar, based on a survey of probable developments in aircraft technology. From about 1943, with the decline of the threat from German bombers, Air Ministry and War Office Staffs had been thinking about the future air defence of the UK (ADUK). An August 1944 Chiefs of Staff paper on 'Air Defence of Great Britain during the ten years following the defeat of Germany' outlined defence priorities for the postwar environment. It was followed in late 1945 by the Cherry Report, which suggested a number of technical and organisational improvements to the radar reporting and control system.[17] As Gough points out, however, the Labour government elected in July 1945 'was disinclined to allot a high priority to defence other than to ensure a rapid and orderly return to a normal civilian life for most of those still in the Services'. The new administration's first Defence White Paper in February 1946 proposed that future development should concentrate on research (rather than the stockpiling of obsolescent and obsolete equipment), that maximum use should be made of existing equipment stocks, and that there should be only a limited introduction of the most modern equipment, while maintaining a reasonable war potential. In addition, finances and manpower were to be greatly reduced.[18]

At a meeting at Fighter Command in December 1945 to discuss future radar requirements, TRE personnel were briefed on Fighter Command's likely needs. Clearly, Service chiefs and users had also learned much from their wartime engagement with the boffins, for this briefing included a specification for FC's 'ideal' radar, together with an outline development programme in five phases spread over ten years.[19] These plans contrasted with the situation on the ground, where the Chain Home (CH), Chain Home Low (CHL) and Ground Controlled Interception (GCI) stations, which had constituted the backbone of ADUK during the war, were being rationalised and run down. A Chiefs of Staff plan drawn up in 1946 suggested that effort should be concentrated on a defended area between Portland Bill (on the south coast of England) and Flamborough Head (on the east), with the rest of the UK constituting a shadow area where skeletal arrangements for radar cover could be activated in the event of war. Throughout 1946 and 1947, in fact, the 'general level of activity within the [control and reporting] system ... was very low indeed'.[20]

The November 1946 Fighter Command memorandum on future requirements for ground radar specified a series of operational requirements for a radar system, and a three-stage development plan which envisaged completion of the proposed new radar by 1957. The memo certainly stimulated TRE into action. The establishment responded in January 1947 with an assessment of the feasibility of the proposed system, including an appraisal of its organisational requirements, as well as its technical dimensions.[21] TRE and RAE staff were involved in negotiations with the Provisional International Civil Aviation Organisation (PICAO); visits from members of this body were important in shaping the work of TRE on ground radar as 'a rationalized programme on the vital subject of Radio Aids to Navigation and Air Traffic Control'.[22] At the meeting of PICAO in Montreal in December 1946, TRE staff pressed for the adoption of the Gee navigation system in Europe, and considerable effort was put into such a scheme over the next few years, culminating in the establishment of an experimental air traffic control system at London Airport in 1948.

REMOBILISATION AND REARMAMENT: RADAR DEVELOPMENT 1948–53

After the Czechoslovakian coup and the Berlin blockade of 1948, the Soviet Union was clearly identified as the 'enemy', although the debate among Chiefs of Staff over strategy and the role of nuclear weapons, and the inter-Service politics concerning the wish of the Army and Navy to retain a war-fighting role, continued unabated into the mid-1950s.[23] The creation of a unified Ministry of Defence on 1 January 1947 institutionalised scientific advice on military matters in the form of the Defence Research Policy Committee (DRPC)—the highest-level site of negotiation between science and the military—initially under the Chairmanship of Sir Henry Tizard.[24] The 1947 Defence White Paper was still concerned with the gradual rundown of the Services, although the commitment to research was evident, for example, in the Air Ministry estimate of £4.4m for the development and acquisition of radio, radar and electrical equipment. This figure was unchanged in the 1948 White Paper, but in June 1948 the Air Ministry issued a draft Air Staff Requirement, OR2020, for a new ground radar for air defence. This was to be constructed as soon as possible to precise specifications, and would complement improvements to the existing air defence system agreed at this same time by the Air Defence of Great Britain Committee. The new requirement called for an increase in the range of the CH system to the maximum possible extent, such that a Meteor aircraft, the fastest then in service, could be detected at 150 miles at heights up to 60,000 feet, with continuous tracking into ground control and interception (GCI) coverage, and significant improvements to display and data coordination. Codenamed ROTOR (the UK part of the plan) and VAST (equipment for use overseas), this plan for the reconstruction and upgrading of the control and reporting system dominated ground radar development at the establishments and in the radar industry for the next several years.[25]

At the programme review held by the Director of Communications Development (DCD) in July 1948, civil air traffic control was deleted from the TRE programme because of 'the relatively higher priority of other work and shortage of scientific staff,' with the air traffic control (ATC) unit at London Airport planned to continue under the jurisdiction of the Ministry of Civil Aviation and TRE to maintain liaison only. Instead, effort was to focus on a new AI system (Mk 16) for the F4/48 and F40/46 fighters, on the further development of AI Mk 9 and H_2S, on guided-weapon work (mainly the Red Hawk beam-following missile) and on the new air defence radar. Contracts were placed with industry for the refurbishment of existing AMES Types 1, 7, 11, 13 and 14 radars for ROTOR/VAST, while TRE itself resumed work on an S-band transmitter.[26] The establishment also began the development of waveguide components with a 10 per cent bandwidth and improved forms of plan position indicator (PPI) display, as well as continuing work on 3-D radar and an experimental X-band (higher frequency) system. The range of projects under way at TRE in January 1949 is indicated in Table 8.1.

The establishment of NATO in 1949, with the accompanying dilution of British strategic autonomy, marked a national watershed. The explosion of the first Soviet atomic bomb in August 1949, and the advent of the Korean War the following year, however, were surprises for the strategists, and their combined effect was to transform defence research policy. First, the calculation that a major war was unlikely for ten years was reappraised. In August 1949 Fighter Command was given a new directive stating categorically that its operational role was the defence of the UK against an

TABLE 8.1: THE TRE RESEARCH AND DEVELOPMENT PROGRAMME, JANUARY 1949 (AN REFERENCES ARE AUTHORITY NUMBERS)[27]

1. Ground radar	
AN72	Aerial mounts
AN73	AMES Type 11 Mk 7
AN505	New ground radar for air defence of GB
2. Airborne radar	
AN88/AN601	H_2S Mk 9 & NBC Mk 2
AN570	Trainer for above
AN99/AN602	Orange Putter (tail warning device)
AN365	Doppler warning derive
AN78	AI Mk 9B
AN79	AI Mk 9D
AN499	AI Mk 9C
AN492	AI Mk 16 (formerly AI for F4/48 & N40/46)
AN483	GL radar for F4/48 & N40/46
AN83	Radar ranging for GGS
AN110	Identification for blind firing
3. Navigation	
AN248	Approach path signals using leader cables Vertical guidance for landing Distance measurement problems in approach and landing
AN354	British GCA
4. Position fixing and homing	
AN271	Responder beacon system in 200 mc/s band
AN290	Gee Mk 3
5. Self contained navigational aids	
AN91	Green Satin (Doppler navigation)
AN303	Blue Sapphire—star sighting system
6. Guided Weapons	
AN389	Red Hawk
AN405	GW research—homing
AN416–418	GW research—simulators
7. Miscellaneous test equipment	

attack.[28] The Chiefs of Staff demanded an acceleration of the re-equipment programme, and the order of priorities within the defence R&D programme was adjusted: guided weapons, in particular, became items of equal priority to nuclear weapons. A second consequence of the events of 1949–50 was therefore massive rearmament, and increased defence expenditure. The 1949 Defence White Paper emphasised that new equipment would be needed, including telecommunications and radar.[29] The outbreak of the Korean War on 25 June 1950 put the country on a defence standing once more, and a three-year defence programme costing £3400 million was announced on 4 August 1950, a sum that had increased to £4700 million by January 1951.

The programme had many ramifications for research and development in the establishments, including radar work. First, the British government approved the production order of the Valiants, the first of a V-bomber force which had been in danger of cancellation before the Korean War. The airborne radar programme was stepped up to include not just development of H_2S Mk 9 and NBC Mk 2 and devices for warning bombers of attack, but modernisation and improvement of the wartime Gee-H and Oboe systems for precision bombing (codename Blue Study).[30] Second, on the ground radar side, the completion of the ROTOR/VAST air defence improvements became of heightened importance. Facing the realisation that the original specifications 'would have occupied a very large research team for many years and an effort from industry which could not be countenanced', the operational requirements for OR2020 were reduced to use power and techniques already available.[31] The establishment formulated and began a new comprehensive experimental programme from January 1950. From July 1950, TRE also sharpened its development of complete equipment. In particular, innovations in S-band receiver components and circuits enabled a prototype system, Green Garlic, to be built with radically increased power and sensitivity. This high-performance system, renamed AMES Type 80 in January 1954, became central to a new proposed Stage 1A—and raised questions over the completion of Stage 1 (ROTOR and VAST). As well as replacements for the chain home-type radars, the Air Ministry also wanted an improved GCI or fighter control radar. Some consideration had been made of adopting the Admiralty's impressive 984 radar, but the Green Garlic prototype could easily be adapted to GCI work.[32] The vanishing distinction between reporting (chain home) and control (GCI) radars was a significant move towards 'comprehensive' unified systems. An intensification of the TRE research programme into turning gear, waveguides, PPIs, moving-target indication, data handling and the theory of radar (following Shannon's 1948 paper on the application of communication theory to radar rangefinding) quickly followed. And third, in the guided weapons field, the successful flight trials of the Red Hawk missile in the spring of 1950 consolidated work on beam following, while work began on control radar for the Blue Sky missile (which later entered service as Fireflash) and continued on X-band semiactive continuous-wave homing under aircraft flight conditions.

Guided weapons work also figured large elsewhere. At the Joint Services Electronics Research Laboratory (SERL) at Baldock, established after the war to act as a focus for basic electronics research, research concentrated on continuous-wave Q-band radar.[33] At RRDE, too, the drift ended as the Soviet atomic bomb and the Korean War led to the acceleration of GW (guided weapons) work. RRDE continued work on their two key defensive radar systems: the tactical control radar Orange Yeoman and the gun-laying fire control Yellow River. From 1950 this programme was expanded with a view to using the two systems not for antiaircraft guns but for guided weapons. Orange Yeoman passed information to Yellow River, which 'put on' the guided weapons Red Duster (Bloodhound) or Red Shoes (Thunderbird) to the target. In 1951, the decision was taken to produce a simplified operational version as quickly as possible, and industry received orders for production: Metropolitan-

Vickers for Orange Yeoman and British Thomson-Houston for Yellow River. In 1953, this use was confirmed when the RAF took over surface-to-air guided weapons (SAGW), and Anti-Aircraft Command was abolished.

These operational and organisational changes had a significant impact on the work of the radar establishments. The burst of renewed activity on airborne radar and the switch of emphasis to guided-weapons work had important consequences for the organisation of research teams and for the recruitment and allocation of staff by establishments. That the individual establishments managed these changes without apparent difficulty is testament to the versatility and responsiveness to the needs of users which characterised them in this period. In a broader political context, however, the new demands raised questions about the overall organisation of the establishments and the division of labour between them. In 1953, RRDE and TRE were amalgamated to form the Radar Research Establishment (RRE), renamed in 1957 the Royal Radar Establishment. Gough explains this reorganisation as stemming from concern at 'higher levels of management that ... there might be duplication and wasted effort', and argues that amalgamation was 'only a matter of time'.[34] This explanation, as far as it goes, is true; however, the organisational change also raised fundamental issues concerned with the relationships between establishments and 'users', and with the trend identified above towards comprehensive integrated radar systems.

CONSOLIDATION BUT NOT UNIFICATION: THE RADAR RESEARCH ESTABLISHMENT 1953–61

The Paris Panel, which examined the coordination of radio and radar research and development in the Ministry of Supply between 1949 and 1951, considered carefully the arguments for and against the amalgamation of TRE and RRDE. In favour were: savings of common services, savings of staff, increased 'technical vitality' through being able to balance research programmes and allocate 'key personalities' to work, savings on local facilities such as workshops, and better coordination through eliminating 'a sense of inferiority in the smaller and of superiority in the larger, and a tendency of jealousy between them'.[35] Arguments against were: that 'a combined establishment would be far too large and would suffer from the disadvantages of over-centralisation', the end of a 'healthy rivalry' and a 'desirable' loyalty to a single Service, and adverse effects on the morale of RRDE staff.[36] The Paris Panel ultimately recommended combination (along with tightening up the role of PDT(RD), and the transfer of RAE's Radio Department to SRDE).[37]

The Research and Development Board of the Ministry of Supply set up a Working Party on the functions and fields of work of TRE and RRDE in October 1952.[38] The Working Party agreed on the reasons why amalgamation must be considered: the growing commitment to guided-weapons defence and the associated transfer of responsibility for GW from Army to RAF meant that RRDE was in 'poor health'. Furthermore, as S.S.C. Mitchell (Controller of Guided Weapons and Electronics) summarised: 'it is clear that we must integrate ground radar for guided weapons (hitherto done by RRDE) with the RAF C&R system (hitherto done by TRE)'—

another aspect of the trend towards unified radar systems. However, sharp differences emerged on the relationship between establishment and user. Barton, for example, argued that RRDE had a 'bad name' because it was associated with the Army, whereas Pollard thought that 'the trouble was that the Universities encouraged men to prefer TRE because of its programme of basic research'.[39] The issue was fundamental: if the establishments were given parity—dividing R&D between establishments on technical criteria (e.g. ground radar to RRDE) then it was a decision 'to abolish the split of work by "user"'—and the close relationship between Service and establishment was and is an often-cited factor behind the success of TRE. This was the reason behind the decision to 'amalgamate' RRDE and TRE (essentially keep the division of sites, but place one person in charge) rather than combine them completely.[40]

The formation of RRE naturally had important consequences for the organisation of research, both at the laboratory and workshop bench and at management level (Fig. 8.1). A reorganisation of the technical departments created six divisions within RRE: airborne radar, ground radar (including ATC requirements), guided weapons, basic techniques, physics and engineering. The integration of the RRDE programme—essentially based on tactical control of AA artillery and GW, fire control of AA artillery and GW and radar ballistic measurements—into the ground radar department created problems of its own, not least because of the continued split-site arrangement (which lasted until the late 1950s). And once again, the establishment was also having to respond to wider political imperatives. Triggered by proposed cuts in defence expenditure during the mid-1950s, a debate over the place of 'basic research' was particularly significant for RRE, which regarded itself as being strong in fundamental work. This was an attraction for applicants from universities, but one which cut both ways at the management level, for the establishment undertook research (for example the use of semiconductors at high frequencies) 'in which there is no prospect of interesting industry'.[41] In September 1953, Owen Wansbrough-Jones, Chief Scientist of the Ministry of Supply, concerned that in 'the squeeze on development projects, research does not go by default,' organised informal meetings from January 1954 to survey and protect basic defence research.[42] The first meeting agreed to formulate criteria under which basic research should be continued,[43] after which (as one member put it) 'it would be necessary to educate the Services so that they came to concur in the way in which the money allocated to basic research was being spent'.

As well as producing criteria and defending the 'freedom' of research establishments to experiment, the group also organised surveys of Admiralty and Ministry of Supply research programmes, as part of the wider campaign to protect defence R&D against cuts. The Ministry of Supply's survey of its research programme was, interestingly, the first it had made 'since [it] existed as a whole'. The review showed RRE to have a larger and more wide-ranging electronics programme than either RAE or SRDE. It also showed the extent of links between RRE, industry and academia in 1953. Industrial links were very strong, not only through the awarding of development and production contracts, but also through extramural research contracts, for example in radar alone: Plessey and Decca for aerials and waveguides, Plessey,

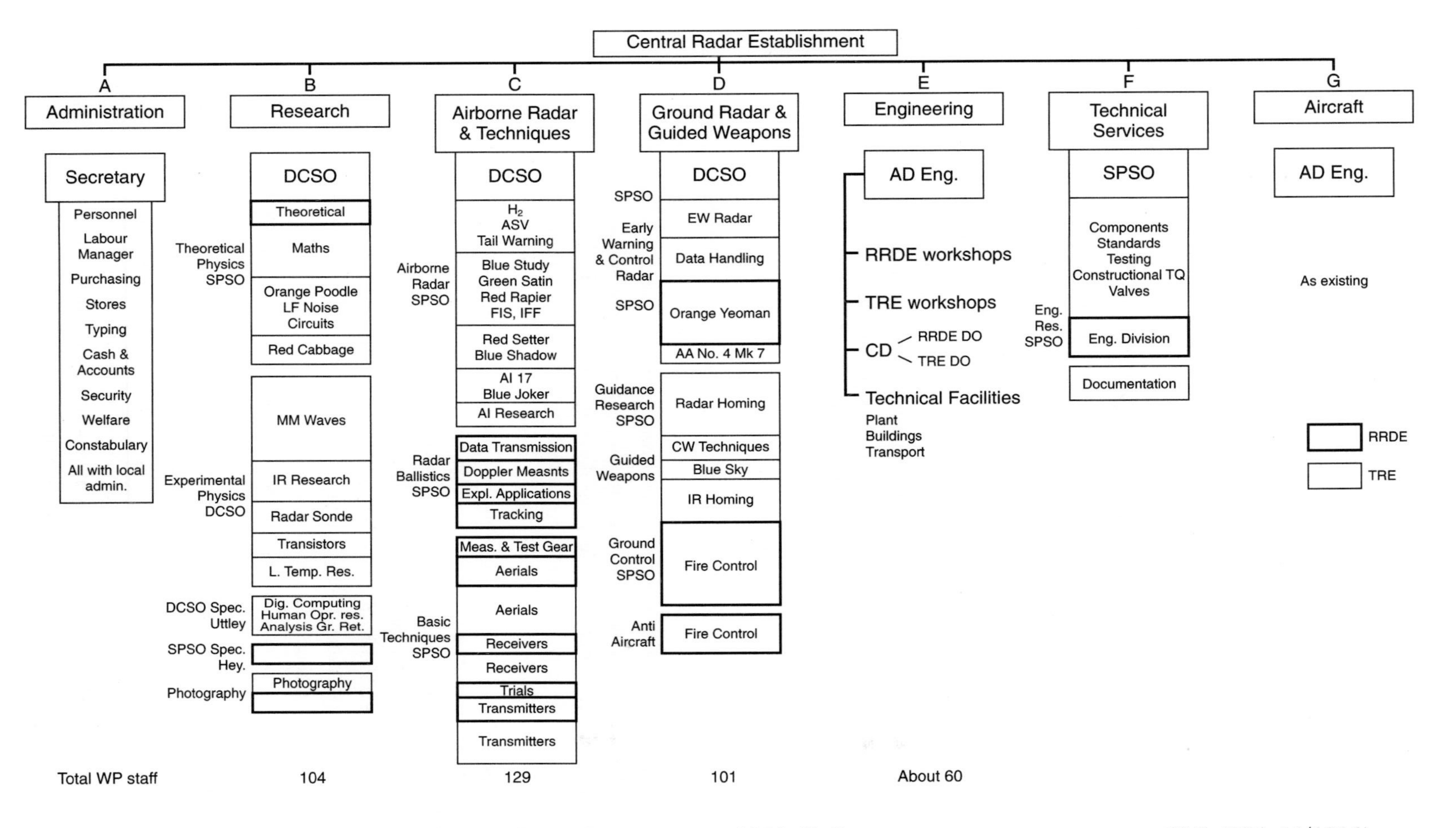

FIGURE 8.1: PROPOSED ORGANISATION OF CENTRAL RADAR ESTABLISHMENT, 1954 (© CROWN COPYRIGHT, REDRAWN AFTER PRO AVIA 54/1716)

Hilger & Watts, Clarke Chapman and Curran for millimetre-wave radar, and Mullard for precision bombing and radar reconnaisance. Academic links were weaker. In the words of Wansbrough-Jones: 'we look most to the Universities for fundamental classified work, which, to be acceptable to them, usually has to be closely linked with work in which they are keenly interested'. Only three examples were listed: Manchester University (twice) for transistors and radio astronomy, and Glasgow University for electronic data storage systems. The arguments over basic science and the first priority status of nuclear weapons were settled by a directive from Harold Macmillan, as Minister of Defence: there was 'to be no reduction in the total effort devoted to basic research without reference to the DRPC'.[44]

Several interlinking factors contributed to a shift in defence policy by 1957: the implications of the hydrogen bomb, the development of ballistic missiles, the unsupportable expense of the Korean rearmament, the fallout of Suez and the dependence on the deterrent. Defence policy changed with the development of thermonuclear bombs after 1952: the capacity to devastate wide areas meant that the defence of point targets, either by fighters or by guided weapons, became untenable.[45] The effect on the British radar programme was fairly slow but profound, and accentuated the previous trend towards integrated systems. Fighter Command, for example, completed its post-H-bomb comprehensive review in April 1956, concluding that major changes to ROTOR were needed. The scheme agreed by the Air Council in June 1956, called the '1958 Plan', was an integrated C&R system based on 'comprehensive master stations' built underground.

However, this was not the only change to ground radar, as a second major change in the threat became apparent, this time from the implications of the carcinotron. The carcinotron, a new form of microwave oscillator developed at Compagnie Générale de Télégraphie Sans Fil in Paris, could radiate strongly over an easily tuneable range of frequencies (unlike the magnetron). Its application in an almost ideal jamming system was only noticed in 1953 in Britain, by the Admiralty's ASRE. Two routes were followed to counter carcinotron-type jamming. First, in a brute-force approach called Blue Riband, RRE drew up designs for a very high-power (klystron) transmitter to be mounted on large dishes. Extramural contracts went to the Jodrell Bank radio astronomy station at Manchester University to research supporting this approach. However, the full version of this expensive route was abandoned in favour of a passive tracking concept from RAE. This system, a correlation radar called Winkle, floodlit air space and used timed reflections to three receivers to calculate position in three dimensions.

Both the integrated C&R system and Winkle formed part of the post-Sandys radar programme. Minister of Defence Duncan Sandys' 1957 White Paper, *Defence: Outline of a Future Policy*, marks a turning point in British postwar defence policy. Sandys abandoned the intensive and expensive post-Korea rearmament programme and, with the lessons of Suez fresh in everyone's minds, declared that future policy would rest on collective defence, and that defence spending must be cut. Sandys proposed a technological fix: more high technology would allow cuts in Service manpower (and thereby, not least, the ending of conscription). The nuclear deterrent

FIGURE 8.2: BLUE YEOMAN, MULTI-BEAM SURVEILLANCE RADAR WITH SECONDARY SURVEILLANCE RADAR ON TOP, USED IN AN EXPERIMENTAL ROLE (© CROWN COPYRIGHT, DERA)

was emphasised, and the V-bombers armed with Blue Steel would, over time, be replaced with the intermediate-range ballistic missile Blue Streak. Further development of supersonic fighters after the P1 (Lightning) was abandoned, with SAGW to be developed instead. Proposals for the new radar defence system mirrored these changes, and a modified plan was approved in December 1958.[46] The components of the modified '1958 Plan' were: Winkle (built by Marconi and RRE), Type 80 radars supporting a new tactical radar called Blue Yeoman (Fig. 8.2, essentially the Blue Riband transmitter on the more modest Orange Yeoman aerial), and automated integrated data handling linking the main control centre to radar stations and fighters.[47]

The carcinotron also threatened airborne radar, such as H_2S, ASV and, in particular, AI.[48] To understand this, we must first follow the postwar development of AI. In the early 1950s, as we have seen, three British AI sets were either in use or under consideration: AI Mk 10 (designed in 1941), AI Mk 17 (designed in 1944 but still not ready) and AI Mk 18 (under development).[49] In this context, with extensive use of American systems, representatives from all three Services 'expressed themselves dissatisfied' with the position of AI R&D in February 1951, and a working party under Cockcroft was reconstituted.[50] To Frederick Brundrett, chairing the DRPC in Sir Henry Tizard's absence, the situation was 'deplorable', and a stopgap measure to manufacture extra AI Mk 10s was imperative. As DRPC minutes put it:[51]

> We were proposing to go into production in 1953 with a set originally designed in 1941 and even if it had been further developed in the interim, it did not meet staff requirements. Further in order to do this, we must necessarily set back by an appreciable time the date by which we could go into production with the future set (AI Mk 18) which would meet requirements. Surely, if there was a case for adopting American interim equipment in order to enable us to reach our ultimate goal more quickly, this was it.

But the Services (and presumably the research establishments) were extremely reluctant to adopt American AI sets. Brundrett noted with exasperation that American navigational aids seemed acceptable, and it was hard to understand 'why it was said to be impossible to accept an American design for an AI set when we appeared to be able to produce a British version of the navigational aid equipment'. In the event AI Mk 10 did not go into production again, but work went slowly on replacements—1959 was the earliest estimated date of production for AI Mk 18, despite GEC having a 'very large and highly qualified team working on the project'.[52] AI Mk 17 finally went into production in 1955. The serious problem with these delays relates to our argument about the systemic nature of radar: AI sets (like all radars) needed to be integrated into a whole weapons system—and there was a danger that, for example, fighters such as the Hunter could be made before the AI was ready. The carcinotron interrupted these programmes further; as S.S.C. Mitchell, Controller of Guided Weapons and Electronics, noted:[53]

> jamming at the power levels postulated ... would deny S and X-band AI sets any range information until the range was less than about 2000 yards. However those AI systems which use an observer would be able to measure the bearing of the jamming aircraft and to attempt a pursuit course interception although this would involve an increased interception time and would not in all cases lead to a successful interception. In general, the fact that carcinotron jamming denies to the AI so much range information on the jamming aircraft and practically prevents the AI from seeing the "quiet" aircraft represents a major threat to AI systems. Under modern conditions of fighter interception any such interference with the AI system is very serious.

From this assessment of the threat and possible responses, it is clear that vulnerability was a key issue throughout this period. This was to have significant implications both for strategy and for the work of the radar establishments.

The Air Council's approval of the modified 1958 Plan was based on a budget of £30m. However, mostly due to rises in computer costs, this estimate soon went awry, reaching £100m within a year. Discussions through the machinery of interdependence with the USA in 1959 on long-range ABM radars resulted in a number of policy decisions with implications for the R&D plan. US plans to site a large early-warning radar station in the UK opened the possibility of a joint station, and in May 1960 the Air Ministry agreed to a slimmer version of the 1958 Plan called Plan AHEAD, which only had one main control centre (at Bawburgh) and three radar tracking stations.[54] Information was fed to Bawburgh, which in turn provided information back to the radars (cybernetic correction data) as well as to fighter airfields and GW complexes. However, Plan AHEAD also needed Treasury approval, and its eventual acceptance depended on two conditions. First, Plan AHEAD had to be the only air defence system. Other projects, such as the long-standing Blue Joker balloon-borne early warning radar developed by RRE and Metropolitan-Vickers, were cut. Second, on the recommendation of the chair of the DRPC and chief scientific adviser to the government on defence matters, Solly Zuckerman, Prime Minister Harold Macmillan ruled in November 1960 that Plan AHEAD must be shared with civil aviation as a joint air defence/air traffic control system. This became the LINESMAN/MEDIATOR system, LINESMAN being Plan AHEAD and MEDIATOR the ATC element.[55]

The Sandys reforms, motivated by the search for economies that produced LINESMAN/MEDIATOR, also embodied a response to the ballistic missile (BM) threat. Sandys announced the siting in the UK of American THOR medium-range BMs, as a stop-gap before Blue Streak. Blue Streak was cancelled in April 1960, however, and after the American Skybolt was also abandoned in 1962 the alternative of submarine-carried Polaris missiles was adopted as the British BM offense. On the defensive side, the advent of the ballistic missile caused considerable alarm. In 1956 Robert Cockburn had been fairly upbeat, telling his DRPC colleagues:[56]

> The general position was that, if we assumed the availability of a small atomic warhead and the invariance of the ballistic trajectory, we could conceive of a defence against long range BMs on the basis of extensions of existing techniques of guidance and little more than existing radar techniques. Thus it was possible to say that there were no technical reasons why interception of a missile was impossible.

Only a year later, however, the attitude of defence scientists and advisers was far less optimistic: radar systems could not quickly discriminate between warhead and decoys, and Sputnik provided a concrete and chilling demonstration of Soviet capability in BMs.[57] Systems such as Plan AHEAD could only deal with the air-breathing missile threat, not BMs, although research into a BM early-warning system by Marconi and RRE had begun from 1954.[58] In November 1958 the UK was given a description of the American Ballistic Missile Early Warning System (BMEWS) and invited to 'cooperate though not to take any part in the development'.[59] However, it seems that this did not extend to research, as Brundrett, now Chair of DRPC, told his committee in March 1959: 'it had been agreed that we should have a joint programme of research against BMs with the US'. Under this agreement, the Americans had suggested that the British 'should concentrate on the electronic problems'. However, a proposal for research leading to 'an experimental discriminating radar' was first delayed because of 'congestion at RRE', and because it was argued that missile research should come first (without it 'it was clear we should not be able to appraise American developments'). A year later the proposal was cut because of a squeeze in money and further uncertainty over re-entry phenomena.[60] An anti-BM defence system was still viewed pessimistically: as Mitchell put it, 'it was far from clear that any defence was operationally feasible and nobody wanted to start serious work on a project unless there was a reasonable chance that it would be brought into service'. BMEWS consisted of three stations: in Alaska, Greenland

and Europe, the latter completed in 1963 at Fylingdales. The UK was 'preferred because partly from a security consideration and partly from being further from enemy bases, it would be less susceptible to ECM [electronic countermeasures]'.[61] The USA and UK undertook intense negotiations, first into the 'extreme difficulty' of the exact site (the further south, the less protection for the USA), secondly into the nature of the microwave radiation hazard from the high-power radars. RRE undertook Project LEGATE—the UK-end data handling—for BMEWS. However, LEGATE was only reported complete in 1966, almost a decade later, and four years after the Cuban Missile Crisis.[62]

One other effect of the Sputnik demonstration of Soviet BM capability was the approval of a large basic research apparatus at RRE: a radio interferometer, under J.S. Hey. Previous efforts to promote this expensive project had failed, but in November 1957 Wansbrough-Jones could claim that 'the arrival of Sputnik had emphasised in a clearer way the need for such an instrument'.[63] The decision had significant impact on university radio astronomy programmes, particularly Jodrell Bank, which was seeking a defence role to lessen the debt on its large steerable radio telescope. When John Carroll of the Admiralty questioned the interferometer project, asking 'whether the cost should be borne on the Defence Vote provided by DSIR for Jodrell Bank or Cambridge', Cockburn responded: 'this facility could have been regarded six months ago as wise provision, but since Sputnik had appeared it had much more direct defence interest. It was not possible to go to a University for information ...' As Wansbrough-Jones defined the problem, the Jodrell Bank radio telescope 'was not under Government control and neither [the radio telescope nor a temporary RRE dish] were designed for the job'. Thus the threat of Soviet BMs led simultaneously to a significant expansion of RRE research and, ironically, to the demise of Jodrell Bank's attempts to mobilise defence concerns to support its radio telescope.[64]

THE LIMITS OF INTEGRATION: DATA HANDLING, AIRBORNE RADAR AND TSR-2

The expansion of research at RRE in the wake of Sputnik was at first accommodated within the organisational structure imposed in the 1953 merger. When G.G. Macfarlane became Director in 1962 a further reorganisation took place in recognition of the on-the-ground integration of the work of the two former establishments, TRE and RRDE.[65] The technical groups were rationalised into three departments, Military and Civil Systems (comprising Ground Radar and ATC, Guided Weapons and Airborne Radar groups), Physics and Electronics (comprising independent Physics and Electronics groups), and Engineering. The organisational changes also reflected significant shifts in technology and strategy, against which establishment heads worked hard to preserve RRE's wider role. Despite the policy shift away from fighters such as the Lightning to guided weapons for UK air defence, for example, RRE continued to argue that a role existed for a complementary independent strike aircraft, and kept up the necessary radar research programmes. The annual reviews of the late 1950s make this point clear. The ground radars were practically immobile and SAGW inflexible: 'it will be preferable to warn off a small number of intruders with long range fighters rather than to shoot them down with SAGW'.[66] Fighters with the assets of a 'long range of operation, mobility and intelligence' would require independence from ground radar: 'the fighter should be able to gather its own intelligence'. Research therefore began, jointly at RRE and GEC's Stanmore laboratory, on continuous-wave AI radar systems to enable the creation of a 'self-contained weapon system'.[67]

Three aspects of RRE's attitude are worth noting. First, there is a significant emphasis on the role of the human operator in RRE's call for fighters rather than GW. Second, an independent fighter was, in one sense, against the trend towards integrated radar systems that we have identified. And third, such a fighter is a close description of the greatest innovative failure of the 1960s: TSR-2.

A significant feature of the RRE programme in the 1950s was a group under Dr A.M. Uttley investigating the role of the human operator as a link in the control chain. This group had originated in wartime TRE, and explored problems of reaction times, response characteristics and pattern recognition in radar users, all with the aim of understanding errors in control operations. Although the group was wound up after Uttley's resignation to take up a senior post at NPL in 1957, RRE nevertheless retained a marked sensitivity to the role of the human operator, and it may be that the work of Uttley's group fundamentally shaped the philosophy of the establishment.[68] Discussing proposed data-handling systems for ASR2232 (the Air Staff requirement for an overseas extension of Plan AHEAD) early in 1962, RRE observed that 'increasing aircraft speeds, and the growing complexity of the air situation, are making the task of safe and efficient control by unaided human beings more difficult ... [both] ... for air traffic and defence requirements'.[69] The specifications for the display and data-handling system called for technology which could be integrated with existing radars at RAF overseas bases, so as to provide tactical control of Bloodhound Mk 2 missiles and interception control of Lightning Mk 3 fighters: a good example of the integration we have discussed. Yet two radically different technical solutions were possible. Both RRE and ASWE (ASRE) succeeded in developing data processing and display systems meeting the operational requirement, one (RRE) employing analogue and the other (ASWE) digital techniques. The difference was nicely captured by the DRPC's Cockburn in 1962, when he noted that 'the important issue was that two experienced establishments faced with meeting similar operational requirements in a similar time scale were depending on different technological solutions'.[70]

Intimately tied to this difference in approach—what Cockburn called a 'transition from well-tried analogue techniques to more promising but risky digital techniques'[71]—were the attitudes of the two establishments to automation: ASWE used automatic data extraction as part of their precocious ADA system, developed since 1953 and described elsewhere in this volume, whereas RRE had chosen manual data extraction and processing for the response to ASR2232.[72] The DRPC had to decide quickly, at a time of cuts and widespread worry about the duplication of defence research (not least from the powerful Public Accounts Committee), whether a common solution could be found. Carroll and Cockburn hastily set up a working party to consider the question. This found that RRE had not pursued manual systems out of ignorance: they were fully aware of ADA, SAGE and STRIDA II automatic tracking systems, but felt that automatic systems could not cope with jamming, insisting instead that the discrimination of a human was needed. ASWE, on the other hand, 'had more faith in the system which they themselves were developing and they were keen to obtain the promised saving of shipboard manpower'—always a key consideration in naval systems. So, behind the difference in technical solutions were radically different attitudes to human operators, jamming, digital techniques, and a natural preference for intramural development.

RRE's emphasis on machine-aided human response was also embodied in its development work on radars for the tactical strike and reconnaisance aircraft, TSR-2. Recognised at the time as being 'probably ... the last military fighting aircraft [to be] developed in the UK,' the TSR-2's importance can be gauged from an impassioned presentation to the DRPC in June 1958:[73]

> this issue, one of the most important ever considered by the DRPC, was vital to the three Services, to the aircraft industry, to the UK balance of payments and to the UK position in NATO. It was vital to the Army in its strike reconnaissance role, to the Navy to strike against submarine bases, and to the RAF abroad. It was important to the aircraft industry in helping to keep it in being and...to assist in exports. It would also demonstrate to NATO that we intended to assist in other ways than maintaining the deterrent.

The TSR-2 electronics were 'complex, varied and highly flexible yet closely integrated and almost entirely new.'[74] Among the R&D responsibilities of RRE were a forward-looking X-band radar (a derivative of the AI Mk 23 Blue Parrot series, intended for the Buccaneer), the rapid photographic processor used with the navigation radar, a sideways-looking Q-band reconnaissance radar, optical line-scan reconnaissance equipment and an associated relay link for in-flight transmission of reconnaissance data to a ground station (Fig. 8.3).[75] The industrial management of the project was complex: although Vickers (now part of BAC) was the main contractor, 21 other firms had direct contracts. Decca Doppler radars provided navigation data, while Ferranti developed the forward-looking radar that would enable the TSR-2 to fly close to the ground, hugging the terrain (an idea imported from Cornell University's Aeronautical Laboratory). EMI's sideways-looking radar provided maps for navigation and reconnaissance.[76] In another innovation all these digital data would be handled by an onboard American Verdan computer, making TSR-2 the first computerised British military aircraft. Not only industry, but also government establishments, had heavily committed resources to TSR-2. By 1961 'roughly half the RRE effort in the airborne radar area' was 'directly involved in TSR-2 work'.[77] This effort was not restricted to the specific subsystems, but also helped shape basic research into techniques at Malvern, especially microminiaturisation to enable the solid-state electronics to be fitted compactly within the aircraft. Clearly, the establishment and the radar industry had a massive investment in the new aircraft.

By 1963 the various TSR-2 radars were undergoing flight trials. Only two years later, however, as tests proceeded apace, the comprehensive programme of innovative work was thrown into disarray with the Wilson government's cancellation of

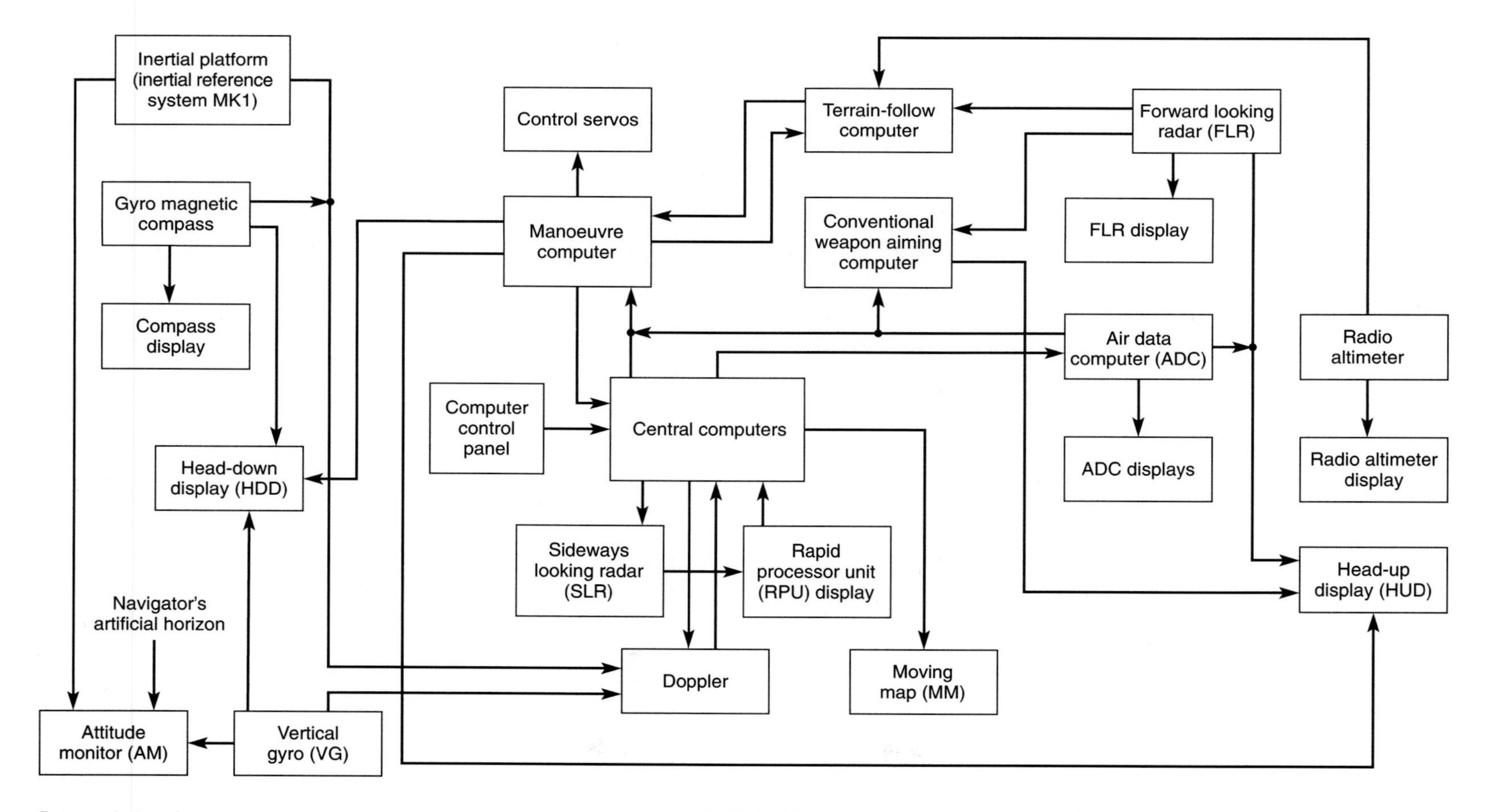

FIGURE 8.3: SYSTEMS DIAGRAM FOR THE ELECTRONICS OF THE AIRCRAFT TSR-2. MANY OF THE SUB-SYSTEMS WERE ADAPTED FOR USE ELSEWHERE AFTER THE CANCELLATION OF THE PROJECT. (© CROWN COPYRIGHT, ROYAL AIRFORCE MUSEUM, REDRAWN)

TSR-2. The news naturally came as a severe blow to the defence industry and research establishments. Trading on the advantages of the systemic and subsystemic nature of the technology, however, the research and development lived on, and each of the TSR-2 radars, suitably adapted, was integrated into other aircraft. The sideways-looking Q-band reconnaissance radar was relocated to the American-built McDonnell F-4 Phantom. The Doppler navigation radar, whose development was almost complete when TSR-2 was cancelled, and which had included a new IF amplifier giving low-level performance down to 150 ft, was used instead in the Nimrod, a Comet airliner modified for reconnaissance and antisubmarine warfare. The Buccaneer provided a home for the linescan reconnaissance system, but the forward-looking radar became a system in search of a platform, moving from TSR-2 to the projected UK variable geometry aircraft before being embedded into the multirole combat aircraft (MRCA), a German–Italian–British project which resulted in the 1980s Tornado. (Remarkably, the other new attack aircraft, the small Jaguar introduced in the early 1970s, had no radar at all.) By the summer of 1966, RRE could note that much of its work now involved 're-packaging and mechanical re-design' of ex-TSR-2 subsystems 'to meet the new environment'.[78]

From the point of view of the establishments, then, the TSR-2 project demonstrated what an integrated programme of technical development could achieve. As events turned out, it also reinforced the political contingencies against which establishments had constantly to work. Although work on the adaptation of TSR-2 radars for other aircraft continued through the late 1960s and into the 1970s, it is clear that the flexibility and pervasiveness of electronics highlighted at the beginning of this chapter protected radar R&D in the rapidly changing military, political and financial contexts of this period. The lessons were not lost on defence science planners. Reflecting on the TSR-2 episode, the DRC (the Defence Research Committee, successor to DRPC) noted that radar electronics were effectively transferable 'building blocks,' and that the 'principle of component research' had become accepted by Whitehall.[79] More than that, 'it would be quite wrong to try and relate the [RRE] research programme to a clear and definite aircraft programme', because '[i]n all areas covered by the Ministry of Aviation it was necessary to maintain expert knowledge and a flexible approach if requirements ... were to be met'. This was significant not just for the way radar technology itself was understood, but because it created an important rhetorical momentum, productive for the defence science community, through the recognition that 'it would be difficult to reconstitute a viable research team either intra or extra-murally' if programmes were stopped.[80]

Basic research at Malvern continued to be taken up in airborne-radar projects of the 1970s. These included significant new techniques, such as digitally processed synthetic aperture radar (a large and very successful programme at Malvern, based on work pioneered in Cambridge radio astronomy), frequency-modulated interrupted continuous-wave (FMICW) radar, and aerial moving-target indicator (AMTI) in reconnaissance and AI for the MRCA.[81] Such techniques also found application in airborne early warning (AEW), deemed essential 'to provide an effective defence against the low-level aircraft and the ... stand-off bomb' and made possible by the decrease in size and weight of electronics.[82] The history of British AEW is troubled. The initial proposal for a replacement of the primitive Avro Shackleton AEW was another Nimrod, or an adapted Comet. A novel approach was suggested: placing two mechanical scanners in bulges in the nose and tail of the aircraft. In 1978 full-power operation of the AEW radar in Comet trials was under way. However, by 1988, after £1 billion had been spent on cost overruns and technical problems, the Nimrod AEW project was cut and American Boeing Sentry aircraft bought, despite a press campaign mounted by the defence industry.[83]

GROUND RADAR AND RADARS FOR OTHER USERS, 1960–80

We have argued that the relationship between establishment and user was an important factor in the development of radar. The establishments were confronted with a variety of users whose needs might diverge widely, of course, and again it is easy to see how the flexible and pervasive nature of radar allowed the technology to be creatively adapted to new situations.

An appraisal of R&D at the newly formed RRE in 1958, for example, pinpointed how army radar differed from other ground radars:[84]

> the problem is very different from that presented in ADUK [Air Defence of the UK] where a large number of desirable targets can be selected from a map. The battle situation over a large area may be fluid and desirable targets such as concentration in the open [or] communication centres, may present only fleeting opportunities for attack. Obviously an important factor in the situation lies in the ability of the Army commander to deny the air space over his forces to enemy reconaissance aircraft, and for this purpose he requires a comprehensive air defence system.

By the early 1960s a number of radars were under development, with the goal of embedding the army in a dense electronic environment: a mobile air-transportable Thunderbird GW system with a new tactical control radar, codenamed Green Ginger, accompanied by radars for light antiaircraft control, night-time movement detectors, a radar for detecting GW launchings, an experimental battlefield surveillance system, and Green Archer, a mortar-detection radar. Contracts were placed with industry, with RRE building some devices and supervising others.[85] Such devices would be useless without robust, lightweight electronics—an area of RRE expertise, for example in microminiaturisation. By the early 1970s these programs had crystallised around three devices: the mortar locator Cymbeline (still in service in the late 1990s, Fig. 8.4), the cannon and rocket locator Zenda (a multinational project from which all but one European partner withdrew in late 1972), and the ZB298 battlefield surveillance system (only in the mid-1990s replaced by the MSTAR radar).[86]

Although the RAF and the Army remained the two foremost Service users of RRE-driven radar, the notion of the 'user' itself underwent considerable change in the 1960s, again in response to wider political imperatives. In the 1960s Malvern made an experimental onsite 45-ft aerial available to NASA for satellite observations (this facility became an operational RAF installation). In a similar vein, collaborations continued with radio astronomers at Jodrell Bank, and with the Radio and Signals

FIGURE 8.4: CYMBELINE, THE MORTAR-LOCATING RADAR DEVELOPED AT RRE IN THE 1970S (© CROWN COPYRIGHT, DERA)

Research Station at Slough, on upper-atmosphere phenomena (crucial to antiballistic missile research). Likewise, from 1960 a joint RRE and Meteorological Office project explored the applications of weather radars.[87] Following the merger of the Ministry of Aviation and the Ministry of Technology in 1967, RRE was absorbed into Mintech, with its emphasis on innovation, production and management. An Industrial Systems Unit was set up within RRE to promote commercial exploitation of ideas and equipment originating in the establishment. Among the products of this initiative was MANAV, conceived at Malvern, partly financed by Esso and produced by Decca for use on oil tankers to assist berthing, coastal navigation and collision avoidance.[88] The trend towards increasing non-military work in the establishments, confirmed in 1968 when the government announced further cuts in defence spending, was, however, halted with the return of a Conservative government in 1970. The abolition of Mintech and the establishment of a Procurement Executive within the Ministry of Defence in 1973 returned the focus of the establishments to defence goals, an orientation that would dominate for the next decade.

Perhaps the most significant impact of the transfer of technology from military to civil users was in the vexed area of air traffic control. As we have seen, from the early 1960s the human operator was seen as a threat to the expansion and sophistication of air travel, so that RRE played only a small role in the resulting automation. To direct and control the information two computers were installed: a Ferranti Apollo at Prestwick, programmed by RAE Farnborough but with display technologies from Malvern, and an Elliott 502 machine, 'Euclid', in London. The computerised ATC system MEDIATOR developed slowly through the 1970s. We have already seen how the decision was taken in the early 1960s that an integrated ground radar system should be built. Integration was understood in two senses. First, on cost grounds, the military LINESMAN would link to the civil air traffic control MEDIATOR through digital data links. Two buildings at West Drayton, one military (L1) and one civil (London ATC Centre) received and distributed raw and processed radar information. On the east coast three air defence radar stations (ADRS) using Type 84 and 85 radars assisted by passive detection systems, at Boulmer, Neatishead and Staxton Wold, fed data to L1. In Ireland a fourth ADRS existed at Bishop's Court. In Scotland, a master radar station at Buchan coordinated information from NATO radars at Saxa Vord in the Shetlands and a British-operated, Danish-owned station in the Faeroes. Other master radar stations were located at Patrington (next to Staxton Wold) and Bawdsey. LINESMAN/MEDIATOR was also to exchange information with the NATO Air Defence Ground Environment (NADGE), which stretched from Turkey to Norway. On the civil side, ATC radars were positioned at Ventnor, Ash, Burrington, Clee Hill and Heathrow. The second sense in which the ground early-warning system was integrated was that Boulmer, Staxton Wold and Neatishead each housed IFF radars to interrogate and identify aircraft, thus distinguishing on the screens at L1 the friendly from the intruder. The final component of LINESMAN was a full automation of the link to HQ Strike Command at High Wycombe through the provision and installation of a recognised air picture dissemination system (RAPDS).[89]

However, there were severe problems with this complex system. Programming LINESMAN/MEDIATOR was much slower and much more expensive than expected (this 'software crisis' was a widespread feature in the mid-1960s). Such delays meant that the system might be antiquated by the time it was brought into service. Furthermore, the L1 building—above the surface and unhardened—presented a vulnerable target. These problems, some publicised by an extremely critical article in the *Observer* in October 1968, stimulated the Ministry of Defence to action.[90] A detailed review commenced in 1969 of technical progress on LINESMAN by a working party under Air Vice-Marshal Moulton. Faced with a situation in which even the basic LINESMAN completion date was receding, the Moulton Report recommended pragmatic and simplified progress in stages, and the establishment of a new LINESMAN Project Management Organisation (LPMO) to oversee installation. However, although LINESMAN was finally declared finished in 1974, a reassessment of the response to the threat from the Warsaw Pact (from massive retaliation to gradual escalation) and progress on the advanced NADGE, meant that British policy makers were already discussing the next system: UK Air Defence Ground Environment, or UKADGE.

In 1972 a new section was formed at RRE in anticipation of formal approval of UKADGE. The new system was 'designed to be less vulnerable than LINESMAN, by the use of dispersed sensors interconnected to control centres and higher management centres'.[91] This section was accompanied by two others, one 'to study the problems of incorporating automatic tracking, emphasis being given to the man/machine interface especially in conditions of heavy clutter or ECM', and another 'to study and oversee the complex data handling task at control and higher management centres'. The form of UKADGE essentially followed LINESMAN: control and reporting centres at Neatishead in Norfolk and Buchan in Scotland (with a stand-in at Boulmer). These received information via datalinks from military reporting post radars (Saxa Vord, Benbecula, Staxton Wold, and Portreath in Cornwall), AEW aircraft, Royal Navy ships and the similar systems of NATO partners. The crucial differences between UKADGE and LINESMAN were in mobility, new techniques and funding. The No 1 Air Control Centre is a mobile command and control facility for the RAF, whereas on the ground radar side the old LINESMAN devices were replaced with transportable systems based on new techniques, notably 3-D radar.[92] This mobility has proved crucial; for example, in the Falklands radar cover was provided for the duration of the 1982 conflict by naval air-defence-capable ships and temporary land systems, and afterwards by a rapidly established Falkland Islands ADGE based on Plessey 3-D radars. Finally, the new UKADGE systems were funded not through the MoD but through NATO by competitive bidding. UKADGE was therefore an extension of a NATO system, rather than one devised nationally.

CONCLUSION: REORGANISATION, RATIONALISATION AND RSRE INTO THE 1980S

In 1976, in a further round of rationalisation, SRDE and SERL were merged with RRE, the resulting establishment being renamed the Royal Signals and Radar Establishment (RSRE). The establishments at Christchurch and Baldock were run down, and staff required to move to Malvern. To a much greater extent than had been the case with the intra-Malvern 1953 merger, staff losses as a result of the moves were a serious cause of concern, compounded by the devolution of extramural research with high staff losses, which had to be made good from RSRE resources.[93] Problems of staff shortages and consequent high workloads continued to trouble RSRE into the early 1980s; in this context, RSRE quickly ceded a GW team working on Blowpipe, fuses and homing projects to Farnborough, leaving behind Rapier, for which long-term development was planned. The establishment was also to retain responsibility for land-based medium surface-to-air missile (SAM) systems, land-based light SAM systems and land-based unit self-defence SAM systems, and would continue to undertake research into homing guidance techniques. It would also take part in international collaborations aimed at the procurement of a new medium SAM to replace Bloodhound/Thunderbird/Hawk.[94]

In the airborne-radar field MRCA continued to occupy an important position, in terms of both development work on its reconnaissance radar pack at RSRE and the AI radar being developed by Marconi-Elliott. A refit of the Nimrod fleet was to continue to completion, as well as the development of an IFF interrogator for

Nimrod Mk II. Staff reductions meant a reduction in research on sideways-looking radar, although the establishment was mandated to continue supervising studies of AI radar and reconnaissance equipment for a new offensive support aircraft. In ground radar, RSRE was expected to continue with the modernisation of Types 84 and 85, the HF200 height-finding radar and passive detection systems, both to extend their lives and to achieve compatibility with the new UKADGE; this work was to be supported by a research programme on clutter rejection and ECCM (electronic counter countermeasures), as well as substantial work on information extraction and display. By the mid-1970s plans were already being laid for the replacement of Types 84 and 85 in the 1990s. Finally, for land warfare, in addition to continuing involvement in the cannon and free-flight rocket detector Cervantes, RSRE was to be responsible for the development of a system designated SCAMPI, the basis of a radar package for SHORSTAS, employing a 'modular approach' offering a choice of radar heads and two displays, 'each with different characteristics to meet the needs of different users'.[95]

This emphasis on the needs of users brings us back to a theme which has figured throughout this chapter: the extremely rich and productive nature of the relationship between establishments and users of radar technology. In 1948, as it came to grips with straitened circumstances and the new demands of peacetime military research, a TRE report had highlighted the nature of this relationship, noting that '[i]n peace as in war there must always be the closest possible co-operation between the scientist and the user to ensure that the user gets as nearly as possible what he wants and that the scientist is not allowed to design an equipment so complicated that it cannot be used'.[96] This could fairly be said to characterise the attitude of the radar establishments for the following 50 years, up to and including RSRE becoming one of the founders of the Defence Research Agency (DRA) in 1990. Indeed, the continuity and stability of the relationships between establishments and users against a rapidly (and occasionally radically) changing organisational and strategic background is one of the most marked features of our account. Mediated by the strong role of industry in radar development and production, the establishments we have discussed always displayed a sensitivity towards users, at first in their traditional Army and Air Force constituencies, and increasingly in the 1960s in industry itself, as government initiatives emphasised the economic aspects of R&D and the need for ever-closer coupling between the work of the establishments and of industry.[97] With a massive increase of radar R&D in industry, pushed along by the establishments, the role of the establishments by this time was increasingly to act as 'technical authority' by providing the expertise needed to study and evaluate contractors' proposals, and to maintain the basic research capacity to keep this expertise up to date.[98]

Part of the reason for the ease with which establishment–industry links could be maintained, and for the adaptability of radar R&D to rapidly changing circumstances, lies, as we have argued, in the nature of the technology itself. As a DRC memorandum on the Ministry of Aviation research programme in 1967 noted: '[i]n almost every field of Defence interests, electronic equipment forms the basis of modern weapon systems'.[99] In this chapter we have tried to demonstrate that sense of pervasiveness, as radar expanded from airborne systems and the wartime reporting and control system to much larger and more comprehensive civil and military air traffic control and navigation systems, including missile guidance, fire control and early-warning systems of increasing variety and complexity. This widespread applicability of the technology, and its openness—its existence as interlocking but relatively independent subsystems with the capacity to be modified and redeployed from system to system and context to context—was important in protecting basic radar R&D, and again allowed for maximum creativity in user–establishment relations.

We have emphasised, too, the themes of integration and centralisation. Radar systems became increasingly comprehensive in the 1950s and after, as Britain became embedded in military alliances with her European and American partners. The 'closed world' of the high Cold War and the harmonisation of strategy provided the larger context within which the establishments carried out their work; but their work was what actually allowed the international networks of electronic surveillance and control that physically constituted the closed world to come into being. Increasingly from the late 1950s, the cost and complexity of radar systems, as well as the desire for standardisation within defence alliances, meant that new radars were often the outcome of international negotiation (for example the NATO IFF system in the 1970s). The accompanying dilution of national autonomy sits uneasily with a study of national research establishments, however, and moves us away from the world of physical systems to consideration of the information provided by the systems we have been discussing. Work on data-display and information-handling systems, to which we have been able to give but little attention in this account, was crucial to the management of the closed world and therefore to the programmes of the establishments. The integration and centralisation of radar networks required new procedures for data transmission and sharing, and the development of computer hardware and software was central to the success of new radar technologies. For this reason automation, too, is an important element of our story. It would be easy to conclude that increasing integration and centralisation imply increasing automation. Yet, as we have seen, there were significant differences of opinion between establishments with respect to the role of the human operator in the command and control chain—and therefore to the kinds of technologies and, ultimately, the kinds of global strategies that might be appropriate to manage the closed world.

An integrated system, embodying the fruits of new research from the establishments and in which automation promised minimisation of human error, should not, however, be equated easily with success. The threat to Britain during the Cold War was considerable, and new techniques could well have overwhelmed the electronic network linking radars to command and control. The carcinotron was a fearsome weapon of electronic warfare, and DRPC discussions suggest that ballistic missiles would always have got through. Moreover, LINESMAN, with its centralised unhardened L1 building, provided an exposed target. The vulnerability of the closed world is perhaps the most significant theme to emerge from a study of postwar radar.

1. For an introduction to the large literature on the history of radar, see *Radar Development to 1945*, edited by R. Burns (IEE, 1988); *Tracking the History of Radar*, edited by O. Blumtritt, H. Petzold and W. Aspray (IEEE, 1994); M.I. Skolnik, 'Fifty years of radar', *Proceedings of the IEEE*, 73, no. 2 (1985): 182–97. On the academic-military-industrial complex in the American context, see S.W. Leslie, *The Cold War and American Science: The Military-Industrial-Academic Complex at MIT and Stanford* (Columbia University Press, 1993).
2. On this systemic aspect of military electronics, see G.E. Bugos, *Engineering the F-4 Phantom II. Parts into Systems* (Annapolis: Naval Institute Press, 1996). More generally on systems in the history of technology, see T.P. Hughes, *Networks of Power* (Johns Hopkins University Press, 1983).
3. On the histories and relationships of the various electronics research establishments, see E.H. Putley, 'The History of the RSRE', *Physics and Technology*, 16 (1985): 5–11, 18. On the reorientation of national scientific establishments to the postwar context, see *The Restructuring of Physical Sciences in Europe and the United States, 1945–1960*, edited by M. De Maria, M. Grilli and F. Sebastiani (World Scientific, 1989); J. Agar, *Science and Spectacle. The Work of Jodrell Bank in Post-war British Culture* (Harwood Academic, 1998).
4. P.N. Edwards, *The Closed World. Computers and the Politics of Discourse in Cold War America* (MIT Press, 1996): 1. For a good overview of the 'new' Cold War history, see J.L. Gaddis, *We Now Know. Rethinking Cold War History* (Clarendon Press, 1997).
5. The 'Sunday Soviets' were informal meetings on Sundays in which Service chiefs, operational users and TRE staff could discuss operational requirements, user feedback and future technological possibilities in a relatively informal setting. See A.P. Rowe, *One Story of Radar* (Cambridge University Press, 1948).
6. J. Gough, *Watching the Skies. The History of Ground Radar in the Air Defence of the United Kingdom* (1993): 58–60. On wartime TRE and its legacy, see also M.M. Postan, D. Hay and J.D. Scott, *Design and Development of Weapons. Studies in Government and Industrial Organisation* (HMSO and Longman, 1964).
7. G.G. Macfarlane and C. Hilsum, 'Francis Edgar Jones', *Biographical Memoirs of the Fellows of the Royal Society*, 35 (1990): 181–99, 185; W.A. Atherton, 'Alec H. Reeves, 1902–1971: inventor of pulse-code modulation', *Electronics and Wireless World*, 94 (1988): 873–4.
8. Born in 1913, Lovell studied physics at the University of Bristol, where he took his Ph.D. in 1936. From 1936 he was a lecturer in physics at the University of Manchester, where he was recruited to the radar project in 1939. H_2S (widely understood to stand for 'Home Sweet Home') used a cavity magnetron to produce a centimetric navigational radar system. His wartime experience and contacts were crucial elements in his development of radioastronomy and the Jodrell Bank radiotelescope after 1945. See D. Saward, *Bernard Lovell. A Biography* (Robert Hale, 1984); B. Lovell, *Echoes of War. The Story of H_2S Radar* (Hilger, 1991).
9. Quoted in B. Lovell and D.G. Hurst, 'Wilfrid Bennett Lewis', *Biographical Memoirs of the Fellows of the Royal Society*, 34 (1988): 453–509, 482.
10. B. Lovell and D.G. Hurst, 'Wilfrid Bennett Lewis' (n. 9 above): 483.
11. Foreword, *TRE Progress Report* (March 1945). In what follows we have drawn heavily on the series of monthly, later biannual, and latterly annual reports of TRE, RRDE and their successor establishments available at the library and information centre, DERA Malvern. We are grateful to Peter Trevitt for access to these and other archival materials.
12. *TRE Progress Report* (April 1945).
13. *TRE Progress Report* (October 1945): 3.
14. *TRE Progress Report* (January 1946).
15. J. Gough, *Watching the Skies* (n. 6 above): 60, 65 (quoting the White Paper *Statement Relating to Defence*). On missile developments see S. Twigge, *The Early Development of Guided Weapons in the United Kingdom, 1940–1960* (Harwood Academic, 1993).
16. J. Gough, *Watching the Skies* (n. 6 above): 60–1.
17. J. Gough, *Watching the Skies* (n. 6 above): 35–41.
18. Cmd 6743; J. Gough (n. 6 above): 42.
19. J. Gough, *Watching the Skies* (n. 6 above): 44–5.
20. J. Gough, *Watching the Skies* (n. 6 above): 47.
21. J. Gough, *Watching the Skies* (n. 6 above): 62–3.
22. *TRE Progress Report* (October 1946).
23. J. Baylis, *Ambiguity and Deterrence: British nuclear strategy 1945–1964* (Oxford University Press, 1995).
24. For the DRPC see: Jon Agar and Brian Balmer, 'British Scientists and the Cold War: the Defence Research Policy Committee and information networks, 1947–1963', in *Historical Studies in the Physical and Biological Sciences*, 28, no. 2 (1998): 209–52.
25. See Second Report from the Select Committee on Science and Technology, HC 213, Session 1968–69 (HMSO, March 1969).
26. Each ground radar system had an AMES (Air Ministry Experimental Station) number. Chain Home, for example, was AMES Type 1.

27. Source for Table 1: *TRE Progress Report* (January 1949).
28. J. Gough, *Watching the Skies* (n. 6 above): 47–8.
29. The Air Ministry estimate for radio, radar and electrical equipment rose to £6 million, of which £4.2 million was for radio and radar alone. Cmd 7631; J. Gough, *Watching the Skies* (n. 6 above): 43.
30. *TRE Progress Report* (July 1950). On the V-bomber programme, see H. Wynn, *The RAF Strategic Nuclear Deterrent Forces: their origins, roles and deployment* (HMSO, 1994).
31. Despite a ministerial decision in July 1951 that contractors should consider the programmes of highest priority, ROTOR and VAST were only slowly completed. Indeed the Korean War had ended before they entered full operational service; ROTOR was not fully operational until mid 1955.
32. The radar was named AMES Type 81, before being renamed AMES Type 80 Mk. 3, an indication of the system's similarity to Type 80.
33. Minutes, DRPC, PRO DEFE 10/38 (20 May 1952).
34. J. Gough, *Watching the Skies* (n. 6 above): 142.
35. *Panel on the Co-ordination of Radio and Radar Research and Development in the Ministry of Supply. Final Report,* PRO AVIA 54/1716 (27 March 1951). Membership was E.T. Paris (chairman), D.W. Bartington, L.H. Curzon and Willis Jackson.
36. TRE had 116 SO, 12 engineers, 186 EO and 98 assistants (scientific). RRDE was smaller: 44, 9, 86 and 74 staff respectively.
37. Note both E.J. Cuckney and Willis Jackson thought that the Paris Panel, re PDT(RD) did not go far enough.
38. *Working Party on the Functions, Fields of Work etc. of TRE and RRDE,* PRO AVIA 54/1716 (21 October 1952). Membership was: Rear Admiral Burghard (DCL, chairman), Cawood (RAE), Richards (CS/TRE), Barton (PDLRD), Bartington (US(R)), Pollard (CS/RRDE), Hanson (DTPA), and Gwynne Jones (TPA1, secretary). On the crucial role of the Ministry of Supply in promoting and coordinating military research, see D. Edgerton, 'Whatever happened to the British warfare state? The Ministry of Supply, 1945–1951', in *Labour Governments and Private Industry. The Experience of 1945–1951,* edited by H. Mercer, N. Rollings and J.D. Tomlinson (Edinburgh University Press, 1992): 91–116.
39. Minutes of WP, PRO AVIA 54/1716 (18 November 1952).
40. RRE was headed from 1953 to 1961 by W.J. Richards, formerly Chief Superintendent of TRE. P.E. Pollard, Chief Superintendent of RRDE at the time of the amalgamation, was transferred to a headquarters post.
41. Minutes, DRPC, PRO DEFE 10/39 (9 November 1954).
42. Minutes, SCI/M(54)1, PRO DEFE 7/300 (5 January 1954). Carroll, Cook, Sargeaunt, Cockburn and Wansbrough-Jones attended the meeting chaired by Brundrett.
43. Wansbrough-Jones' undisputed suggestions were: (a) there were in a particular establishment individuals with the qualifications to do a certain piece of research of great scientific importance with some probabilty of direct interest to defence; (b) a piece of research was manifestly important to national defence and could not be done elsewhere; (c) the facilities needed to do a particular piece of research were so expensive that they could be economically provided only on a national scale; (d) every establishment should have some basic research going to maintain scientific standards, but the work to be done with this object should only be that which comes under (a), (b), or (c) above.
44. *MoD. DRPC. Review of Defence Research and Development. Report to the Minister of Defence,* DRP/P(55)50, PRO AVIA 54/1922 (23 January 1956).
45. *Defence Policy and Global Strategy,* JP(54) Note 10, PRO DEFE 4/70 (20 April 1954). We are indebted to Stephen Twigge for this reference.
46. J. Gough, *Watching the Skies* (n. 6 above): 184.
47. Blue Yeoman was subsequently called AMES Type 85.
48. In 1953, Mitchell noted that, 'there was a very definite limit to the accuracy obtainable by an H_2S system. There would be no improvement until a guided bomb arrived.' Minutes, DRPC, PRO DEFE 10/39 (21 April 1953).
49. By 1957 an AI Mk 23 was also being designed for fighters (Lightning), including single seater fighters. By 1962, AI Mk 18 was operational in the Sea Vixen, providing fire control for Firestreak and rockets. An AI Mk 22 also existed. Contractors for AI Mk 17 were Metropolitan-Vickers, EMIED, E.K. Cole, Scophony-Baird, ERI, J. Stone, Portsmouth Aviation, UFMT and Dunlop Rubber. GEC produced AI Mk 18.
50. Minutes, DRPC, PRO DEFE 10/38 (6 February 1951).
51. Minutes, DRPC, PRO DEFE 10/38 (12 June 1951).
52. Minutes, DRPC, PRO DEFE 10/38 (12 August 1952).
53. Memorandum, *Wide Band Noise Jamming of the Air Defence System. Note by CGWL,* DRP/P(55)11,PRO DEFE 10/34 (7 March 1955). RRE Annual Review no. 2, April 1958 records that improvements were needed for AI mks 18 and 23 in 'an era of higher performance targets, collision course weapons and reduced ground control in a jamming environment'.

54. J. Gough, *Watching the Skies* (n. 6 above): 186–7.
55. J. Gough, *Watching the Skies* (n. 6 above): 189.
56. Minutes, DRPC, PRO DEFE 10/39 (9 October 1956).
57. Gough states that the problem of 'diversionary systems' was accepted from the Tripartite Conference on Defence against BMs, held in 1956. Sharp differences of opinion about the seriousness of this threat appeared in the DRPC (Cockburn v Tuttle).
58. J. Gough, *Watching the Skies* (n. 6 above): 198. Meteor and aurora research was also funded at Jodrell Bank.
59. J. Gough, *Watching the Skies* (n. 6 above): 201.
60. Minutes, DRPC, PRO DEFE 10/355 (10 March 1959); Minutes, DRPC, PRO DEFE 10/355 (26 June 1960).
61. Minutes, DRPC, PRO DEFE 10/277 (18 November 1958).
62. On the negotiations over BMEWS see I. Clark, *Nuclear Diplomacy and the Special Relationship* (Oxford University Press, 1994).
63. Minutes, DRPC, PRO DEFE 10/277 (6 November 1957).
64. Minutes (n. 63 above); see also A.C.B. Lovell, *The Story of Jodrell Bank* (Oxford University Press, 1968): 186 ff.
65. W.J. Richards resigned as Director in 1961. He was succeeded by W.H. Penley, formerly Head of the Guided Weapons Group, for one year (1961–2). G.G. Macfarlane had joined TRE at Dundee; he had become Head of the Theoretical Physics Group after the war, then in 1959 was appointed Deputy Director of the National Physical Laboratory, from which post he returned as Director of RRE in 1962.
66. *RRE Annual Review*, no. 1 (April 1957).
67. *RRE Annual Review*, no. 2 (April 1958).
68. *RRE Annual Review*, no. 1 (April 1957).
69. *RRE Half Yearly Progress Report* (January 1962).
70. Minutes, DRPC, PRO DEFE 10/417 (14 March 1962).
71. Minutes (n. 70 above).
72. ADA (Action Data Automation) had been installed on HMS EAGLE and was soon to be extended to other ships. The RRE equipment was built by Decca, and is described elsewhere in this book.
73. Minutes, DRPC, DRP/M(58)8, PRO DEFE 10/277 (17 June 1958).
74. G. Williams *et al.*, *Crisis in Procurement: a case study of the TSR-2* (Royal United Service Institution, 1969): 25. See also D. Edgerton, *England and the Aeroplane. An Essay on a Militant and Technological Nation* (Macmillan, 1991): 90–2.
75. The TSR-2's radar equipment was completed by a sideways-looking X-band navigation radar and a doppler equipment, both of which were the responsibility of the main aircraft contractor, with RRE acting in an advisory capacity. *RRE Annual Review*, no. 5 (April 1961).
76. This system stemmed from gaps in cover: 'With the developed NBS, the V-force will thus be equipped with radar which we believe will be adequate to deal with strategic targets whose geographical location is known. Further equipment is required for high altitude reconnaissance of targets whose position are not known in advance and, for this purpose, it is planned to use the long fixed sideways-looking aerial technique.' *RRE Annual Review*, no. 1 (April 1957).
77. *RRE Annual Review*, no. 5 (April 1961).
78. *RRE Half-Yearly Progress Report* (July 1966).
79. Memorandum, *MoD. DRC. The Ministry of Aviation Research Programme. Note by the Air Force Department*, PRO DEFE 10/626 (6 February 1967).
80. Minutes, DRC, DR/M(66)9, PRO DEFE 10/624 (7 December 1966).
81. *RRE Annual Report* (1973/74).
82. Memorandum (n. 79 above).
83. P. Gummett, 'The government of military R&D in Britain', in *Science, Technology and the Military*, edited by E. Mendelsohn, M.R. Smith and P. Weingart (Kluwer Academic Publishers, 1988): 481–506, 501.
84. *RRE Annual Review*, no. 2 (April 1958).
85. Marconi (Green Ginger), Cossor and Ferranti (Green Ginger peripherals), Elliott and EMI (SSGW detection), Decca (battlefield surveillance) and Plessey (radar camouflage). RRE built the LAA radar and Green Archer, and collaborated over SSGW detection. For an industrial perspective see D. Martin, *THORN EMI: 50 Years in Radar* (THORN EMI, 1986).
86. *RRE Annual Report* (1972/73).
87. *RRE Annual Review*, no.4 (April 1960).
88. *RRE Programme 1971/72* (final draft June 1971). Of £190,800 work for outside customers, £91,000 came from Esso and DTI for MANAV. A further £36,900 was made by sales of electronic crystals. On Mintech, see D. Edgerton, 'The "White Heat" revisited: the British Government and technology in the 1960s', *Twentieth Century British History*, 7 (1996): 53–82, esp. 65–73.

89. *RRE Annual Report* (1973/74).
90. J. Gough, *Watching the Skies* (n. 6 above): 281.
91. *RRE Annual Report* (1973/74).
92. *RSRE Annual Report* (1979/80).
93. *RSRE Annual Report* (1979/80). The establishments at Christchurch and Baldock did not complete their move to Malvern until 1980.
94. *RSRE Annual Report* (1977/78).
95. *RSRE Annual Report* (1977/78).
96. Foreword, *TRE Progress Report* (October 1948).
97. P. Gummett, 'The government of military R&D in Britain' (n. 83 above): 492–3.
98. On more recent developments in the organisation of UK military R&D, see Council for Science and Society, *UK Military R&D* (Oxford University Press, 1986); P. Gummett, 'United Kingdom', in *European Defence Technology in Transition*, edited by P. Gummett and J.A. Stein (Harwood Academic, 1998): 261–88.
99. Memorandum (n. 79 above).

CHAPTER 9

Naval Command and Control Equipment: The Birth of the Late Twentieth Century 'Revolution in Military Affairs'

ERIC GROVE

INTRODUCTION

One of the major themes of the late 1990s strategic debate is the 'digitisation of the battlefield'—the use of electronic sensors and command and control equipment to collect and digitally distribute information to allow precise engagements of distant enemy targets. It is held by some that this information-led warfare has been the basis of a modern 'revolution in military affairs', whose potential against less well equipped enemies was demonstrated in the Gulf War of 1991. What is less well known is that the foundations of this revolution were laid half a century ago, in the beginnings of the digitisation of naval warfare to cope with the challenges of the progress of contemporary maritime threats. Moreover, this naval response was led by British naval scientists and research establishments. This work in the United Kingdom introduced the Americans to the concept of the electronic mechanisation of tactical data handling and the use of data links. Not only has such equipment become the defining technology of modern navies but, as noted above, related concepts lie at the heart of the advanced warfare techniques of the next century.

The British contribution to the development of air defence techniques using radar and relatively simple manual action information organisation has recently been well documented.[1] This had been revolutionary enough, allowing the prewar pessimism that fighter aircraft would prove relatively useless at sea to be proved wrong. The synergy of fighters and action information organisation gave naval commanders a powerful antiair-warfare instrument vital for the maintenanace of sea control in areas of high air threat, and enabling maritime-based air arms to take on and defeat even powerful land-based air forces. The advent of the jet aircraft, however, promised to stress such systems 'beyond their elastic limit'.[2]

High-speed jet aircraft arriving in waves or a high-density stream would saturate even the improved defences of 1945. In high-altitude engagements the future generation of jet interceptors, with both a high rate of climb and associated high wing loading, would require especially accurate individual direction, given their

FIGURE 9.1: DR RALPH BENJAMIN,1950 (COURTESY OF DR R. BENJAMIN)

relatively poor manoeuvrability. Even existing piston-engined low-flying attackers required improvements in the integration of ships deployed as radar pickets and airborne early-warning platforms with the main units of the fleet. To cope with the new threat every available fighter had to be used, individual aircraft intercepting at maximum range to allow multiple engagements. The threat of electronic jamming was also increasing, necessitating not only antijamming properties in new radars, but also the integration of all available electronic data, both active and passive, into a single plot. Only by electronically mechanising the data acquisition and plotting activities would these formidable problems have any chance of solution.[3]

THE REVOLUTION

The key figure in the development of such a system was Ralph Benjamin (Fig. 9.1), a German Jewish refugee forced to flee Nazi persecution. A first-class honours graduate of Imperial College, London, he joined the Royal Naval Scientific Service in 1944, being allocated to radar development at the Admiralty Signals Establishment (ASE, renamed in 1948 the Admiralty Signals and Radar Establishment, ASRE).[4] In order to integrate radar pickets with the main force, Benjamin developed the concept of digitally encoding information compactly and using a high-frequency data link, which would have a longer range than the microwave video links originally suggested for this role. The data links would transmit over the horizon the positions of targets defined by simple grid coordinates.

To display the information, Benjamin developed the concept of using a cathode ray tube display with a set of independent electronic markers that could be placed on the indicated positions of the relevant contacts, whose characteristics (track number, identity, height) would be displayed. A Polish-born scientist at ASRE, Sacha Woroncow, developed the mechanism that replaced the mechanical rotating deflection coil used on then-current radar plan-position indicators with a fixed-coil system controlling both up-and-down and side-to-side movement of the marker spots.

The markers could be coded with appropriate symbols to display their contacts' characteristics; alternatively (or in addition) information on a contact could be displayed separately when a cursor was placed upon it. A joystick was used to control the display, although a fixed rolling ball or even an early 'mouse' was considered. A joystick was, however, easier to engineer. At first only a couple of markers were used to extract information on selected radar echoes and to transmit target designations elsewhere. Benjamin, however, foresaw that multiple markers could produce a sanitised synthetic picture which could be either superimposed on the raw radar display of fighter controllers to extract target information more easily from it, or be displayed as a separate tactical display used by controllers coordinating the air defence of an entire task group.[5] The interlaced marker system was described in an ASRE Technical Note 'Some Suggestions for the Design of Target Indication Equipment', coauthored by Benjamin and Peter Wallis and issued on 17 March 1947.[6] The data extraction and display technique was patented by the Admiralty in the name of Benjamin and Wallis in September 1947.[7]

The Division Head at ASRE, D.S. Watson, gave Benjamin 'full support and a free hand' in designing and building a system to demonstrate the potential of these ideas. In this process he was assisted by two naval colleagues, Commander David S. Tibbits and Lt Bobby Woolrych. Woolrych, described by Benjamin as 'a brilliant, imaginative, enthusiastic, highly intelligent and very unconventional young naval officer', was especially important in work on operational concepts.[8]

THE COMPREHENSIVE DISPLAY SYSTEM

Thus encouraged and assisted, Benjamin became Technical Project Leader for the development of the world's first electronically mechanised combat data system for action information organisation (AIO), the Comprehensive Display System (CDS). In a 1953 ASRE Technical Note Benjamin described CDS as follows:

> The Comprehensive Display System is designed primarily to abstract from suitable sources (mainly radar) the information required for the control of fighters (or very long range missiles) in air defence . It deals with the processes of compilation, classification, storage and selective distribution of tactical information. The various functions are carried out by a combination of human operators, electronic circuits and electromechanical devices. The system is designed to produce a flexible organisation and to achieve, with maximum economy, a performance adequate for coping with fairly large numbers of independently operating high-performance aircraft.[9]

The demands of following a very large number of contacts were a considerable challenge. The original intention was to produce a 96-track capability but in 1952 this was reduced to 48.[10] Contemporary target indication technology allowed one operator to cope with only two strong contacts from a fast-rotating radar. Moreover, the motions of the expected targets would have defeated the automatic following capabilities of a surveillance radar with the required range turning at only six revolutions per minute. These problems were solved by the adoption of the technique of 'rate aiding', in which operators had the task of monitoring the alignment of contacts, taking action only if they changed direction or speed. Speed was also added to position in the contact's recorded characteristics, which aided classification and provided parameters to the computers used for threat evaluation and fighter direction (see below). Both rate aiding and the computers used the principle of switched-capacitor computation, at least a quarter-century before its more general adoption.[11]

To help solve the tracking problem caused by slow antenna rotation, identification friend or foe (IFF) techniques were used to aid the labelling of contacts and to resolve tracking ambiguities. Tracks from raw radar data were injected by detectors into the system by ringing them on a screen using a joystick. The contacts were then shared equally among six trackers, each at a cathode-ray tube display workstation, and responsible for keeping up-to-date plan position information on up to eight tracks. The contacts were then analysed by four analysers, each responsible for 12 tracks. Their workstations had analysis displays, which included nine small cathode-ray tubes to allow height to be detected and displayed, and two plan position indicators to display radar and IFF data to help classify targets and assess formation size. The analysers then injected information into the code store. The store, the 'heart' of the system, was an 'assembly of electromechanical and electronic devices' maintaining 'an up-to date record of all available information on all targets being tracked in the system'.[12] The store contained identity categories, track references, plan position information, height and formation size characteristics of all tracks.

The trackers and analysers were supervised by the Radar Directing Room Officer and his assistant, who resolved points of potential confusion. The stored information was passed to the tactical operators, notably the Fighter Direction Officers. Users of the equipment could choose to display only markers in a given category, e.g. specific attackers and the fighters allocated to dealing with them, or non-engaged hostile contacts and the available defending fighters. It was found better—so as not to clutter the displays—for fighter controllers to use their facility to 'hook' contacts with their joystick control cursors and read a full description on an ancillary display. Alternatively, track numbers could be selected using rotary switches. Once chosen, the given track was highlighted on the display with its other parameters displayed on the ancillary display, as if it had been 'hooked' by the joystick. Alphanumeric symbols were used with the cathode-ray tube tracing the pattern of the symbols, a method devised by Benjamin himself. More mechanical 'totes', based on railway indicator board technology, were used to display the status information of defending aircraft and all identified hostile contacts. Each aircraft in the air was indicated with its track number.[13]

A key component of CDS was the intercept computer. This worked from the movement of the markers on the screen. A new course and speed could be injected into the system and a 'time' knob turned to let the markers advance to where they might be at the selected moment. Having thus found the nearest approach the course of the fighter could be further adjusted to make the markers coincide, thereby obtaining the required intercept vector. Also computed was the need for the fighter to manoeuvre into a firing position astern of the hostile contact. The displacements in the positions of both aircraft during this final manoeuvre were represented by offsets on the starting positions of the 'future position' markers. These allowed fighters to be ordered to turn into their firing positions when the markers came together on the screen.

As Benjamin later wrote with justifiable pride, this intercept computer:

> was almost certainly the first time-shared, programme controlled analogue computer. Like the rate aiding it employed switched capacitor arithmetic. By using the same components for computing X and Y, for target and fighter, for straight flight and for manoeuvre-equivalent displacements (and for two simultaneous interceptions), most of the equipment duplication, tolerance and calibration problems of more conventional analogue computers were avoided.[14]

The fast-moving nature of jet-age maritime combat required more speedy and efficient threat evaluation and weapon assignment (TEWA) than was possible from unaided human operators, and a computer was developed for this task also. This sorted the targets in order of urgency as they turned on to their attacking courses, and matched each with a defensive asset in terms of the latter's flexibility. As Benjamin later described it:

> After assigning any "only possible target" for a weapon, and any "only possible weapon" for a target, the remaining targets were taken in order of urgency, and each assigned to the least flexible weapon capable of doing the job, e.g. preferably a gun or a missile of limited coverage, or an airborne fighter with little fuel left.[15]

This maximised the flexibility to cope with the evolving threat.

The TEWA computer was a decision-making aid only: the Direction Officer could veto or alter its decisions if necessary. In order to diminish workload, however, the Direction Officer 'was encouraged to modify the general weights, rather than specific decisions. Modifications of either type were made "user friendly"—and were known as "gold braided modulation". When there were only a few seconds to spare, to ensure a target was engaged before its predicted weapon release, a warning light indicated that, unless the recommended assignment was overridden promptly, it would be transmitted to the relevant weapon system.'[16] It proved a little difficult to persuade officers skilled in the still-novel art of fighter direction to leave so much of their art to a machine and concentrate on "higher level decisions: changes in the readiness state of fighters on the flight deck: the establishment and deployment of combat air patrols; the relative priorities for the use of fighters vs missiles, or of fighters of one type vs those of another'.[17]

The Direction Officer, in effect, had to delegate command to the CDS, guiding and modulating it as required. This required changes in professional culture, perhaps helped a little by the return of the older generation of direction officers to civilian life. The new generation of FDOs both needed CDS and were better prepared to exploit it.

A contract to produce CDS was drawn up with Elliott Brothers and two former workers at ASE, C.A. Laws and M.V. Needham, produced two demonstration systems. Elliotts were very strong in electromagnetic technology and the demonstrators used this approach for the various CDS subsystems, such as potentiometers, motors and relays. This created very serious maintenance problems, and ASRE worked on more electronic solutions that would be not only more maintainable, but also more reliable because of redundancy. Stimulated by the work of H.W. Pout of ASRE on the Sea Slug missile guidance radar, Benjamin 'stressed modularity, redundancy and reconfigurability throughout the design.'[18] P. Dax of ASRE designed a central monitoring position for fault finding, based on a sequence of binary tests. The concept was expanded by Benjamin and A.D. Brown to cover the complexities of parallel units. Faulty units at sea were replaced for repair in the maintenance annex in spaces alongside the CDS equipment room, this process being facilitated by similar binary analysis designed by Dax. It is this ASRE design that was eventually manufactured by Pye and put into service.

Thanks to careful design CDS proved to be an extremely reliable system in service. Although it sometimes degraded at the margin, with one tracking or intercept position workstation out of use, the operational availability of the entire system was never seriously compromised.[19]

To provide radar data for CDS a new and advanced three-dimensional set, Type 984, was developed by ASRE. Benjamin suggested the use of a slanted stack of scanning 'pencil' beams, each of which scanned a narrow range of elevation and azimuth angles, while the whole assembly was rotated to sweep 360°. Maximum range was 180 miles and height 40,000 ft. The stacked overlapping beams gave clear echoes and good accuracy, both in elevation and azimuth (the individual beams were used by CDS analysers to give height information on their small cathode-ray tube displays). The initial work at ASE was carried out by a team under the direction of D.S. Watson, which included, as well as Benjamin, B.W. Lythall, an antenna expert who, together with D. Roessler, developed a novel microwave lens. According to Benjamin, 984 was 'probably 10 years ahead of other radars in 3D resolution'.[20] Its development after 1947 was supervised by J. Snowdon, 'a superb project manager ... leading one of the strongest teams ever assembled at ASRE'.[21]

Capability came at a price, and 984 was a large piece of equipment, the antenna casing having a volume of over 250 cubic feet, enough to hold a cocktail party within it! CDS, 'a complex man-made organism' was also bulky, weighing about 25 tons and occupying about 2000 cubic feet in volume. It consisted of some 30 consoles, 45 individual display units and 60 cabinets containing 5000 valves.[22] There were three major spaces, two operations rooms (upper and lower) and a radar direction reporting room; in the first installation aboard the carrier *Victorious*, the lower operations room and the radar direction reporting room were placed side by side. The much-rebuilt *Victorious* commissioned with the 984/CDS combination in 1958, and the light fleet carrier *Hermes* commissioned with a similar fit the following year. This reduced CDS system was also installed in the carrier *Ark Royal*, when she refitted in 1959.

Installation of the full 984 and CDS system had been proposed for the planned 18,500-ton guided missile cruisers. However, these were cancelled in 1956, and the system was too large for the fleet-escort guided-missile destroyers (DLGs), the first pair of which was laid down in March 1959. Fitting had been considered for these too, but this would have meant dispensing with these ships' 4.5-inch guns, thus sacrificing much of their effectiveness in limited-war and policing functions. Their more limited radar capability made it all the more important that the integrated air-situation picture, compiled in the carrier's CDS, be transmitted to the reduced (but CDS-based) guided-missile destroyer's display systems. The destroyers in turn could act as stand-off 'picket' ships, filling the gap below the horizon of the carrier's 984 coverage (when Airborne Early Warning was not available or adequate). All the DPT ships were able to complement and integrate their joint radar and tracking capacity.

Digital Plot Transmission (DPT) had always been an integral part of the overall CDS concept. Benjamin had described the details of a digital system in an ASE Technical Note in June 1947, and patented it at the end of the year.[23] The concept was

> the sequential digital transmission of the Cartesian coordinates defining the CDS track marker positions, together with the associated height identity, track number and assignment information ... The received synthetic picture was aligned with the local radar picture, using targets seen on both, in order to compensate for errors in the ships' navigational relative-position estimates. The assignment of track numbers to newly-detected targets, and the prime reporting responsibilities for all AA targets, were allocated to ships on a sector basis, and appropriate protocols were developed for multi-ship tactical integration.[24]

The system worked in the HF band and had a range of over 100 miles. W.P. Anderson and N. Godel were the leading figures in its development at ASRE. The analogue information from CDS was digitally encoded on a RJT(1) encoder and decoded by an RJR(1) for use by the analogue display equipment in the receiving ship.[25]

Both CDS and DPT gave the UK something of a lead over the USA. The latter had been kept informed of British thinking on radar data extraction, DPT and synthetic marker and symbol displays from the beginning, Benjamin and Tibbits having first visited the USA for discussions on the subject in 1946.[26] The visitors 'seemed to cause a considerable stir at the Navy Department' and at the Bell Laboratories. At Bell the visitors met Claude Shannon, who was already working on similar theoretical lines but was unable to discuss his work. In any case, the British were way ahead in practical application.[27] The Americans were presented with one of the first CDS demonstration models but decided to apply the concept using digital means, with their naval tactical data system (NTDS). A Bureau of Ships Technical Requirement for the system was issued in 1955, and the Chief of Naval Operations signed a Characteristics and Operational Requirement the following year. It began tests ashore in 1959 and sea trials in September 1961.[28] It was first fitted in the carrier *Oriskany* and the DLGs *King* and *Mahan*, the smaller ship fittings being a sign of the more advanced technology. Nevertheless, this was still some three years after the in-service date of CDS, and a convincing demonstration the Royal Navy had given the Americans in 1958 of the power of computerised AIO.

The operations room and radar room crews of HMS *Victorious* had become highly proficient in working CDS before their carrier was recommissioned by a pioneering use of simulators. They were thus able to deal most effectively with a heavy coordinated 'attack' by 23 USN aircraft on *Victorious*, designed to test the new system. The Americans agreed that the carrier's Scimitars had 'shot down' 22 of the 23 attackers, the remaining one of which had turned back. The US analysts could not believe that this had been achieved with so few fighters, most of which had been launched to deal with the attack rather than flying standing patrols. The Americans were equally amazed to learn that the British aircraft were not carrying radar, and that the Fighter Direction Officers were both relatively inexperienced and were controlling multiple fighters.[29]

The Americans and British had already joined with the Canadians to develop common standards for tactical data links. A CANUKUS Naval Data Transmission Working Group was set up in 1954, and three years later its proposals were finalised at a meeting in Ottawa.[30] Three tactical international data exchange (TIDE) links were proposed, designated P (HF, high-speed), Q (HF, low-speed) and R (UHF); the name TIDE was suggested a little whimsically by the British after the brand of washing powder common to all three countries: the TIDE links would 'clear up a messy tactical picture'.[31] The letter designations were then changed to numbers. DPT was clearly Link I, so the more complex Link P, designed with NTDS in mind, became Link II. The Roman II, however, soon became the Arabic 11. Thus was born the basic data link that still forms the standard for the all NATO navies. Link Q formed the basis for the simpler links 13 and 14, and R became Link 12.[32] Only Link 14 entered service under that number; Link 13 became Link X (after the 'Brand X' detergent compared unfavourably in advertisements with Tide) and the X soon became a number. Such was the origin of Link 10, used later for a time as a cheaper alternative to Link 11.[33]

LATER SYSTEMS

The more complex and expensive TIDE Link 11 was part of the specification of the all-digital system that replaced CDS in RN service. This was developed by Benjamin's research division at ASRE from the mid-1950s onwards.[34] The resulting action data automation (ADA) was based on the DAA data-processing system. The 984 radar fed through its own detection computer into a general-purpose digital computer suite, which contained three Ferranti high-speed Poseidon computers, developed in association with ASRE. (The latter establishment was combined with the Gunnery Establishment as the Admiralty Surface Warfare Establishment, ASWE, in 1959—this in itself was a sign of the changing nature of naval warfare.) Each Poseidon had a fast-access memory of 8192 24-bit words; computer 2 had a programme store of the same length; computers 1 and 2 stores of half the length, making a total programme store capacity capacity of 16,384 k words.[35] The Poseidons occupied ten cabinets. This computer suite also received the inputs of available data links and electronic warfare sensors, and gave its output to 15 JGA workstations (plus one spare); three of the latter had facilities to receive information from airborne early-warning aircraft (AEW) to assist the direction officers. One of the JGAs was in the general direction room; the other 14 operational displays were in the operations and air defence room, where they serviced four interception officers. Each JGA could display both marker pictures and totes, thereby dispensing with the need for large display plots and mechanical totes. In general, ADA/DAA offered more capability—notably automatically generated vectoring orders—and generally greater capacity, with some savings in space and manpower. Semiconductors and printed circuits replaced valves.[36]

The prototype Poseidon was completed in 1961, but problems were experienced with the programme store, which had to be redesigned by its contractor, Mullard. The first—and only—production ADA was commissioned in HMS *Eagle* on completion of her major rebuild on 14 May 1964. Her system could interact with the CDS in *Victorious* or *Hermes* and with any assets (so far just US) equipped with Link 11 and could transmit her plot to accompanying carriers or to the first four DLGs, all of which were now in commission. The Royal Navy had fully entered the digital age.

Already a more compact digital AIO system was under development at ASWE, work having begun in 1961. This went to sea in the second batch of DLGs, the first pair of which were commissioned in 1966. It was designated Action Data Automation Weapon System, or ADAWS. Based on two Poseidon computers, it integrated the ship's sensors, sonar and EW as well as radar into a complete tactical picture. Link 11 was fitted. ADAWS was improved into ADAWS 2, based on the new Ferranti FM1600 computer and Plessey Mk 8 digital autonomous displays. Improved computer capacity added fire-control functions to the previous picture compilation task. Only one ship received ADAWS 2, the DLG *Bristol*. The other ships of this class were cancelled, along with the proposed carrier CVA01, HMS *Queen Elizabeth*, which was to have carried ADAWS 3. The subsequent Type 42 destroyers were equipped with ADAWS 4 (Fig. 9.2), their more austere nature being marked by the move to Link 10; ships such as *Bristol* were used as gateways between Links 11 and 10. Further versions of ADAWS were ADAWS 5, fitted to the Leander class frigates fitted with the Ikara antisubmarine missile (to facilitate integration of these ships' main armament with the force air picture), and ADAWS 6, developed for the Invincible class through-deck cruisers/carriers.[37]

An altogether cheaper and less complicated AIO system was also developed in the 1960s, the computer assisted action information system (CAAIS). As its name suggests, it was intended to assist operators in less demanding tactical situations than fully automating a fast-moving and intense air battle. This led to its rather unfortunate nickname 'Can't Automate Anything Insufficient Software', but a CAAIS ship was in a much better tactical situation than a non-computer vessel. The system offered automatic tracking of surface and air contacts and automatic picture compilation and weapons assignment. The combination of picture-keeping and weapons-direction functions into a single FM1600B computer greatly reduced track capacity, however. The DBA(1) variant began retrofitting into frigates in 1974; the same year, the slightly improved DBA(2) entered service with the Type 21 frigates. Later versions were fitted to *Hermes* as an interim through-deck cruiser, DBA (3); the Hunt class minehunters, DBA(4); and the Seawolf-equipped modernised Batch

Figure 9.2: The Operations Room of a Type 42 destroyer equipped with the ADAWS 4 AIO system. Such equipment is now the foundation of modern naval warfare and owes its origin to the Comprehensive Display System developed by ASE/ASRE fifty years ago. (© Dave Reynolds, The Military Picture Library)

3 Leanders and Batch 1 Type 22 frigates DBA(5); in this version an improved FM1600E provided 128 k of memory and a capacity for 60 tracks, a greater track capacity than the old CDS.[38]

By the 1980s modern command and control techniques were being exploited in innovative ways, such as in the Falklands War, when the Type 42 destroyers and Type 22 frigates were able to network their ADAWS and CAAIS systems using Link 10 to allow the synergistic combination of the different capabilities of their sensors and weapons. No ship not equipped with computerised AIO and data links could by then be considered fully modern. CDS and DPT had spawned a real revolution, one that would soon spread ashore.

THE LEGACY

In his memoirs Benjamin makes the telling point that his 1947 patents, taken out to protect the Admiralty from outside patents being enforced against the Crown, 'are to this day the basis for all military Command-and-Control Information Systems, worldwide, and for all civil, air, sea or road traffic control systems and they are also the basis of the "mouse" cursor technique used on all modern personal computers. However I have no knowledge of the Crown having sought or received any royalties from them'.[39] It is hard to disagree with this justifiably proud claim—or to wonder at the potential revenue lost! Ralph Benjamin and his colleagues in ASE/ASRE/ASWE were responsible for what can only be described as some of the fundamental technologies of the second half of the twentieth century. This achievement is far too easily ignored. It is thus especially vital that, as information technology becomes ever more dominant in both military and civilian applications, the seminal work of this little known Admiralty establishment be belatedly given the recognition it richly deserves.

1. Notably in the work of the Naval Radar Trust. See D. Howse, *Radar at Sea: the Royal Navy in World War Two* (Macmillan, 1993); *The Development of Radar Equipments for the Royal Navy 1935–45*, edited by F.A. Kingsley (Macmillan, 1995); *The Application of Radar and Other Electronic Systems in the Royal Navy in World War Two*, edited by F.A. Kingsley (Macmillan, 1995).
2. R. Benjamin, 'The post war generation of tactical control systems', *Journal of Naval Science*, 15, no. 4, November 1989: 263–75. This key article has been one of the foundations of this chapter. Benjamin has also written an autobiography which usefully supplements his article: *Five Lives in One: an insider's view of the intelligence world* (Parapress, 1996).
3. R. Benjamin, 'The post war generation of tactical control systems' (n. 2 above): 263. See also *Department of Research Programme and Planning, Admiralty: report on the defence of seaborne forces from future air attack, an ASE appreciation of the electronic requirements*, PRO ADM 213/485 (1947).
4. R. Benjamin, *Five Lives in One* (n. 2 above): 1–21.
5. R. Benjamin, 'The post war generation of tactical control systems' (n. 2 above): 263; R. Benjamin, *Five Lives in One* (n. 2 above): 32–4.
6. XRC4/47/4.
7. 'Improvements in Cathode Ray Tube Display Systems'.
8. R. Benjamin, *Five Lives in One* (n. 2 above): 34.
9. R. Benjamin, *An Outline of the 48-track Comprehensive Display System and its Variants*, ASRE Tech. Note DX-53–3, PRO ADM 220: 647.
10. R. Benjamin, *An Outline of the 48-track Comprehensive Display System* (n. 9 above).
11. R. Benjamin, 'The post war generation of tactical control systems' (n. 2 above): 264.
12. R. Benjamin, 'The post war generation of tactical control systems' (n. 2 above): 264.
13. R. Benjamin, 'The post war generation of tactical control systems' (n. 2 above): 264.
14. R. Benjamin, 'The post war generation of tactical control systems' (n. 2 above): 265.
15. R. Benjamin, 'The post war generation of tactical control systems' (n. 2 above): 266.
16. R. Benjamin, 'The post war generation of tactical control systems' (n. 2 above): 266.
17. R. Benjamin, 'The post war generation of tactical control systems' (n. 2 above): 267.
18. R. Benjamin, 'The post war generation of tactical control systems' (n. 2 above): 267.
19. R. Benjamin, 'The post war generation of tactical control systems' (n. 2 above): 268.
20. R. Benjamin, *Five Lives in One* (n. 2 above): 47; see also R. Benjamin, 'The post war generation of tactical control systems' (n. 2 above): 269–70. The handbook for 984 can be found at PRO ADM 238/476. Other relevant documents are in PRO ADM 220.
21. R. Benjamin, 'The post war generation of tactical control systems' (n. 2 above): 271–2.
22. PRO ADM 20/647. See also *The Facilities and Organisation of the Comprehensive Display System*, ASRE Tech. Note DX-52–11, PRO ADM 220/630.
23. XRC4/47/10 (7 June 1947); Patent 33/573/47 (December 1947).
24. R. Benjamin, 'The post war generation of tactical control systems' (n. 2 above): 268.
25. *A Summary of Data Links for Naval Tactical Operations*, ASWE Tech. Report no. 3, PRO ADM 220/1312.
26. R. Benjamin, *Five Lives in One* (n. 2 above): 36–40.
27. R. Benjamin, *Five Lives in One* (n. 2 above): 36–40.
28. N. Friedman, *Naval Radar* (Conway, 1981): 77, 80.
29. R. Benjamin, *Five Lives in One* (n. 2 above): 265.
30. N. Friedman, *The Naval Institute Guide to World Naval Weapons Systems 1997–98* (Naval Institute Press, 1997): 28.
31. N. Friedman, *The Naval Institute Guide* (n. 30 above): 28.
32. PRO ADM 220/1312.
33. N. Friedman, *The Naval Institute Guide* (n. 30 above): 28.
34. R. Benjamin, *Five Lives in One* (n. 2 above): 57.
35. R.A. Ballard, *The General Purpose Computer Poseidon: Its Facilities and System of Operations*, ASWE Tech. Note DX-60–6, PRO ADM 220/944; *Handbook for Data Processing System DAA, Action Data Automation*, CBO 4889, PRO ADM 239/586.

36. R.A. Ballard, *The General Purpose Computer Poseidon* (n. 35 above); *ASWE Progress Report*, PRO ADM 220/1324 (June 1962).
37. Details of early ADAWS can be found in contemporary *Jane's Weapon Systems*; fuller details of later systems in N. Friedman, *The Naval Institute* (n. 30 above).
38. Contemporary *Jane's Weapons Systems*; N. Friedman, *The Naval Institute* (n. 30 above).
39. R. Benjamin, *Five Lives in One* (n. 2 above): 34.

CHAPTER 10

Laser Research and Development, 1960–80

RICHARD MILLS

INTRODUCTION

Lasers have had a profound influence on the conduct of modern warfare. According to Lambeth:

> Laser-guided bombs largely dictated the outcome of the 1991 Gulf War by shutting down Iraq's integrated air-defence system, keeping its air force out of the fray, and destroying its fielded armour. Although their individual probability of kill turned out to be considerably lower than the initially touted figure of 80–90%, they were nevertheless decisive in their overall operational effects. These munitions have already offered an increase in destructive power by as much as a factor of 1,000 over unguided bombs.[1]

In addition to their use as target designators, lasers are employed in a number of other military applications, including rangefinding, fibreoptic communications, navigation and training aids. They have also found many important civil uses in such diverse fields as cutting and welding, scientific research, surgery, communications, navigation, surveying, printing, pollution detection, compact disc players, supermarket checkout systems and light shows. Lasers are, therefore, the physical manifestation of an important 'generic technology' because they are applicable across many industries. They are also the physical manifestation of an interdisciplinary technology arising from the conjoining of the disciplines of physics and electronics. Indeed, the field of research associated with lasers is often referred to as 'quantum electronics'.

Lasers differ from other natural and manmade sources of light, such as the sun or light bulbs, in that they emit an intense and highly directional beam of monochromatic radiation. There are many different types of laser, and these are normally classified according to the nature of the active medium. Lasers can be compared using a variety of parameters, such as size, weight, cost, output power, efficiency, pulse repetition frequency (PRF) and wavelength. In terms of their size, weight and cost, lasers differ radically between the little, low-power semiconductor lasers used in compact-disc players, at one extreme, and huge high-power lasers that fill an entire room and are used for research on beam weapons and nuclear fusion, at the other. Like other light sources, lasers are relatively inefficient, the best converting only about 20 per cent of the input energy into light. The remaining input energy is usually dissipated in the form of waste heat. Therefore, high-power lasers require special cooling systems to prevent them overheating. Not all lasers can generate a continuous beam of radiation. Consequently, their PRF is an important characteristic. Lasers can emit radiation in the visible, infrared, ultraviolet or X-ray regions of the electromagnetic spectrum. However, they tend to be fixed-wavelength devices,[2] i.e. their wavelength cannot be tuned and is determined by the nature of the active laser medium.

A comprehensive history of laser research in British government defence research establishments[3] has never before been written. Two important historical studies of the development of laser technology have been produced by Bertolotti.[4] However, both are written from the 'internalist' perspective and consequently fail to place the evolution of the technology within a broader social, political and economic context. This limitation is also shared by various historical accounts of the invention of specific devices written by the inventors themselves or the institutions to which they belonged.[5] More recently, Forman,[6] Eppers,[7] Seidel[8] and Bromberg[9] have sought to rectify this omission, with studies that have examined the growth of laser technology in the USA. Some research on the growth of laser technology in the UK has been undertaken by Collins, who studied the transfer of knowledge required to construct a particular type of carbon dioxide laser.[10] He found that the primary mechanism by which this technology was communicated was not through publications, but rather personal contact, telephone conversations and personnel transfer. Collins then mapped the pattern of communications between different laboratories within the UK, and discovered that the resultant network was dominated by a defence research establishment, which served as the principle conduit through which this technology was transferred. This study clearly suggests that defence R&D establishments played a crucial role in promoting the growth of laser technology within the UK. However, it should be noted that Collins only examined one particular type of laser a few years after its invention, and so his findings may lack more general applicability.

The task of writing a history of the development of a technology within an organisational context is complex and challenging. This approach chosen in this chapter is to evaluate the laser programmes undertaken by the defence R&D establishments in terms of the objectives they were intended to support at the time. No definitive statements of these objectives could be located. Evidence from a 1968 enquiry by the House of Commons Select Committee on Science and Technology on defence research[11] and the 1980 Strathcona Report on the research and development establishments[12] suggest that in practice the defence research establishments supported a number of broad and overlapping goals. A composite list of these consists of:

1. Creating the technological base needed to support the development of future weapons systems.
2. Providing advice on future Staff Targets, Staff Requirements and specifications, where to order equipment from, and worthwhile development programmes.
3. Undertaking development work, including project acceptance and evaluation.

4. Providing, where appropriate, large capital equipment for common use by industry.
5. Generating technologies of wider social benefit.

Virtually every one of the government's defence R&D establishments has been involved in laser work at one time or another. A detailed analysis of all their laser programmes is clearly beyond the scope of this chapter. Instead, the approach taken has been first, to establish how government defence R&D policy developed with respect to lasers. The way in which this policy was implemented and its consequences will then be explored in more focused studies of the laser R&D programmes in RRE, SRDE, SERL and RSRE, formed as a result of their amalgamation in 1976. These establishments were selected because they are a representative sample of the defence research establishments in that they undertook both fundamental and applied laser work, and because they were where most of the laser work was concentrated.

This chapter is divided into six sections. The first deals with the period leading up to the invention of the first laser and, in particular, seeks to establish the role of the UK defence R&D establishments in that process. In the second section UK policy towards laser R&D for defence purposes will be outlined. The next three sections will then examine the laser programmes pursued by RRE, SERL and SRDE. Each of these sections will also indicate how the work initiated at these establishments was later taken forward at RSRE. Conclusions will then be drawn.

INVENTION

The invention of the laser was predicated upon a number of prior inventions and scientific breakthroughs.[13] It is generally agreed that the first of these was the publication of a paper by Albert Einstein in 1916, in which he suggested that atoms can release energy through a process called 'stimulated emission'.[14] Atoms are capable of possessing different amounts of energy, and these energy levels are called 'states'. When an atom in a lower energy state absorbs energy it is said to become 'excited', and moves to a higher energy state. Conversely, when an atom drops down from an excited state to a lower energy state it emits electromagnetic radiation in a process known as 'spontaneous emission'. However, Einstein suggested that if an atom in an excited state absorbs energy of the correct wavelength it will be stimulated to emit electromagnetic radiation of the same energy and drop into a lower state, emitting further radiation of the same wavelength in the process.

Experimental proof of the existence of stimulated emission was first achieved by the German physicist Rudolf Ladenberg during the late 1920s.[15] However, the importance of the phenomenon was not appreciated at the time because stimulated emission was regarded to be an unusual event. It required energy at precisely the right wavelength and, under normal conditions, any energy that is produced is quickly absorbed by the atoms in the lower energy states. The problem that confronted the scientists was how to achieve a situation where atoms in an excited state would predominate—a condition known as 'population inversion'.

The Second World War triggered the development of radar technology, which helped open up the microwave portion of the electromagnetic spectrum. A number of groups began to consider ways of generating very short-wavelength microwave radiation through the process of stimulated emission. This was first achieved in 1954 by a team led by Charles Townes at Columbia University.[16] Their device worked by generating a beam of ammonia molecules and then separating those in the excited state using a focusing device. The excited molecules were then directed into a microwave-resonant cavity. When spontaneous emission occurred from some of the excited ammonia molecules this would trigger stimulated emission in others. The resultant signal would then be amplified in the resonant cavity. Townes called the device a maser (microwave amplification by stimulated emission of radiation).

The maser developed by Townes was not capable of sustained operation: it would cease to operate once the ammonia molecules dropped into the lower state. To overcome this limitation a more complex energy scheme was required. In 1955 Alexander Prokhorov and Nikolai Basov of the Lebedev Physics Institute in Moscow proposed the first two schemes for gas masers that employed three energy states, and hence were known as three-level masers.[17] One scheme was to excite molecules from the ground state to a higher energy level with an external radiation source. Population inversion now existed between the top level and an intermediate level. As molecules then fell back to an intermediate level they experienced stimulated emission. The other was to remove molecules from the lowest energy level reducing its population below the level in the slightly higher intermediate level. This would create population inversion between those low-lying states, leading to maser action. The following year Nicolas Bloembergen, working at Harvard University, published the first proposal for a three-level solid-state maser.[18] Operation of such a maser was first achieved in December 1956 by Derrick Scovil, George Feher and Harold Seidel, working at Bell Laboratories.[19] They used gadolinium ions in a salt of lanthanum ethyl sulphate as the active medium. However, their maser only operated as an oscillator. Alan McWhorter and James Meyer, at the MIT Lincoln Laboratory, were the first to operate a maser as both an amplifier and an oscillator, in the spring of 1957.[20] Their maser was based on potassium cobalticyanide doped with chromium. It was at this time that scientists at RRE also began to take an interest in the possibilities these new devices offered.[21]

Maser research at RRE appears to have been initiated in 1957 following a visit to the USA by the head of the establishment's Physics Department, R.A. Smith. Like their US counterparts, staff at RRE thought that masers could be used as very low-noise amplifiers. They might, therefore, have important applications in radar systems and, as RRE was a radar establishment, this was an area in which they ought to become involved. A small team was established in the Physics Department and began by constructing spectrometers to investigate the paramagnetic resonance phenomena of candidate maser materials. They then attempted to repeat the work of Scovil and colleagues by constructing their own gadolinium ethyl sulphate maser. Having gained experience with this type they then constructed masers based on potassium cobalt chromicyanide and potassium chromicyanide. The last of these operated as both an oscillator and an amplifier in March 1958, and was the first maser to operate in the UK. It was subsequently demonstrated to the Malvern meeting of the Physical Society a few months later. Further success was to follow later that year, when the RRE maser group achieved a world first by operating a ruby maser at a temperature of 60 K, rather than at the temperature of liquid nitrogen (about 2 K), as had normally been the case. To support this research a maser crystal-growth programme was also initiated at RRE and extramural contracts placed with UK firms to supply maser crystals and components.

During 1957 Charles Townes, who was still based at Columbia University, and his brother-in-law Arthur Schawlow, who was working at Bell Laboratories, published their famous paper entitled 'Infrared and optical masers'.[22] In it, they outlined how the maser principle might be applied to a device that could generate radiation at infrared and optical wavelengths.

Publication of the Schawlow–Townes paper triggered a race to complete the world's first operational laser. However, no evidence was found to suggest that scientists and engineers in Britain had any great interest in constructing a device of their own. The only British scientists who did take an interest in the possibility of constructing a laser at this time were John Sanders from Oxford University and Oliver Heavens from Royal Holloway College, London.

In October 1958 Sanders had received a preprint of the Schawlow–Townes paper from a former Oxford student who was now in charge of a group at Bell Laboratories concerned with satellite communications. The student suggested that Sanders spend a sabbatical at Bell, attempting to construct a laser. Sanders accepted, but with less than a year available he decided on a cut-and-try attack by exciting pure helium in a discharge tube and searching for lasing by varying the discharge parameters. However, he did not allow the length of his discharge tube to exceed 15 cm—the maximum distance over which a pair of laser mirrors could be aligned with existing incoherent light sources. Unfortunately, the gain in the gas mixture was so low that lasing could not be achieved.

Following the publication of the Schawlow–Townes paper, Charles Townes had decided to concentrate his efforts on demonstrating laser action with a potassium vapour laser which was optically excited with a mercury arc lamp. However, during September 1959 he began a two-year leave of absence from Columbia University to serve as Vice President of the Institute for Defense Analyses. To help his graduate students with their laser project during his period of absence, Townes invited Oliver Heavens to spend a sabbatical year at Columbia. Heavens was an expert in the physics of highly reflecting surfaces and therefore possessed knowledge very relevant to the development of lasers. By this time, the Columbia potassium vapour laser project had run into serious difficulties, so Heavens redirected the efforts of the group towards the construction of a caesium vapour gas laser which would be optically excited by a helium lamp. However, they failed to obtain laser action with this device before Heavens had to return to the UK.

The race to operate the world's first laser was won in May 1960 by Theodore Maiman, who was working at Hughes Research Laboratories in Malibu, California.[23]

His laser employed a rod of synthetic ruby as the active medium, with excitation being achieved by a helical-shaped flashlamp placed around the rod. This laser emitted a pulsed beam of radiation in the visible red region of the electromagnetic spectrum.

Later that year Peter Sorokin and Mirek Stevenson, who were working at the IBM T.J. Watson Research Center, demonstrated laser operation with flashlamp-pumped cryogenically cooled crystals of uranium-doped calcium fluoride.[24] This was followed in early 1961 by the demonstration of a flashlamp-pumped cryogenically cooled laser using crystals of samarium-doped calcium fluoride.[25] These devices emitted pulsed beams of radiation in the infrared and red regions of the spectrum, respectively. Their operation represented an important advance because the lasers operated through transitions between four energy levels, rather than the three involved with Maiman's ruby laser. Four-level lasers require less power to achieve laser action, and can therefore be more easily operated in continuous mode.

During December 1960 laser operation was also demonstrated in a gas. This was achieved by a team based at Bell Laboratories consisting of Ali Javan, William Bennett and Donald Herriott. In this laser the active medium was a mixture of helium and neon gas excited by a electrical discharge.[26] The laser was a four-level device and emitted a continuous beam of radiation in the near-infrared region of the spectrum.

Thus, by the end of 1960 several different types of laser had been invented in the USA. However, in Britain it appears to have been largely ignored. Staff at RRE had certainly been active in the maser field since 1957, and the possibility of constructing infrared and optical masers had been postulated in one of the most prestigious American physics journals, triggering a race by groups in that country to build such a device. Yet there is no evidence to suggest that staff at any of the British defence R&D establishments took any interest in lasers until after they had been demonstrated abroad.

LASER R&D POLICY

Laser research started in the UK defence R&D establishments during 1961.[27] This work appears to have been initiated because these devices were believed to have potential defence applications in such diverse areas as metrology, cutting and welding, communications, target designation, night vision, identification-friend-or-foe systems, optical radar and beam weapons.[28] However, these ideas were extremely speculative and unproven. Therefore, it was believed that further research was necessary before the possible future applications of the laser could be realistically assessed.

The decision to initiate laser work appears to have been made by senior scientists within the establishments. No evidence was found to suggest that it was driven by senior policy staff within the establishment's parent departments. Indeed, lasers were not discussed by the Defence Research Policy Committee (DRPC) until 1963.[29] This appears to have reflected the relative responsibilities of the defence R&D establishments and the departments they served. According to Sir George Macfarlane, former

Director of RRE and Controller (Research) in the Ministry of Technology:

> I think the people who worked in headquarters were primarily concerned with development. The research programme in the Aviation and Defence establishments was really determined by the scientists who worked there. This didn't mean that they weren't under some sort of control from their headquarters—they were in the sense that once or twice a year there was a review of their research programme. But, headquarters Directors carrying out the review of the programme were very generous in their attitude to it primarily because experience showed that best results were obtained by allowing the research scientists freedom to pursue their own initiatives providing they were working in a field which was deemed to have aviation or defence applications.[30]

During the next two years the amount of laser R&D work conducted in the UK gradually increased as laser programmes were established at a number of universities, companies and government laboratories. However, this was dwarfed by the sheer scale and speed with which laser R&D expanded in the USA. One commentator from the NATO Supreme Headquarters Allied Powers in Europe (SHAPE) Air Defence Technical Centre summed up the position in 1962 as follows:

> On one side of the Atlantic an entire industry seems to be building up around a new defence-technical device, while on the other, European, side just the first spade work is being laid on the applications of that very same device ... it would be surprising if one would find more than one or two dozen groups working on lasers in all of Europe. France is still the most active country in laser research ... There are a small number of projects going on in the United Kingdom; there is a cautious start in Germany, and a very few scattered efforts under way in the rest of Europe. In general, the subject in the European laser program is still fundamental research studying the phenomenology of the laser. Hardly any group is yet thinking in terms of hardware development or even military systems.[31]

By 1963 lasers did begin to receive serious attention from senior policy makers within the MOA and the MOD. That year the Chief Scientist of the MOA, Sir Robert Cockburn, visited the USA, during the course of which he was briefed on laser developments in that country.[32] When he returned to the UK he invited the Directors of RRE, RAE and SRDE to make proposals for a considerable increase in their laser research programmes. In response, they suggested an increase in intramural staff up to a level of about 45 white paper grades; an extension of the crystal growing facility at RRE; and the development of a range for laser damage assessment. It was also proposed that this programme should be supported with an extramural programme running at about £100,000 per annum. These recommendations were endorsed and Sir Robert asked RRE to hold a symposium on the military applications of lasers later that year.[33]

Between 1964 and 1967 the level of effort devoted to laser R&D in defence R&D establishments continued to increase, reaching a high point in about 1967. By this time there were over 70 government scientists working in the defence R&D establishments on various aspects of laser technology and its applications, and the associated extramural R&D programme amounted to over £700,000.[34] This should be compared with the figures of the UK's civil laser R&D programme, which involved less than half that number of white paper grades and a negligible amount of extramural contract work. Moreover, although many government establishments had a laser group, most of the effort was concentrated at three defence R&D establishments: RRE, RAE and SERL. A 1968 survey by the Ministry of Technology estimated that these three establishments accounted for nearly 75 per cent of the intramural effort and about 75 per cent of the extramural funding.[35] Thus, most laser work in the UK was funded by the defence budget, and the government establishments that were principally involved were also defence establishments. However, despite the increased size of the UK laser programme it continued to remain a fraction of that of the USA. For example, minutes of a Defence Research Committee (DRC) meeting during 1965 noted that UK effort on high-power lasers for use as damage weapons was 10 per cent of that committed by the USA.[36]

By the mid-1960s, however, the early enthusiasm for the laser appeared to be on the wane. Lasers were being marketed commercially as scientific instruments and for a few very specialised applications. However, as a laser scientist from SERL acknowledged, these were 'not world shattering, as may once have been hoped'.[37] A similar point was made by a Ministry of Technology Paper written in the same year which observed that 'predictions of widespread use of lasers in engineering, communications and medical fields have not been fulfilled. The situation is still one where the research worker and teacher are the main users'.[38] The failure of the laser to reach people's initial expectations led some to refer to this device as 'a solution in search of a problem'.[39]

The above might be taken to suggest that laser technology was developed through 'technology-push'. However, this would appear to be oversimplistic. In a pure technology-push model no consideration is given to its applications, yet, as has been shown, applications of the laser were considered at an early stage. Unfortunately, its initial exploitation was held up by the lack of lasers with the appropriate characteristics. As a laser scientist from NPL argued in 1965:

> Most of the effort in Government laboratories has been devoted to developing the many forms of laser and improving their performance. There are very few people working on laser applications who would say this is undesirable, most of the applications new being studied are in fact held up, not by want of funds or effort, but by want of the right sort of laser.[40]

There were several major limitations with the early lasers. One was the characteristics of the radiation that could be generated. For most of the applications that were envisaged, the existing lasers lacked the right combination of output power, PRF, wavelength and spectral purity. The other was their high cost, short life and low reliability.[41]

Evidence suggests that this disappointment, combined with competing demands on resources from other areas of technology, led to a reduction of effort in the laser field between 1967 and 1970.[42] In 1967 a decision was reached to significantly reduce the level of effort on laser R&D at RRE and redefine the responsibilities of RRE and RAE in this field.[43] During the same year intramural work on semiconductor lasers at SERL was terminated.[44]

It is difficult to estimate the level of effort devoted to laser R&D during the 1970s because the relevant policy files have either been destroyed or archived in such a way as to make them extremely inaccessible. However, the available evidence suggests that the early 1970s saw a sudden resurgence of effort directed at the development of lasers and investigation into their applications. The laser programme was relaunched at RRE and, in response to interest from the SBAC and EEA, the MOD held a two-day classified symposium on lasers at the establishment on 4 and 5 May 1971.[45] Although definitive evidence is hard to obtain, the resurgence of interest in the laser would appear to have arisen from a belief that new advances in this technology now allowed it to be exploited in military applications. Certainly, by the early 1970s the laser was beginning to find major defence applications in rangefinding and target designation, with others under consideration.[46]

Given the lead established by scientists and engineers in the USA, why did British government defence procurement agencies choose to sponsor the growth of laser technology at all? They could have simply waited for the necessary R&D to be conducted in the USA and procured the equipment from US firms.

The reason why this approach was not adopted was the fact that, at the time, the UK government had an explicit policy of maintaining a national indigenous technology base in this field and, as far as possible, procuring all defence equipment from British firms. This is illustrated by the minutes of one the meetings of the CVD Optical Maser Committee:

> At the moment no good large rubies are grown in this country, and it is necessary to purchase any that are required from abroad ... The meeting felt that an effort should be made to ensure that there was a source of large rubies within the country.[47]

Similarly, members of the CVD Technical Committee observed that:

> There is a shortage in this country of suitable flash tubes for lasers and CVD will have to support some development which may in the first instance involve the copying of US devices.[48]

There were, however, attempts to promote international collaboration with the UK's NATO allies. Details of new fundamental technological advances in other countries were regularly published in the unclassified literature, and classified information related to laser applications was exchanged through multilateral and bilateral arrangements with Australia, Canada and the USA through the TTCP. Discussions related to lasers were also held on a bilateral basis with NATO countries, including France. Under the aegis of an Anglo-American Steering Committee, an agreement was reached during 1966 in which both countries committed themselves to spend £0.5 million over two years in a collaborative R&D programme to investigate the potential use of high-power lasers in beam weapons applications.[49]

Despite these initiatives, senior policy staff within the MOA and MOD remained extremely ambivalent about research collaboration with the USA.[50] On the one hand there were a number of benefits to be gained from collaboration. The USA was leading the world in terms of laser technology. Consequently, if the UK collaborated with the USA it would gain access to a much larger and more rapidly expanding technology base. Exchange of information was also expected to help avoid waste of effort by pursuing unprofitable lines of research. Even if the UK and USA did not decide to proceed with collaborative development, the knowledge gained would be of value in taking decisions regarding the purchase of foreign equipment. Furthermore, although much information on laser developments in other countries was published in the unclassified literature, this was heavily censored. Thus, unrestricted access to information on foreign military laser work could only be gained through international agreements.

On the other hand, defence research collaboration was also perceived to have a number of drawbacks. First, such agreements sometimes excluded collaborative development. Consequently, if any usable results did emerge from the research, 'the Americans with their vastly superior development resources, would be in a much more favourable position to exploit them, and will almost certainly leave us standing'.[51] Second, such agreements sometimes precluded, without US consent, the possibility of joint development and production with third countries. Although this restriction applied equally to the USA, it bore upon the UK to a much greater extent because the US had little need for collaboration.

In addition to collaborating with the USA, the UK also sought to respond to the larger US effort in this field by concentrating on the study of those lasers of particular interest to the UK, rather than examining every possible material that might be made to lase.[52] In particular, there was an attempt to focus upon acquiring the technologies necessary to develop possible applications. According to a DRC paper from 1965:

> It can be argued that because of the superiority of American resources the UK has no real chance of making an independent initiative in the laser field ... But, this argument takes too little account of the fact that much of the current research work is essentially laboratory-type investigation which does not call for expensive facilities. It is consequently quite possible that a breakthrough would be achieved from a fairly modest programme of work within the resources of European countries. It can also be argued that it will be to the ultimate advantage of the UK not to try to keep pace with the very wide range of activities going on in the US but instead to concentrate, so far as possible with our European allies, on a more modest programme aimed at simple, peripheral weapons which we and the Europeans stand some chance of being able to develop in competition with the Americans.[53]

When laser work was first initiated in the defence R&D establishments there was considerable overlap between their programmes. Only a limited degree of coordination of their extramural R&D programmes was provided by the CVD organisation. During 1962 an informal division of responsibility between RRE and SERL had been agreed, whereby SERL would work on gas lasers because of its expertise on the development of vacuum tubes, and RRE would concentrate on the development of solid-state lasers because of its expertise in crystal growth.[54] The following year this agreement was extended to cover RAE and SRDE: RRE was to be responsible for fundamental research on solid lasers, for crystal development, and for radar-like applications; RAE would be responsible for potential applications of the laser as a beam weapon; SRDE would be responsible for applications of the laser in visual aids and communications systems; and SERL would continue to be responsible for fundamental research on gas and semiconductor lasers.[55]

Despite of this division of responsibility, effective coordination of the laser R&D programmes of the different establishments appears to have continued to pose a problem for the rest of the decade. A 1968 paper prepared by a senior member of staff from the Ministry of Technology noted that there was a danger of 'needless duplication', and consequently a need to centralise certain areas of laser work.[56] This view was endorsed by a meeting of senior policy staff from the relevant government departments. In response, they decided to reaffirm the division of responsibility set out in the 1963 agreement. However, the development of the high-power carbon-dioxide laser invalidated the idea of having a single centre for high-power laser work because such performance could now be achieved with more than one type of laser.[57] It was also decided to form a Laser Committee to oversee all government-sponsored laser R&D. The role of this new committee was to provide a forum for exchanging advice and information on lasers and their applications, and had three supporting panels concerned with crystal lasers, gas and semiconductor lasers, and non-linear optics.

Efforts were also made to promote spinoffs. Fundamental advances in laser technology were usually reported in the open unclassified literature and unclassified conferences. This information was therefore readily accessible to anyone with an interest in the field. It was also common practice for the designs of lasers, laser components and laser-based systems developed within the defence research establishments that were of general or specific civil interest to be transferred to the National Research and Development Corporation (NRDC) for licensing to British industry.[58] In addition to these formal technology transfer mechanisms there was also a large amount of unstructured interaction between different members of the UK laser community. One important vehicle for this was the 'Laser Club', which was formed in 1962 and consisted of laser scientists and engineers drawn from government laboratories, firms, research associations and universities.[59] It met several times a year at different government research establishments, and those who were interviewed claimed it played an extremely important role in promoting contact between different members of the UK laser community. However, its value began to decline because of the growing need for commercial secrecy of its industrial members, and by the end of the decade it had ceased to exist.[60]

LASER R&D AT RRE

One of the RRE scientists who had been working on masers during the 1950s was Peter Forrester.[61] Between June 1959 and January 1961 he had managed to obtain a leave of absence from RRE to work at Bell Laboratories on microwave devices. Towards the end of his time there Theodore Maiman's paper was published announcing the operation of the first laser. Jim Gordon, who was managing Forrester's group, suggested that they ought to try to make a laser. This they achieved before Forrester's time at Bell ended.

When Peter Forrester returned to the UK at the end of January 1961 he had planned to resume working on masers. However, R.A. Smith said he thought that RRE ought to begin working on lasers. Forrester did not want to become involved in laser research because optics was not his field of expertise. As he had already made a working laser, however, Smith insisted that he and another RRE scientist begin a small programme of laser research.[62] A formal statement of the rationale for establishing a laser group at RRE could not be located. However, Forrester suggested that lasers were going to be of great technological importance and that RRE, which considered itself to be a premier electronics establishment, ought therefore to be active in this field.[63] Their initial objective was merely to construct a working laser of their own.[64] This they achieved at the end of July 1961, using a ruby laser.[65] Following this successful demonstration a Laser Division was formed at RRE, initially consisting of about 10 scientists, and work on lasers stepped up.[66]

The formal objectives of the Laser Division were initially to develop lasers that operated at wavelengths other than that of ruby (particularly in the infrared region), and to increase the efficiency of the present lasers, leading to devices capable of continuous or high PRF operation.[67] These objectives appear to have been selected because RRE chose to focus this research on investigating the possible use of lasers in optical radar systems. This type of application would require a laser with several characteristics. First, it would have to be capable of high PRF operation. Second, the light it emitted would have to be at a wavelength where atmospheric transmission was satisfactory and for which suitable detectors existed. The output power of the beam would also have to be adequate.

These objectives were pursued in a very exploratory manner. Two general routes were followed. One was to attempt to improve the performance of ruby lasers; the other was to search for new candidate laser materials that might not have ruby's limitations.

Attempts to improve the quality of existing ruby lasers led the group to explore the use of external resonators as a method for controlling the unstable output. In doing so they built a rotating wheel with a great number of slots: if this was spun at high speed it was found that the laser could pulse through it.[68] According to Peter Forrester they were one step away from discovering Q-switching.[69] This is a process by which the output power of a laser can be greatly increased by temporarily suppressing one of the laser mirrors, so that the radiation that hits it is not reflected into the resonant cavity. This inhibits stimulated emission and therefore allows the inverted population to build up to a high level. The mirror is then reactivated, stimulated emission is allowed to occur on a large scale, and a giant pulse of energy is released. Unfortunately, the RRE team failed to make the necessary mental connection and in 1961 were beaten by Robert Hellwarth and Fred McClung from Hughes Research Laboratories.[70]

A similar missed opportunity occurred in the case of mode locking.[71] This technique allows the generation of even higher-power pulses than Q-switching by breaking up a single pulse into a sequence of very sharp, equally spaced pulses using a suitable resonator. The RRE scientists realised that by looking at the output spectrum of ruby with a fast amplifier beats could be seen in the ruby laser signal, and that these were mode beats. Unfortunately, they failed to take the next mental step and discover that mode locking was possible. Mode locking was first demonstrated by Hargrove and colleagues in 1964, using helium–neon lasers.[72]

The Laser Division also attempted to improve the performance of ruby lasers by improving the efficiency of the excitation process. The advantages of using linear flash tubes located within an elliptical cavity, instead of a helical flash tube inside a spherical cavity, had already been demonstrated abroad. A number of linear flash tubes and elliptical cavities were constructed at RRE and their efficiency compared with helical flash tubes in spherical tubes. The flash linear tubes had to be constructed at RRE because there were no UK suppliers for this component, and the only US firm that made them was unwilling to supply RRE. This encouraged CVD to place a contract for the supply of flash tubes with English Electric Valve (EEV) Ltd.

Another area where the domestic supply of solid-state laser components presented problems was good-quality ruby rods.[73] The first RRE ruby lasers used rods obtained from the USA. Ruby laser rods were then obtained from some UK and French sources, but these proved inferior to the American rods. In response, RRE placed a contracts with Thermal Syndicate and GEC to provide better-quality rubies. Commercial polishing of ruby laser rods also proved unsatisfactory, and in response a polishing shop was established at RRE. Unfortunately, its completion was significantly delayed.

A number of additional laser materials were also investigated. The first were calcium fluoride doped with samarium and calcium fluoride doped in uranium.[74] These materials had already been shown to lase by Stevenson and Sorokin at IBM Research Laboratories. Crystals were grown by RRE and supplied by a number of UK firms through CVD contracts.

Some effort was also devoted to investigating lasers based on other media. However, these were low-priority background activities undertaken when staff were available. A helium–neon laser was constructed because at this time these were the only ones capable of high PRF or continuous operation. However, work on gas lasers ceased following the 1962 agreement on the division of responsibility for laser work with SERL.[75] Some consideration was also given to the possibility of achieving laser operation in liquids. An extramural research contract was placed with Woolwich Polytechnic to investigate candidate liquids. Unfortunately, this did not yield any promising candidate lasers and work on this type of laser was terminated.[76] Staff at RRE also considered the potential of glass lasers.[77]

During 1962 a second group, consisting of about 10 scientists, was formed within the Laser Division to investigate laser techniques required to realise some of the device's potential applications.[78] This group started a number of laser-ranging experiments in 1963, using a Q-switched ruby laser and a helium–neon laser.[79] An RRE report on potential laser applications was prepared.[80]

Staff at RRE continued to focus on investigating solid lasers for radar-like applications[81] and attempted to improve the performance of ruby lasers. Unfortunately, the quality of ruby laser rods produced by UK firms was disappointing, and these lasers proved incapable of operating at high PRF. Consequently, the search for new laser materials continued. Effort initially focused on neodymium-doped calcium tungstate lasers. They had already been demonstrated by Johnson and Nassau at Bell Laboratories in 1961, and were the first type of laser to operate in a continuous mode. However, they would only operate at 7 K, and therefore required a cryogenic cooling system. Neodymium-doped calcium tungstate laser rods and flash tubes were developed through CVD contracts, and a number of lasers constructed at RRE.

During 1965 emphasis shifted to lasers that were capable of generating a continuous beam at room temperature. Attention was focused on neodymium-doped calcium fluoride, which was just capable of continuous operation at room temperature. The following year scientists at RRE started to become interested in neodymium-doped yttrium aluminium garnet lasers. This type of laser had been demonstrated by Joseph Geusic and his co-workers from Bell Laboratories in 1964,[82] and was regarded as more promising than neodymium-doped calcium tungstate because it was believed to have a lower threshold. To support this work extramural contracts to develop laser rods and flashlamps were placed with British firms through CVD, and a number of neodymium-doped yttrium aluminium garnet lasers assembled at RRE.

The use of semiconductor lasers as a pump source instead of flashlamps was also investigated. To improve the quality of laser crystals an RRE polishing shop was improved. Unfortunately, the completion of this conversion was delayed considerably. Work was also undertaken on the development of Q-switches and detectors for use with lasers.

Studies of potential applications continued.[83] Ruby, neodymium-doped calcium tungstate and neodymium-doped yttrium aluminium garnet laser rangefinders were constructed (Fig. 10.1) and trials undertaken at the RAE range at Larkhill. Trials were also undertaken jointly with RAE and Hawker Siddeley Dynamics, of a semiactive guidance system. This employed a neodymium-doped calcium tungstate laser designator and a modified Firestreak missile homing head. In addition to rangefinder trials, one of these rangefinders was also used to measure emissions of toxic gases from the CDEE. Some effort was also devoted to the development of a laser microwelding system.

During 1966, effort on laser research at RRE was reduced significantly. No formal explanation for this decision could be found. However, the evidence suggests that it was due to a feeling of disappointment at the failure of the laser to rapidly find applications, combined with competing demands on resources. The remaining effort was largely deployed on the continued development of high-PRF neodymium-doped calcium tungstate lasers for possible laser-radar applications. To support this work staff at RRE continued to develop new flashlamps and Q-switches. One of the lasers was licensed to industry for commercial production. Scientists also demonstrated frequency doubling with these lasers, and grew non-linear optical crystals to support this work. They also demonstrated mode locking in a neodymium-doped yttrium aluminium garnet laser in 1969.

FIGURE 10.1: LASER RANGEFINDER, RESEARCH MODEL *C.* 1962 (SCIENCE MUSEUM INV. NO. 1979–0054, SCIENCE MUSEUM/SSPL)

Trials of experimental neodymium-doped calcium tungstate and neodymium-doped yttrium aluminium garnet laser rangefinders continued. In 1966 RRE issued a specification for an experimental laser rangefinder and TV sight to be test flown on a Varsity aircraft.[84] The system was to be based on an RRE-supplied neodymium-doped calcium tungstate laser, and the selected contractor was required to design and supply the optics, which were to include a device to point the laser beam and TV sightline and stabilise the system against airframe movements. Approximately a year later the contract was won by Ferranti and the equipment was delivered in September 1967. However, shortly afterwards responsibility for airborne laser rangefinders was transferred to RAE, and the equipment was converted to new tasks.

Laser work resumed as a major effort at RRE in 1970, and responsibility for this work was transferred from the Electronics Group to the Physics Group.[85] One of the first major activities of the new laser group was to re-establish facilities for making measurements on pulsed lasers. This entailed both refurbishing existing equipment and purchasing new items of equipment. The group was also initially smaller than planned, owing to a shortage of staff.[86]

The primary aim of the laser research programme remained the investigation of solid-state lasers for radar-like applications.[87] Some work continued on the refinement of neodymium-doped yttrium aluminium garnet lasers, but the specific direction of the research changed. Until this point it had been to identify a laser material that could operate in a continuous mode at room temperature and produce a beam of light that could be detected. Instead, they now began to look for cheaper materials than neodymium-doped yttrium aluminium garnet, with the same attributes.[88] Neodymium-doped glass and several solid-state materials were considered, and a limited amount of effort was also devoted to the evaluation of dye lasers. However, the latter was discontinued after a short time owing to the redistribution of staff. Towards the mid-1970s the group began to turn its attention towards the identification of eye-safe solid-state lasers.[89] Two approaches were identified. One was to use a neodymium laser and then convert the radiation to a different wavelength using stimulated Raman conversion. However, this approach was abandoned because of the amount of development work required. The other approach was to use holmium-doped yttrium aluminium garnet and holmium-doped yttrium lithium fluoride.

Work also continued on the assessment and development of new electro-optical components. A significant success was achieved in 1970, when a cheap reliable glass acoustic–optic Q-switch was developed for a high-PRF laser that both met the requirement and absorbed 30 per cent less energy than comparable US systems. This design was transferred to industry and the device manufactured under licence. It was also used in an experimental laser forward-looking system.[90]

A number of potential applications were studied during this period. Doppler laser-radar techniques were investigated for the measurement of wind velocity and turbulence. These employed a carbon dioxide laser supplied by SERL. Although operational difficulties were still to be overcome, it was demonstrated that these lasers could measure target velocity.[91]

In response to a request from the Army, a neodymium-doped yttrium aluminium garnet handheld laser rangefinder was developed in a collaborative project between RRE and RARDE.[92] Because of their existing work in this area the establishments were able to respond very quickly to the request, developing the device in only seven months. RARDE supplied the optics, and the rest of the equipment was developed by RRE. The design was then transferred to Barr and Stroud Ltd for commercial exploitation, and 12 production models were ordered. RRE then conducted further studies aimed at reducing the cost of the system and extending its applications. This rangefinder programme was portrayed as 'outstandingly successful' in the RRE Annual Report.[93] However, the former manager of the Barr and Stroud laser group has suggested that this programme was not as successful as it could have been.[94] He suggests that this stemmed from the institutional rivalry between RRE and RARDE, lack of engineering expertise in the former establishment, and the availability of a competitive product from the Norwegian firm Simrad. RARDE, which was responsible for assessing the system, suggested that the rangefinder was too heavy. In response, RRE proposed constructing the device from magnesium alloy. However, lacking the necessary engineering expertise, they failed to appreciate the extent to which the use of this material would complicate the manufacturing process, because magnesium alloy is extremely flammable. At the same time Simrad was marketing a better-engineered competitive product, which seized the market. No further orders for the Barr and Stroud handheld laser rangefinder were received, and consequently the device was not a commercial success. Following on from the above work RRE (and later RSRE) began to investigate the possibility of constructing eye-safe handheld laser rangefinders. A holmium-doped yttrium lithium fluoride, self-contained, eye-safe, handheld laser rangefinder was developed by RSRE and tested. Trials with the equipment showed that, under conditions of good visibility, the performance of the equipment was comparable to that of neodymium-doped yttrium aluminium garnet-based systems,[95] but the performance of the eye-safe laser was superior when visibility was reduced. Unfortunately, the rangefinder was never procured.

During this period RRE, in collaboration with MVEE, began to investigate possible laser rangefinders for a future UK main battle tank (later to become the MBT80 programme).[96] An experimental pulsed carbon-dioxide laser rangefinder was built and compared with ruby and neodymium-doped yttrium aluminium garnet systems. These studies suggested that carbon-dioxide laser rangefinders offered potential improvements in performance over ruby and neodymium-doped yttrium aluminium garnet models, being almost unaffected by smoke, dust and haze, and generating laser radiation at an eye-safe wavelength. As this work was completed effort was transferred to the development of a heterodyne carbon-dioxide waveguide laser rangefinder which could measure not only range, but also the velocity of the target. This type of laser rangefinder was also expected to be smaller and more rugged than a neodymium-doped yttrium aluminium garnet laser rangefinder (which was already in use), and to cost much less.

In addition to the work on lasers themselves, a small group at RRE led by Roy Pike became involved in conducting research in the photon counting and statistics of laser radiation. The group quickly became one of the world leaders in this field, and developed and patented a system called the Malvern Correlator, which can be used to measure fluid flow in a non-invasive manner.[97] The Malvern Correlator rapidly found many defence and civil applications, in areas such as the prediction of air turbulence ahead of an aircraft, or the design of engines. At the beginning of the 1970s the patent was transferred to the NRDC for commercial exploitation. This was taken up by Malvern Instruments Ltd, who achieved great commercial success with the product. For their work in this field RRE won a MacRobert Award from the Confederation of British Industry.

LASER R&D AT SERL

SERL was established in 1945 by the Admiralty to undertake vacuum and discharge tube R&D to satisfy the needs of the UK's Armed Forces. SERL was also the official research laboratory of the Admiralty's CVD organisation, and through this body has strong links with universities and industry. During the 1950s, the remit of SERL was extended to include other active electronic components when it formed a Semiconductor Devices Division and a Nuclear Devices Division. A further extension to the work of SERL occurred during the 1960s, when it became actively involved in gas and semiconductor laser R&D.

Research on gas lasers began at SERL during 1961.[98] Some of the laboratory's existing work had been terminated, and after a review it was decided that lasers offered the greatest potential future in defence applications. An Optical Maser Division was formed under Harry Boot and began work on gas lasers. One reason for this was because the establishment's considerable expertise lay in high-vacuum techniques and discharge tubes, both of which were important enabling technologies in the development of gas lasers. Another important factor was that RRE wanted to focus on solid-state lasers because of their expertise in solid-state masers. The objectives of the Optical Maser Division were to construct a fully engineered helium–neon laser with reasonable life; to identify gas mixtures which could exhibit laser action; and to investigate possible applications for these devices.

In 1962 no commercial source of gas lasers existed within the UK, so the first objective of the gas laser programme was to ensure that a satisfactory device was available for experimental purposes. Members of the Optical Maser Division succeeded in operating their first helium–neon laser in 1962, assisted by John Sander's laser group at Oxford University, which was funded through a CVD contract. This was the first gas laser to operate in the UK and was essentially a copy of the Bell model. Improved models of the helium–neon were developed based on advances made in the USA. A small production line was set up within the establishment and 106 models were manufactured, 22 of which were supplied to other government laboratories. SERL received many enquiries from industry regarding the manufacturing techniques for these devices, and held a demonstration during 1963. In the same year SERL also transferred the design drawings to the NRDC for commercial exploitation. Elliott Automation Ltd and G. & E. Bradley both took out licences and began to manufacture lasers based on this design.

There appear to have been three major limitations with the early helium–neon lasers constructed at SERL, as far as its staff were concerned. These were the laser's limited and variable life, its lack of stability and its low output power. Work was undertaken to address all these issues. Lasers with substantially longer lives were developed, a highly stable device was constructed in collaboration with NPL, and significant increases in output power were achieved using a helium–neon laser amplifier. In 1964, however, work on helium–neon lasers was terminated and effort diverted to the consideration of other types of gas laser.

The second principal objective of the Optical Maser Division was to identify new types of gas laser. In particular, it would appear that staff at SERL were seeking to identify gas lasers with a higher output power than helium–neon. They were also looking for gas lasers that generated radiation at different wavelengths.

In 1963 two of the Division's members, L.E.S. Mathias and J.T. Parker, succeeded in operating the world's first molecular gas laser when they discovered laser action in nitrogen.[99] This was a significant breakthrough, because this type of laser could generate a beam of much higher output power than had hitherto been possible with a gas laser. It also emitted radiation in the near-infrared region of the spectrum, where atmospheric transmission is good. Unfortunately, its practical applications were limited by the fact that it was a pulsed device with a low PRF, so the mean power was limited. Mathias and Parker also succeeded in operating the first carbon-monoxide laser later the same year.[100] This produced radiation in the visible region of the spectrum, but the beam was also pulsed and had low peak power.

After examining lasers that operated in the visible and near infrared, they then moved on to look at molecular lasers that operated in the far infrared. At the time far-infrared lasers were expected to have applications in several potential areas where the good atmospheric transmission in mist and fog would be an important factor, and in specialist areas such as spectroscopy and plasma diagnostics. A collaborative research programme was established with RRE and NPL, and in 1964 they were the first group to report laser operation in water vapour.[101] To ensure that a far-infrared source would be available within the UK a sealed-off water vapour laser was designed for industrial development. Members of the team then examined other potential far-infrared laser sources. However, collaboration with NPL appears to have ceased, and that group continued to work on far-infrared lasers independently.

By 1966, work on molecular lasers at SERL had become focused on carbon-dioxide devices. Between 1964 and 1965 Kumar Patel, who was working at Bell Laboratories, first demonstrated a continuous carbon-dioxide laser and then managed to steadily increase the output power of the device by the addition of other gases.[102] By 1966, staff at SERL regarded these as the most powerful and efficient continuous lasers. Furthermore, they were also known to emit radiation at a wavelength where atmospheric transmission was good. A low-power carbon-dioxide laser was designed and samples supplied to a number of government establishments for exploratory work. A large amount of effort was also devoted to finding ways of increasing the output power of the carbon-dioxide laser. Early research suggested that the output power was proportional to the length of the laser tube, and not its diameter. Therefore, the output power could only be increased by constructing longer lasers. The length could be reduced by folding the laser back on itself. Nevertheless, this approach made carbon dioxide lasers extremely bulky. In response, staff at SERL also began to look at methods of increasing the output power per unit length. A number of different approaches were explored, including the use of pulsed excitation, higher gas pressure and increased gas flow rate.

From the mid-1960s staff at SERL also undertook work on ion lasers. During 1964 ion lasers were discovered in the USA. These appeared to have potential because they could produce higher output power than atomic lasers such as helium–neon, and could operate in the blue/green region of the spectrum, which suggested possible applications in holography and antisubmarine warfare. The design of an argon-ion laser was eventually transferred to NRDC for commercial exploitation, and was eventually licensed by Ferranti Ltd.

In addition to the construction of lasers the group also began to consider possible applications for the device. One team spent a considerable time attempting to make a working ring-laser gyroscope for use in submarine navigation systems. Unfortunately, the device could never be made sufficiently stable to equal the performance of the gyroscope in the ship's inertial navigation system, and the project was eventually abandoned.

Techniques for using argon-ion lasers in holographic techniques for the manufacture of integrated circuits were also developed, patented, and the licence transferred to the NRDC.

Scientists and engineers at the establishment also developed very high-power carbon-dioxide lasers for exploring cutting and welding applications and its potential use as a beam weapon. After trials at SERL the design of the first of these was eventually licensed to Ferranti for commercial manufacture. A second improved model with a higher output power was developed for the Welding Institute. This was followed by a third model with even higher output power, which was used to provide technical advice to the Welding Institute on the design of a 2 kW laser, and to CVD to aid design studies at Elliots Automation Ltd and the United Kingdom Atomic Energy Authority (UKAEA) Culham Laboratories of a 10 kW welding laser for the Navy. Work on lasers for this application ceased in about 1973, however, when responsibility for this type of work was transferred to Culham.

During the late 1960s scientists at the establishment also began to seek to design carbon-dioxide lasers suitable for optical radar applications. This encouraged them to attempt to make these lasers compact and sealed-off, so they did not depend on external gas supplies.

During the early 1960s staff at SERL also began to undertake research on semiconductor lasers.[103] SERL's Semiconductor Devices Division was already growing gallium arsenide to support investigations into the possible use of this material in transistors and light-emitting diodes.[104] When lasing was discovered in solid-state and gaseous media, a couple of the members of the Division began to consider whether it was also theoretically possible to make a semiconductor laser. They had almost come to the conclusion that it was not feasible, when the operation of gallium-arsenide semiconductor lasers was reported by groups at IBM and General Electric in the USA.[105]

The initial response of the Division to this discovery was to build some gallium-arsenide semiconductor lasers of their own, in order to understand their properties and methods of construction. However, like the early US models, they could not be used in a continuous mode; produced a low-power beam of the order of 1 W; and had to be cooled to 77 K with liquid nitrogen. One member of the Division succeeded in operating a gallium-arsenide laser at room temperature by passing a very high current through the material. However, the efficiency of this device was only 1 per cent.[106]

After this initial exploratory work, the laser research of the Semiconductor Devices Division was focused on the development of gallium-arsenide lasers with higher output power. After the initial thrust towards high power, the emphasis of the gallium-arsenide laser programme appears to have changed in the mid-1960s towards developing lasers capable of operation at room temperature. The rationale for this would seem to have been the desire to eliminate the need for cooling systems, which were complicated and expensive, and therefore thought to be highly undesirable for practical applications. This intramural work was supported by an extramural R&D programme funded through CVD. A number of these contracts were placed with universities and Standard Telecommunications Laboratories (STL), the R&D arm of Standard Telephones and Cables (STC).

In addition to this research programme, the Semiconductor Devices Division also undertook pilot production of two standard types of gallium-arsenide semiconductor lasers. One was a low-power device with a junction area of 0.4 × 0.4 mm, and the other was a higher-power model with a junction area of 2 × 1 mm. During the early 1960s there were no commercial manufacturers of semiconductor lasers, and they therefore had to be produced in-house. This production line not only served SERL's internal needs, but also the requirements of the other government research establishments and their contractors. For a number of years it was the only source of semiconductor lasers in the UK.

Several applications of the semiconductor laser were seriously considered at that time. It was hoped that they could be used to 'pump' glass and solid-state lasers because they possessed superior electrical energy-to-light conversion efficiency to flashlamps, and a joint research programme to investigate this possibility was initiated with RRE. Other applications that were considered were: an optical guidance system for the Rapier surface-to-air missile; a tank identification-friend-or-foe system and laser fusing with RRE; an aircraft laser altimeter and a communications link for a guided bomb with RAE; an antitank missile flare with RARDE; and a ship-to-ship communications system. Some feasibility contracts were placed with industry. However, the only one that appears to have moved to full-scale production was the use of semiconductor lasers in combat simulation aids.

During 1967 research on gallium-arsenide lasers ceased in SERL as commercial supplies became increasingly available from STC. This freed resources for the laboratory's expanding silicon semiconductor device programme. However, members of the establishment (and later staff at RSRE) continued to monitor and place extramural contracts for semiconductor laser R&D throughout the 1960s, 1970s and 1980s.

During the 1970s laser work at SERL was directed towards investigating gas lasers suitable for target ranging and tracking systems, and for beam weapons applications.[107] Interest in the possible use of carbon-dioxide lasers for rangefinding had developed at SERL during the late 1960s. In 1970 staff at SERL became aware of A. Jacques Beaulieu's invention of the transversely excited atmospheric pressure (TEA) laser at the Defence Research Establishment Valcartier (DREV) in Canada.[108] In this type of laser a high voltage is applied perpendicular to the direction of the gas flow, which is at atmospheric pressure. TEA lasers were perceived to have a number of potential advantages over other types, including higher energy/pulse, high efficiency and a simple mechanical design because no Q-switch was required. A research programme on TEA lasers was rapidly initiated, and the team began to look at addressing one of their early limitations, which was the fact that the discharge was not uniform. This had the effect of limiting the energy obtained per unit volume, and the mode purity. A major advance was achieved when two members developed a method for overcoming this problem using an auxiliary trigger electrode.[109] This led to the output power being more than doubled.

The improved performance of the TEA laser compared to conventional types of gas laser led staff at SERL to suggest that they 'change radically the thinking on carbon dioxide laser applications'.[110] SERL supplied a number of low-PRF TEA carbon-dioxide lasers to RRE for comparative assessment with neodymium-doped yttrium aluminium garnet and ruby devices in rangefinding applications. RRE also requested SERL to undertake preliminary development of a high-PRF laser for a guidance and tracking system.

In 1971, staff at SERL began to study the feasibility of constructing a sealed-off low-PRF TEA carbon-dioxide laser. A problem encountered with these devices was that they only had very short lives, and this was believed to be due to the dissociation of the carbon dioxide into carbon monoxide and oxygen. This problem was overcome by the staff at SERL adding appropriate concentrations of carbon monoxide and hydrogen to the normal gas mixture. Consequently, during the early 1970s they were able to demonstrate the world's first successful sealed TEA carbon-dioxide laser (Fig. 10.2).[111] A CVD contract was placed with Marconi Elliott Avionics Systems (MEAS) Ltd to develop this laser. MEAS also mechanically redesigned the device to improve performance. Some of these lasers were eventually supplied to RRE and Ferranti Ltd for incorporation into rangefinding systems. Meanwhile, staff at SERL continued research on this type of laser with the aim of reducing its length and increasing its peak power. In furtherance of this objective, between the early 1970s and 1990 they developed further significant refinements.[112]

Interest in the possible use of lasers in radar systems also led staff at SERL to begin to investigate waveguide carbon-dioxide lasers, so called because the gas mixture is confined at high pressure in a very small-diameter tube. These lasers were extremely compact and rugged. They were also expected to offer increased tunability and thus be capable of providing both range and velocity information. To support this work a CVD contract was placed with Ferranti Ltd. In response to a request from RRE, staff at SERL developed a sealed-off version of a waveguide carbon-dioxide laser for incorporation into a ranging system, and for the investigation of laser-radar applications.

A small amount of effort was also devoted to studying the use of the TEA laser configuration with other gases. Hydrogen chloride and deuterium fluoride were considered. However, this was not a priority project and appears to have been discontinued after a short period.

During the early 1970s a major research programme to evaluate the potential use of high-power lasers for use as beam weapons began at SERL in collaboration with several other government defence R&D establishments. One type of laser that was considered for this type of application was the electron-beam (e-beam) pumped laser, which had already been demonstrated in the USA and the Soviet Union. Staff at SERL believed e-beam lasers had an advantage over TEA lasers in that they were scalable to very high powers, and could generate pulses of longer length. Their first e-beam laser used carbon dioxide as the active gas medium. A number of single-shot experimental e-beam lasers were constructed at SERL. Staff at the establishment then constructed a double-pulsed system, which was fully developed by MEAS under

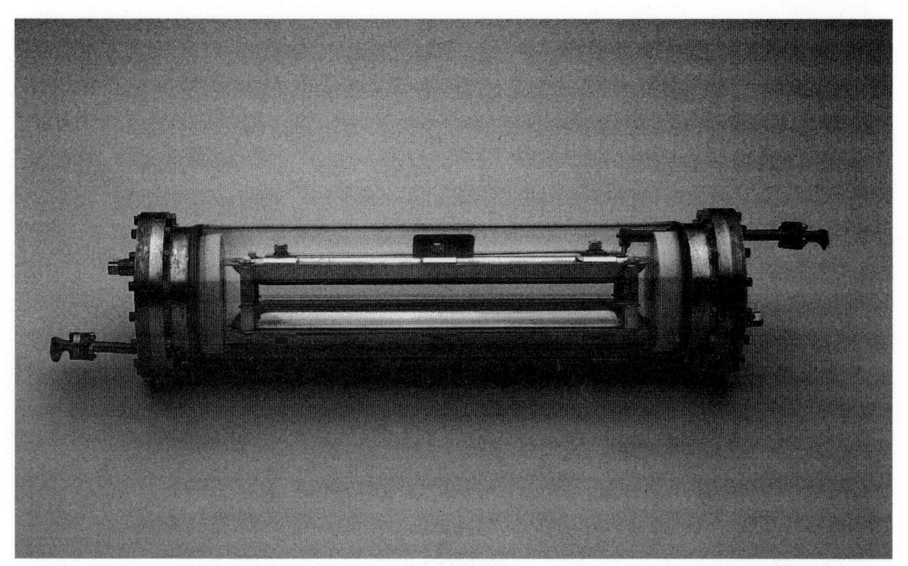

FIGURE 10.2: FIRST SEALED-OFF CO_2 PULSED TEA LASER, *c.* 1973 (SCIENCE MUSEUM INV. NO. 1982–0140, SCIENCE MUSEUM/SSPL)

a CVD contract and had an output energy in excess of 1 kJ. SERL also constructed several high-PRF e-beam lasers of increasing powers (Fig. 10.3). The last of these produced a beam with a mean power in excess of 30 kW, making it the most powerful laser of this type in the Western world. Ultimately it was transferred to the Naval Research Laboratory in Washington as a UK contribution to an international technology transfer agreement. Intramural research on e-beam excited chemical lasers was also conducted at SERL.

During the course of their work on e-beam lasers staff at SERL achieved several important advances. One was to develop a novel electron gun design.[113] This was originally developed by GEC under a CVD contract, but later modified by SERL. A second was the development of improved designs to separate the electron gun from the resonator cavity.[114] Third, a phenomenon called laser-induced medium perturbation (LIMP) was discovered.[115] This fundamental effect limits the quality of the laser output beam.

In addition to examining e-beam lasers, SERL also considered the use of gas-dynamic and chemical lasers for high-power applications. The reason for this was that e-beam lasers can only operate in a pulsed mode, whereas the other lasers could potentially be used to generate continuous beams. CVD contracts were placed with Rolls-Royce to study both types, and this work was supported by theoretical studies conducted within the establishment and universities funded through CVD. This programme was successful from a technical point of view. However, it became clear that the way forward was to use pulsed lasers, so further work on gas-dynamic and chemical lasers was discontinued and effort focused on e-beam lasers. This work continued for about another year, until the end of the 1970s, when the MOD took

FIGURE 10.3: A HIGH-POWER, HIGH-REPETITION-RATE, ELECTRON-BEAM SUSTAINED CARBON DIOXIDE LASER CONSTRUCTED BY SERL DURING THE 1970S. IT WAS THE MOST POWERFUL LASER OF THIS TYPE IN THE WESTERN WORLD. (© CROWN COPYRIGHT, DERA)

a policy decision not to continue it because of the high cost of further developing the technology.

LASER R&D AT SRDE

As with RRE and SERL, laser R&D at SRDE was initiated in the early 1960s. By 1963 the establishment had initiated a broad programme of research covering studies of laser materials, new types of laser, the characteristics of laser radiation and potential applications.[116] However, following the 1963 division of responsibility agreed between the establishments, this work became increasingly focused on the study of the application of lasers in night-vision and communications systems.[117] Despite the breadth of the SRDE laser programme, the overall level of intramural effort involved appears to have been small compared to that of the largest laser groups in the UK. For example, in 1967 official papers put the number of white paper grades working on lasers at SRDE at 'about 5', compared to 30 at RRE and 14 at RAE.[118]

Staff at SRDE used solid-state, gas and semiconductor lasers in their research. They initially assembled ruby, neodymium-doped glass and neodymium-doped calcium tungstate lasers themselves, using some components supplied by firms. Experiments with laser amplification and laser diode pumping were undertaken. Electro-optical, rotating prism and dye Q-switches were also constructed, and the shifting of the laser's output was investigated using non-linear optical materials. However, SDRE's role in the development of lasers appears to have been limited, and declined as the decade progressed. The reason for this appears to have been the far greater level of resources committed by RRE and SERL, and their division of responsibility for fundamental research on solid-state, gas and semiconductor lasers. This left little scope for SRDE to make a significant impact in the field. The establishment relied increasingly upon outside sources for its solid-state lasers, and when it began to use gas and semiconductor lasers these were supplied by SERL.

Research into the potential applications of the laser was also undertaken. A study to investigate laser radiation damage on military optical systems was initiated. However, the 1963 division of responsibility gave RAE the responsibility for studying beam weapon applications of lasers, and in consequence the SRDE project was terminated. The 1963 agreement did, however, give SRDE responsibility for investigating night-vision and communications applications of the laser. SRDE constructed experimental night-viewing systems using ruby and neodymium-doped glass lasers in conjunction with an infrared image converter. Field trials were undertaken. Investigation of the use of lasers for short-range communications purposes covered the application of gallium-arsenide laser sources, modulation methods and atmospheric propagation studies. An experimental system was demonstrated and a contract placed with Plessey Ltd for its development.

At the end of the 1960s SRDE was reorganised and its laser programme became even more specifically tied to the study of night-vision and communications applications.[119] The latter work became associated with its broader interest in fibreoptic communications for use in battlefield communications systems, optical data buses on aircraft, and optical data links to missiles. Fibreoptic communications had a number of potential advantages over conventional copper cables in terms of data capacity, lightness, strength and resistance to electromagnetic pulses encountered in a nuclear environment. A considerable amount of work was sponsored in this area during the 1970s. However, the Ptarmigan battlefield communications system, which was then under development, eventually used copper cables rather than fibreoptics because of the higher risk associated with the latter.

CONCLUSIONS

Laser research was initiated at the defence R&D establishments because the laser was seen as a device which might have widespread military applications. However, these ideas were extremely speculative and unproven. Therefore, further research was necessary to clarify the capabilities and limitations of these devices and their possible uses. The performance of a laser could be measured in a number of ways, such as its output power, PRF, stability and wavelength. To be of use in a specific application it was necessary for a laser not just to have sufficient performance in one of these areas, but to have the right combination of performance characteristics. The first lasers lacked these for most of the applications that were envisaged. In response, scientists and engineers sought to identify new types of laser that would possess such characteristics. As these were discovered new applications became feasible, and effort then focused on refining and tailoring particular types of laser. Thus, the growth of laser research was driven by a mixture of technology push and need pull, with technology push being dominant at the early stages and need pull being more important later on.

The decision to initiate laser research was taken by senior staff within the R&D establishments, not by the departments that controlled them. This was because the responsibility for determining the content of the research programmes of the establishments appears to have lain primarily with their directors. The departments did, however, have an oversight role over the research programmes, providing guidance as to major changes in resource allocation and negotiating divisions of responsibility for various aspects of the work. As the results of the research programme were exploited in development programmes, the department became far more actively involved. Thus, the establishments had considerable latitude in exploring new areas of research. However, it was necessary to obtain the support of their departments if these technologies were to be fully exploited in new weapons systems.

Despite the early lead established by the USA in the laser field, the UK still sought to build up and sustain its own laser defence technology base. The reason for this appears to have been that it aimed to maintain an independent capability, both in terms of possessing the relevant knowledge and also of having an indigenous manufacturing capability. To achieve this a major programme of work was funded at the defence R&D establishments. Defence expenditure provided the largest source of funding for laser R&D in the UK, and the government establishments that provided most of the effort were all primarily concerned with defence work.

Virtually every non-nuclear defence research establishment has had a laser R&D programme at some stage since 1960. This was because lasers are a generic technology with widespread potential defence applications. Some establishments became involved in laser work because of their expertise in electronic components, whereas others became interested because of the potential applications of the laser. The particular types of laser work undertaken by the different establishments, therefore, appear to have been determined by the core technical activities of the relevant establishments.

The UK's defence R&D effort in the laser field was concentrated on those types of laser that appeared most promising for defence applications. Staff in the defence R&D establishments sought to undertake some original work and some significant world 'firsts' were achieved, particularly in the gas laser field. In this area SERL gained an international reputation. However, the speed with which laser technology developed in the USA ensured that a considerable amount of the UK's effort was devoted to understanding and indigenising advances achieved overseas. To further this, international defence research collaboration was promoted through a variety of mechanisms. However, there were perceived to be limitations to what international defence research collaboration could achieve, and thus it was not regarded as a panacea. Once the defence R&D establishments acquired technology, this was then transferred to industry for exploitation in new weapons systems, usually through the placement of extramural R&D contracts; the CVD organisation played an important role in this respect.

Despite the large amount of effort devoted to laser work very little of it was ultimately exploited in operational weapons systems. The main areas of application have been in rangefinding and target designation. Here the laser has helped revolutionise the conduct of air interdiction and close air support operations. However, there is evidence to suggest that much of the technology used in these systems was licensed by British firms from their US counterparts.[120] Yet when staff at SERL developed a world-class capability to design compact carbon-dioxide lasers for these types of application they were never exploited by the MOD, a decision that remains surprising given the operational benefits they would have conferred. In other cases lasers have not been exploited in defence applications because of performance limitations, the perceived risk involved and resource limitations.

Scientists and engineers in the defence research establishments did provide advice to their parent departments. However, the quality of this technical advice was dependent upon their level of technical excellence. Thus, in some cases they were able to recommend superior technological solutions, whereas in others their advice merely reflected received wisdom in the field.

The record of the establishments in undertaking project development work also appears mixed. Evidence from one former industrial manager of a laser R&D programme suggests that, despite the fact that some of the establishments attempted to undertake development, in at least one instance they were not fully aware of the manufacturing issues associated with the system because of their lack of engineers. However, there are other instances where the establishments were involved in development and the lasers were transferred to industry and very successfully commercially exploited.

In terms of the provision of large capital equipment the defence R&D establishments did not play a major role. Most laser work was laboratory based and did not require the provision of major plant or facilities.

Laser R&D also yielded spinoffs. Some of these arose from work of a general nature. Advances in the design of lasers were usually disseminated in the unclassified literature and unclassified conferences, and the designs transferred to the NRDC for commercial exploitation. Other spinoffs occurred because the work was application specific but dual use, such as that on laser cutting and welding systems. This work was also licensed to industry for commercial exploitation.

The achievements of the defence R&D research establishments would probably have been even greater if their individual laser programmes had been better coordinated. During the 1960s there were occasions when collaborative laser research programmes were undertaken between the establishments. However, when laser programmes were first started overall coordination between the different establishments was relatively weak. Consequently, significant overlap developed in some areas of research concerned with these devices and their applications, and the situation was only resolved by the establishment of a series of coordinating committees.

There is also evidence to suggest that there was some occasional short-termism in the management of the laser programme in particular establishments. This is perhaps best illustrated by the decision to terminate work on solid-state lasers at RRE, only to restart it a few years later. This appears to have stemmed from a sense of disappointment at the failure of lasers to yield significant defence applications, and pressure on resources.

1. B.S. Lambeth, 'The technological revolution in air warfare', *Survival*, 39 (1997): 68–9.
2. Dye lasers are an exception in that they are tuneable over a limited range of wavelengths.
3. In this chapter the term 'defence R&D establishment' will be taken to refer to the establishments currently managed by DERA. Consequently, laser research undertaken at the Atomic Weapons Establishment, Aldermaston will not be included.
4. M. Bertolotti, *Masers and Lasers: an historical approach* (Adam Hilger, 1983); M. Bertolotti, 'Twenty-five years of the laser, the European contribution to its development', *Optica Acta*, 32 (1985): 961–80.
5. See, for example, the articles in Vol. 20, no. 6 of the *IEEE Journal of Quantum Electronics*; and the papers in *Laser Pioneers*, edited by J. Hecht (Academic Press, 1992).
6. P. Forman, 'Beyond quantum electronics: national security as the basis for physical research in the United States 1940–1969', *Historical Studies in the Physical and Biological Sciences*, 18, no. 1 (1987): 149–229.
7. W. Eppers, 'The role of the military in laser research and development', *Society of Photo-Optical Instrumentation Engineers*, 41 (1985): 31–5.
8. R.W. Seidel, 'From glow to flow: a history of military laser research and development', *Historical Studies in the Physical and Biological Sciences*, 18, no. 1 (1987): 111–47; R.W. Seidel, 'How the military responded to the laser', *Physics Today*, 41, no. 10 (October 1988): 36–43.
9. J.L. Bromberg, 'Research efforts that led to laser development', *Laser Focus* (October 1984): 58–60; J.L. Bromberg, 'The birth of the laser', *Physics Today*, 41, no. 10 (October 1988): 26–33; J.L. Bromberg, *The Laser in America 1950–1970* (MIT Press, 1991).
10. H.M. Collins, 'The TEA Set: tacit knowledge and scientific networks', *Science Studies*, 4 (1974): 165–86; H.M. Collins and R.G. Harrison, 'Building a TEA laser: the caprices of communication', *Social Studies of Science*, 5 (1975): 441–50.
11. *Defence Research*, Second Report from the Select Committee on Science and Technology, HC 213, Session 1968–69 (HMSO, March 1969).
12. Lord Strathcona, *Steering Group on Research and Development Establishments* (Ministry of Defence, 1981).
13. An excellent history of the invention of the laser is provided by J.L.Bromberg, *The Laser in America* (n. 9 above).
14. A. Einstein, *Mitt. Phys. Ges., Zurich*, 16, no. 18 (1916): 47. This work was subsequently translated into English in *Sources of Quantum Mechanics* edited by B.L. Van Der Werden (North-Holland, 1967).
15. R. Ladenberg, 'Untersuchungen über die anomale Dispersion angeregten Gase', *Zeitschrift für Physik*, 48 (1928): 15–50.
16. J.P. Gordon, H.J. Zeigler and C.H. Townes, 'Molecular microwave oscillator and new hyperfine structure in the microwave spectrum of NH_3', *Physical Review*, 99 (1955): 1253–63.
17. N.G. Basov and A.M. Prokhorov, 'Possible methods of obtaining active molecules for a molecular oscillator', *Soviet Physics. Journal of Experimental and Theoretical Physics of the Academy of Sciences of the USSR*, 1 (1955): 184–5.
18. N. Bloembergen, 'Proposal for a new type of solid state maser', *Physical Review*, 104 (1975): 324–7.
19. H.E.D. Scovil, G. Feher and H. Seidel, 'Operation of a solid state maser', *Physical Review*, 105 (1957): 760–3.
20. A.L. McWhorter and J.W. Meyer, 'Solid-state maser amplifier', *Physical Review*, 109 (1958): 312–8.
21. Information on the maser work sponsored by RRE is drawn from the following sources: Interview with Peter Forrester (8 September 1992); *RRE Half Yearly Progress Report* (January 1957): 121; *RRE Half Yearly Progress Report* (July 1957): 103–7; *RRE Half Yearly Progress Report* (January 1958): 117–22; *RRE Half Yearly Progress Report* (July 1958): 101–4; *RRE Half Yearly Progress Report* (January 1959): 88–92; *RRE Half Yearly Progress Report* (July 1959): 97–100; *RRE Half Yearly Progress Report* (January 1960): 105–8; *RRE Half Yearly Progress Report* (July 1960): 115–7; *RRE Half Yearly Progress Report* (January 1961): 108–10; *RRE Annual Review* (1959): 4–7; *RRE Annual Review* (1960): 7–9.
22. A.L. Schawlow and C.H. Townes, 'Infrared and optical masers', *Physical Review*, 112 (1958): 1940.
23. T.H. Maiman, 'Stimulated emission in ruby', *Nature*, 187 (6 August 1960): 493–4.

24. P.P. Sorokin and M.J. Stevenson, 'Stimulated infrared emission of trivalent uranium', *Physical Review Letters*, 5 (1960): 80–2.
25. P.P. Sorokin and M.J. Stevenson, 'Solid state optical maser using Sm2+ in CaF_2', *IBM Journal of Research and Development*, 5 (1961): 56–8.
26. A. Javan, W.R. Bennett Jr and D.R. Herriott, 'Population inversion and continuous optical maser oscillation in a gas discharge containing a He-Ne mixture', *Physical Review Letters*, 6 (1961): 106–8.
27. See the following sections on laser R&D in specific establishments.
28. See A.H. Boot, *Applications of Lasers*, Minutes of the 112th Meeting of the CVD Technical Committee (22 May 1963); I.L. Davis, *Some Laser Applications*, Minutes of the 113th Meeting of the CVD Technical Committee (July 1963: 8–13); Minutes of the ERC, ERC.495 M.20 (12 May 1964).
29. DPRC Minutes of Meeting, DRP/M(63)5, par. 2 (13 March 1963): 6.
30. Interview with Sir George Macfarlane (17 August 1992).
31. *Proceedings of NATO–SADTC Symposium on the Technical and Military Applications of Laser Techniques*, Vol. 1 (The Hague, April 1962): 4.
32. Ministry of Aviation, *Minutes of the Chief Scientist's Joint Monthly Meetings, Minutes of the First Meeting Held on Thursday 31 January 1963*, JMM/M(63)1 (1963): 1–2; Ministry of Aviation, *Minutes of the Electronics Research Council Infra Red Committee*, ERC 396 (1963).
33. The symposium was held on 10–11 October 1963 and some of the papers were eventually issued in an unclassified RRE memorandum: G.L Hutchinson, *Lasers—Present State of the Art*, RRE Memorandum no. 2063 (February 1964).
34. DSRL, *Lasers*, BN/33/013 (13 May 1968).
35. DSRL, *Lasers* (n. 34 above).
36. Minutes of the DRC (25 May 1965): 6.
37. D.M. Clunie, 'The Physics Exhibition 1966: Lasers', *Journal of Scientific Instruments*, 43 (1966): 509.
38. B. Snowdon, *Laser Development*, Ministry of Technology, EA 12/7 (1966): 1.
39. J.M. Burch, 'The application of lasers in production engineering', *The Production Engineer* 44 (1965): 431–49.
40. J.M. Burch, 'The application of lasers' (n. 39 above): 441.
41. B. Snowdon, *Laser Development* (n. 38 above).
42. See the sections on the work in specific establishments.
43. See the section below on laser R&D at RRE.
44. See the section below on laser R&D at SERL.
45. *Laser Symposium Proceedings, 4th & 5th May 1971* (unpublished, 1972).
46. DSRL, *Lasers* (n. 34 above).
47. *Minutes of the 7th Meeting of the CVD Optical Maser Working Party Held on 22 September 1964* (1964): 1.
48. *Minutes of the 115th Meeting of the CVD Technical Committee on 19 February 1964*, CVD/D64/50 (1964): 5.
49. B. Snowdon, *Laser Development* (n. 38 above).
50. DRC Paper, *Co-operative Research on Laser Damage Weapons*, PRO DR/P(65)29 (21 May 1965); Minutes of the DRC, Paper DR/P(65)29 (1965).
51. DRC Paper, *Co-operative Research on Laser Damage Weapons* (n. 50 above): 4.
52. DRC Paper DR/P(64)34, par. 117 (29 December 1964).
53. DRC Paper DR/P(65)29, p. 4, par. 10 (21 May 1965).
54. See *SERL Classified Journal*, 57 (March 1962): 1.33.
55. DSRL, *Lasers* (n. 34 above).
56. DSRL, *Lasers* (n. 34 above).
57. Ministry of Technology, Meeting on Laser Research (1 October 1968).
58. See the following sections on laser R&D at specific establishments.
59. Interview with Professor Jim Burch (22 January 1992); telephone conversation with Mr Peter Bateman (5 May 1993).
60. Professor Jim Burch (n. 59 above).
61. Interview with Peter Forrester (8 September 1992).
62. This section will draw upon the following sources: *RRE Half Yearly Progress Report* (July 1961): 123–4; *RRE Half Yearly Progress Report* (January 1962): 154–7; *RRE Half Yearly Progress Report* (July 1962): 166–7; *RRE Half Yearly Progress Report* (January 1963): 179–81; *RRE Half Yearly Progress Report* (July 1963): 123–6; *RRE Half Yearly Progress Report* (January 1964): 126–31; *RRE Half Yearly Progress Report* (July 1964): 120–7; *RRE Half Yearly Progress Report* (January 1965): 1283–8; *RRE Half Yearly Progress Report* (July 1965): 118–24; *RRE Half Yearly Progress Report* (January 1966): 97–102; and *RRE Half Yearly Progress Report* (July 1966): 103–9; *RRE Half Yearly Progress Report* (January 1967): 108–11; *RRE Half Yearly Progress Report* (July 1967): 111–4; *RRE Half Yearly Progress Report* (January 1968): 112–5; *RRE Half Yearly Progress Report* (July 1968): 122–3;

RRE Progress Report (1969): 150–85; *RRE Annual Report* (1970): 209–10; *RRE Annual Report* (1971): 10; *RRE Annual Report* (1972/73): 27; *RRE Annual Report* (1973/74): 25; *RRE Annual Report* (1974/75): 19; *RRE Annual Report*, 1975/76: 25; P.A. Forrester, K.F. Hulme, 'Laser rangefinders', *Optical and Quantum Electronics*, 13 (1981): 259–93; Peter Forrester (n. 61 above); Interview with Professor John Midwinter (23 September 1997).

63. Peter Forrester (n. 61 above).
64. *RRE Half Yearly Progress Report* (July 1961): 123.
65. Peter Forrester (n. 61 above); *RRE Half Yearly Progress Report* (January 1962): 154.
66. *RRE Half Yearly Progress Report* (January 1962): 154.
67. *RRE Half Yearly Progress Report* (January 1962): 154.
68. Peter Forrester (n. 61 above).
69. Peter Forrester (n. 61 above).
70. F.J. McClung and R.W. Hellwarth, 'Giant optical pulsations from ruby', *Journal of Applied Physics*, 33 (1962): 828–9.
71. Peter Forrester (n. 61 above).
72. L.E. Hargrove, R.L. Fork and M.A. Pollack, 'Locking of He-Ne laser modes induced by synchronous intra-cavity modulation', *Applied Physics Letters*, 5 (1964): 4–5.
73. *RRE Half Yearly Progress Report* (July 1962): 166.
74. *RRE Half Yearly Progress Report* (July 1962): 166–7.
75. *RRE Half Yearly Progress Report* (January 1962): 156; *RRE Half Yearly Progress Report* (July 1962): 167.
76. *RRE Half Yearly Progress Report* (January 1962): 156; *RRE Half Yearly Progress Report* (July 1962): 167; *RRE Half Yearly Progress Report* (January 1963): 181.
77. *RRE Half Yearly Progress Report* (January 1962): 156.
78. *RRE Half Yearly Progress Report* (January 1963): 179.
79. *RRE Half Yearly Progress Report* (July 1963): 125–6.
80. I.L. Davies and R. Meredith, 'Some notes on the military applications of lasers', RRE Memo. 2002 (*c.* 1960).
81. *RRE Half Yearly Progress Report* (July 1963): 123–5; *RRE Half Yearly Progress Report* (January 1964): 126–30; *RRE Half Yearly Progress Report* (July 1964): 120–7; *RRE Half Yearly Progress Report* (January 1965): 123–8; *RRE Half Yearly Progress Report* (July 1965): 118–24; *RRE Half Yearly Progress Report* (January 1966): 97–101; *RRE Half Yearly Progress Report* (July 1966): 103–9.
82. J.E. Geusic, H.M. Marcos and L.G. Van Uitert, 'Laser oscillations in Nd-doped yttrium aluminium, yttrium gallium and gadolinium garnets', *Applied Physics Letters*, 6 (May 1965): 175–7.
83. *RRE Half Yearly Progress Report* (July 1963): 125–6; *RRE Half Yearly Progress Report* (January 1964): 130–1; *RRE Half Yearly Progress Report* (July 1964): 127; *RRE Half Yearly Progress Report* (January 1965): 128; *RRE Half Yearly Progress Report* (July 1965): 124; *RRE Half Yearly Progress Report* (January 1966): 101–2; *RRE Half Yearly Progress Report* (July 1966): 109.
84. See R.J. Marquis, 'Light from darkness', *GEC Review*, 10, no.3 (1995): 161–75; and W. Blain, *UK Military Laser/EO Systems Research in the 1960s–1970s* (unpublished, 1987).
85. Peter Forrester (n. 61 above); *RRE Annual Report* (1970): 209–10.
86. RRE Programme (June 1971).
87. Peter Forrester (n. 61 above); *RRE Annual Report* (1970): 209–10.
88. *RRE Annual Report* (1970): 209–10.
89. *RRE Annual Report* (1974/75): 19–20.
90. *RRE Annual Report* (1970): 209–10.
91. *RRE Annual Report* (1971): 10; *RRE Annual Report* (1972/73): 27.
92. *RRE Annual Report* (1973/74): 25.
93. *RRE Annual Report* (1974/75): 19.
94. Interview with Gordon Hamilton (16 January 1997).
95. See P.A. Forrester and K.F. Hulme (n. 62 above): 284.
96. *RRE Annual Report* (1973/74): 25; *RRE Annual Report* (1974/75): 19; *RRE Annual Report* (1975/76): 25.
97. See chapter 13.
98. Information on the gas laser work at SERL is drawn from the following sources: *SERL Classified Journal*, 56 (July 1961): 1.33; *SERL Classified Journal*, 58 (September 1962): 1.14; *SERL Classified Journal*, 59 (May 1963): 1.8–1.10; *SERL Classified Journal*, 60 (February 1964): 1.13–1.14, 1.19; *SERL Classified Journal*, 61 (May 1964): 1.12–1.13, 1.17; *SERL Classified Journal*, 62 (November 1964): 1.21, 1.26–1.28; *SERL Half Yearly Report*, 63 (May 1965): 2.1, 3.2; *SERL Half Yearly Report*, 64 (November 1965): 12–4; *SERL Half Yearly Report*, 65 (May 1966): 13–6, 20–1; *SERL Half Yearly Report*, 66 (November 1966): 12–3, 20–2; *SERL Half Yearly Report*, 68 (November 1967): 12–7, 22; *SERL Half Yearly Report*, 69 (July 1968): 12–6, 19; *SERL Half Yearly Report*, 70 (December 1968): 13–6, 19; *SERL Half Yearly Report*, 72 (December 1969): 19–22; *SERL Half Yearly Report*, 71 (July 1969): *20–3, 25–6*; G.P. Wright, *SERL: 1945–1976* (HMSO, 1986); Interview with Les Large (27 February 1997); Interviews with Alan Crocker and Don Stark (6 March 1997).

99. L.E.S. Mathias and J.T. Parker, 'Stimulated emission in the band spectrum of N_2', *Applied Physics Letters*, 3 (1963): 16–18.
100. L.E.S. Mathias and J.T. Parker, 'Visible laser oscillations From CO', *Physics Letters*, 7 (1963): 194–6.
101. A. Crocker, A. Gebbie and H.A. Kimmitt *et al.*, 'Stimulated emission in the far infra-red', *Nature*, 201 (1964): 250–1.
102. C.K.N. Patel, W.L. Faust and R.A. McFarlane, 'Laser action on rotational transitions of the vibrational band of CO_2', *Bulletin of the American Physical Society*, 9 (1964): 500; C.K.N. Patel, 'Selective excitation through vibrational energy transfer and optical maser action in N_2-CO_2', *Physical Review Letters*, 13 (1964): 617–9; C.K.N. Patel, P.K. Tien and J.H. McFee, 'CW high-power CO_2-N_2-He laser', *Physical Review Letters*, 13 (1964): 290–2.
103. Information on the semiconductor laser work at SERL is drawn from the following sources: *SERL Classified Journal*, 59 (May 1963): 1.5–1.6; *SERL Classified Journal*, 60 (February 1964): 1.6, 1.8–1.10; *SERL Classified Journal*, 61 (May 1964): 1.4, 1.7–1.8, 1.10; *SERL Classified Journal*, 62 (November 1964): 1.2, 1.9–1.13, 1.18; *SERL Half Yearly Report*, no. 63 (May 1965): 1.1–1.3, 1.5, 1.7–1.8; *SERL Half Yearly Report*, no. 64 (November 1965): 2–6, 10; *SERL Half Yearly Report*, no. 66 (November 1966): 1–3, 6, 11; *SERL Half Yearly Report*, no. 68 (November 1967): 4; *Gallium Arsendie Lasers*, edited by C.H. Gooch (Wiley-Interscience, 1969); G.P. Wright, *SERL:1945–1976* (n. 98 above); Interview with Mike Coupland (13 August 1992); Telephone conversation with David Wight (4 May 1998).
104. Interview with Mike Coupland (n. 103 above); G.P. Wright, *SERL:1945–1976* (n. 98 above): 13.
105. M.I. Nathan, W.P. Dumake and G. Burns *et al.*, 'Stimulated emissions of radiation From GaAs p-n junction', *Applied Physics Letters*, 1 (1962): 62; R.N. Hall, G.E. Fenner and J.D. Kingsley *et al.*, 'Coherent emissions from GaAs junctions', *Physical Review Letters*, 9 (1962): 336; Interview with Mike Coupland (n. 103 above).
106. R.F. Broom, 'Room temperature operation of GaAs *Lasers*', *Physics Letters*, 4 (1963): 330.
107. Information on the work of SERL on gas lasers during the 1970s has been drawn from interviews with Alan Crocker, Don Stark and Viv Roper (6 March 1997); interview with Viv Roper (21 May 1997); *SERL Half Yearly Report*, 73 (July 1970): 23–6; *SERL Half Yearly Report*, 74 (December 1970): 22–5; SERL Annual Report, 75 (December 1971): 36–40; *SERL Annual Report*, 76 (December 1972): 29–33; *SERL Annual Report*, 77 (December 1973): 33–8; *SERL Annual Report*, 78 (December 1974): 31–8; *SERL Annual Report*, 79 (December 1975): 35–42.
108. A.J. Beaulieu, 'Transversely excited atmospheric pressure CO_2 lasers', *Applied Physics Letters*, 16 (1970): 504–5.
109. H.M. Lamberton and P.R. Pearson, 'Improved excitation technique for atmospheric pressure CO_2 lasers', *Electronic Letters*, 7 (1971): 141–2.
110. *SERL Half Yearly Report*, 74 (December 1970): 22.
111. D.S. Stark, P.H. Cross and H. Foster, 'A compact sealed pulsed CO_2 TEA laser', *IEEE Journal of Quantum Electronics*, QE-11, 9 (September 1975): 774–8.
112. D.S. Stark and M.R. Harris, 'Platinum-catalysed recombination of CO and O_2 in sealed CO_2 TEA laser gases', *Journal of Physics E: Scientific Instruments*, 11 (1978): 316–9; D.S. Stark, A.Crocker and G.J. Steward, 'A sealed 100-Hz CO_2 TEA laser using high CO_2 concentrations and ambient-temperature catalysts', *Journal of Physics E: Scientific Instruments*, 16 (1983): 158–61; D.S. Stark, A.Crocker and N.A. Lowde, 'A semiconductor-preionised sealed TEA laser operating at high CO_2 concentrations and repetition rates up to 100 Hz', *Journal of Physics E: Scientific Instruments*, 16 (1983): 1069–71; D.S. Stark and A.Crocker, 'A sealed high-CO_2 high-PRF semiconductor-preionised TEA laser without a solid catalyst', *Optics Communications*, 48, no. 5 (1984): 337–42.
113. J.D.L.H. Wood and P.R. Pearson, 'A repetitively pulsed carbon dioxide laser with mean power output in excess of 30 kw', *Paper Presented at the Third International Symposium on Gas Flow and Chemical Lasers* (Marseille, September 1980).
114. J.D.L.H. Wood and P.R. Pearson, 'A repetitively pulsed carbon dioxide laser' (n. 113 above).
115. V.G. Roper, H.M. Lamberton and E.W. Parcell *et al.*, 'Laser induced medium perturbation in a pulsed CO_2 laser', *Optics Communications*, 25 (1978): 235–40.
116. *SRDE Progress Report* (April–September 1963): 81–3, *SRDE Progress Report* (October 1963 – March 1964): 58–61.
117. *SRDE Progress Report* (April–September 1964); *SRDE Progress Report* (October 1964 – March 1965); *SRDE Progress Report* (April–September 1965); *SRDE Progress Report* (October 1965 – March 1966).
118. G.G. Macfarlane, 'Lasers and their applications', paper submitted to Def. Res. Comm., DR25/67, 15 May 1967, PRO DEFE 10 626.
119. This section draws upon material from the following sources: Advisory Council on Science and Technology, *Optoelectronics: building on our investment* (HMSO, 1988): 81–3; *SRDE Establishment Report* (1972/73): 15; *SRDE Establishment Report* (1973/74): 15; *SRDE Establishment Report* (1974/75): 18; Interview with Professor Alic Gambling (11 August 1992); Interview with Professor John Midwinter (23 September 1997).

120. According to industry sources the ruby laser technology used in the Barr and Stroud rangefinder for the Chieftain tank was licensed from Hughes UK Ltd and that employed in the laser for the Ferranti Laser Ranger and Marked Target Seeker (LRMTS) incorporated in to British combat aircraft was licensed from Westinghouse. See M. Moss and I. Russell, *Range and Vision: the first hundred years of Barr and Stroud* (Edinburgh); R.J. Marquis, 'Light from darkness' (n. 84 above); W. Blain, *UK Military Laser/EO Systems Research in the 1960s–1970s* (n. 84 above): 149–229.

CHAPTER 11

Chemical and Biological Warfare and Defence, 1945–90

GRADON CARTER AND BRIAN BALMER

INTRODUCTION

At the end of the Second World War the UK had a substantial stockpile of chemical warfare agents and weapons and a token biological warfare weapon system. The former were based on the agents of the Great War of 1914–18 and weapons which were largely developed in the late 1930s and 1940s. The latter was an antilivestock and economic weapon, both archaic in concept and devised to fulfil an urgent political requirement in the Second World War to possess the ability to retaliate in kind. Neither chemical or biological warfare was used in the Second World War (apart from some Japanese activity in biological warfare), but at its end it was clear that remarkable new chemical warfare agents, which had remained undiscovered by the Allies, had been developed to operational readiness by the Germans. It was also evident that the potential of biological warfare, although now evident, had hardly been explored.

In the face of a new postwar threat from the USSR and the Warsaw Pact, and the significant new advantages provided by the nerve agents, the UK planned to modernise its chemical warfare capability. It also proposed to acquire an essentially antipersonnel biological warfare capability. Equally, it became necessary to provide the means of defence against these two methods of warfare for the military and civil population. How such plans developed and how some were subsequently abandoned is best told if chemical and biological warfare are considered separately, albeit that most of the research took place within the two major establishments at Porton Down, near Salisbury in Wiltshire.

The complexities of relating 45 years of diverse research in two major defence fields set against a changing backcloth of political uncertainties, financial stringency and the peculiar emotive considerations that have always attended the subject, are considerable. We also examine something of essentially civil research conducted at Porton, and the once eminent advisory boards connected with chemical and biological warfare.

CHEMICAL WARFARE AND CHEMICAL DEFENCE

At the end of the Second World War, the UK's capability for chemical warfare was manifest in a complex array of facilities. There were seven Ministry of Supply Agency Factories, which had been designed, built and managed for the War Office, and later the Ministry of Supply, by Imperial Chemical Industries. Process research for such agent production was based at the Chemical Defence Research Establishment at Sutton Oak, near St Helens. The research, development and trials focus was the Porton Down Chemical Defence Experimental Station, the home of UK chemical warfare and chemical defence since 1916. A London-based Ministry of Supply headquarters, the Controllerate of Chemical Defence Development, managed all research, development, trials, production and weaponising. Further, a considerable chemical warfare and chemical defence staff existed in the three Services, notably in the Army and Royal Air Force.

The UK capability in 1945 was based on three agents: mustard gas, phosgene[1] and a tear gas, usually bromacetone phenone. The associated munitions and weapons included a small range of artillery shell and mortar rounds, aircraft bombs of several sizes, and several aircraft spray devices. In addition, a few archaic and quaint devices for contaminating terrain had survived from the 1930s, despite obsolescence. The total agent production in the Second World War had been 40,719 tons of mustard gas and 14,042 tons of the other agents. Most agents had been charged into munitions and weapons. Comprehensive details are readily available and there is little point in reproducing these here, but some of the stockpile of chemical munitions and devices was impressive, e.g. 2,364,000 25 lb base ejection shells and 711,640 aircraft bombs, all charged with mustard gas, had been acquired by D Day. If chemical warfare had been used, the stockpile would have been replenished by output from the Agency factories. Their total capability was 900 tons of agents a month, with the means to produce 3680 tons each month within six months' notice. The capability of the Commonwealth, in South Africa and Canada, could have added a further 5330 tons each month.

Notwithstanding this considerable capability, the postwar situation in 1945 was complicated. The UK was bound by the terms of its 1930 ratification of the 1925 Geneva Protocol[2] to a virtual no-first-use policy. This was a considerable limiting factor, which diminished somewhat during the Second World War but which had been paramount in the 1930s and would be again in the 1950s. In fact, this and defence economies would become the major decisive factors affecting the development of these weapons after the Second World War.

- Chemical warfare had not been used during the Second World War. Was its utility waning? Was it clear why it had not been used? Was there a future for chemical warfare? In reality, the Services had no great enthusiasm for being plunged into chemical warfare.
- The UK's practical experience of using chemical warfare and being exposed to its hazards was limited to the Great War, and was thus dated. The UK had no Chemical Corps. The Royal Engineers quasi-corps role in chemical warfare had been disbanded early in the war.

- The nerve agents had been developed and stockpiled by Germany: they provided a dramatic advantage to users and seemingly intractable problems of defence to the vulnerable.
- A future chemical warfare threat arose from the USSR, which was known to have rebuilt one of the German nerve agent production plants in the USSR and undoubtedly possessed nerve agent-based munitions.

Despite these complications, the UK's chemical warfare capability required modernisation to match the threat from the USSR. Such a modernisation had to be based on the nerve agents. The implications for research and development in the UK were vast. There would have to be new process chemistry research, a new process chemistry pilot plant, and methods of charging munitions with a much more toxic agent than hitherto. That achieved, a full-scale production and weaponising plant was needed. Also, new doctrines of use needed to be evaluated and planned.

On the defensive side the problems seemed much more long term and difficult. Earlier chemical warfare agents had distinctive smells or evoked some more-or-less immediate sensory irritation. Recognition could be taught. The recognition of agents by smell had been an important part of Service antigas training since the Great War, but was eventually abandoned because of health and safety concerns. Speedy recourse to the respirator was then the mainstay of defence. True, liquid mustard gas presented a greater problem, but oilskin capes, goggles and associated equipment provided effective countermeasures. Nerve agents, by contrast, were odourless and colourless, exerted highly toxic effects very rapidly, through the skin, respiratory tract and eyes. A single drop of nerve agent splashed on the skin or in the eye could, untreated, be fatal. Service training with actual nerve agents to demonstrate effective detection and decontamination would not be possible, as the risk of fatal accident was unacceptable. Beyond the problem of detection lay those of preventing and curing symptoms. Both demanded new basic research in toxicology, pharmacology and pathology, for the means of prophylaxis and therapy could only emerge when the detailed mechanisms of action were identified. Physical protection also required research, even though the existing military and civilian respirators were effective. Although respirators provided protection for the eyes and the respiratory tract, protection of the body needed to be improved and based on new concepts. Servicemen could not be protected from nerve agents by the chemically impregnated battledress and oilskin capes of the Second World War. Decontamination, too, needed to be extensively revised. For decades decontamination had been by liquid bleach or bleaching powder and, for skin, also by antigas ointments, which were both prophylactic and therapeutic. Nerve agent decontamination of personnel eventually became centred upon a form of aluminium silicate used for cleaning cloth, known as Fuller's earth.

The long-term research and development needs were thus formidable, particularly in a period of postwar austerity and shortages. The problems, both for the research community and the Services, were exacerbated by the need to maintain secrecy about the UK's offensive contingencies and, for decades, to retain an almost equal level of secrecy about many defensive aspects. The place of secrecy in chemical warfare and defence was curious, often untoward and protracted. It was designed originally to inhibit foreign knowledge of the agents and weapons of national interest. It was also to inhibit public knowledge and thus allay concern. Ultimately it became almost a catch-all, political and near-automatic recourse to protect manifold sensitivities about chemical and biological warfare matters. Inevitably, as official documents reach the Public Record Office, such sensitivities are sensationally resurrected by the media, often to the detriment of current defence. For, it is worth emphasising in chemical and biological warfare, obsolescence is virtually unknown. The mustard gas of 1918 was no less effective in the Iran–Iraq war during the 1980s. In the public domain in the UK, no official acknowledgement of the existence of nerve agents was made for some years. By 1950, media citations derived from US sources began to cause unease, particularly when they suggested that the UK was not only actively studying the nerve agents, but doing so in collaboration with the USA. The late 1950s and 1960s saw massive US media coverage of official domestic publicity, particularly on the American development of nerve agents.

In the USA there was always greater official openness about chemical warfare. Of the circa 40 English-language books on chemical and biological warfare published between 1918 and 1969, most originated in the USA. Even when, later in the mid-1950s, the UK had decided to abandon the modernisation of its chemical warfare capability, to forsake and destroy the Second World War stockpile, and to restrict research to chemical defence and assessment of chemical warfare as it might be used against the UK, little was exposed to parliament or the public. The untoward secrecy which had surrounded chemical warfare topics in the UK since 1920 continued. Concern in parliament and in the country caused the Ministry of Defence to mount a week of 'open days' at the Microbiological Research Establishment in 1968, and at the Chemical Defence Experimental Establishment in 1969. Generally, ministers were not keen to comment. Not until the late 1960s were muted ministerial public statements made on the now solely defensive nature of the UK's policy. The dates of the first *unequivocal* ministerial statements are uncertain, but on 2 July 1968 John Morris, Minister for Defence (Equipment), told the Commons that in respect of both chemical and biological warfare 'the programme of research at Porton is defensive'.

The decision to abandon offensive chemical warfare had been primarily economic. How could the UK justify the continued expense involved in developing a method of warfare of which UK use, except in direct retaliation, had been effectively prohibited by ratification of the Geneva Protocol? This logic was supported by the Duncan Sandys policy of reliance on nuclear weapons as deterrent, and as the ultimate and major weapon in our armoury. Further, at that time there were no constraints on nuclear first use.

Since the mid-1950s the need for the reacquisition of the discarded UK chemical warfare capability has been officially studied at intervals. In 1961 two committees studied the prospects of chemical and biological warfare: first, an operational assessment by the Chiefs of Staff and, second, the Todd Panel, established by Sir Solly Zuckerman, investigated the potential of chemical and biological warfare and the scope of future UK research in the area. A year later these were integrated into a single paper by a subcommittee of the Defence Research Policy Committee (the most senior defence research advisory body in government), which recommended 'a modest expansion of effort on the offensive aspects of BW and CW in tactical situations, and further examination of the strategic potentialities of BW'.[3] In 1963 the Prime Minister, Harold Macmillan, concerned about the international situation and appraisals of the potential military utility of chemical warfare in a limited war, eventually authorised a five-year exploratory programme of offensively related research and a limited production of agents. Staff Requirements and Staff Targets were prepared between 1963 and 1965, notably for aircraft spray tanks to disseminate the new British nerve gas VX[4] and for improved artillery shells for sarin, the nerve gas originally developed in wartime Germany (isopropyl methylphosphonofluoridate, also known as GB). Nonetheless, no deployment of weapons that might be developed in the research programme was contemplated. Although with the formation of a new government under Harold Wilson, there was recognition of the necessity for reapproval of the plans, Cabinet discussion was repeatedly deferred, reputedly because of the Rhodesia crisis and the 1966 general election. At the same time there was a massive reduction in funding for the general defence programme. Chemical defence was considerably affected. Any resurgence of offensive chemical warfare plans fell by the wayside, and by 1968 had been deferred indefinitely. In later years the need for a chemical warfare capability was raised again, but proposals were never put to the Cabinet.

Perhaps the last occasion when considerations were reputedly made at the highest level was in 1985, in response to a 1984 Joint Intelligence Committee assessment of the chemical warfare threat from the USSR. Here, according to the journalist Duncan Campell, a proposal for UK nerve agent production was an option presented to an ad hoc highly secret Cabinet committee chaired by the Prime Minister, Margaret Thatcher.[5] Reputedly this option, favoured by Mrs Thatcher, was successfully opposed, as was a further option to encourage the USA to develop new chemical weapons and base them in Europe. The validity of Duncan Campell's report is unknown. However, his assessment that arms control concepts presented a more attractive option for the UK are certainly substantiated by the efforts that had been made by the UK since 1970 in this field, and which culminated in the UK's signature of the 1993 Chemical Weapons Convention and its ratification in 1996. The Convention came into force on 29 April 1997, and is nationally enforced in the UK by the 1996 Chemical Weapons Act.

This brief history of the policy associated with chemical warfare was not generally available to the public until the mid-1990s, although it could probably have been largely deduced from Public Record Office documents. Here the history is a scene-setting preamble to a review of the research that subsequently took place at Porton, Sutton Oak and Nancekuke.

CHEMICAL WARFARE OBJECTIVES 1945–56

The ability to produce agents on a large scale is a basic need in chemical warfare. For most involved nations, this provided an initial stockpile and then an essential ability to maintain the level of the stockpile after any deployment, by continued production. The chemistry involved in synthesising the nerve agents at laboratory scale was quickly absorbed by Porton and Sutton Oak. The procedures used by Germany at several scales for the production of sarin and tabun (a G-agent, also known as GA) at Munsterlager, and later at Raubkammer, had been determined by on-site UK intelligence exploitation and the considerable help of senior German staff. The German full-scale 1000 tons/month plant at Dyhernfurth in Silesia was in Soviet hands.

Initially some German work and trials were repeated at Raubkammer by the West. Concurrently, samples of nerve agents were hastily transported to Porton and Sutton Oak. Also, 71,000 German aircraft bombs charged with tabun were brought to the UK in 1945 as a contingency for possible retaliatory use against the Japanese. These were to remain at RAF Llandwrog in North Wales until their disposal in Operation Sandcastle in 1955–56.[6] By the end of 1946 research and trials at Porton had provided a considerable assessment of the utility of sarin and tabun. Essentially, these agents were so highly toxic and lethal that they seemed to render phosgene, the cyanides and cyanogen chloride obsolete. They also provided usefully severe harassing effects with non-lethal doses from effects on the iris, with dimming of vision and difficulty in adjusting the eyes to focus on different distances [accommodation]. Now, it was clear, new munitions and weapons were required in order to disseminate the nerve agents effectively. The modernised UK chemical warfare capability had to be based primarily on one of the nerve agents, with mustard gas in a persistent, incapacitant and terrain-denying supporting role. Research and development on new munitions and weapons for such roles was of great importance. In the event, the only postwar chemical munition that reached the production level was a new 1000 lb bomb for mustard gas.

Research on optimising processes for production of the nerve agents started at Sutton Oak in 1946, with the techniques used by the Germans. Sutton Oak, however, was in a built-up area of St Helens and, for safety reasons, production could not be safely studied beyond the 1 lb/hour scale. As a 1 ton/week pilot plant had been identified as essential, this was clearly not sufficient to meet requirements. The prime need was for a safer and more isolated site, where any accidental release of agents would not be likely to result in exposure of the civil population. The wartime Agency factory sites were also largely unsuitable. After much exploration of 14 potential sites, the disused wartime airfield at RAF Portreath, near Nancekuke, in Cornwall, was chosen. It was taken over from the Air Ministry by the Ministry of Supply on 28 July 1949 and opened as the Chemical Defence Establishment, Nancekuke, in 1951 under the former Superintendent of Sutton Oak. The years 1951–1953 were spent on building laboratories and plant at Nancekuke, winding down at Sutton Oak, which finally closed in 1954, and transferring about 40 staff. This was followed by the acquisition of about 200 further Nancekuke staff during the early 1950s. The cited dates of opening and closing of Nancekuke and Sutton Oak are diffuse because of the overlaps during the transfer of equipment and staff. Sutton Oak has now been demolished and the cleared site is devoted to light industry. The Nancekuke site was also cleared, and by January 1979 had reverted to Royal Air Force use.

Despite some initial study of the processes for tabun, which was the only nerve agent that the Germans had manufactured in quantity, and some initial enthusiasm for a further nerve agent GE, it was ultimately established that precursors for sarin did not pose the technical difficulties envisaged initially. Further, field trials at Porton had established that sarin was the best of the existing nerve agents for UK requirements. By the time the technical programme got under way in 1951 there were two major items:

- The design, erection and proofing of the 1 ton/week pilot plant for sarin.
- The design and planning of a 50 ton/week production-scale plant for sarin.

The pilot-plant design had been done at Sutton Oak. Work started on building this plant at Nancekuke in August 1951, and was completed by 1953. In the latter part of 1953 the pilot plant was operational but using ethyl chloroacetate as simulant[7] for sarin, to familiarise operators with equipment without the toxic hazard. The first pilot-plant run to produce sarin began in January 1954, and such work continued intermittently until January 1956. This process research led to the accumulation of 20 tons of sarin.

The 1951 programme envisaged a 10 ton/week plant as an intermediary stage between the pilot plant and the production plant, but this plan was discarded in an attempt to recover lost time. The design and planning of the production plant was completed but because of the 1956 decision to abandon plans for modernising the UK's chemical warfare capability, it was never built. After 1956, Nancekuke's and Porton's studies on nerve agents were done solely to assess how potential enemies could make agents and the nature of the chemical hazards to the UK and its Services against which defence was required. Probably the conclusion of studies that were instigated before 1956 was in understanding long-term stability and corrosion, to determine the best conditions for storing precursors and nerve agents in bulk, and for nerve agents charged into munitions. Subtle and slow chemical reactions between some agents and the metal of the munition casing caused instability, which could result in the build-up of pressure and dangerous leaking of the agent. These necessarily long-term investigations were finished in 1966. The only other offensively motivated activity was the design and construction of a charging machine for 25 lb shells. This was installed at Nancekuke and tested with simulants. It was never used to charge nerve agent into munitions. Most of the 20 tons of sarin which had accumulated from pilot-plant studies were either destroyed in 1967–68 or used in studies of corrosion and stability.

In so far as, at Porton, a programme of offensively relevant research work may seem to have continued, it is important to understand that much activity at Nancekuke and at Porton can only be categorised as essentially offensive or defensive by the context of contemporary national policy, and not by the intrinsic nature of that activity. To properly assess a potential enemy's optimum means of producing chemical warfare agents, turning them into weapons and deploying them against the UK, it may well be necessary virtually to rehearse all these activities on a small scale. With the passage of time, or even at the time, such activity may be construed, either reasonably or unreasonably, as offensive research. Real offensively motivated research can rarely be unequivocally identified, and is more often associated with development at a significant scale. Collateral authoritative data on policy are inevitably needed. In the UK, political decisions about chemical warfare capabilities were always most secret and decisions were not much revealed, let alone widely promulgated. When details from the UK's chemical warfare and defence past enter the public domain, there is often some misunderstanding by investigators and the media. Sometimes nothing capable of distortion and misuse escapes would-be commentators. In 1996, when some details of the guidelines for Ministry of Defence document reviewers was released to the Campaign for Freedom of Information, these were selectively quoted to project an imagined one-time deliberate defiance by Whitehall of the 1956 Cabinet Defence Committee decision to abandon the UK's chemical warfare capability.[8] Inspection of written technical data is unlikely to reveal what policy existed at the time of writing. Most older documents, written since 1956, which are sometimes taken to show some alternative decision to that of 1956, give this impression from their vernacular, which continues the language of earlier years and is often misconstrued.

It is worth repeating, therefore, that despite the 1963 decision to re-enter offensively-related research and a limited production of agents, and despite the reputed wishes of Margaret Thatcher in 1985 referred to earlier, in practical terms the 1956 decision has remained inviolate ever since. By 1960 the UK no longer possessed an offensive chemical warfare capability. The country had totally discarded the capability to produce chemical warfare agents, and by 1960 had destroyed the stockpile left over from the Second World War. There were, of course, no munitions or weapons containing nerve agents in the stockpile, and the German tabun bombs had been disposed of by deep-sea dumping in Operation Sandcastle. Since the late 1960s the UK's research and development programmes have been exclusively in support of a chemical defence capability. Further, the UK is bound by the 1993 Chemical Weapons Convention. The Chemical Weapons Act of 1996 is the domestic legislation which legally enforces the Convention.

During the 1950s extensive development work was done at Porton on weapons, mainly for the proposed equipment of the Services with a nerve-agent-based capability. To this end, the main thrust was for 20 and 25 lb artillery shells, several antitank projectiles, 30 mm Aden gun[9] ammunition, mortar rounds and a naval shell. Some progress was made with a 1000 lb bomb, a 1000 lb cluster bomb[10] and 60 lb aircraft rockets. Nevertheless, the only reserves of chemical weapons built up after the Second World War were 10,000 1000 lb mustard gas bombs, charged at the Ministry of Supply Agency factory at Randle in 1954 using largely wartime stocks of mustard gas.

After 1956 the earlier offensive developments were quickly rounded off and written up. The mustard gas bombs were destroyed. Perhaps the last major activities of the era were the trials at Obanakoro in Nigeria, held between 1951 and 1955 to evaluate the performance of experimental nerve-agent weapons in the tropics. The Obanakoro trials did not, as some have asserted, demonstrate that the UK possessed nerve-agent munitions. The munitions used were handcharged specifically for the trials, and did not come from any in-service stock.

Work was now limited to an understanding of how chemical warfare might be used against the UK. In this context the V-agents,[11] which required a considerable shift in research, became a priority. The V-agents, methyl phosphonothiolates, are generically nerve agents although they differ chemically from the older nerve agents, the G-agents,[12] which were methyl phosphonofluoridates. The manifestations of the UK change of policy are reflected in much documentation of the period, albeit in a restrained manner. One category reported in the Directorate of Chemical Defence Research and Development Progress Report for the half-year ended 30 June 1958 is titled 'Operational evaluation of toxic agents, weapon research and development'. By the next Progress Report for the half-year to 31 December 1958, this title had been changed to 'Operational evaluation of toxic agents'. Synthesis and semitechnical scale preparation, chemistry, physicochemical and stability studies, toxicological work, the development of analytical methods and field activity and the methods whereby V-agents could be disseminated, evoked a decade or more of detailed assessment. On the explicit defence problems, those of body protection were considerable, the nature of the V-agents evoking the need for more complete skin protection and more effective means of decontamination. Detection, especially by automatic means, posed further problems, and innumerable minor aspects of V-agents were studied in the 1960s until the UK had made a proper assessment of their considerable military significance and the hazard to which the Services could be exposed in future wars. Meanwhile, the evaluation of the original G series of nerve agents also continued.

The V-agents were the most toxic chemical warfare agents studied at Porton up to the mid-1950s. This particular nerve agent series had emerged from research on insecticides at Imperial Chemical Industries in the early 1950s. One, codenamed VX, emerged as the most significant of the series. Whereas UK interest was limited to research, in the USA VX was produced on a multiton basis and made into a weapon, notably as a binary agent. Developed purely for safety reasons, the binary concept of chemical agents is based on the charging into munitions of two precursor compounds. These are held separately in the munition and when the munition, such as a shell, is fired from a gun, special design features enable the rapid efficient mixing of the two precursors to produce the agent. Research on the V-agents, and mainly VX, continued throughout the 1960s. Inevitably this diminished as data were acquired, and also because of other priorities; as there was now no UK interest to acquire any chemical warfare capability, studies were inevitably self-limiting.

By the end of the 1960s VX was seen as a standard chemical warfare agent within the capabilities of other nations. As such, in the context of research at Porton, it became merely one of the agents against which defensive countermeasures were assessed. At Nancekuke, virtually all the work on V-agents had been completed by 1969. Much collaborative V-agent research was done with the USA under the Mutual Weapons Development Programme. No 'mutual' weapons arose for the UK, but the projects undertaken by the then Chemical Defence Experimental Establishment on the V-agents attracted later criticism because they were deemed to have contributed to the USA's chemical warfare capability. As far as the UK was concerned, the projects provided data for the assessment of how V-agents might be developed by the UK's enemies, and also much-needed money. Unlike in other international collaborative research, the USA was essentially a customer.

CHEMICAL DEFENCE 1956–90

This title embraces not only chemical defence *per se*, but any aspect of the study of how chemical warfare might be used against the UK. The nature of chemical defence research during this period was diverse, changing at intervals in its qualitative and quantitative aspects. Many factors combined to cause changes, e.g. Service priorities, the subordination of the Establishment, political fashions in the administration of government science, financial resources, the situation in Northern Ireland and the development of riot-control agents, specific threats such as that from the USSR (and later from Iraq), arms control, and the effects of the devolution of biological defence research to the older Porton establishment. All such factors exerted effects either constantly or intermittently over a period longer than 30 years. The overall picture of this period cannot therefore be even summarised here. Nor would it be useful to concentrate on a mere three or four highlights of the period, e.g. the development of the S10 respirator, prophylaxis and therapy for nerve agent casualties, or technology opportunities in agent detection. We have instead decided to list, almost verbatim, the major research programme headings taken from documents circa 1960, 1970, 1980 and 1990. To these, some comments and notes have been added. The totality provides at least a qualitative résumé of the main research of the period.

TABLE 11.1: DIRECTORATE OF CHEMICAL DEFENCE RESEARCH AND DEVELOPMENT: PROGRESS REPORT FOR THE HALF-YEAR ENDED 30 JUNE 1960

I Chemical, toxicological, physiological etc.; research on CW agents

a. Study of new types of agent
    i. Lethal agents
    ii. Incapacitating agents
b. Chemistry and toxicology of anticholinesterase agents
    i. Laboratory-scale synthesis of V-agents
    ii. Semitechnical-scale preparation of V-agents
    iii. Chemical, physicochemical and stability studies
    iv. Analytical methods for V-agents
    v. Toxicology of anticholinesterase agents
c. Physiological studies
d. Biophysics and related studies—skin penetration

II Operational evaluation of toxic agents

a. Long-distance travel of BW agents
b. Evaluation of V-agents
    i. Dispersion studies
    ii. Impaction on the body and penetration of clothing
    iii. Persistence on and pick-up from surfaces
c. Meteorological research
d. Laboratory and field techniques

III Protective measures against biological and chemical agents

a. Respirators
    i. New Service respirator
    ii. Rubber research
    iii. Charcoal research
    iv. Filter research
b. Collective protection
c. Protective clothing and materials
d. Detection of CW agents
e. Decontamination
f. Therapy and prophylaxis

IV Smoke weapons and apparatus; riot control

a. Screening smoke
b. Coloured smoke
c. Riot control
    i. Research on agents
    ii. Anti-riot weapon and devices

V Pest control

a. Dispersion of insecticides
b. Shark repellents

VI Miscellaneous projects; work for other departments etc.

a. Wound ballistics
b. Respirators, breathing apparatus, filters etc.
c. Hazards from propellants, lubricants, exhaust fumes etc.
d. Miscellaneous

VII Radiological defence

a. Transmission of thermal radiation
    i. Basic study of transmission through the atmosphere
    ii. Field investigation of transmission of thermal radiation
    iii. Thermal radiation measures
b. Radiological decontamination
    i. Adhesion of fall-out to surfaces
    ii. Decontamination techniques
c. Examination of bomb debris
d. Chemotherapy

The diversity of research and development is well demonstrated, but note that there are many military topics beyond chemical defence. These are worthy of comment, for some continued for decades and some continue today. Item II(a), on the long-distance travel of BW agents, was done at the then Chemical Defence Experimental Establishment, rather than at the Microbiological Research Establishment, because this work with the fluorescent microscopic particulate powder zinc cadmium sulphide was essentially a meteorological study. This provided the first initimations of the hazard to the UK of clandestine strategic biological warfare attack. Before this period the term 'strategic' had been applied to biological warfare bombing attacks. After this period it became associated with a much larger area coverage by standoff long-line dissemination of biological agents by aircraft spray. Item IV emphasises Porton's role in smoke research and development, which continued from 1916 to the 1990s. The early work on riot control agents, notably CS (2-chlorobenzylidene malonitrile), became a major commitment during the 1970s and 1980s. Item V(a) reflects another long-held role that peaked during the Second World War. The dissemination of insecticides for both military and civil purposes was a former long-term preoccupation at Porton.[13] Item V(b) was, by contrast, short-lived. Item VI(a) represents work that continues at present and has a long history at Porton. The study of trauma and wound ballistics at Porton has not been historically recounted in any depth. It began sporadically in about 1937. Many scientific and medical papers from CBD Sector on such topics continue to appear in learned journals. Item VII represents tasking by the Atomic Weapons Research Establishment, which lasted for a decade. This diversity of research and development arose because Porton was uniquely placed, equipped and staffed for such contingencies. Although the format of the 1970 report on the research programme, at what had by now become the Chemical Defence Establishment, is different, the diversity remained.

A few comments on the 1970 programme (Table 11.2) may be helpful. Items I(b) and (c) included new research on glycollates (a class of psychotomimetic compounds), carbamates and indole analogues (novel psychotomimetic compounds from one of the major classes of alkaloids). Most of the work in Item I(d) was on glycollates. Item III(a) involved work on new protection-suit fabrics. Item IV on riot control agents has grown vastly in response to Service requirements—notably in Northern Ireland—and the work on safety of CS[14] use was engendered by the Himsworth Committee.[15] Item V is largely rather more academic research initiated by the Chemical Defence Establishment, rather than the Services. By 1980, the emphasis had subtly shifted again (Table 11.3).

Note that biological warfare topics are included in Item I(a) and that microbiological defence forms a major new item. This reflects the 1979 acquisition of the responsibility for biological defence research by the older Porton establishment and the closure of the Microbiological Research Establishment.

By 1990, the now Chemical and Biological Defence Establishment programme had been streamlined to respond to newer concepts of customer relationships with the establishment (Table 11.4).

TABLE 11.2: REVIEW OF CDE RESEARCH PROGRAMME BY CHIEF SCIENTIST (ARMY), 30 APRIL 1970

I Identification and evaluation of the threat from chemical agents
a. Search for leads to compounds likely to constitute a CW threat
b. Synthesis of compounds of high or novel biological activity
c. Toxicological and pharmacological studies
d. Evaluation of highly active compounds in man
e. Preparation and production of compounds

II Assessment of factors likely to affect the conduct of military operations in a CW environment
a. Methods for the dissemination of chemical agents
b. The fate of chemical agents after dispersion
c. Operational analysis
d. Field studies

III Research into the defensive response to the CW threat
a. Protection of the individual
b. Aids to the relaxation of individual protection
c. Counters to breaches of individual protection
d. Testing facilities for defensive equipment

IV Evaluation of improved riot control agents and means of dissemination
a. Search for more effective irritant chemicals
b. Investigation of methods for the dissemination of irritant compounds
c. Assessment of the effectiveness and safety of irritant compounds (experimental devices)
d. Safety aspects of the civil use of CS (Himsworth Committee)

V Self-initiated research
a. Basic research on dispersion of powders
b. Sterospecificity in relation to pharmacological action
c. Metabolism and distribution of anticholinergics in the body
d. Mechanism of action for penetration of skin
e. Antimaterial research

VI Investigation of new materials and improved compositions for screening smokes
a. Chemical changes in smoke compositions on storage
b. Reactions on combustion of smoke mixtures
c. New materials for smoke mixtures

A few points may need explanation. Chemical hardening refers to the protection of Service equipments against contamination from chemical warfare agents, and is analogous to the better-known concept of nuclear hardening. Item VI was primarily concerned with the simulation of chemical and biological warfare environments to facilitate better training in the Services. At this period the establishment was also preoccupied with lessons learnt from the Gulf conflict, the United Nations Special Commission on Iraq, and support for arms control negotiations leading to the Chemical Weapons Convention. Also, Defence Support Agency status was about to be acquired by the establishment. By 1994 newer plans had been confirmed, and

TABLE 11.3: CHEMICAL DEFENCE ESTABLISHMENT PROGRAMME: 1980 REVIEW

I Chemical defence—protection
a. Assessment of the current CB defensive systems against the likely threat
b. Personal protection
c. Activated charcoal

II Chemical defence—detection
a. Assessment of the effectiveness of present systems and the need for improvements
b. Examination of methods for the detection of chemical agents
c. Training

III Chemical defence—contamination control
a. Assessment of the factors governing contamination control
b. Methods for the measurement, monitoring and control of contamination

IV Dispersion
a. Operational analysis and assessment
b. Dispersion of chemical agents
c. Dispersion of agents for IS and COIN operations
d. Dispersion of simulants for chemical defence training

V Chemical defence—medical countermeasures
a. Assessment of the efficacy of medical treatment and its impact on combat effectiveness in a chemical environment
b. Treatment for poisoning by nerve agent
c. Treatment of poisoning by agents other than nerve agents

VI Response to chemical injury
a. Assessment of the present or potential chemical threat

VII IS, COIN and anti-terrorism
a. Trialling and training

VII Response to physical injury
a. Injuries due to penetrating missiles
b. Injuries from non-penetrating missiles
c. Injuries due to explosive effect
d. Injuries from translational movement

IX Health and safety aspects of chemicals
a. Definition of possible health hazards from compounds
b. Measurement, monitoring and control of health hazards
c. Development of techniques for safety evaluation

X Microbiological defence
a. Assessment of the threat of biological warfare
b. Identification and detection of micro-organisms

TABLE 11.4: THE STREAMLINED CHEMICAL AND BIOLOGICAL DEFENCE ESTABLISHMENT PROGRAMME

I	Assessment of the hazard
II	Detection, identification, monitoring and reconnaissance
III	Protection
IV	Contamination management, i.e. decontamination and chemical hardening
V	Medical countermeasures
VI	Training and operational effectiveness
VII	Response to physical injury
VIII	IS and COIN and antiterrorism

in April 1995 the Chemical and Biological Defence Establishment became part of the Defence Evaluation and Research Agency. There were, of course, several sets of evaluations during the period 1957–90, on respirators, foot and hand protection, the NBC suit, the casualty bag, facelets,[16] training simulants, the NAIAD[17] alarm and detector, and the chemical agent monitor and other devices. All these were associated with Staff Requirements and Staff Targets for equipment procurement by the Services, and again are too numerous to list here.

At the time of writing (1997), the future of chemical defence is closely tied to measures to deal with the proliferation of chemical weapons and to the 1993 Chemical Weapons Convention. Whatever non-proliferation measures are developed, and despite the 1997 coming into force of the Chemical Weapons Convention, the specific threats that still exist indicate that ultimately the need for effective chemical defence measures remains.[18] In this context, the UK Services are probably the best equipped and trained in the world.

BIOLOGICAL WARFARE AND BIOLOGICAL DEFENCE

In contrast to chemical warfare and defence, the UK's official interest in biological warfare and defence dates only from the mid-1930s. Interest was aroused mainly by intimations of foreign—especially German—activity. By 1936 this interest was formalised in a Committee of Imperial Defence subcommittee 'to report on the practicability of the introduction of biological warfare and to make recommendations as to the countermeasures which should be taken to deal with such an eventuality'. Until it was exacerbated by additional intelligence during the early stages of the Second World War, official interest was limited to theoretical assessment of the hazard and to the organisation of defensive measures. The war eventually saw the prewar subcommittee reconvened as the War Cabinet Committee on Bacteriological Warfare. In 1940, after the fall of France, official studies of biological warfare left the theoretical world and entered the world of practical experiment to determine whether biological warfare was militarily feasible and, if it was, to determine the ways in which the UK could retaliate in kind in the event of the use of biological warfare by the Axis powers.

A most secret group was formed through the Medical Research Council at the then Chemical Defence Experimental Station at Porton in October 1940, under Dr

Paul Fildes (later Sir Paul Fildes). The group was euphemistically entitled the Biology Department, Porton, and, although situated within the senior establishment, was largely independent and autonomous. During the Second World War the Biology Department established, by realistic field trials on Gruinard Island and at Penclawdd, the feasibility of using anthrax spores in small, bursting submunitions, designed for a cluster bomb, the 'N' bomb, to infect and kill sheep through inhalation. The projected acquisition of an Allied antipersonnel biological warfare capability based on such cluster bombs was halted by the end of the war. The USA never produced anthrax spores at its Vigo plant in Haute Terre, Indiana, and the completed cluster bomb was never field trialled with an anthrax spore charge. Even the 1000 lb cluster bomb, which remained the most probable biological weapon projected for development in the UK after the Second World War was never developed. However, the urgent political requirement for the means to retaliate in kind had been fulfilled by the development of cattle cakes contaminated with anthrax spores and intended to be disseminated by aircraft over pastures for ingestion by cattle (Fig. 11.1). An operational stockpile of 5 million such cakes was produced by 1943 for retaliation in kind. Except for two boxes of cakes which were retained as curiosities, all the stockpile was destroyed very soon after the end of the war. The retained cakes were destroyed in 1972. The anthrax spores in these cakes were still viable in 1955. Essentially, the end of the war had seen the establishment of the feasibility of biological warfare by the UK, Canada and the USA by all means short of actual use in war. Moreover, the defence requirements were now all too obvious. The hazards from a biological warfare attack were immense.

BIOLOGICAL WARFARE OBJECTIVES 1945–56

The end of the war saw the contaminated cattle-cake stockpile destroyed: there was no further operational need for this somewhat unmilitary antilivestock weapon. Fildes, now knighted, returned to the Medical Research Council; other colleagues had also departed from the Ministry; a minimal staff remained at Biology Department, Porton. As has been described earlier, when dealing with chemical warfare, a partly similar paradoxical situation existed.

- As with chemical warfare, the no-first-use aspect of the UK's ratification of the Geneva Protocol was a constraining factor on the development of an offensive capability after the war.
- Biological warfare remained largely unproven: it had certainly been used by Japan in China and Manchuria in the 1930s and 1940s, but at the time uncertainty prevailed about the validity of limited intelligence.
- The UK perception of biological warfare and most putative agents was extremely limited. By the end of the Second World War UK research had virtually been limited to anthrax and, to a much lesser degree, to botulinum toxin.
- The specific biological warfare threat in 1945 was virtually undefined. An assumption was made that a capability would be acquired by the USSR: there were certainly indications of offensive work extending back to prewar years. The

FIGURE 11.1: LARGE-SCALE BATCH CULTIVATION OF ANTHRAX SPORES AT THE BIOLOGY DEPARTMENT, PORTON, DURING THE SECOND WORLD WAR, USING A MEDIUM LARGELY CONSISTING OF MARMITE, MOLASSES AND SIMPLE SALTS (© CROWN COPYRIGHT, DERA)

USSR emerged as the UK's principal potential enemy and thus constituted the prime biological warfare threat. Intelligence assessments of the Soviet biological warfare threat of the postwar years are probably not plentiful in the public domain. Where they can be found they lack detail, but attest, on a rather nebulous basis, to a long-standing programme. Much had come from German Second World War intelligence. The most significant report of this type was Colonel Walter Hirsch's 'Soviet BW and CW Preparations and Capabilities' (circa 1946–50). This large (671 pp) translated report from the Chemical Corps HQ in Washington DC was declassified on 6 June 1974, but its availability status is currently unknown.

Postwar decisions about biological warfare policy were considerably more intractable than those on chemical warfare. Nevertheless, the Biological Warfare Sub-Committee of the Chiefs of Staff Committee and the Defence Research Policy Committee recommended that research and development should be directed to:

- a biological weapon for strategic use
- the means of agent production and storage; and
- defensive measures.

It was recommended that the effort on biological warfare be increased 'to a level that may make it possible to bring into service by 1957 biological weapons comparable in strategic effects with the atomic bomb'.[19] Comparison of biological and atomic warfare was a considerable preoccupation. Points of emphasis included the potential to increase the lethality of biological warfare agents on an effective dose/weight-of-agent basis. Anthrax spores, although demonstrably effective, were not highly infectious: something in excess of 10,000 inhaled spores were needed to kill most species. Generally, animal lungs needed to receive a large number of inhaled spores (more than 10,000) before some lung clearance threshold could be passed and a systemic disease evoked. This phenomenon had been observed in humans within the skin and hide industry: although factory air was commonly contaminated with anthrax spores, there were few manifestations of inhalation infection. Conversely, a few spores could evoke skin lesions and even lethal systemic effects if they entered a skin abrasion and the disease remained untreated. It seemed likely that other agents, unresearched to any extent at Porton but examined by the USA, could be effective at doses a thousand times lower. *Brucella* species appeared to be highly infectious after inhalation, and had been studied in the USA at Camp Detrick during the Second World War, but not by the UK. Also, the production of agents was comparatively easy and less costly than atomic weapons. Further, there was some possibility that biological warfare could be complementary to atomic weapons. It was also noted that 'biological warfare need not remain a method of warfare repugnant to the civilised world ... a certain amount of informed guidance of the public might well result in it being regarded as very humane indeed by comparison with atomic bombs'. The view was also held that research must continue, 'preferably on a basis of Anglo-American cooperation'.

The prime concern was the need for a long period of systematic and fundamental research into the basic biological, chemical and physical parameters that underpinned biological warfare and defence against it. This reflected the view that biological weapon research and development would be largely a US role: the UK would examine fundamentals. Much more needed to be known about putative agents beyond anthrax spores. Also, field trials with agents needed to be carried out: not the ad hoc trials done during the war on Gruinard Island and Penclawdd, but ones capable of providing an understanding of the factors that controlled the survival of aerosolised pathogens in the field. Fildes' deputy, Dr D.W.W. Henderson, agreed to lead the postwar successor to the Biology Department, Porton, and was appointed Chief Superintendent of the renamed facility, the Microbiological Research Department at Porton, in 1946. The period from 1946 to 1951, when the new department remained physically within what had by 1948 become the Chemical Defence Experimental Establishment, was characterised by seven main activities:

- the design and planning of a new establishment adjacent to the Chemical Defence Experimental Establishment;
- the design and erection of a pilot plant to study the basics of large-scale production of bacteria;
- an expansion of fundamental research;
- the recruitment and training of new staff;

- the mounting of a major biological warfare trial, Operation Harness, at sea, off the Leeward Islands;
- the devolvement to the Chemical Defence Experimental Establishment of research and development for biological warfare munitions and weapon design; and
- the setting up of a new and eminent body of independent advice, the Biological Research Advisory Board.

BIOLOGICAL WEAPONS—OFFENSIVE RESEARCH

In this section we focus on the pilot plant and on Operation Harness and the successive sea trials which were held between 1948 and 1955 (space does not permit an account of the other major activities before 1956). Henderson and the Biological Research Advisory Board decided that before any full-scale production plant could be designed and built, it was necessary to have a pilot plant to study the fundamentals, using not pathogens but convenient non-pathogenic bacteria which could be regarded as simulants (e.g. *Serratia marcescens, Bacillus globigii* and some *Escherichia coli* strains). The pilot plant, also known as Experimental Plant No. 1, was designed in collaboration with staff at the Chemical Defence Research Establishment at Sutton Oak. Once processes had been developed at the pilot plant, it was intended to design and build Experimental Plant No. 2 adjacent to the new Microbiological Research Department building. This was specifically for pilot-plant or larger-scale production of pathogens, i.e. biological warfare agents. It is important to emphasise that the projected plant (Experimental Plant No. 2) for pathogens was never built, nor indeed finalised in its design. Confusingly, the designation Experimental Plant No. 2 was later used for a small continuous-culture plant added on to Experimental Plant No. 1 in 1960. This was intended to be used for pathogens but was never used for such purposes. It was dismantled in 1970, as alternative laboratory-scale apparatus for producing pathogens was eventually constructed within the Microbiological Research Establishment building. The location of the Experimental Plant No. 1, within the boundary of the Chemical Defence Experimental Establishment, was dictated by site and services availability. The plant was finally handed over by the Ministry of Works in May 1949. It was based on six 112 litre batch fermentors and was to evaluate in such vessels processes which had only ever been accomplished in a research laboratory in 300 ml vessels. To add further to the complexity, accurate temperature control, the provision of sterile air and the ability to contain and sterilise all effluent was to be explored. Downstream concentration and processing of the products were further needs. This sophisticated plant was also justified by its ability to accommodate the Medical Research Council's need for a pilot plant to produce antibiotics. The Council's small plant at Clevedon, in a former Royal Navy School site, was inadequate. It was agreed that the first work in conjunction with the Medical Research Council would be to produce the antibiotic licheniformin from *Bacillus licheniformis.* Licheniformin was active against *Mycobacterium tuberculosis,* but was eventually found to be highly damaging to the kidneys and never passed into general use as an antibiotic in tuberculosis treatment. This early collaborative effort with the civil sector involved not only the Ministry of Supply but also Imperial Chemical Industries, Glaxo, The Distillers Company and the Boots Pure Drug Company. The use of the plant for civil research was continued over the years, and it is extant today under the aegis of the Centre for Applied Microbiology and Research.

The trials at sea, which began with Operation Harness in 1948–49, continued until 1955. Like the erection of the vast new Microbiological Research Department at Porton in 1948–51 at a cost of £2.25 million, these trials underline the importance attached to biological warfare by the UK during a period of considerable postwar austerity. Gruinard Island was no longer considered suitable for use in peacetime. Trials at sea possessed obvious advantages: restrictions due to wind direction would not be a limiting factor and land contamination could be avoided; safety and security were assured; aerosols of pathogens would be destroyed by sunlight and diluted to extinction in the air and in the sea.

The size of the operations can be glimpsed by the total number of 475 men from Porton, the Royal Navy, the Army and the USA involved in Operation Harness. These trials demonstrated that effects could be produced in the field under realistic conditions and with a variety of dissemination devices, using the agents anthrax, tularaemia, brucellosis, plague and Venezuelan equine encephalomyelitis (see Table 11.5).

Generally, aerosols survived considerably less well in the field than experiments under enclosed holding conditions in the laboratory suggested. It came as something of a surprise that, apart from anthrax, the half-life of most disseminated agent aerosols was short, i.e. seconds rather than minutes. In the relative absence of

TABLE 11.5: A SUMMARY OF THE SEA TRIALS 1948–55

Title	Date	Location	Agents used
Operation Harness	1948–49	Off the Leeward Isles	*Bacillus anthracis* *Brucella suis* *Brucella abortus* *Francisella tularensis*
Operation Cauldron	1952	Off Stornoway in the Isles of Lewis	*Brucella suis* *Yersinia pestis*
Operation Hesperus	1953	Off Stornoway in the Isles of Lewis	*Brucella suis* *Francisella tularensis*
Operation Ozone	1954	Off the Bahamas	*Francisella tularensis* *Brucella suis* Venezuelan equine encephalomyelitis
Operation Negation	1954–55	Off the Bahamas	*Brucella suis* *Francisella tularensis* Vaccinia virus*

*Vaccinia virus was used as a simulant for variola virus, the agent of smallpox.[20] Other simulants were also used in the set of trials, e.g. *Bacillus globigii* and *Bacillus pumilus.*

sunlight, the half-life extended to 5 minutes. The sea trials finished in 1955, essentially because of the decision to abandon the efforts to acquire a UK biological warfare capability. Although they might well have continued for purposes relevant solely to biological defence, their cost was prohibitive at a time of increasing defence costs and economies. They had, however, provided an immense amount of knowledge of what happened to several agent aerosols under realistic conditions, and provided further compelling evidence of the military utility of biological warfare. They also emphasised that particularly intriguing aspects of agent viable decay needed to be studied.

We now turn to the abandonment of the policy to acquire a biological warfare capability in the postwar years, which had been caused by a combination of events unconnected with biological warfare *per se*.[21] Surprisingly, no specific date appears to be associated with a critical decision. In 1947, just after the establishment of the Board, BW research enjoyed a priority on a par with atomic weapons and guided missiles.[22] By 1953 this had shifted primarily to defensive research, with offensive research being confined to 'long-range' possibilities.[23] In 1955 the overall position of BW had been lowered in the military R&D programme. [24] A further policy statement by the Chiefs of Staff in 1958 declared that 'neither BW nor CW had significant strategic value and there was negligible risk of the use of these methods against the UK'.[25] Between these dates, changing policy on biological warfare can only be described as ambiguous and uncertain.

It seems almost certain that biological warfare was eventually subsumed in chemical warfare-related decisions of 1956, although not identified by name in key documents. This is a curious situation. No Cabinet committee decision specific to biological warfare appears to have been made. The discussion of chemical warfare appears to have incorporated biological warfare, but this is not certain. In any event, decisions that inhibited the future development of a biological warfare capability had been taken much earlier. Given the very highest levels of decision making about biological warfare during and just after the Second World War, the apparent absence of continued high-level decision making is odd. However, manifestations of biological warfare's decline in the UK can be seen as early as 1949, when it became apparent that whereas 'the development of the means for bulk production of pathogens shall be given top priority .... In view, however, of the state of national finances, ... it has been decided to postpone construction of this plant [The Experimental Plant No. 2]'.[26] Experimental Plant No. 2 was seen as a key element of an offensive programme and several advisers construed this decision as a judgement on the status of the entire programme. The decision also went through several reversals over the next few years. We will return to the independent scientific advice pertaining to this decision later.

By 1951, financial stringency was being stressed in the tripartite meetings[27] held with the USA and Canada. The following year it was emphasised there that the 1925 Geneva Protocol had a dampening effect because chemical and biological weapons, however good, could not be integrated with early and immediate war plans. In 1953, economies and the minimal scale of UK offensive research were again being emphasised, and in 1954 and 1955 further intimations of a lessening interest in development of chemical and biological weapons were given. By the late 1950s the tripartite meetings were being clearly told that UK research and development was now limited to defence and to assessments of how biological warfare could be used against the UK.

On the chemical warfare side, decisions were followed by several resurgences of interest and planning for some development. By contrast, the overall decline in the importance of British biological warfare research prevailed until 1961, when the two assessments of chemical and biological warfare mentioned earlier were carried out.

At the Chemical Defence Experimental Establishment biological weapon development was rounded off and written up and no biological weapons were developed. Instead, the work there was fundamental, involving studies of improved methods of dissemination from experimental munitions. The secret and protracted change of policy towards defence only had little effect at the Microbiological Research Department and the later Microbiological Research Establishment. Since the end of the Second World War, very few of its staff had been concerned with offensive matters in any case. No biological weapons were under development within the Department itself, and research had always been fundamental and long term. Such research was now relevant only to assessing the ways in which biological warfare might be used against the UK, and to defence. It was fortunate that the new Microbiological Research Department's building had been finished in 1951, otherwise this too might have been abandoned for essentially economic reasons. However, the Microbiological Research Establishment flourished.

The further development in the prestige and stature of the establishment was due largely to several factors:

- It occupied a vast, luxurious, modern building specifically designed for its role (Fig. 11.2). This was regarded as the best-designed and equipped microbiological institute in the world.[28]
- It was led from 1946 to 1964 by a purposeful, powerful and enthusiastic director, acknowledged as a microbiologist of stature who was capable of inspired leadership. He had a dislike of ministers and bureaucracy, but always recognised his duty and responsibilities. His relationship with the US biological warfare community was excellent and this played a major part in promoting collaborative efforts.[29]
- The establishment was associated from 1946 to 1977 with the powerful Biological Research Advisory Board under the aegis of the Defence Scientific Advisory Council. This was chaired successfully by Lord Hankey, Sir Charles Dodds and Sir David Evans, and it included, especially in its first two or three decades, a considerable number of the UK's most eminent independent scientists and medical men. We shall deal with this Board in a later part of this chapter.
- The research programme was multidisciplinary and diverse, attracting a wide variety of good scientists by its facilities and the research programme. The publication of scientific papers and involvement with the greater national and international scientific community was encouraged. In retrospect the diversity of research was amazing. The establishment soon attained world-leading status in

FIGURE 11.2: THE NEW MICROBIOLOGICAL DEPARTMENT AT PORTON IN 1951(© CROWN COPYRIGHT, DERA)

research on the chemical basis of pathogenicity, aerobiology,[30] virus epidemiology, experimental pathology, immunology, the identification of pathogens, biotechnology, continuous culture and microbial safety.

The interaction of these factors led eventually to the emergence of the establishment as a virtual national microbiological institute. By the time that Henderson retired in 1964, there was indeed a wish that it should become such a national civil institute, while retaining its biological defence role. Unfortunately, this mixed role was not acceptable and in 1964, on the dissolution of the Ministry of Aviation (as the Ministry of Supply had been renamed five years earlier), the establishment entered the Ministry of Defence. This move was opposed by Henderson, by the Biological Research Advisory Board and by many staff.

BIOLOGICAL DEFENCE 1957–90

The calibre of the research at the establishment was of the highest. It centred on the innumerable facets of microbiology and medicine which impinged on biological warfare defence, but the number of workers on highly specific aspects of defence, e.g. the study of the factors controlling microbial aerosol survival, and the evaluation of methods for near-real-time detection and early warning, was limited to a relatively small proportion of the total staff. Unlike the applied way of life at the older Porton establishment dealing with chemical defence problems, the Microbiological Research

Establishment received only one formal Service manifestation of a need for equipment in the form of work for Staff Targets and Staff Requirements. In the more than 30 years of its postwar research the Microbiological Research Establishment was largely concerned only with the Naval General and Air Staff Target 3083 for an automatic and near-real-time biological agent aerosol detector, which was later abandoned. This foundered because of financial constraints and other Service priorities, although the basic need remained an imperative and is the subject of continued research at the Chemical and Biological Defence Sector of the Defence Evaluation and Research Agency at Porton.

The majority of research contributed to the biological defence requirements, and although the yearly programmes were agreed with the Services and monitored by the Biological Research Advisory Board and the formal processes of the Ministry of Defence, the basic product of the Microbiological Research Establishment was fundamental data. These were not necessarily highly specific to biological defence, but increased understanding of microorganisms and their behaviour, and this new knowledge enabled assessment of the utility and hazards of biological warfare. Most of the establishment's work was of great academic interest and relevant to civil medical, veterinary and biotechnological uses of microbiology. This can be seen in a simplified representative (1967) establishment research programme (Table 11.6). This academic aspect provided a considerable impetus for the establishment to become a national civil microbiological institute, while retaining its defence responsibilities. However, this was not to be fulfilled.

In the more applied area, the establishment continued its interest in field trials. The possibility of further trials using pathogens in the field or at sea had gone: the last such use was in Operation Negation in 1955. Clearly pathogens could not be disseminated in the open air in the UK, and there were no longer isolated areas of Empire for such a purpose. Work consequently turned to the use of non-pathogenic simulants. These had been used since 1941 and provided a useful means of studying the movement, behaviour and decay of microbial aerosols in the field. Data could be extrapolated to pathogens, which could be studied safely in apparatus within the laboratories or in the large bursting chambers then available at the senior Porton establishment, one of which had been modified for microbiological use (Fig. 11.3). These twin 22-ft diameter test spheres or bursting chambers, constructed circa 1950 to provide safety containment in which vapours or aerosols produced by bursting munitions could be studied, were being destroyed at the time of writing (1997) as redundant facilities. Equally, data on the infectivity of microbial aerosols for laboratory animals could be obtained by the several means developed for laboratory use. On a quantitative basis, the concentration of airborne simulants could be determined from sampling devices located miles away from the sources of dissemination. If viable simulants at a concentration, say, of 30 per litre of air, were found ten miles from a source, knowing the wind speed and number of litres of air inspired by humans per minute, one could determine the effects of a biological warfare attack by extrapolation from laboratory determinations of the amount of agent that needed to be inhaled to evoke a disease.

TABLE 11.6: SIMPLIFIED 1967 RESEARCH PROGRAMME, MICROBIOLOGICAL RESEARCH ESTABLISHMENT

Growth and survival of micro-organisms
Bacterial growth and microbial products
- Continuous culture
- Microbial products
- Nucleic acids

Growth of cells and viruses
- Cytology
- Tissue culture
- Virus growth and concentration
- Production and activity of antibody

Survival
- Suspensions and freeze drying
- Survival of airborne organisms
- Field studies[31]

Pathogenesis and spread of infections
Microbial genetics
- *Pasteurella*
- Studies with antibiotic-producing fungi

Immunology
- Bacterial cell surfaces
- Animal cells
- Other studies

Pathogenesis of virus diseases
- Pathogenesis of viral encephalitis
- Vervet monkey disease

Epidemiology of virus diseases
- Studies in Sarawak[32]
- Field studies on British mosquitoes
- Laboratory studies on mosquitoes

Defence measures
Early warning and rapid detection
- Early warning
- Rapid detection

Vaccines and sera for prophylaxis or therapy
- *Brucella* vaccine
- Virus vaccines
- Prophylactic or therapeutic antisera

FIGURE 11.3: BUILDING THE TEST SPHERES AT THE CHEMICAL DEFENCE EXPERIMENTAL ESTABLISHMENT, PORTON, IN THE EARLY 1950S. IN 1998, THESE WERE BEING DISMANTLED AS UNWANTED FACILITIES. (© CROWN COPYRIGHT, DERA)

To such ends, initial trials with simulants on the Porton range evolved over the years to embrace the surrounding countryside, and later large areas of the central south of England. Here the simulants were disseminated as long-line sources from a naval vessel traversing the sea off Portland and Lyme Bay. The aerosol, moving inland with the wind, was tracked to enable the deployment of mobile sampling teams through the countryside. In other trials aircraft were used to disseminate the simulants. These trials had their origins in those conducted by the senior Porton establishment in the mid-1950s, when a non-biological and microscopic fluorescent particulate zinc cadmium sulphide was disseminated from ships or aircraft as a long-line source of many miles. These, and the subsequent trials using live simulants conducted by the Microbiological Research Establishment in the 1960s, demonstrated the feasibility and hazards of strategic attack on large areas of the UK. Whereas the fluorescent simulant could not suffer biological decay, i.e. die, the later trials with live simulants, where viability could be assessed, were conclusive in demonstrating that the use of actual agents in the same manner would result in massive casualties.

Although these aspects of the strategic attack preoccupied the establishment for many years, the possibilities inherent in several forms of sabotage-type attacks were not neglected. Work in the Second World War on such aspects had been minimal and occasionally bizarre. In the mid-1960s trials were carried out on the introduction of simulants into the ventilation systems of public buildings, exploration of the vulnerability of railways, and ultimately of the London Underground system.[33] Thus, the military utility of biological warfare in strategic, tactical and sabotage modes was firmly established. Inevitably, the defence requirements were obvious and considerable. The prime defence need was for a detector which would indicate the presence of an alien microbiological aerosol in the ambient air. The difficulties were immense, principally because air naturally contains a constantly variable concentration of microscopic particulates. Many of these are biological in nature and many are bacteria, fungal spores etc. The requirement was therefore to discriminate between the natural and the unnatural, quickly and precisely. Many potential systems were evaluated and tested in laboratory studies and field trials from the 1960s. These include those based on antigen–antibody reactions,[34] physical characteristics, and those methods involving some chemical reaction with enzyme systems. The trials on early warning and detection demanded considerable interaction with laboratory studies and studies of the normal air flora.

Underpinning all trials was the relatively new science of aerobiology.[35] The Microbiological Research Establishment was a world leader in this preoccupation with biological warfare and defence facilities. The topics addressed by Porton cannot readily be summarised here: most have been published in the open literature. Many relevant once-classified reports and annual reports from the Microbiological Research Establishment up to about 1967 are in the Public Record Office. However, it is worth dealing briefly with the microthread technique, which enabled the Microbiological Research Establishment to study the factors that controlled viable decay of aerosolised agents in the field but without any dissemination of such agents in the open air. The significance of such an opportunity in biological defence was immense. Essentially, ultrafine spider threads are wound on to small frames and exposed to aerosols of the pathogen or simulant. The microscopic particles adhere to the threads and provide a 'captive aerosol'. The frames, with more than 2000 trapped aerosol particles minimally stuck to threads, are a valid simulation of the true airborne state and can subsequently be exposed to a variety of conditions indoors or in the open, under conditions of safety: the particles are not readily dispersed from the threads by wind but may be effectively eluted for the determination of viability. By 1966 field trials designed solely for viability testing had been largely abandoned in favour of the microthread technique.

The survival of airborne bacteria had been studied at the establishment over the years in three ways. First, large rotating drums were used which could maintain small aerosol particles airborne for several days, but which did not produce results similar to those in the field; second, the microthread technique was a highly versatile and safe technique giving a good simulation of realistic field conditions; third, the field trial, which was costly and inefficient in that usable days were dependent largely on meteorological conditions. Trials were, however, the most fully representative of a biological warfare attack, particularly those done with simulants at night. From the data derived it would now be concluded that the decay rate of most aerosolised bacteria in the open was greater than occurred in the laboratory; the rate in the field was also greater in daylight than at night. This did nothing to diminish the hazard of biological warfare, but rather suggested that agents could be more self-limiting and thus more controllable. It also suggested that strategic attacks would be mounted at night for maximum effectiveness. Further, conditions of bacterial growth and storage had a profound influence on the biological stability of microbial aerosols. By the decade in which the establishment closed, another main factor controlling viable decay had been determined—largely through the application of the microthread technique—known as the open air factor. This germicidal factor, which had many peculiar characteristics, was a highly reactive product of ozone and olefins: it was essentially a partly manmade pollutant. Its effect on airborne microorganisms resulted in the inactivation of coat constituents and of nucleic acids. Also, the germicidal effect was due less to a high toxicity than to a readiness to condense on to aerosol particles.

Unhappily, the 1979 closure of the Microbiological Research Establishment led to a massive diminution in UK work on aerobiology, which is now slowly recovering after the build-up of equipment and expertise within CBD. Studies on the open air factor had virtually ceased by 1978, but not before it was shown that its effects could be reduced or nullified by the use of certain additives in disseminated bacterial suspensions.

Over the years from 1964 factors of increasing complexity began to impinge on biological defence. Successive defence cuts had little effect, as the new Director and the Biological Research Advisory Board regarded the integrity of well-established multidisciplinary teams within the establishment as something to be maintained, both to fulfil defence roles and to increase the proportion of civil research. Consequently, defence cuts were increasingly offset by civil sector funding of innumerable activities, both long and short term. Such activities did much to reduce opprobrium and unease about the establishment's activity, which were at times considerable, although temporarily nullified by a week of open days in 1968. Such opprobrium continued and is still active today, especially as more official documents on biological warfare and defence appear in the Public Record Office. The opprobrium, constant modifications of defence policy and the spectre of further defence economies led to the assumption by many that the way forward to a stable future was subordination to one of the civil ministries or one of the research councils.

However, less parochial forces were beginning to appear, and to combine to alter perceptions. These included the 1969 declaration by President Nixon that the USA's biological warfare capability would be destroyed and the Geneva Protocol ratified (the USA ratified in 1975). This political decision, not based on any proper scientific assessment, had a stated but unusual rationale which promptly de-emphasised the significance of the biological hazard and threat. In the 1960s and 1970s there was much international concern about arms control of chemical and biological weapons. The UK became a codepository of the 1972 Biological Weapons Convention, and ratified it in 1975. The Convention, too, became a factor diminishing perceptions of the significance of biological warfare. Ultimately, the need to continue the UK's existing level of biological defence came under close scrutiny. From

1971, factors moved increasingly to diminish biological defence research. Eventually, the Ministry of Defence decided that the future of the Microbiological Research Establishment was under consideration, and that in 1979 biological defence would become the responsibility of the then Chemical Defence Establishment, at a fairly token level.

The great debates over the future of the former Microbiological Research Establishment after 1979 cannot be dealt with here, but ultimately, upon its closure in 1979, the establishment became the Centre for Applied Microbiology and Research (CAMR) under the Public Health Laboratory Service. Later, the centre was removed from the Public Health Laboratory Service and in April 1994 was established as a Special Health Authority, an independent public-sector body under the Microbiological Research Authority and the Department of Health. An interim threat of privatisation was thus thwarted. A Defence Microbiology Division was set up within the Chemical Defence Establishment just before the closure of the Microbiological Research Establishment, with a small number of staff from the defunct establishment, minimal equipment and no proper facilities. Its original role was to provide a 'watchtower' and focus of advice on biological warfare, and to provide a group with microbiological expertise relevant to biological defence, notably on the identification of agents. It also had an almost immediate remit to attend to the assessment of residual contamination of Gruinard Island, and then its decontamination.

The Defence Microbiology Division spent some years establishing its laboratories and activity was limited, but by the mid-1980s official recognition of advances in biological sciences, such as the burgeoning of recombinant DNA technology[36] and biotechnology, led to funding for an increased role. Intelligence assessments of the biological warfare threat from the USSR had been largely ignored outside the intelligence community, but intelligence about the 1979 incident at a USSR biological warfare agent production facility in Sverdlovsk focused increased attention. The Sverdlovsk matter is complex. There is abundant literature from the 1980s which generally examines the nature of the anthrax outbreak among the citizens in April 1979, which had its origin in an accident and the liberation of an anthrax spore aerosol from a biological warfare agent production and storage facility. What interest should have been focused on was the nature and role of the facility in the context of the Soviet biological warfare threat. It was instead largely involved in a protracted discussion of the alternative, naturally occurring disease story disseminated by the Soviet authorities. Ultimately, a combination of growing evidence on Soviet activities and the relevant advances in biological science forced the reluctant realisation that the 1960s assessment of the threat from biological weapons had been ill-conceived, even if few were prepared to admit this. During the late 1980s and 1990s an enhanced biological defence programme was approved by the Chiefs of Staff. The wisdom of this move was underlined *post hoc* by a Russian admission that the former USSR had maintained an extensive clandestine biological warfare programme until at least 1992, while it was a codepository and ratifier of the Biological Weapons Convention. The cessation of this programme and the destruction of its weapons remains an enigma.

Further, in recent times the proliferation of biological weapons and offensive research in the world, notably by Iraq, suggested that the earlier rationale for a reduced biological defence programme and the closure of the Microbiological Defence Establishment may indeed have been a seriously ill-advised error. By 1989, when the Chemical Defence Establishment became part of the Defence Scientific Staff, 15 per cent of its staff were deployed on the biological defence programme. The Gulf Campaign in the early 1990s further underlined the relative paucity of the biological defence programme. Thereafter, considerable moves were initiated to put the programme on a proper basis, notably by the planning of the large new facilities for microbiological research within the present Chemical and Biological Defence Sector, Porton, and the appointment of extra staff. Notwithstanding such developments, the breakup of the integrated mixture of facilities and knowledge which had been accumulated at the Microbiological Research Establishment will take some years to restore.

CIVIL RESEARCH

We now look briefly at the civil research conducted at the several establishments after the Second World War. Again, it is convenient to look at chemical and biological matters separately, and single out a few representative highlights: we can do no more than give a brief resumé, which is far from comprehensive.

Chemical Matters

The earliest postwar civil research is not always readily identifiable: much was done on an informal basis, and for many years activities were conducted for British scientific and technological areas often on the 'old-boy' basis, and without much thought for written agreements or indeed proper funding. An impression can best be gained by a representative list, with some brief comments. Civil research is barely identifiable in periodical establishment reports until the 1950s: here they are often inextricably mixed with intramural accounts. One of the earliest topics in the 1950s was in the context of Colonial Office interest in antilocust research, when African Locust Research Council personnel worked under local supervision at Porton and in Africa on insecticides and their dissemination. Assistance was given to the Post Office Research Station in the development of instrumentation for insecticide distribution determination. The links between Porton and work on insect pest control continued into the 1980s. Another early topic involving much civil sector liaison lay in particulate filter research and testing. Charcoal research was another major civil spinoff from defence, notably in conjunction with Messrs Sutcliffe Speakman. Work was done in 1951 for the Ministry of Civil Aviation on the hazard of the jettisoning of fuel tanks from civil aircraft. Research on smog was carried out in the mid-1950s. Rubber research, a fundamental in respirator facepiece construction, was a further long-term project which had major civil applications and involved close collaboration with the British rubber industry. Induced precipitation (rainmaking) was studied in the field under the aegis of the Meteorological Office for some years in the 1950s. Other esoteric work covered the hazards arising from the use of leaded petrol in cooking stoves used in confined places. Work on the protection of agricultural workers from insecticides was carried out for the Ministry of Agriculture for several years. Liaison with civil firms in the chemical and pharmaceutical industries, which had been extant since the Great War of 1914–18, continued, notably on the identification of new compounds of putative chemical warfare significance and on new potentially therapeutic and prophylactic compounds. This too had relevance beyond defence. The potential of the establishments' mass spectroscopy facilities was made available to a wide variety of civil contacts.

At Nancekuke there was extensive collaboration with the chemical industry up to the time of its closure in 1980. In the mid-1950s several tons of an extremely toxic organophosphorus insecticide, 'Amiton' were made for ICI. In 1976–80 Nancekuke's charcoal cloth plant provided material to the Charcoal Cloth Co. and Siebe Gorman Ltd. An advisory service was instituted in the 1970s in support of civil authorities: this was particularly relevant to chemical accidents, and played an important part in 1974 when a leak of arsine occurred on the freighter SS *Asia*. By the 1970s, interaction with the civil community became focused with the appointment of an Establishment Repayment Liaison Officer post to positively seek work from industry and academics. This later became the modern business development and marketing area of the Chemical and Biological Defence Establishment, when it took on trading fund status. When the establishment entered DERA in 1995, this became even more important.

Microbiological Matters

The beginning of civil research at the Microbiological Research Department and later Establishment is not readily identified. Inevitably it had its roots in personal contacts, notably those enabled by the prestigious nature of the Biological Research Advisory Board. Monetary aspects are even more obscure initially, especially during the Henderson era. In 1951–52, scientists from the Department of Scientific and Industrial Research, the Medical Research Council and the Courtauld Institute of Biochemistry were actually working in the establishment on civil topics. Work was also taking place for the Public Health Laboratory Service, the School of Pathology at Oxford, the Lister Institute, the School of Medicine at Leeds and the Medical Research Council. In contrast, only one extramural research contract—i.e. where a civil contractor pursued research for the establishment—existed.

By the mid-1950s the Production Unit was satisfying demands made upon it, both internal and external, for a variety of organisms. Actinomycin had been prepared for Cambridge University; other microbial products had been supplied to the Royal South Hampshire Hospital and Dr G.P. Gladstone of Oxford. 1956–58 saw an increase in work for civil bodies, in accordance with government policy. A member of The Distillers Company was now on loan to the establishment. Bacterial cells were now being supplied to the civil sector in kilogram quantities. An increasingly large amount of freeze-drying work was also being done for civil customers. The Production Unit produced virus for 600,000 doses of Asian flu vaccine: this involved the use of 8000–9000 eggs each week for nearly five months. Each annual report provides an impressive list of microbial products for civil outside bodies, but never any indications of the costs involved, or whether all of these were recovered from the customer. In 1962–63 aflatoxin was made for the Tropical Products Institute. By 1963–64 the Vaccine Unit was committing half its output to the civil sector. The Experimental Plant (EPI) was fully employed in meeting civil demands. From 1965, the annual reports pay increasingly detailed attention to liaison with industry: vaccines for civil use were becoming increasingly important. Collaborative work with the Research Institute for Animal Virus Diseases at Pirbright was started. Perhaps the largest activity in this era was the establishment's new Virus Epidemiology Unit, partly funded by the Ministry of Overseas Development, which carried out studies in Sarawak. By 1968, the objective of 25 per cent support from non-defence and civil funds had been achieved. The philosophy was that the establishment had the minimum of expertise for a biological defence programme, and that excision of this by defence cuts would damage the defence capability. As an alternative civil support was obtained, and if the full capability was not required for defence, the resultant spare capacity was deployed on civil work. It was also explained that other government departments had a need for establishment expertise. It was thus fulfilling a national and broad-based need for major microbiological production. Louping Ill[37] vaccine was now being made available; later, tickborne encephalitis vaccine was made for use in central Europe.

By 1968, the list of microbial products for civil use was immense, but regrettably the sponsors were rarely cited in annual reports. The establishment produced a vast range of microbial products from *Escherichia coli*, e.g. penicillinase, glutamine synthetase, RNA polymerase etc. Also, bacteriophages, leucocidin, cell-wall lipopolysaccharide etc. were produced from a wide range of fungi, bacteria and viruses (Fig. 11.4). The enzyme L-asparaginase was being produced for clinical studies of its use in the treatment of leukaemia, and a significant programme was under way for the Department of Health and Social Security on herpes B virus in laboratory monkeys. Other customers included the Science Research Council, the Medical Research Council, the Atomic Energy Authority, the Unilever Research Laboratories and the Wellcome Foundation. Microbial ore leaching of uranium and dental vaccines were further interests. In 1962 the establishment began issuing separate reports on microbial products: these continued until 1970, when they were supplemented by civil programme reports, both as quarterly technical and cost reports as part of the overall annual report, and on specific topics, such as a series on enzymes for cancer therapy. Ultimately, in 1962, the Biological Research Advisory Board set up its Microbial Products Committee to advise on policy and scope of production, under Professor Wilson Smith and then Professor W.T.J. Morgan of the Lister Institute. In 1972 this became the Civil Programme Advisory Committee under Professor D.G. Evans. By the time of establishment closure the large civil research programme was yielding yearly work valued at £118k, based on major activities such as assorted microbial products, asparaginase and other therapeutic enzymes, herpes B virus studies, whooping cough vaccine, tickborne encephalitis virus vaccine, carboxypeptidase enzymes for cancer therapy, microbial toxins for

FIGURE 11.4: DEVELOPING FERMENTERS FOR THE LABORATORY-SCALE CULTIVATION OF PATHOGENS IN THE LATE 1960S (© CROWN COPYRIGHT, DERA)

pharmaceutical use, malaria vaccine, viral haemorrhagic fever in the southern Sudan, and safety testing of insect viruses. A Special Pathogens Unit was also continuing its valuable international responsibilities to the World Health Organization in studies on especially dangerous tropical virus infections in humans such as Marburg disease, Ebola fever, Lassa fever and Legionnaires' disease. When the Microbiological Research Establishment closed in 1979, the future for civil research was transferred to the Centre for Applied Microbiology and Research.

The new Defence Microbiology Division pursued no specific civil research but had inevitably to concentrate its limited staff and facilities on its minimal biological defence role. The focus for civil research at Porton had now passed to the Public Health Laboratory Service's Centre for Applied Microbiology and Research.

THE INDEPENDENT ADVISORY BOARDS

Studies in the history of military research and development tend to focus on the tangible products of science and technology. This emphasis ignores the equally important intangible products of scientific research, in particular expert scientific advice.[38] As a consequence, the role and nature of scientific advice as a 'product' of defence research establishments is rarely considered by historians. The onetime two major establishments at Porton had long been associated with bodies containing a mixture of Service officers, civilian staff and, most importantly, professionally eminent scientists acting in their individual capacities to advise and comment on the research programmes, both at the local level and, when necessary, to the responsible minister. In this respect, it is important to note that these people provided advice to the establishments—and more importantly—passed on, throughout the government, advice based on the work of the establishments.

These advisory bodies have their roots in the ten committees which, at several periods from 1916 to 1918, gave guidance on chemical warfare. The most important of these was the tenth and last, the Chemical Warfare Committee, which had been suspended after the Armistice.[39] A new Chemical Warfare Committee was set up in 1920 on the advice of the Holland Committee, and in 1931 its title was changed to the Chemical Defence Committee.[40] After the outbreak of the Second World War, the Committee was largely replaced by a new Chemical Board, which continued until 1946 when the Ministry of Supply replaced its 1939 Advisory Council of Scientific Research and Technical Development, under whose wing the Board had existed, with a new Scientific Advisory Council under Sir Frank Smith. Its remit was to:

> consider and initiate new proposals for research and development and to review research and development in progress at Ministry of Supply Establishments in relation to the most recent advances in scientific knowledge, to advise on scientific and technical problems referred to then, to make recommendations regarding the most effective use of scientific personnel for research and development and to report to the Minister of Supply.[41]

The Council consisted of independent members and officials and operated through a number of subordinate boards and committees, including (in 1959) boards for advising on chemical defence, guided weapons, radar and signals, weapons and explosives, and agricultural defence. Those that concern us here are the Chemical

Defence Advisory Board and the Biological Research Advisory Board. It is convenient to consider these separately. The discussion of the Biological Research Advisory Board will be more detailed in order to illuminate some facets of the role and nature of the expert advisers.[42]

The Chemical Defence Advisory Board had a fairly heterogeneous membership. Scientific members included organic and physical chemists, biochemists, physicists, physiologists and meteorologists. There was also industrial representation, together with departmental representation initially from the Admiralty, War Office, Air Ministry, Home Office and Ministry of Supply. Its ostensible role was purely advisory:

> It was not the responsibility of the Board to tell CDEE what to do, but rather to comment and advise on research matters. [Advice] also went to the upper echelon where it helped to form policy directives.[43]

The Board had, at various times from 1946 until it was disbanded in 1994, several Committees:

a. Chemistry Committee
b. Manufacturing Processes Committee
c. Physics and Physical Chemistry Committee
d. Offensive Equipment Committee (later the Offensive Assessment Committee)
e. Defensive Equipment Committee
f. Biology Committee (This gave rise to the separate Applied Biology Committee which eventually became the Medical Committee. The original Biology Committee became the Life Sciences Committee.)
g. Medical Committee, with subgroups for:
   Internal Security and Anti-Terrorist matters
   Chemical Defence
   Trauma Research
   Ethical considerations.

There was also an interim short-lived Biological Defence Group and Biotechnological Committee.

At a later stage the Board possessed only three committees:

a. Biomedical Committee
b. Microbiology and Biotechnology Committee
c. Physical Sciences Committee.

At a subsequent stage the Board dropped 'Advisory' from its title and still later became the Chemical and Biological Defence Board. In April 1994 the Board was replaced by the Defence Scientific Advisory Council's Chemical, Biological and Human Technology Board. All other Council Boards were similarly reorganised at this time.

The Chemical Defence Advisory Board was initially Chaired by Captain J. Davidson

Pratt of the onetime Chemical Manufacturers Association. Independent members over the years have included elite scientific and medical researchers, e.g. both the first and the second Lord Adrian, Sir Cyril Hinshelwood, Professor Sir James Frazer, Sir Harry Melville, Sir Rudolph Peters, Sir Robert Robinson, Sir Ian Heilbron, Lord Todd, Professor H.J. Emeleus, Professor R.H.S. Thompson, Professor G.R. Cameron, Professor Sir Charles Lovatt Evans and Sir John Cadogan. The part played by these men, many Fellows of the Royal Society, was significant and extended for many years until the 1970s, when the proportion of Fellows of the Royal Society associated with the Board decreased, although there was a brief resurgence to four Fellows in the early 1980s. It is worth providing here a list of the Chairmen of the Board, with their approximate dates of tenure:

1946 Captain J. Davidson Pratt
1949 Professor (later Lord) Todd
1952 Sir Rudolph Peters
1957 Professor H.J. Emeleus
1961 Professor J.H. Gaddum
1962 Professor H.J. Emeleus
1963 Professor R.B. Fisher
1969 Professor R.H.S. Thompson
1976 Professor N.B. Chapman
1981 Professor R.N. Hazeldine
1984 Dr B. Atkinson

A study of the independent membership of the Board from 1946 until its reorientation and replacement in 1994 reveals a profoundly interesting and senior scientific and medical advisory body. Members were undoubtedly well able to provide scientific advice and views, which did much to foster the research programmes in chemical warfare and chemical defence and civil work for nearly half a century. Their guidance over the years added to the status of the relevant establishments at Sutton Oak, Nancekuke and Porton. The concept of this type of elite interface between government and science is examined in greater detail in the next section.

The Biological Research Advisory Board (BRAB) was established in 1946 to provide independent scientific advice on 'biological problems with special reference to micro-biological research carried out in the Ministry of Supply and extra-murally'.[44] It was accountable primarily to the Advisory Council on Scientific Research and Technical Development of the Ministry of Supply (MoS), but provided technical advice to the Inter-Services Sub-Committee on Biological Warfare (ISSBW). The Ministry of Supply, in turn, was 'responsible for carrying out the technical programme for BW research and development at the Microbiological Research Department and other establishments'.[45] The Board existed until 1977 when it was dissolved, before the closure of the Microbiological Research Establishment in 1979, in anticipation of the Chemical Defence Board absorbing its role.

The ISSBW comprised independent scientists, Ministry of Supply, Home Office and Ministry of Health representatives, together with Admiralty, War Office, Air

Ministry staff and medical representation.[46] Soon after the end of the Second World War it was reconstituted as a subcommittee of the Defence Research Policy Committee. Its terms of reference stated that it was to be:

> the focus for the discussion and formulation of all aspects of BW policy, both offensive and defensive, and [it] reports to the Chiefs of Staff. On technical matters it is advised by the Biological Research Advisory Board and for foreign intelligence it relies on the Joint Scientific Intelligence Committee.[47]

In addition, the ISSBW supplied 'advice on policy and technical matters relating to civil defence against BW' to the Home Defence Committee. It received general guidance on research policy from, and provided advice on 'the policy and technical aspects of Biological Warfare' to, the Defence Research Policy Committee (DRPC). The DRPC was 'responsible for advising the Minister of Defence and the Chiefs of Staff on matters connected with the formulation of scientific policy in the defence field'.[48]

The initial membership of BRAB was intentionally small, consisting of the Chairman, four scientific members and representatives of the MoS (including the Chief Superintendent of the Microbiological Research Department).[49] The original independent members were Lord Hankey, Professor E.C. Dodds (later Sir Charles Dodds), Sir Paul Fildes, Sir Howard Florey (later Lord Florey) and Lord Stamp.[50] The Board included three scientists also appointed to the ISSBW.[51] It is worthwhile noting that by 1950 the size of the Board had grown to 19 members and remained at about this level through to 1967, when the records close. The Chairmen were: 1946, Lord Hankey; 1952, Sir Charles Dodds; 1966, Sir David Evans.

As with the Chemical Defence Advisory Board, the status of the Chairmen was undoubted. Lord Hankey FRS (1877–1963), although not a scientist, had been elected to a Fellowship of the Royal Society and had been concerned with defence since the turn of the century, with biological warfare since the early 1930s, and had been Secretary of the Committee of Imperial Defence and Secretary of the War Cabinet. Hankey retired from the Board at the age of 80. He can be considered the founding father of biological warfare and defence in the UK.[52] Sir Charles Dodds FRS (1899–1973) was Director of the Courtauld Institute of Biochemistry and President of the Royal College of Physicians.[53] Sir David Evans FRS (1909–84) was at various times Head of the Biological Standards Control Laboratory at the National Institute for Medical Research, Professor at the London School of Hygiene and Tropical Medicine, Director of the Lister Institute, Director of the National Institute for Biological Standards and Control and President of the Society for General Microbiology. Study of the independent membership of the Board reveals the senior status of this body by the identification of 30 Fellows of the Royal Society who served as members.

The advice formulated by BRAB members was passed in two directions, first, to Porton on the conduct of the BW research programme, and second, to the policy-making hierarchy, drawing on the results of BW research. One key difference between

BRAB, CDAB and the other advisory boards is pointed out by the Scientific Advisory Council. Both the chemical and biological advisory boards:

> had an easier task because they were dealing with more or less unified fields. In the case of the Weapons and Explosives Advisory Board the position was rather different because that Board was concerned with an extremely wide range of disciplines and techniques.[54]

Although BRAB had a remit to provide technical advice, this varied in nature and substance. Its members frequently provided advice and made recommendations on quite detailed technical matters, such as the conduct of field trials with pathogens and simulants. They were consulted on issues concerning the potential threat from BW, for instance, to suggest the agents likely to be used in a BW attack against the UK.[55] BRAB also advised on more managerial aspects of the BW research programme, primarily on matters of staff recruitment. This was a recurrent problem that took up a substantial portion of their discussions.

The committee was well aware that they had no remit to discuss policy issues, and was often reminded of this by the chairman. Nonetheless, all of their advice was proffered in the context of the changing policies towards biological weapons outlined earlier. It was not always possible to exclude such political considerations entirely from what was ostensibly the realm of technical debate. A full history of scientific advice would not make for engaging or illuminating reading. We have chosen instead to focus on BRAB's changing self-perception, and then to examine in some detail the anatomy of several key recommendations made by the Board. By doing this, we aim to draw out some general features of this advisory board.

Members of BRAB tended to regard the Board as an extremely specialist unit consisting of 'eminent members in a circumscribed field'.[56] Minutes of meetings record, at several points, their close relationship with staff at the Microbiological Research Department (MRD).[57] At one point the Chief Scientist, Sir Owen Wansbrough-Jones, even went as far as to state that the members had 'virtually assumed the role of supervisors of each section'.[58]

This position is in marked contrast to some of the statements made by members of the Board about those bodies receiving their advice. Sir Paul Fildes, for example, reported in 1952 that he had been involved in disagreements with MoS officials which he attributed to his 'lack of knowledge of changing policies'. [59] Later in the same meeting Fildes added that 'it was quite unsatisfactory that advisers should have advice rejected ... without some slight explanation'.[60] Regardless of whether or not the bulk of BRAB advice was actually being rejected at this time, it is evident that this was the perception of at least one influential member of the committee.

As the status of BW fell, so did the links with the formulation of policy. In 1955 a key link between BRAB and policy makers disappeared with the dissolution of the ISSBW. Concrete evidence of further divorce from the formulation of policy came when the Todd Panel, evaluating the role of CBW, was established in 1961.[61] BRAB had not been informed about this and expressed their 'astonishment that an outside body such as the Todd panel could be set up without any word to the BRAB, which was presumably the foremost authority on BW'.[62]

At the following BRAB meeting, Fildes expressed dissatisfaction with the position of the Board.[63] Because they no longer had direct contact with the 'users', he felt that their advice was now ignored. Sir Howard Florey also:

> criticised the remoteness of the final advisors from the source of first hand information, and felt strongly that the Board might expect to be told what decisions had been reached and on what grounds.[64]

It is evident that changes in the status of biological weapons research and policy were equally reflected in the attention paid by higher authorities to the field's expert advisers. Although the members of BRAB noted their frustration at having advice ignored, another fate for their counsel was simply to be overtaken by events. Nowhere is this more evident than with the advice proffered by BRAB over Experimental Plant No. 2. As we have seen, between 1947 and 1955 several recommendations were made by BRAB to proceed with the construction of a large-scale plant for the manufacture of pathogenic organisms.

Discussion surrounding Experimental Plant No. 2, mentioned earlier, focused in part on costs. The initial budget, estimated at £1 million, was a matter of concern to several BRAB members, the amount being comparable to the annual expenditure of the Medical Research Council.[65] The discussion over this item was not entirely about the economics of the proposal. As a backdrop to the financial considerations, the Board also saw the decision as contingent on two broader questions of policy. First, the question of how heavily the UK was to rely on the USA for its supply of pathogens. One implication of building the plant would be to grant a degree of independence from the USA in this area. A second, related, consideration was that the plan was thought to offer 'the only possibility' for obtaining full-scale production of BW agents by a target date of 1957 for an operational air requirement. Without this target date, it was suggested by the Board that the new plant could be delayed until experience was gained from Experimental Plant No. 1.[66] The latter option was strongly endorsed by Lord Stamp, who argued that policy should revert to that of the Second World War, when all mass production of pathogens was to be undertaken by the USA.

In all this debate the BRAB members acknowledged that actual policy decisions were beyond the scope of their responsibilities. It nevertheless highlights the way in which broader policy deliberations interacted with concrete decisions, in this case over whether or not to acquire a pathogen plant. In short, Experimental Plant No. 2 was perceived as an embodiment of national BW policy. Discussions as to whether it was 'technically' to be recommended could not readily be separated from broader political considerations such as the status of the UK programme *vis à vis* the USA.[67]

As the estimated costs continued to rise, the plant was deferred by the DRPC in 1950.[68] Shortly afterwards, at the 16th meeting of BRAB, a central place was accorded to the plant when it was acknowledged that an indefinite delay in building Experimental Plant No. 2 would engender major alterations to the entire MRD programme.[69] The decision by the DRPC was then reversed later in the same year, although with the proviso that other aspects besides the plant itself, such as stores, a workshop, laboratories, changing rooms and offices, would be reduced in order to make significant savings.[70]

By 1952 the plant had still not been constructed, although the USA was reported to be proceeding rapidly with its own plant. As a consequence, it appears that the UK plant's future was still in the balance. The Board was reminded by Florey that BRAB had vacillated over its decision to recommend the plant, and therefore another reversal of policy should not be taken lightly.[71] Fildes argued that the new plant would help ascertain whether countries other than the US could produce operational quantities of pathogens. In other words, the success or failure of the UK would mirror Soviet abilities. It was also pointed out that there were significant differences in the aims of the two production programmes, the UK having a particular interest in continuous-flow production. The Board concluded by reaffirming the necessity of the plant for BW research; nevertheless, it appears that this was a decision that was not taken easily.

Despite the recommendation to the contrary, it was announced at the 25th meeting of the Board that building work on Experimental Plant No. 2 would not be started in 1952.[72] The position was to be reconsidered in the 1953 spending estimates.[73] In connection with this postponement, it was also mentioned that the plant would have to fit in with other MoS defence commitments. BRAB continued to recommend the construction of the plant over the next two years, and by November 1954 a new specification for buildings and services had been drawn up.[74] A year later the plant looked likely to become a victim of the reduced status of biological weapons research. At its 33rd and 34th meetings BRAB:

> discussed the project for the provision of Experimental Plant No. 2 in the light of a policy directive by the Minister of Defence, which lowered the general position of Biological Warfare in the research and development programme. The opinion was expressed that it would be entirely wrong to stop production research on disease-producing organisms, that the plant would have wider use than the culture of pathogens, and that the plant was an essential piece of apparatus for pursuing microbiological studies.[75]

Delays caused by economics and policy changes meant that other events eventually overtook the decision on whether to proceed with the plant. The annual report of the MRE for 1956–58 noted that an 'entirely new' apparatus had been developed for producing quantities of around 100 litres or a kilogram of dry-weight cells per day.[76] It was also noted that the plant components were now available entirely from commercial sources, thus reducing the need for specialist engineering. Plans were made for the construction of a 100-litre continuous-flow plant: these were approved and building finished in August 1962.[77]

Despite the above comments, BRAB advice did enter into policy considerations on occasions, even when the Board had voiced concerns over its waning influence on policy makers. From 1956 onwards a dominant theme in BRAB's discussions was the threat from the widespread dissemination of biological agents, the so-called large-area concept. The report by the Chairman of BRAB for 1957, for example, stated that: 'the concept of large area coverage over, say, 200 by 200 miles offers a potent exploitation of the insidious character and low dosage which are peculiar to BW'.[78]

The perceived threat of large-area coverage led to a variety of outdoor trials and provided a new rationale for research on microbial detection methods to provide early warning. Open-air trials, as described earlier, involved spraying zinc cadmium sulphide fluorescent particles from an aircraft across large areas of Britain, followed later by trials with live simulant organisms.[79] In 1959, in its annual report, the Board 'reaffirmed its belief that large area attack with BW agents constituted a major threat to the country's defences and its profound dissatisfaction with the low priority that had been accorded to the threat'.[80] The overall level of concern about large-area attack expressed by BRAB eventually found its way into the deliberations of the subcommittee of the DRPC in 1962. This subcommittee, as mentioned earlier, had been charged with preparing a paper, based on the findings of the Todd Panel and operational assessments of CBW, for the Chiefs of Staff.[81] In the introduction to their report the subcommittee justified a renewed interest in the potential of CBW. Their reasons included the increased effectiveness of CW agents, the future potential of incapacitating agents, and the vulnerability of the civilian population. The report also—mirroring the concerns voiced by BRAB—noted that a key reason for renewing attention in the area was 'the realization that large populated areas could be subjected to clandestine BW attack by aircraft or ships operating at distances of many miles from the target area' and the severe difficulty of detecting such an attack.[82] This is by no means to claim that BRAB advice was decisive in the later recommendation of the subcommittee to renew research on the offensive aspects of BW. Nonetheless, it shows that at least some of BRAB's concerns were assimilated into important policy deliberations.

Advisory committees can generally be regarded as an elite interface between government and society.[83] From this perspective, both BRAB and CDAB may be construed as an elite interface between the government (and the Services) and another elite, a particular research constituency (Porton Down). The Boards acted as conduits, passing and processing expertise between scientists and policy makers. In this capacity they acted as a key route by which abstract policy declarations and the concrete activities of defence scientists reciprocally influenced each other.

Evidence from the records of BRAB activities suggests that the relationship of the Board to researchers was closer than the links with government and the Services. In part, this could be attributed to a certain professional solidarity among a particular section of the scientific community. This was reinforced by the wartime experience of some early members of BRAB, who had been closely involved with the wartime effort in BW and with the prewar Committee of Imperial Defence Sub-Committee on Bacteriological Warfare. The extreme secrecy of the field and the gradually diminishing status of biological weapons research may also have contributed to the quality of these relationships.

An initial reading of the documents and minutes of the various committees responsible for chemical and biological warfare reveals many recommendations and decisions reproduced verbatim along the chain of advice. On this reading, one could readily construe the advisory process as if scientific advice were something passed from committee to committee like batons in a relay race. The detailed evidence demonstrates the subtle ways in which technical advice, initially as a product of defence research, enters the political arena as one resource among many.[84] At times, BRAB advice may simply have been ignored; at other times it was assimilated into the policy-making process—although not necessarily with the results recommended by the committee. In general, the intangible nature of scientific advice makes it difficult to narrate its history or to judge its full influence. Nonetheless, by including advice in our account of chemical and biological warfare, we hope to illustrate that artefacts, weapons and other defence equipment are not the only items of interest to have emerged from the research performed at Britain's defence establishments since 1945.

1. Phosgene is essentially an industrial gas, first synthesised in 1812 and used by all sides in the Great War as a lung-injurant: carbonyl chloride.
2. This was the earliest (1925) significant arms control agreement to prohibit the use of chemical and biological warfare in war, but not possession of the capability.
3. Technical Supplement to the *Report for the Year 1962*, appendix to SAC566, PRO WO 195/15491 (1962). See also J. Agar and B. Balmer, 'British scientists and the Cold War: the Defence Research Policy Committee and information networks, 1947–1963', *Historical Studies in the Physical Sciences*, 28, no. 2 (1998): 1–40.
4. VX gas is one of the most potent of the V-agents and the one most studied.
5. D. Campbell, 'Thatcher goes for nerve gas', *New Statesman* (11 January 1985).
6. Most accounts of Operation SANDCASTLE are poor. A useful account is given in Reg Chambers Jones, *Bless 'Em All: aspects of the war in north west Wales 1939–1945* (Bridge Brooks, 1995). A further detailed text, published since this chapter was written, is Roy Sloan, *The Tale of Tabun: Nazi Chemical Weapons in North Wales* (Carreg Gwalch, 1998).
7. A simulant in this context is an inert chemical compound or harmless microorganism capable of being used to imitate actual chemical or biological warfare agents.
8. The 1956 decision is reported in *Cabinet Office: Defence Committee 1946–1963 Minutes and Papers*, DC(56) 6th Meeting, PRO CAB 131/17 (10 July 1956).
9. An Aden gun is a 30 mm antitank light cannon of the 1950s.
10. A cluster bomb contains submunitions or 'child bombs' which are released after deployment of the 'mother' bomb to cover a relatively wide area.
11. V-agents are a series of nerve agents with greater military utility than the early G series.
12. G-agents are essentially the nerve agents. The term arises from the German and later the Allied code letters for such agents, e.g. GA, GB, GE, GF.
13. An account of insecticide work up to and including the Second World War is in the Ministry of Supply Monographs nos 9.600, 9.601 and 9.602, PRO WO 88/739, 740 and 741.
14. CS is a lachrymator or tear gas widely used in riot control. The initials stand for Corson and Staughton, who first synthesised it in the USA in 1928.
15. 'Report of the enquiry into the medical and toxicological aspects of CS (orthochlorobenzylidene malononitrile)' Part 1 and Part 2, The Himsworth Committee 1961 and 1971, respectively Cmnd 4713 and Cmnd 4775.
16. A facelet is an oronasal mask of charcoal cloth capable of providing some initial protection against inhalation of chemical agents.
17. Nerve Agent Immobilised Alarm and Detector, now an in-service item of equipment.
18. See G.S. Pearson, 'Prospects for CB arms control: the web of deterence', *Washington Quarterly* (Spring 1993): 145–62.
19. DRPC, *Final Version of Paper on Future of Defence Research Policy*, PRO DEFE 10/19 (30 July 1947).
20. Vaccinia is the virus used in smallpox vaccine. As a virus of somewhat low pathogenicity it could, under some circumstances, be used as a simulant for its related smallpox virus. At that time smallpox virus was regarded as a putative agent: but given the almost universal use of vaccination throughout the civilised world, the status of smallpox as an agent could not have been great. Now, at a time when vaccination has diminished it might be greater, but for the fact that there are only two or three smallpox virus stocks left in the world. Essentially smallpox is no longer extant as a disease.

21. B. Balmer, 'The drift of biological weapons policy in the UK 1945–65', *Journal of Strategic Studies*, 20, no. 4 (1997): 115–45.
22. Chiefs of Staff Committee, Biological Warfare Sub-Committee, Division of Responsibilities, BW(47)32 (3 November 1947), PRO WO 188/667.
23. G.B. Carter and G.S. Pearson, 'North Atlantic chemical and biological research collaboration: 1916–1995', *Journal of Strategic Studies*, 19, no. 1 (March 1996): 74–103. The ambiguous term 'long range' is deemed by the authors to mean fundamental research—a view supported by the BRAB minutes. See, for example, BRAB 27th Meeting, AC12127/BRBM27, PRO WO 188/668 (6 December 1952).
24. *Report by the Chairman on the Work of the Board during the Year 1955*, Item 3, AC13646/BRB139, PRO WO 195/13460 (7 November 1955).
25. DRPC, *Chemical and Biological Warfare*, DRP/P(62)33, PRO DEFE 10/490.24144 (10 May 1962).
26. Chiefs of Staff Committee: Biological Warfare Sub-Committee,BW(49) 40(Final) (28 February 1950); *Report on Biological Warfare*, Report to the Chiefs of Staff (1949).
27. In the context of this chapter, the Anglo-American-Canadian meetings on toxicological warfare held from 1947 to 1964, latterly becoming quadripartite with the entrance of Australia.
28. Details of the building have appeared in a variety of texts: papers, Establishment brochures, etc. over the years. A paper by J.C. Knight, 'Engineering services in a laboratory', *Journal of the Institute of Heating and Ventilation Engineers*, 23 (May 1955): 33–78 was one of the earliest and most comprehensive accounts by the Ministry of Work's Senior Engineer. Curiously, the building is not identified.
29. See especially H.A. Druett, 'David Willis Wilson Henderson 1903–1968' in *The Dictionary of National Biography 1961–1970*, edited by E.T. Williams and C.S. Nicholls (Oxford University Press, 1981): 504–6, which gives references to obituaries.
30. Aerobiology is essentially the science of how microorganisms behave in the airborne state. The term first appeared in 1937 and entered the biological warfare vernacular in the early 1950s.
31. Field trials were also involved in early warning and rapid detection in the Defence Measures category.
32. This civil research was initiated by the Overseas Development Ministry and the Medical Research Council.
33. Much publicised in the media, e.g. B. Balmer, 'The day germ warfare came to Tooting Bec', *The Independent* (28 March 1995). A contemporary account appears in BRAB Minutes, 55th Meeting, SAC 784, PRO WO 195/15709 (1963).
34. The antigens of microorganisms evoke the body's production of antibodies, which may then be used to detect antigens and thus microorganisms.
35. This term is quite old but was first popularised by the US Navy Biological Laboratories in Oakley, California, in its researches on experimental airborne infection.
36. Recombinant DNA technology, first developed in the 1970s, enables the production of designed novel microorganisms, leading to 'genetic engineering'.
37. This is a viral disease of sheep encountered in parts of Scotland.
38. Some contemporary analyses of science have even claimed, counter to this emphasis on artefacts, that scientific advice is the *primary* reason why the State supports civil and military science. See C. Mukerji, *A Fragile Power: scientists and the state* (Princeton University Press, 1989).
39. An account of the Great War Advisory Boards is set out in *The History of the Ministry of Munition Volume XI: The Supply of Munitions Part II Chemical Warfare Supplies* (undated but ca. 1921). This is undoubtedly in the public domain but no reference can be ascribed.
40. The substitution of 'Defence' for 'Warfare' was also made in the title of the Chemical Warfare Experimental Station at Porton, a title which had been extant since 1929. The senior Porton Establishment has had eight changes of title since 1916.
41. The Chemical Defence Advisory Board's *Terms of Reference and Constitution* are in the PRO at WO 195/9062 and for the Biological Research Advisory Board at WO 195/9087.
42. A similar analysis could have been made of the Chemical Defence Advisory Board. The choice here reflects the broader research programme of one of the authors (Brian Balmer). We are grateful to Zahid Karim for research assistance in obtaining certain primary documents relating to CDAB.
43. Minute 47, SAC 684/CDB, M53 CDAB, PRO WO 195/15609 (6 June 1963).
44. *BRAB Constitution and Terms of Reference*, AC9091/BRB1, PRO WO 195/9087 (11 July 1946).
45. Chiefs of Staff Committee, Biological Warfare Sub-Committee, *Division of Responsibilities*, BW(47)32, PRO WO 188/667 (3 November 1947).
46. Chiefs of Staff Committee (n. 45 above).
47. Chiefs of Staff Committee (n. 45 above).
48. Chiefs of Staff Committee (n. 45 above).
49. This small and intimate Board was, at its inception dwarfed by the Chemical Defence Advisory Board chaired by Professor A.R. Todd (later Lord Todd), which had nine independent members and a further membership within its six subordinate Committees.

50. *BRAB Constitution and Terms of Reference* (n. 44 above). Lord Florey and Lord Stamp have not been mentioned earlier. Florey was, with Sir Alexander Fleming, one of the pioneers of penicillin during the Second World War. Stamp had been one of Fildes' senior staff at Biology Department, Porton and spent a large part of the war in Canada and the United States on liaison matters and particularly field trials. See Hansard: House of Lords 229/30 (5 February) wherein Trevor Stamp describes how he came to Porton.
51. Fildes and two others: *BRAB Constitution and Terms of Reference* (n. 44 above).
52. Some acknowledgment of Hankey's role in biological warfare is found in Stephen Roskill's three-volume biography, *Hankey: man of secrets* (Collins, 1974).
53. Sir Charles Dodds' obituaries give little credit to his role as Chairman of BRAB. Security concepts of the past would have imposed constraints on a wider knowledge of his fourteen years service.
54. Advisory Council on Scientific Research and Development, 133rd Meeting, AC13757/M133, PRO WO 195/13753 (5 July 1956).
55. BRAB, *Review of Work Done by the Board During the Year 1947*, Item 8, AC9554/BRB33, PRO WO 188/667 (19 November 1947).
56. BRAB 19th Meeting, Minute 141, AC11260/BRBM19, PRO WO 188/668.21 (December 1950).
57. The Microbiological Research Department (MRD) at Porton Down changed its title to the Microbiological Research Establishment (MRE) in 1957.
58. Minute 286, BRAB 41st Meeting AC14413/BRBM41, PRO WO 195/14405 (18 July 1958).
59. BRAB 27th Meeting, AC12127/BRBM27, PRO WO 188/668 (6 December 1952).
60. BRAB 27th Meeting (n. 59 above).
61. Membership: Sir Harry Melville, Dr F.J. Wilkins, Professor E.R.H. Jones, Professor E.T.C. Spooner, Major-General J.R.C. Hamilton, Professor P.B. Medawar.
62. BRAB 49th Meeting, Minute 329, SAC243/BRBM49, PRO WO 195/15168 (15 April 1961).
63. BRAB 49th Meeting, Minute 333 (n. 62 above).
64. BRAB 49th Meeting, Minute 333 (n. 62 above).
65. *Report on Plant for Experimentation on Bulk Production of Bacteria*, Amendments to AC10522/BRB70, PRO WO 188/668 (28 September 1949).
66. *Report on Plant for Experimentation on Bulk Production of Bacteria* (n. 65 above).
67. Members of the DRPC were more frank in this regard, describing the plant as a 'bargaining counter': see DRP [50] 14, Annex 2, PRO DEFE 10/26 (1 February 1950). Nonetheless, just as the BRAB concerns are not 'purely technical', it would be equally simplistic to delineate the DRPC concerns as 'purely political'.
68. DRPC were a little more circumspect about Experimental Plant No. 2 than BRAB. The chairman noted that 'while in the long run much information for civilian purposes might well be obtained from a plant of this kind, its potentialities for war purposes were still very speculative'. The DRPC were asked to 'consider very seriously' whether the plant was worth the £1,300,000 estimated cost. It was also noted that 'the necessity for proceeding with the construction of this plant at the present time arose entirely from the 1957 planning date'. DRPC 17th Meeting, PRO DEFE 10/37 (22 November 1949).
69. BRAB 16th Meeting, Minute 122, AC10875, PRO WO 188/668 (16 March 1950).
70. BRAB 19th Meeting, AC11260/BRBM19, PRO WO 188/668 (21 December 1950).
71. BRAB 24th Meeting, AC11817/BRBM24, PRO WO 18/668 (29 March 1952).
72. BRAB 25th Meeting, Minute 171, AC11872/BRBM25, PRO WO 188/668 (14 May 1952).
73. *MRD Annual Report 1951–1952*, AC12140/BRB105, PRO WO 195: 12136 (1952): 1.
74. AC12980/BRB130, PRO WO 195/12976.
75. *Report by the Chairman on the Work of the Board during the Year 1955* (n. 24 above).
76. *MRE Annual Report 1956–1958*, Paragraph 39, PRO WO 195/14357 (1958).
77. *MRE Annual Report 1961–62*, Item 36, SAC457/BRB210, PRO WO 195/15382 (1962).
78. BRAB, *Report by the Chairman 1957*, AC14169/BRB150, PRO WO 195/14164 (1957).
79. *Concentration, Viability and Immunological Properties of Airborne Bacteria Released From A Massive Line Source*, MRE Field Trial Report no. 3 (Trials Conducted October 1963 – April 1964) PRO DEFE 55/78 (January 1966).
80. BRAB Report 1959, Item 3, AC14820/BRB172, PRO WO 195/14811.
81. DRPC, *Chemical and Biological Warfare*, DRP/P(62)33, PRO DEFE 10/490.24144 (10 May 1962).
82. DRPC, *Chemical and Biological Warfare* (n. 81 above).
83. P. Gummett, *Scientists in Whitehall*, chap. 4 (Manchester University Press, 1980).
84. P. Gummett, *Scientists in Whitehall* (n. 83 above); S. Jasanoff, 'Contested boundaries in policy-relevant science', *Social Studies of Science*, 17 (1987): 195–230; T.F. Gieryn, 'Boundaries of Science' in *Handbook of Science and Technology Studies*, edited by S. Jasanoff *et al.*, (Sage 1995).

Chapter 12

Defence Physiology

John Ernsting

INTRODUCTION

It is perhaps ironic that the mechanisation of twentieth century warfare has vastly increased the range of physical demands on fighting men. From the beginning of the century aircraft have tested the ability to survive at increasing altitudes, and submariners have routinely patrolled great depths. Men have thereby encountered rapidly changing pressures—either under water or in the sky—extreme vibration, and very rapid shifts between radically different climatic zones. Such unprecedented experiences put the human frame under severe test.

Physiologists were already making very considerable contributions at the beginning of the century to the effectiveness and safety of men working in hazardous environments. Thus the great physiologist J.S. Haldane revolutionised deep diving by discovering and eliminating the carbon dioxide narcosis which had resulted in so many divers becoming unconscious at depth, and developing the concepts of stage decompression which did so much to eliminate decompression sickness.[1] The first centrifuges for studying the effects on aircrew of high acceleration were introduced in Germany and the USA before the Second World War. It was, however, that war which saw the clear recognition of the importance of understanding the effects of the environment within which military personnel operate, upon their wellbeing, performance, and indeed survival. It was also recognised how much the effectiveness of weapons depended on the relationship between the human operator and the machines he controls. Indeed, the Second World War saw the emergence of a new science variously called applied psychology, human factors engineering, ergonomics or biotechnology.

During the 25 years following the war, the importance and true value of physiology as applied to service personnel and weapon systems slowly gained recognition within Britain's Service ministries. This recognition was not gained easily, and the establishment of procedures and organisations whereby the appropriate physiological research was conducted and applied during the development of weapons systems and military operations gave rise to considerable argument and conflict among those responsible for defence policy. Satisfactory arrangements were, however, in place in many areas by the early 1960s, and over the following 30 years applied physiological research played its true role in enhancing the performance and safety of service personnel. It also saw the final acceptance of the value of applied psychology and ergonomics (human factors) to the efficiency and effectiveness of defence systems.

The various international military research cooperative programmes within NATO and between individual nations within the NATO Alliance, such as the Advisory Group for Aerospace Research & Development (AGARD), the Technical Co-operation Programme (TTCP), and the Anglo-French programme, led to international recognition of the considerable contributions of British applied physiological research to the efficiency and wellbeing of the members of the defence forces. Close links were formed and maintained with sister military medical research establishments in Canada and the USA. These were enhanced by the exchange of military and civilian research staff between these organisations.

The Legacy of the Second World War

Although fighting the Second World War within a unified command, each of the three Armed Services—Navy, Army and Air Force—nonetheless had its distinct policies, traditions, style and organisation. Each was represented by its own ministry within the overall War Department. Accordingly, each Service had its own network of physiological organisations with its own separate programme and chain of command. Within the Navy, the Royal Navy Medical School, which had been established originally at Greenwich in 1912 and which moved to Clevedon, Somerset, in 1939, was responsible primarily for training Service medical officers and ratings in naval medical matters. It concentrated on research on clinical topics such as the manufacture of vaccines. The Royal Navy Physiological Laboratory at Alverstoke, Hampshire, conducted extensive research on underwater and hyperbaric physiology. The Army's physiological research was undertaken by various units of the Directorate of Biological Research. The Royal Air Force Physiological Laboratory, originally established in 1918, which had moved from RAF Hendon to the Royal Aircraft Establishment in 1939, conducted aviation physiological research for the Royal Air Force and the Royal Navy throughout the war.[2]

The staffing and control of these establishments differed considerably. The Royal Navy Medical School and the Royal Air Force Physiological Laboratory were staffed at the senior level predominantly by serving medical officers, and were directly controlled by their respective Directors General of Medical Services. They each employed some 15–20 professional staff and a further 20–30 support staff. The Royal Naval Physiological Laboratory, employing 30 staff, was, in contrast, principally a civilian scientist organisation to which Royal Navy medical officers were attached. The physiological research units of the Army Directorate of Biological Research comprised a mixture of civilian scientists and medical officers of the Royal Army Medical Corps.

From the beginning of the war the Service ministries recognised the value of external committees of experts to advise—and in some circumstances to direct—the personnel research required by the Armed Forces. Each, however, had its own committees. The Air Ministry established the Flying Personnel Research Committee in March 1939 to advise the Secretary of State for Air on 'the medical aspects of all matters concerning personnel which might affect safety and efficiency in flying'. The Medical Research Council played a major role in the formation and operation of the personnel research committees for the Royal Navy and the Army, but not of that for the Royal Air Force. The Royal Navy Personnel Research Committee was formed in 1943, whereas the Military Personnel Research Committee provided advice on research for the Army, some of which was conducted in the Medical Research Laboratories in Hampstead, London. The latter became the Army Personnel Research Committee in 1953.

Personnel Research Establishments, 1945–90

Despite the cessation of hostilities in 1945, the pace of physiological research increased as new weapons emerged in the postwar years, nuclear warfare threatened and the Cold War intensified. A characteristically baroque arsenal of institutions emerged and was periodically reorganised. In 1948, the Royal Navy Medical School moved from Somerset to a site adjacent to the Royal Naval Physiological Laboratory in Alverstoke. The research conducted at the school changed from clinical matters to environmental physiology, such as survival at sea and the problems associated with closed environments on ships (described as the 'citadel' concept of protection against nuclear, biological and chemical (NBC) threat). With the planned expansion of the nuclear submarine programme in the late 1960s the research facilities at the school were expanded, and in particular a controlled gaseous environment laboratory (the Environmental Medicine Unit) was built. In addition, the Naval Radiological Protection Service and the Naval Emergency Monitoring Team were co-located at the school, which in April 1969 was renamed the Institute of Naval Medicine. The Institute of Naval Medicine continued to develop and expand until the early 1980s, when it had a total staff of 160, of whom 30 were engaged in physiological and human factors research. Underwater and hyperbaric medical research continued to be conducted principally at the Royal Naval Physiological Laboratory, which was incorporated into the Admiralty Marine Technology Establishment and subsequently, in 1985, into the Admiralty Research Establishment. In the early 1990s, physiology as a subspecialty of naval medicine was abolished in favour of occupational medicine. The Medical Research Council's Royal Navy Personnel Research Committee continued to play an important role in advising on and monitoring the physiological and medical research conducted for the Navy.

The Army's physiological research effort, as part of a much larger human factors research element, was incorporated into the Army Operational Research Group at West Byfleet in 1948. Shortly thereafter this was renamed the Army Operational Research Establishment (AORE). At much the same time the Army set up the Clothing and Equipment Physiological Research Establishment (CEPRE) at Farnborough (lodging, so to speak, with the Royal Aircraft Establishment). The principal task of this unit was to monitor the work and products of the Colchester-based Stores and Clothing Research and Development Establishment. Army research was frequently compelled to make do with borrowed and makeshift resources: in the early 1960s, for example, important acclimatisation studies were carried out in an elderly superannuated Aldershot barrack block converted into a temporary climatic chamber, effective for generating the necessary heat but erratic in terms of precision and control.

By 1965 the physiological and psychological research wings of the Army's Operational Research Establishment had become increasingly isolated from the rest of an establishment which was largely exercised with more recondite military strategy and war gaming. Accordingly, it was decided to merge these two wings with CEPRE at Farnborough to form the Army Personnel Research Establishment (APRE). In fact, not until 1972 was the new organisation brought together physically at Farnborough, in facilities on the old CEPRE site, and these were only finally completed in 1980. The new buildings which eventually arose provided greatly enhanced research facilities. They included a large hot chamber, capable of accommodating the Army's largest armoured fighting vehicle (AFV), a cold chamber able to reproduce the most severe conditions likely to be encountered by British military personnel, a large physical fitness laboratory, and a nutritional laboratory for long-term feeding studies. At about the same time, in 1978, the Applied Physiology Division of APRE was reorganised, placing the Royal Army Medical Corps (RAMC) personnel in a new section, the Army Occupational Health Research Unit (AOHRU), alongside the existing A and B Divisions of APRE. The staffing of APRE continued to increase, until in the mid-1980s it consisted of 82 civilian and 25 military personnel, of whom 35 were in the senior scientific and professional grades.

Unlike the Navy and Air Force personnel research establishments, control of APRE bypassed the Director General of Medical Services. It was vested instead in the Chief Scientist of the Army. Undoubtedly this chain of command brought its own benefits: financial support and ready access to soldiers to serve as experimental subjects. The omission of the Director General of the Army Medical Services from the command and control structure, however, reduced the potential input from the Army Medical Services. There was, moreover, no career structure for research-minded medical officers of the Royal Army Medical Corps (RAMC), and the staff of APRE was drawn predominantly from the scientific civil service.

The Army command organisation, responsible for strategic planning and for operations, was the main source of requests for physiological research particularly where environmental conditions (for example heat or cold) or other factors, such as protective clothing and equipment, were likely to impose a functional limitation on human performance. Requests for initiating major research projects were subjected to rigorous screening procedures, once their practical feasibility had been assessed, taking into account the financial, ethical and manpower implications of the proposed research. Such screening, a lengthy process, customarily involved the Chief Scientist (Army), the Adjutant General's Department and the Army Personnel Research Committee (APRC).

In 1945 the Royal Air Force Physiological Laboratory moved into new buildings in the south-east corner of the Royal Aircraft Establishment and was renamed the Royal Air Force Institute of Aviation Medicine (IAM). Here it remained a lodger, controlled administratively by a variety of Royal Air Force Commands (Strike

Command from 1972). Initially the facilities which were available to simulate the various features of the flight environment were limited to decompression and low-temperature/decompression chambers. It was vital, however, to be able to investigate the effects of flight on human subjects, and from 1945 the Institute 'owned' and conducted physiological research in a succession of aircraft, including a Piston Provost, a Meteor T Mk 7, a Canberra B Mk 6, a Hunter T Mk 7 and a Hawk T Mk 1. These aircraft were flown principally by medical officer pilots on the staff of the Institute. A large and versatile climatic facility was commissioned in 1952 and a man-carrying centrifuge with a 9.1 metre radius arm came into operation in 1955. A high-performance hypobaric chamber, together with a therapeutic hyperbaric chamber, was installed in a new building in 1963. The final additions to the Institute's capital facilities were in 1968 a combined climatic and hypobaric chamber capable also of generating high radiant heat loads, and in 1970 a 46-metre-long decelerator track capable of generating decelerations of over 40 *g*, and an advanced helmet impact test facility. A new three-floored building to house the expanding psychology and special senses research groups was occupied in 1971. The Institute had a staff of approximately 200, of whom one-third were military and two-thirds were civilians; 75 were in the senior scientific and professional grades.

Although for administrative matters the Institute was incorporated within an RAF Command, the setting of its tasks and reviews of its activities were the responsibility of the Air Ministry, and subsequently the Ministry of Defence. The Director General Medical Services (RAF) was primarily responsible for approving the fundamental research programmes of the Institute, which generally arose from the staff itself, whereas the operational requirements staff and the procurement offices were principally concerned with the Institute's applied research, which included the assessment and approval of human-related aircraft systems. The latter two activities were co-ordinated after 1962 with those of the Human Engineering Division of the Royal Aircraft Establishment by the Aircrew Equipment Research and Development Committee of the Ministry of Defence. More formal reviews of the Institute's work and activities were established by the Air Force Board in 1968 with the formation of the IAM Tasking Advisory Board. Soon after the Second World War the RAF Medical Branch recognised the importance of maintaining a cadre of medically qualified physiologists, and established the speciality of Aviation Physiology, the members of which were engaged full time in research and teaching at the Institute. A second, more applied speciality, Aviation Medicine, was formed in 1968; in 1975 the two were combined as Aviation Medicine, which had a total strength of 16 medical officers, of whom nine were consultants, all employed at the RAF IAM. The independent Flying Personnel Research Committee, with direct access to the Secretary of State, continued to advise on physiological and psychological matters and to monitor the research conducted by the Institute and elsewhere. It was disbanded in 1979.

The tasks of many of these separate organisations were comparable. Each Service was concerned with the development of clothing suitable for novel experiences of warfare and with enabling service personnel to survive extremes of temperature. Pressure suits were a particular problem—for both submariners and aviators. Methods of escape from crashed aircraft and immobilised ships and submarines were another generic problem. At the same time the problems raised by the individual technologies did have their distinct characters, and the Services addressed very specific problems. They also each had their own strong civilian links—to the emergence of the North Sea oil industry in the case of the Navy and the civilian aircraft industry, particularly through Concorde, for the Air Force. These civilian links are well illustrated by the provision of treatment facilities by the Royal Navy for diving accidents involving civilian personnel, and the support to civil aviation provided by the Royal Air Force. Thus, the RAF Institute of Aviation Medicine provided all the physiological, life support equipment assessment and human factors expertise required by the Civil Aviation Authority from 1945 to the mid-1980s.

The MoD Personnel Research Establishments had close links with academia throughout the postwar period. Individual members of the staff, particularly the RN and RAF medical officers at the establishments, taught applied physiology at a wide spectrum of universities within the UK, and many of them were appointed to visiting lectureships at universities as wide apart as London and Glasgow. There was a distinct divide with respect to the placing of extramural contracts with universities and civilian research institutes, between the Royal Navy and the Army on the one hand, and the Royal Air Force on the other. Thus, through their respective Medical Research Council Personnel Research Committees, the Royal Navy, and especially the Army, placed significant extramural physiological research contracts with MRC units and universities. The Royal Air Force and the Flying Personnel Research Committee established only very few extramural contracts for physiological research, the philosophy of the RAF IAM being to conduct the necessary research in-house rather than to contract it externally.

NAVAL PHYSIOLOGICAL RESEARCH

Survival at Sea

The Second World War's Battle of the Atlantic, in which convoys were pitted against submarines, quickly revealed the inadequacies of the lifesaving equipment provided in both merchant and war ships, which resulted in an appalling loss of life.[3] In 1941, at the request of the Medical Director General (RN), the Medical Research Council (MRC) set up the Committee on the Care of Shipwrecked Personnel, under the chairmanship of Sir Sheldon Dudley, whose terms of reference were to study the difficulties facing Royal Navy and Royal Air Force personnel after their ship or aircraft had been lost and they were adrift in boats or rafts. Special attention was to be given to their requirements for food and water, the effects of exposure to heat and cold, and the value of stimulants. This appears to have been the first attempt at a concerted effort to understand the underlying physiological problems confronting a shipwreck survivor. The MRC's committee report *A Guide to the Preservation of Life at Sea after Shipwreck*, published as a pamphlet in 1943,[4] was based on the scientific interpretation of data compiled from interviews with many survivors and reports by naval medical officers responsible for their care. This short pamphlet was widely circulated during the war and remains a milestone on the subject. The value of the committee's work was acknowledged by a request from the Admiralty to the MRC to form a permanent advisory committee to further the research in this and allied areas. Thus, in 1943 the Royal Naval Personnel Research Committee (RNPRC) was born[5] and led international research in this area for many years.

In 1945, the Admiralty set up a committee under A.G. Talbot to investigate all aspects of naval lifesaving.[6] The major finding of this committee was that approximately two-thirds of all naval casualties at sea (between 30,000 and 40,000 officers and men) were through 'drowning', many after reaching 'lodgement' in the water; it emphasised the importance of environmental protection for survivors. Research initiated by the committee led to the introduction in the 1950s of two new revolutionary items of equipment, the inflatable liferaft and the inflatable lifejacket, which were universally adopted and have saved many thousands of lives.

A good understanding of the needs of survivors emerged in the immediate postwar years as a result of the pioneering work and direction of the RNPRC Survival-at-Sea Subcommittee. The subsequent research performed in this country by both the RN and the RAF under the direction of the Defence Services Life Saving Committee, which was set up in 1966 to look after the interests of all three Services, has resulted in the provision of excellent marine lifesaving equipment, including specialised immersion suits. Much of this pioneering physiological research was due to the remarkable insight of Professor R.A. McCance of the Department of Experimental Medicine, University of Cambridge, who was Chairman of the RNPRC Survival-at-Sea Subcommittee until he retired in 1970. This research at Cambridge was assisted by Hervey and Keatinge, who continued to make significant contributions to the understanding of the problems of survival at sea, including hypothermia,[7] in later years. The development of the inflatable tented liferaft owes much to McCance and his colleagues, who defined the physiological criteria on which the naval architects formulated their designs. Subsequent material and physiological testing and evaluation in both the Arctic and the tropics refined the design principles, which have since saved countless lives.

In association with the liferaft development trials, experiments were conducted in Cambridge to determine the minimum daily water requirements for shipwreck survivors and the most efficient survival rations. Prior to these experiments it had been assumed that, as the major cause of death in survivors adrift in open boats in tropical waters was water deprivation, all available storage space should be devoted to potable water. In 1951, Hervey and McCance demonstrated the value of replacing some of the water with sugar.[8] They showed that a daily ration of 100 g of barley sugar improved water balance by 200 ml, which had considerable benefits on available storage space as well as on survival. Studies of the physiological effects of extreme water deprivation and the control of urinary volume led to McCance's team discovering the principle of osmotic diuresis. Finally, controlled experiments by Hervey showed that the drinking of sea water, however little, always increased plasma hypertonicity and must accelerate death in these circumstances.[9]

The development of personal survival suits in the years since the war has been largely led by military physiological research, with industrial technological advances attempting to meet the required criteria. An analysis of Second World War maritime casualties by McCance and his colleagues highlighted the importance of whole-body cooling in the outcome of many shipwreck survivors. This led to research in this area, initially by naval medical officers seconded to Cambridge who later continued their research through support from the RNPRC. The early pilot studies by Glaser and Hervey were continued and expanded by Keatinge.[10] Subsequently, human cold-water immersion studies, carried out by Golden at the Institute of Naval Medicine, followed in the 1970s by animal studies at the University of Leeds, provided a new insight into the mechanisms of 'circum rescue' death which resulted in the modification of rescue techniques.[11] In the 1980s, experiments by Tipton and Golden highlighted the dangers of the responses associated with the initial few minutes of immersion in cold water—the 'cold shock' response—and the possible causes of swim failure in cold water.[12] This work led to the design by Tipton of a simple but highly practical underwater breathing system to facilitate escape from submerged helicopters, which was adopted by the companies operating in the North Sea oil industry.

Diving

The role of British military research in the field of diving has never been given the recognition it so richly deserved, largely because of the inherent attitude of the 'silent service' to publicity. This point is nicely made by K.W. Donald in his introduction to what is possibly the first book in the English language on underwater medicine, by Stanley Miles, published in 1962. He states:

> In 1945 Great Britain led the world in these types of activities [self-contained diving] from both the operational and scientific aspects. However, since these days the diving authorities in this country have had little to do with the world-wide publicity of the underwater advances which have popularised independent diving as a sport and recreation. The attitude of the Royal Navy, perhaps unfortunately, remains strictly professional and practical. The contrasting names for nitrogen narcosis in France ('rapture of the deep') and in Britain (the 'narks') are a delightful example of this.[13]

The failure of the existing decompression tables to prevent a significant incidence of decompression sickness when used for ascent from depths greater than 40–50 metres, especially when the time at depth was extended, led Hempleman, at the RN Physiological Laboratory, in 1952 to propose a new approach to the construction of these tables. Extensive chamber trials to depths of 55 metres conducted at the Laboratory during the 1950s, and subsequent sea trials in 1957, demonstrated the success of his approach.[14] Hempleman also developed very successful decompression procedures for men working in compressed air at pressures up to 3.8 ATA (28 metres depth).[15] These procedures greatly reduced the incidence of decompression sickness in tunnel workers. Royal Navy and US Navy interests in diving at great depths (submarine rescue and escape) and commercial diving operations resulted in extensive international collaborative studies involving the RN Physiological Laboratory, the RN Medical School, the US Navy and the French Navy in the 1960s and 1970s.

In the early 1960s it was realised that the long decompression procedures required to avoid decompression sickness made saturation diving the only economic method of carrying out prolonged dives to depths greater than 50 metres. The RN Physiological Laboratory developed safe decompression procedures for divers who had spent days working at depth.[16] As this work was more relevant to the developing North Sea oil industry than to the military, much of the supporting funding came from non-military sources. The subsequent development of robotic techniques, in the late 1970s, for much of the deep engineering work reduced the requirement for human dives to these depths.

The narcotic effects of nitrogen when breathing air at depth were studied by Bernard and his colleagues in the early 1960s, and the value of substituting helium for nitrogen at depths greater than 90 metres was firmly established. Bennett subsequently described and investigated the impairment of performance and the muscle tremor that occurred when oxygen–helium mixtures were breathed at depths of 180–240 metres. This condition, which was termed the high-pressure nervous syndrome, was researched intensively at the RN Physiological Laboratory, with studies ranging from 10 hours at 460 metres to several days at depths between 300 and 660 metres over the decade from 1970 to 1980.[17] Methods of suppressing the syndrome were evolved and applied to deep diving.

Submarine Escape

During the Second World War 77 Royal Navy submarines were sunk. The loss of many lives stimulated research into the problem of escape from sunken submarines. The Admiralty appointed a committee under the chairmanship of Ruck-Keen to look into the question of submarine escape. Two important findings emerged: first, the major hazard was within the submarine prior to ascent, with three-quarters of all submarine casualties occurring during the period of flooding; and second, many men escaped successfully without recourse to breathing apparatus. Submarines were equipped with the Davis Escape Apparatus, which consisted of a mouthpiece connected to a flexible bag containing a carbon dioxide absorber and which was filled with oxygen from a cylinder mounted below the bag.[18] The atmosphere of incapacitated submarines submerged for some time could contain as much as 5 per cent carbon dioxide. With the increase in partial pressure associated with flooding, toxic levels of carbon dioxide were achieved very rapidly. Although the Davis Escape Apparatus provided protection against carbon dioxide toxicity, the high concentration of oxygen in the closed circuit produced oxygen poisoning. The combination of oxygen toxicity and unfamiliarity with the correct use of the apparatus led to an unnecessary number of deaths, whereas many who had not used this equipment appeared to escape successfully. As a consequence, it was decided that research should commence on 'free ascent' as the preferred method of escape.

The physiological problems associated with escape from a submarine at depth are caused, above all, by the rapid alterations in barometric pressure. These affect the organs that contain free gas, such as the lung, gut and ear, and change the gas tensions in the tissues during compression and ascent. The former may result in a fatal pulmonary barotrauma, ruptured viscus, or rupture of an eardrum; the latter in debilitating or possibly fatal decompression sickness. Pulmonary barotrauma could largely be avoided by training submariners to breathe out continually during ascent at a rate sufficient to prevent overexpansion of the lungs while retaining sufficient alveolar gas to satisfy respiratory requirements. However, finding the optimal rate of ascent to minimise the absorption and facilitate the elimination of nitrogen by the body tissues proved to be a more difficult problem. Exposure time to raised pressures had to be minimised. Ideally the requirement was for a 'bounce dive', with minimum bottom time and a rapid ascent, but at a rate which avoided the risk of barotrauma.

This problem was tackled by a number of researchers at the Royal Naval Physiological Laboratories, Alverstoke, through mathematical modelling and chamber experiments using an animal model (the goat), followed by human volunteers. Between 1945 and 1948, Donald, Davidson and Shelford showed that goats could be decompressed safely from 76 metres sea water while breathing air.[19] The bottom time was 3–5 minutes and the rate of ascent was similar to that of compression, i.e. 0.3 m/s. Incurable decompression sickness occurred if the time at depth was extended to 7 minutes at 76 metres or 3 minutes at 91 metres. However, increasing the oxygen concentration to 34 per cent removed signs of decompression sickness. Further experiments showed that men could be decompressed without discomfort at a rate of 1.8 m/s from 91 metres, but the preferred rate continued to be 0.3 m/s. Subsequently it was found that trouble-free ascents could be made from 100 metres at 1.2 m/s. In 1949, Dr Cam Wright, who personally carried out most of these early human simulated escapes from 76 and 91 metres, found he could exhale throughout an ascent at 1.2 m/s. Nevertheless, there was considerable doubt that subjects could ascend from 91 metres without inspiring and drowning, even if the rate of ascent was increased.[20]

The inflatable submarine escape survival suit which was under development at this time had an integral hood and provided 10 lb of buoyancy, which produced an ascent rate of 1.2–1.5 m/s and gave thermal protection to the survivor on reaching the surface. The sinking of the submarine HMS *Truculent* in 1950 following a collision in the Thames estuary, when all 50 crew members managed to escape but only 10 survived, stimulated the early introduction of the suit. In 1952 a 100-foot escape training tower was built at the headquarters of the submarine fleet in Gosport near Portsmouth, HMS *Dolphin.*

Although buoyant ascent training was started in 1953, trials of this method of escape were not conducted at sea until nearly 10 years later. Preliminary compression trials conducted in HMS *Tireless* in 1962 were followed by escapes in open water from HMS *Tiptoe* from 71 metres, with rates of ascent up to 2 m/s.[21] In 1963 the Royal Naval Personnel Research Committee established a working party to advise on the conduct of escapes from depths down to 183 metres in the open sea. Barnard and Eaton conducted a series of simulated escapes from depths of 91 to 152 metres using goats and then, when these had suggested safe rates of compression, bottom times and ascent rates using human subjects.[22] Eight simulated escapes were performed by volunteers from a depth of 152 metres with bottom times varying between 20 and 30 seconds. These latter were probably the most dangerous escape exposures carried out by humans. One subject developed decompression sickness but responded fully to treatment. These laboratory experiments led in 1965 to a series of highly successful open-water escapes from HMS *Orpheus*, from a depth of 152 metres.

Laboratory studies continued using goats, simulating escapes from depths up to 44 metres. The beneficial use of oxygen at 18 metres in the treatment of the decompression sickness found in these experiments improved the therapeutic procedure for divers suffering from this troublesome problem. Eventually, in 1969, dives to 290 metres were achieved using goats, followed by successful simulated escapes by humans from 190 metres. In July 1970 successful escapes were made in the Mediterranean off Malta from HMS *Osiris* by a number of the training staff from the Escape Training Tank, ably led by Cdr M.R. Todd RN.[23] The magnitude of this achievement can perhaps be better comprehended by noting that in a free ascent from 183 metres, the individual briefly passed from 1 to 19 atmospheres and back again in the space of about 95 seconds, during which he was initially exposed to horrendous noise and very cold water, followed by an ascent through inky-dark water. The rate of compression was such that eardrums were frequently ruptured, but in the turmoil of the situation this was rarely noticed at the time and, if noticed, it was regarded as a minor inconvenience by the individual!

The mechanism of pulmonary barotrauma continued to be a cause of concern in trainees undergoing escape training in the 30-metre escape training tower at HMS *Dolphin*. Methods of detecting lung abnormalities that could predispose to barotrauma on ascent were developed at the Institute of Naval Medicine, and were applied to all submarine candidates prior to the commencement of training.[24]

Underwater Explosions

The observation early in the Second World War that underwater explosions were particularly lethal prompted research into the mechanisms responsible for the fatal damage to the organs of the body. Pioneering research on the mechanisms of damage, on the lethal deterrent and safe ranges from a variety of charges, and methods of providing protection, were conducted by H.C. Wright. Working at the Royal Naval Physiological Laboratory, Wright conducted extensive studies of underwater blast using animals and human divers in the period immediately following the Second World War. He had a very great—probably the greatest—personal experience of the effects of underwater blast. On one occasion, having dived to a depth of 15 metres, at a distance of 640 metres from a 90 kg charge, he had to be rescued from the water unconscious and suffering temporary paresis. Between 1950 and 1955 he was ably supported by an Admiralty physicist, A.H. Bebb. Their work resulted in mathematical equations which provided lethal ranges in open water, and the deterrent range for shallow-water explosions.[25] The results of this unique research were subsequently used extensively by the US forces in Vietnam to determine safe distances for deterrent underwater charges in the Mekong Delta, and by industry in the development of underwater cutting techniques using plastic explosive.

Underwater Ejection

The introduction of jet aircraft to aircraft carriers in the 1950s was associated with an increased loss of aircrew following 'ditching' accidents. Unlike the more buoyant earlier aircraft, the heavier jet-powered aircraft sank very rapidly. Various options were considered to facilitate escape from a sinking aircraft, with the Institute of Aviation Medicine favouring some modification of the ejection seat to enable it to be used under water without harming the pilot. At a meeting of experts held on 12 October 1954, it was considered that underwater ejection was not feasible and that an alternative method of escape should be sought. The following day, a Lt B. McFarlane RN successfully ejected under water from a Wyvern aircraft.

The physiological problems associated with underwater ejection included protection of the seat occupant from the effects of underwater explosion of the gun cartridge; protection from the drag of driving man and seat through water at high velocity; and pulmonary barotrauma resulting from the rapid decrease in barometric pressure with ascent to the surface. Early successful trials by Rawlins, using divers, evaluated the feasibility of jettisoning the canopy of partially flooded cockpits and performing free ascents from various depths down to 15 metres. These trials were followed by ejections, using a mannequin, during which seat accelerations and velocities were recorded along with the shockwaves on the mannequin. Eventually, after a series of experiments using mannequins and anaesthetised sheep, human volunteers—led by Surgeon Lt Davidson RN—demonstrated that the system was effective and safe (Fig. 12.1).[26]

The introduction of the highly efficient rocket-powered ejection system by the Martin Baker Aircraft Company in the late 1960s made the requirement for an underwater ejection system redundant for jet aircraft, but a system based on compressed air fitted to the propeller-powered Gannet remained in use by the Royal Navy until these aircraft were eventually phased out in 1978.

Submarine Habitability

The advent of the nuclear submarine programme in the mid-1960s stimulated research into the effects on humans of prolonged exposure to raised ambient levels of carbon dioxide and carbon monoxide. A special environmental laboratory, where volunteers could be accommodated for weeks and months and in which the atmosphere could be carefully controlled, was constructed in the Institute of Naval Medicine to study these problems. The first part of a three-phase investigation aimed at defining the maximum safe level of carbon dioxide for continuous exposure in humans commenced in 1973, with studies of the physiological and biochemical changes produced by 0.5 per cent and 1.5 per cent carbon dioxide. Simultaneously, haematological studies were carried out on the crews of Polaris submarines at sea. Research on the effects of continuous exposure to low levels of carbon monoxide commenced in 1974. This resulted in a reduction of the maximum permissible concentration in submarines from 25 ppm to 15 ppm.[27] Disturbances of mineral metabolism and the urinary excretion of calcium were found in the long-duration exposures to carbon dioxide. In view of these findings, and of the absence of sunlight

FIGURE 12.1: AN EXPERIMENTAL UNDERWATER EJECTION FROM THE FUSELAGE OF A SEA VIXEN AIRCRAFT PERFORMED BY SURGEON LIEUTENANT COMMANDER JOHN RAWLINS (LATER SURGEON VICE ADMIRAL) IN THE DEEP TANK AT THE ADMIRALTY HYDRO-BALLISTICS ESTABLISHMENT, GLEN FRUIN (© CROWN COPYRIGHT, MOD)

in submarines, studies were made of the levels of vitamin D in their crews. In addition, microbiological and other environmental contaminants, as well as possible radioactive contamination, were also carefully monitored at sea.

Cold Injury

The published literature on cold injury is remarkable in that many papers are published following conflicts in cold climates, but interest gradually wanes until the next conflict. The Second World War proved no exception, and the now-classic descriptions by Surgeon Commander Ungley in 1945 of non-freezing cold injury (NFCI) in shipwreck survivors, which he termed 'immersion foot', provided a new insight into this troublesome problem, which had previously been regarded as a form of freezing cold injury (frostbite).[28] The unusually high incidence of NFCI in the Royal Marines and Parachute Brigade personnel during the Falklands campaign in 1981 regenerated research interest in this topic. A comprehensive questionnaire survey of the troops returning from the campaign revealed that some 80 per cent had suffered some degree of NFCI. A technique for screening and classifying the severity of the condition had to be rapidly devised, and a clinic was instituted by Golden at the Institute of Naval Medicine to advise on the treatment and disposal of those suffering from the more severe effects.[29] Research was conducted into the nature of the cold sensitivity caused by this form of injury.

ARMY PHYSIOLOGICAL RESEARCH

Heat Physiology

The Army's requirement for troops to serve in hot climates and—more importantly—to be able to operate with optimum efficiency on first arrival, stimulated a broad-based programme of research into heat acclimatisation and its induction by artificial means. Gone were the days when a leisurely journey by troopship to distant overseas stations allowed the physiological adaptations of acclimatisation to take place naturally. The Army was faced with the need to move large bodies of troops quickly by air from temperate to hot climates, and to expect them to be fit to fight from the moment of landing. During the immediate postwar years the Army relied largely on moving volunteer soldiers to the hot climates concerned, but the expense, inconvenience and, above all, the lack of effective climatic control led to the concentration of research within the UK.

The physiological adaptations of acclimatisation to heat were fully explored by the Medical Research Council Laboratories (MRC) at Hampstead, where the measurement guide of acclimatisation—the predicted four-hour sweat rate (P4SR)—was developed.[30] The Army continued to take advantage of the MRC expertise in the induction of artificial acclimatisation, and it also used the climatic chambers at RAF IAM for its accurate measurement, but large field trials were needed to demonstrate the operational effectiveness of artificial acclimatisation and these were undertaken by AORE in the early 1960s. For the definitive research study (Exercise Mixed Grill, 1963), an infantry company of 100 men was divided into two equal groups, one half to exercise for two hours a day in the newly provided Aldershot hot chamber, and the other half to undertake the same rigorous physical exercise without the added effect of heat. The physiological adaptations, principally by way of increased sweat output, were demonstrated conclusively in the heat-treated group (and also, but to a much lesser extent in the exercise-only group) and the complete company was then flown to Aden where, in the rugged hills of the Jebel Khoraz, a strenuous six-day exercise was staged to compare the military effectiveness and capability of the two groups. Although a small but significant superiority in performance was demonstrated for the heat treated group, the most important difference lay in the casualty experience: the unacclimatised group suffered three times as many heat casualties as the artificially acclimatised.[31]

The benefit of prior acclimatisation and, to a lesser extent, that of a high standard of physical fitness, had been demonstrated beyond doubt by these experiments. Thoughts turned to improved (and cheaper) methods of inducing artificial acclimatisation, its maintenance and measurement. Alternative schemes not requiring the use of a hot chamber, but where the individual's own metabolic heat was to be the stimulus to the necessary physiological adaptations, were investigated. Extra clothing, coupled with exhausting work in a gymnasium, proved ineffective, and the next variant on the theme was to employ a suit impermeable to water vapour, thereby inhibiting entirely the cooling effect of the evaporation of sweat.[32] Although the changes induced during the training period were promising, the full picture of adaptation was not reproduced when the subjects were tested in a hot climate.

Further research on methods of acclimatisation, both active and passive, continued into the early 1970s, by which time the Army had ready for use a portable gas-heated and inflated air dome. However, by then the strategic requirement for the rapid movement of troops to hot climates had come to a halt and the research ceased.

Arising from the problems first encountered by British troops in the Arabian Gulf in 1960, there began a long programme of research into the effects of clothing and the environment inside armoured fighting vehicles (AFV) such as tanks and armoured cars. The heat burden on the crew was necessarily exacerbated by engine-generated heat, protective clothing, and the installation of NBC (nuclear biological chemical) air-filter packs. The thermal loads and stresses imposed by the various clothing assemblies were investigated by Army Personnel Research Establishment (APRE), both for the Stores and Clothing Research and Development Establishment (SCRDE) at Colchester and, in the case of the NBC assemblies, for the Chemical Defence Establishment (CDE), Porton Down. The high temperatures encountered in the AFV crew compartments, coupled with the need to wear NBC assemblies, instigated studies of the use of air-ventilated and liquid-conditioned suits originally developed for use by aircrew. The value of this work to the Army's programme of AFV development led directly to the establishment in 1970 of a subsection of APRE within the Fighting Vehicles Research and Development Establishment (FVRDE) at Chobham.

The scientific design of clothing raised further questions, such as its optimum size. Formal anthropometric surveys of Army personnel were conducted by the APRE during the late 1970s[33] using the techniques developed by the Royal Aircraft Establishment for the 1970/71 survey of RAF aircrew. These surveys provided information for the design of clothing and crew workspace, and also contributed to the international anthropometric database. The National Institute for Medical Research conducted an ergonomic study in the late 1960s under contract to the Army Personnel Research Committee, the aim of which was to specify the dimensions and ergonomic characteristics of the smallest spaces for crew members of fighting vehicles (the 'Capsule Investigations').[34] The studies were subsequently extended to include requirements for reclining postures. The results were provided to industry and the Military Vehicles Establishment. Subsequently, the APRE conducted ergonomic surveys of the workspaces of AFVs as they were developed. The APRE established a Human Factors Unit at the Military Vehicles Engineering Establishment (Chertsey) (MVEE) in 1972.[35] This greatly facilitated the delivery of appropriate and timely advice on human factors to MVEE. A second Human Factors Unit was established at Fort Halstead.

In 1964 and 1965 AORE (later APRE) set up an ambitious research programme to fulfil a long-felt need, namely to produce a reliable evidence-based guide to the quantities of drinking water required by soldiers deployed in hot countries. The need for the study was accentuated by the absence of any agreed logistic scale for Army use, and also by the alarming prevalence of the view among self-appointed military experts that men may be trained by progressive deprivation to manage on a much-reduced water ration. The research project, essentially one of practical applied physiology in the field, aimed to predict the daily water requirements of fully acclimatised fit young combatants in terms of climate and physical activity. The experimental work was conducted in four hot overseas locations: Tripolitania, Swaziland, Bahrain and Malaya. In each place a group of 35 soldiers had their water balance measured twice a day over a period of 11 days: 6 of rest and 5 of hard physical work. The results, published in 1968, produced a recommended scale of supply indicating (for example) that the mean daily requirement, necessary to maintain a positive water balance, for men working in open country where the mean daily wet bulb globe temperature index was 32°C was 16 litres/24 hours.[36]

The requirement for combat forces to operate efficiently in cold environments led in the late 1970s to renewed interest in the effects of cold, especially cold/wet climates, and methods of providing adequate protection against these effects.[37] The difficult problem of providing good protection in these environments led to the development of new types of protective clothing and sleeping bags, the effectiveness and acceptability of which were determined by laboratory and field trials conducted by APRE during the 1980s. Before adequate cold chambers were available to APRE it was often necessary to graft experimental work on to exercises taking place in naturally cold countries—typically Norway in winter. This unsatisfactory state of affairs was rectified by the construction of APRE's new cold chamber in 1973.

The threats posed by nuclear, biological and chemical (NBC) warfare posed particular challenges to protective clothing. The Army Personnel Research Establishment determined the physiological penalties of various types of personal NBC equipment and associated procedures developed for ground personnel by the Chemical Defence Establishment and the Defence NBC School. The establishment determined the respiratory and thermal loads and the general comfort of the NBC equipment. The impairment of performance imposed by the assemblies was also investigated. These studies were conducted in the laboratory and in field trials. They yielded valuable information which aided further improvements in the design of equipment and procedures.[38] They also provided the basis for Commander's *Guides on the use and physiological penalties of NBC equipment.*

The energy costs of the various tasks of the combat soldier when wearing various clothing assemblies and operating various weapons were determined by Army physiologists as new operational procedures and equipment were introduced.[39] Standard respiratory techniques for measuring energy expenditure were adapted to the special conditions, and new techniques, including telemetry of data, were developed for use in the field. Fundamental studies of the efficiency of various techniques for operating military equipment were conducted in parallel. These measurements of energy expenditure provided estimates of nutritional requirements.

Nutrition and Physical Fitness

From 1967 to 1971 APRE was concerned with the concept of packed rations and the weight of food that had to be carried by soldiers on extended patrols where there was no prospect of resupply. Two experiments were carried out, one in Scotland and the other in Malaysia, in which the effect on performance of a diet supplying rather less than half the measured energy expenditure was assessed. The results of both the studies showed that, for well-led fit young soldiers, rations providing for only 50 per cent of the energy expended were compatible with the maintenance of full military efficiency. The first experiment was concerned primarily with the capacity for hard physical work, whereas the second explored vigilance and more skilled military tasks.[40]

The ability to determine the degree of physical fitness of combat personnel is of great importance to the Army. Studies conducted by APRE in the 1960s which attempted to determine the relationship between the mass of body fat, as indicated by measurements of the thickness of skin folds, and physical fitness revealed the lack of an accurate method of assessing physical fitness. The research did, however, yield a large body of information on the value of skinfold thickness as a measure of the fat content of the body.[41] Reliable methods of assessing the physical fitness of combat personnel were developed by APRE during the 1970s in collaboration with the USA and Canada. The Army Personnel Research Establishment played a leading role in the formulation of NATO standard tests for the measurement of physical fitness. These tests were subsequently used by APRE to determine the levels of physical fitness required for specific Army tasks.[42]

Vibration and Noise

The AFV raised a host of physiological problems even beyond temperature control. The poor ride qualities of many AFVs led to the need, in the mid-1960s, for guidelines on acceptable exposure to vibration. The Army Personnel Research Establishment formulated guidelines on the basis of published work.[43] Studies of the vibration spectra specific to fighting vehicles were then carried out with the Institute of Sound and Vibration at Southampton. Surveys by APRE of the noise levels in AFVs showed them to be excessive in many vehicles. Improved methods of providing protection against the deleterious effects of exposure to noise were assessed in parallel with the adoption of a hearing conservation programme. Measurement of noise levels in new AFVs were made as the vehicles were developed. The establishment also contributed to NATO standards defining acceptable levels of impulse and continuous noise. In the late 1980s the use in AFVs of active noise reduction, pioneered by the Royal Aircraft Establishment in aircraft to improve reception of speech, was investigated by APRE and found to be valuable.[44]

AVIATION PHYSIOLOGICAL RESEARCH

Altitude and Altitude Protection

The planned high-altitude role of future fighter, bomber and photoreconnaissance aircraft such as the Lightning and the V-bombers stimulated in the late 1940s the development of pressure suits to provide protection in the event of loss of cabin pressure and/or escape at high altitude. Flying at altitudes above 50,000 feet commenced in the UK in 1951, in order to test the performance of jet engines at these extreme altitudes. A number of USAF capstan partial-pressure suits had been purchased from the USA to support these flights. These suits comprised a full pressure

FIGURE 12.2: A SUBJECT WEARING THE PARTIAL PRESSURE HELMET–SLEEVED JERKIN–G TROUSER ENSEMBLE DEVELOPED BY RAF IAM TO PROTECT AIRCREW AT ALTITUDES UP TO 100,000 FEET. THE ENSEMBLE IS INFLATED TO A BREATHING PRESSURE OF 110 MMHG. (© CROWN COPYRIGHT, MOD)

helmet and a close-fitting fabric suit encompassing the trunk and limbs.[45] On exposure to high altitude, oxygen was delivered at the appropriate pressure to the helmet, and the tension in the fabric suit was increased by the inflation of tubes connected by inelastic loops to the suit (the capstan principle). A simple but ingenuous partial-pressure helmet was developed by the Royal Air Force Institute of Aviation Medicine (RAF IAM) and the Mechanical Engineering Department of the Royal Aircraft Establishment (ME Dep, RAE), initially for use with the capstan suit (Fig. 12.2).[46] The RAF IAM performed a full evaluation of the system up to a pressure altitude of 60,000 feet and trained flight crews in its use. It was worn by Walter Gibb and his navigator when they gained the world altitude record of 63,668 feet in a Canberra in 1953. Although it was originally intended that RAF aircrew should also use the capstan suit, evaluation by RAF IAM demonstrated that the suit greatly encumbered the wearer and that it did not integrate with UK escape systems. RAF IAM advanced the concept of a minimal-coverage suit which would not encumber the aircrew and which would provide 'get-me-down' protection.[47] This proposal was accepted by the Air Staff and gave birth to the 'pressure-jerkin–G-suit' combination, which was then developed by RAF IAM and industry.[48] The pressure-jerkin system replaced the capstan suit in test flying in 1956. Two versions of a partial-pressure helmet with open pressure visor were developed, and the helmet–jerkin–G-suit assembly was accepted for the Lightning. A sleeved version of the pressure jerkin was designed for use in the Saunders-Roe P117 interceptor, which required 'get-me-down' protection from 100,000 feet (Fig. 12.2). The performance of this system was confirmed in a series of decompressions to a pressure altitude of 100,000 feet conducted by a team from RAF IAM in the hypobaric chamber at the Canadian Defence Medical Research Laboratory at Downsview, Ontario, in 1956. During this series of decompressions one subject, John Ernsting, achieved an altitude of 140,000 feet.[49]

Laboratory studies at RAF IAM in 1955 demonstrated that the newly developed (by RAF IAM and the Chemical Defence Establishment) type P/Q oxygen mask would seal very effectively at high pressure and, when combined with the pressure jerkin and G-suit, would provide 'get-me-down' protection from 56,000 feet.[50] This combination was adopted for use in the Royal Air Force. The pressure-helmet–jerkin–G-suit assembly was introduced into service in 1959, and the RAF Aviation Medicine Training Centre was established initially at RAF Upwood to issue the equipment and to train aircrew in its use. The decision in 1962 not to continue with full pressure-suit development led to the mask–jerkin–G-suit assembly being adopted for the Vulcan B2, Victor B2 and Canberra PR 9. A later development of the assembly—the combined partial-pressure suit which also provided immersion protection—was adopted for use in the RN Phantom in the late 1960s.

The operational requirement issued in 1950 that the V-class aircraft and the Canberra PR 9 should be able to remain at high altitude for several hours following loss of cabin pressure initiated the development of a UK full-pressure suit. This was required to provide a pressure altitude within the suit of 25,000–28,000 feet, and to maintain thermal comfort in spite of exposure to temperatures of –40°C. The lead in the design and ground assessment of the prototypes was undertaken by ME Dep. RAE with the contractor. RAF IAM assessed the suits at altitude and at low temperatures. RAE produced several important improvements to the design of the suit, increasing mobility when pressurised and introducing a flexible headpiece which was stowed around the neck in the uninflated condition.[51] Development of a UK full-pressure suit ceased in 1963 as a result of the decision that the V-force was to operate at low level.

The effects of the mild hypoxia induced by breathing air at altitudes between 5000 and 8000 feet was researched in depth by Denison at RAF IAM in the early 1960s.[52] He found that it produced a significant impairment in the performance of novel tasks. This finding led to the recommendation that the cabin altitude in Concorde should not exceed 6000 feet. The transient profound hypoxia induced by rapid decompression while breathing air and followed by oxygen was also studied extensively by Ernsting and Denison.[53] The requirements for the delivery of oxygen in these circumstances were defined in detail and formed the basis of the regulations on the use of oxygen in civil and military high-altitude aircraft in which air is breathed routinely in flight. Nicholson and Ernsting investigated the incidence of permanent brain damage in non-human primates exposed to a variety of altitude profiles. This work led to a definition of the maximum size of the cabin windows in Concorde.[54]

The pressure demand equipments available in the 1950s, which were of American design, imposed a very significant resistance to breathing and failed to control the composition of the gas delivered to the user. During the early 1960s the RAF IAM formulated the 'ideal' physiological requirements, and proposed improvements to MoD specifications which were a reasonable compromise between the ideal and that which was technically feasible. This process of raising standards was facilitated by the unique position of RAF IAM as the organisation which measured the performance of oxygen systems on humans and the excellent working relationships with Human Engineering Division (HE Div.) RAE, the research and development branches of the Ministry of Defence and industry. As each new generation of oxygen equipment was specified the physiological requirements were brought closer to the ideal until in the mid-1970s a very satisfactory performance was reached.[55] These physiological standards were then agreed with the USAF School of Aerospace Medicine and subsequently incorporated into ASCC Air Standards and NATO STANAGs.[56]

The serious operational penalties arising from the need to replenish the oxygen store of a combat aircraft between sorties and the logistics of supplying liquid oxygen led to the investigation in the UK and the US of methods of generating oxygen on board the aircraft. The mid-1970s saw the adoption of pressure-swing adsorption using synthetic molecular sieves as the method of choice for the production of oxygen-rich breathing gas from engine bleed air. The prototype molecular sieve oxygen concentrator system (MSOCS), developed in the UK by Normalair-Garrett Ltd, was assessed in the laboratory at RAF IAM and in flight in the IAM Hunter T7. These in-depth evaluations supported the successful bid by this UK manufacturer to supply the MSOCS for the USAF B-1B aircraft.

A US manufacturer produced the MSOCS for the USMC AV-8B, and the same system was to be fitted to the RAF Harrier GR5. The US system had a number of features which were unacceptable to the MoD, and RAF IAM proposed and developed a breadboard system using the US oxygen concentrator which overcame these disadvantages.[57] The IAM modifications to the system were adopted for the Harrier GR5. The development version of the Harrier GR5 system was assessed by RAF IAM and test flown in the IAM Hunter T7.[58] The IAM provided extensive advice to British Aerospace with respect to the specification and design of the MSOCS for the European Fighter Aircraft.

Sustained Accelerations

Prior to the availability of the man-carrying centrifuge in 1955, RAF IAM continued its wartime practice of conducting research on sustained accelerations in aircraft. Most of the in-flight work conducted in the 10 years before 1955 related to the development and testing of UK-designed anti-G suits and G valves. Several prototype

G valves were tested by a joint team of RAE engineers and IAM pilots using the IAM Spitfire Mk 9 and the Meteor T7. Anti-G suits were developed and tested in parallel to a form which was to be employed by the Royal Air Force and Royal Navy, with only slight modifications, over the next 30 years.[59] An alternative approach to enhancing G tolerance which was explored between 1946 and 1955 was the prone position, which greatly reduced the vertical distance between the heart and the brain. A Meteor Mk 8 was fitted with an extended nose containing a prone-position cockpit. The IAM medical pilot team, led by Ruffel Smith, flew some 99 sorties in the aircraft in 1955–56.[60] The prone position was comfortable except in turbulent flight, and gave a significant increase in G tolerance. There were, however, major restrictions of upward and rearward vision, and the project was discontinued.

The commissioning of the man-carrying centrifuge in 1955 enabled IAM to explore the mechanisms responsible for blackout and loss of consciousness, and to study the cardiovascular responses to +G and –G accelerations. The gross disturbances of the gas-exchange function of the lungs produced by sustained accelerations were extensively researched by David Glaister during the 1960s.[61] In the early 1970s he investigated the increase in G tolerance associated with reclining the pilot, and introduced the use of positive-pressure breathing to ease breathing on exposure to high G in the reclined position. Glaister found during these studies that pressure breathing enhanced the G tolerance of the seated subject, and proposed that pressure breathing, together with an anti-G suit, could be used to raise G tolerance without the intense fatigue produced by the standard straining manoeuvres. The great value of pressure breathing was amply confirmed in a joint RAF IAM-USAF School of Aerospace Medicine (SAM) study in 1972.[62] Flight trials of pressure breathing with G (PBG), conducted in the IAM Hunter between 1975 and 1980 and in the IAM Hawk in the mid-1980s, when the system was flown by RAF instructor pilots at a Tactical Weapons Unit, provided overwhelming evidence of the value and acceptability of the system.[63] In parallel, RAF IAM and USAF SAM conducted an integrated programme of development which demonstrated that PBG with counterpressure to the chest and G trousers would maintain full vision in the seated subject exposed to 9 G for several minutes.[64] This system was adopted by the USAF in the F-16 in order to reduce the incidence of G-induced loss of consciousness. It was also adopted for the European Fighter Aircraft and the USAF F-22.

The history of the development of ejection seats in the UK is almost solely that of the work of the Martin Baker Aircraft Company from 1945, when the requirement for an ejection seat was first raised. Throughout the period under review there was a very close symbiosis between the company and RAF IAM, with IAM providing the physiological and medical advice.[65] A succession of medical officers from IAM also acted as subjects on the ejection tower in joint studies to determine the maximum acceleration and jolt to which the ejectee should be exposed. Research of a more basic nature, such as the distribution of ejection load within the body, was conducted at RAF IAM.[66] Peter Howard of IAM was the subject for the first human ejection using a rocket-assisted seat in 1962.[67]

Neurosciences

During the 1960s neuroscience began to have an impact on the practice of aviation medicine, and this led to the formation of the Division of Neurosciences at RAF IAM. One area of interest was the advent of aircraft operating around northern and southern latitudes and crossing time zones at almost the same rate as the earth's rotation. The circadian rhythms of the aircrew had to adapt to new time zones and this, together with the irregularity of their work and rest times, led to serious problems with their sleep.[68] These problems were particularly acute in long-haul civil aircrew, and in 1967 the Flying Personnel Research Committee set up the Flight Deck Workload Study Group to study the matter.[69]

In support of the Flight Deck Workload Study Group, the Institute brought together a range of interdisciplinary skills, including medicine, physiology and mathematics, and this team sought to establish the principles behind the circadian desynchrony and sleep disturbance created by the increasing complexity of civil worldwide operations. The group was supported by the Civil Aviation Authority and was afforded excellent facilities with the civil airlines, particularly the antecedent of British Airways. In due course the group collaborated worldwide in studies with other centres of aviation medicine, so that the sleep of aircrew could be studied at various locations.[70]

By the late 1980s it had become clear that there were two main issues affecting the wellbeing of long-haul aircrew. These were the effects on performance of the continuous irregularity of work and rest on sleep, and the effects of the duration of duty and the time of day when the duty period occurred. It was evident that it would be possible to develop models by which the projected work and rest patterns of aircrew schedules could be assessed to establish whether they would be compatible with acceptable sleep and performance.[71]

The 1960s saw increasing pressure for a greater understanding of the use of medication by aircrew, so that they could benefit from advances in therapeutics which were available to the general population. A unit was therefore set up within the Neurosciences Division to investigate these issues and, over the years, made several recommendations for the safe use of medication by aircrew. The work of this group continues to have a marked influence on therapeutics as it relates to aircrew. Antihistamines free of sedation were introduced in the late 1970s, and studies carried out on antihypertensive agents have led to aircrew using these drugs without restrictions on their flying. Another area was the management of sleep disturbance. Although the understanding of the nature of sleep disturbance in aircrew was enhanced by the work of the Flight Deck Workload Study Group, it was still necessary to provide a means of alleviating the problem. However, at that time hypnotics caused residual effects on performance the next day. There were two factors involved: first, hypnotics were very long acting, and second, inordinately high doses were being used routinely in general medicine. Studies were carried out on these two issues and led to the identification of a hypnotic which could be used by aircrew at an appropriately low dose. The work was completed by 1980, and led to the introduction of temazepam for the use of aircrew under well-specified conditions.[72]

The possibility that hypnotics could now be used safely by aircrew was particularly opportune with the advent of the South Atlantic Campaign. The Royal Air Force was faced with a difficult task because of the distance of over 7000 miles from the UK to the Falklands, and the absence of any forward airfields. The only option was to use Ascension Island, which was almost midway between the two. Support for the task force from Ascension required return flights of up to 6800 miles, and this posed problems for the aircrew, in particular the sleep disturbances inevitable in maintaining a long-range air operation over several months.

Studies carried out at the Institute during the early 1970s on reinforcing the Far East from the UK had suggested that, with some augmentation of the crew, the duration of individual flights could be extended to 30 hours, even though this involved the loss of sleep over two consecutive nights. However, there was little information available on how frequently such flights could be operated over several weeks. Studies on civil operations provided some idea of workloads which would still preserve an acceptable sleep pattern, but it was evident that duty hours thought to be compatible with an acceptable sleep pattern would be far exceeded. This would lead to considerable sleep difficulties and so, from the early stages of the South Atlantic Campaign, temazepam was used between missions to ensure sleep during the rest periods, which occurred at various times of the day or night. The transport crews attained 150 hours flying per month, with duty hours much more, and the aircrew were able to cope with six such long-range flights in periods of less than a month.[73]

Advice on human factors, particularly in the relation to the problem of sleep, played an important part in maintaining the air operation during the South Atlantic Campaign. The use of a hypnotic ensured adequate sleep under difficult circumstances. The information gained during this campaign was relevant during the Middle East operation to relieve Kuwait. There, again, the workloads of the transport crews were exceedingly high, but the judicious use of a hypnotic helped to maintain a high-workload transport operation.

Motion (Air) Sickness and Spatial Disorientation

Wastage and disruption of flying training due to air sickness was studied by Tom Dobie in the early 1960s. He was responsible for the introduction of desensitisation therapy, which involved ground-based and airborne phases with incremental exposure to increasingly provocative motion stimuli.[74] Initially established at a flying unit in Flying Training Command, both the ground and airborne phases of the programme were transferred to RAF IAM in 1981. The flying phase was then conducted by a dedicated Medical Officer Pilot using the RAF IAM Hunter T7 aircraft. The use of active rather than passive motion by Reason and Benson, and linear as well as angular stimuli by Stott in the ground-based phase increased the rate and extent of adaptation before the flying phase of therapy was begun.[75] The programme has over the years returned more than 80 per cent of chronically motion sick aircrew to flying training, and subsequently to operational flying.

At the same time as the introduction of desensitisation therapy, Reason developed a neural mismatch theory of motion sickness which provided a common aetiology for all situations where sickness is engendered by motion (which include flight simulators and space flight), and a mechanism for adaptation to unfamiliar motion environments.[76] Over the ensuing years the theory has been refined by experiments conducted at IAM by Benson, Stott and Golding using novel angular and linear motion devices.

Although the occurrence of illusions and errors in aviators' perception of aircraft orientation has been known since the early days of powered flight, heightened awareness in the RAF of the incidence of spatial disorientation and its role in the aetiology of aircraft accidents was dependent upon interviews and a survey carried out by G. Melvill Jones in the late 1950s.[77] This study led to the detailed clinical investigation—in particular the assessment of vestibular function—of aircrew who had come under medical care because of in-flight disorientation incidents, or accidents in which spatial disorientation was implicated. The principal outcome of the clinical evaluations carried out by O'Connor and Benson was an elucidation of the relative importance of psychological and neuropsychiatric dysfunction in spatial disorientation.[78] In addition, it provided a better understanding of the dissociative sensations—the so-called 'breakoff phenomenon'—and its prevalence among helicopter aircrew, as well as the pilots of high-altitude, fixed-wing aircraft in whom the condition was first described.

With the greater awareness of spatial disorientation came the recognition of deficiencies in the aeromedical training of aircrew on the topic. In 1973—74 an AGARD Working Group, led by Benson, defined training requirements and the need for rotational devices that would familiarise aircrew with some of the sensory limitations that are responsible for disorientation in flight.[79] Implementation of the recommendations of the report was achieved at IAM by the design and construction of a spatial disorientation familiarisation device (SDFD), which was installed and used at the RAF Aviation Medicine Training Centre, RAF North Luffenham, for more than 20 years.

Vibration and Vision

The establishment of guidelines for human exposure to vibration, and its effects on performance, which were embodied in ISO 2631 (1974)[80] and in greater detail in BS6841 (1987)[81] was, in part, dependent upon work carried out at RAF IAM using unique facilities, namely a 2 m stroke vertical motion device and a high-torque turntable. These were capable of producing oscillatory (vibratory) stimuli over a wide frequency range with low harmonic distortion. Experimental work has been in three major areas. The first was concerned with the dynamic properties of the human body and the transmission of translational (linear) and rotational (angular) vibration to the head. This essential biodynamic data, provided by the work of Guignard, Rance, Barnes and Stott, has underpinned the guidelines on the subjective tolerance of vibration as a function of magnitude, frequency spectrum and duration of the stimulus.

The second area of research concerned the effects of vibration on vision and the role of vestibular and oculomotor mechanisms. Research by Barnes, Benson, Stott and Viveash[82] on the visibility of displays when either the display or the observer is vibrated, and how it is dependent on physical variables relating both to the display (e.g. luminance, size, contrast) and to the mechanical stimulus (e.g. frequency, amplitude, axis of motion), has elucidated the role of retinal and vestibular information in the control of eye movements. Mathematical models of oculomotor systems developed by Barnes allow predictions of the likely visual deficit in specific vibration environments. In addition, knowledge of the differing dynamics of the visual pursuit and vestibular/ocular reflexes has pointed to the need for head-mounted displays to be space stabilised if visual acuity is to be preserved in the presence of head vibration at frequencies above 2 Hz.

The problems of vision during high-altitude flight were studied intensively by Whiteside at IAM in the early 1950s.[83] In parallel with this, RAF IAM developed the helmet-mounted visor to replace the Second World War goggles that protected the eyes.[84] Basic research defined the optical properties required of antisolar-glare visors. Protection of the eyes against the hazards arising from a bird strike in the canopy area and wind blast on ejection at high speed led to the development in the 1960s of dual visor systems using polycarbonate.[85] Brennan, at IAM, set and assessed the optical properties of these visor systems. He also developed very effective corrective flying spectacles, which were fully integrated with aircrew headgear. Brennan also played a central role in the specification and assessment of the optical characteristics of the night-vision devices developed for aircrew use. The IAM integrated many of these devices into aircrew helmets and supported directly much of the in-service flying with NVGs.

The effects on vision of the very high-intensity light created by the detonation of a nuclear weapon were studied both in the laboratory and at the nuclear weapon tests at Christmas Island by Whiteside.[86] He also tested an electromechanical shutter developed jointly with the Atomic Weapons Establishment and RAE. RAF IAM advised on the standards of protection against nuclear flash blindness, and assessed the performance of various potential protective devices. Research on the biological effects of laser radiation, particularly on the eye, began at IAM in 1963, and the Institute rapidly became the authority within MoD on the threshold levels for effects on vision and retinal damage.[87] This experimental knowledge formed the basis of UK and international standards.

Aircrew NBC Equipment

During the 1960s both IAM and RAE were intermittently involved in the assessment of various items of NBC protective equipment for use in flight. A major step forward in the protection of the skin was the proposal by Simpson of the RAE that the activated charcoal fabric should be worn beneath the flying coverall or immersion suit. The outcome was the aircrew inner coverall, designed by RAE and assessed physiologically by IAM. This garment was adopted by the RAF and RN and by several NATO air forces, including the USAF, USMC and Canadian Forces. The work intensified with the issue of NGASR 562 for NBC protective assemblies for aircrew, and attempts by industry to produce acceptable aircrew respirators. The unacceptable features of the low-altitude aircrew respirator produced by industry led RAE to design an over-the-helmet respirator using principally available components of aircrew equipment and a breathing and ventilating system developed by IAM.[88] The Aircrew Respirator No 4 overcame virtually all the disadvantages of the respirators developed by industry. It was not, however, compatible with weapon sights and NVGs. Simpson proposed an under-the-helmet respirator with its visor plate fitting into the facial opening, thus ensuring compatibility between helmet, visors, NVGs and sights.[89] A joint RAE–IAM team, working from 1975 to 1979, developed the Aircrew Respirator NBC No 5 (AR 5) in-house and assessed it in the laboratory, in the RAF IAM aircraft, and in field trials in RAF Germany.[90] This respirator was then produced by industry. The AR 5 supply systems were specified by IAM. The AR 5 entered service with the RAF in 1980 and was subsequently issued to RN aircrew and adopted by the US Marine Corps and the Canadian Forces. The RAE–IAM team also developed and, in association with CDE, proved don–doff procedures for contaminated aircrew NBC assemblies, and designed the associated contamination control areas for collective protection. One outcome of this joint endeavour was that in the war with Iraq the RAF fast-jet aircrew had fully proven and acceptable CB protective equipment to wear in flight. Aircrew of the other air forces (except those using the AR 5) did not.

CONCLUSION

The 45 years spanned by this history of physiology in the MOD saw the growth and consolidation of applied physiological research in support of the development of weapon systems and their use by the armed forces. Whilst each of the three personnel research establishments was successful in its own field, the very wide spectrum of challenges set by military aviation in the post-war period and the wise direction of the RAF Institute of Aviation Medicine by Air Vice-Marshal Stewart who commanded the unit from 1946 to 1967 led to the Institute being widely recognised as one of the leading human physiology research establishments in the United Kingdom. The close symbiosis which was developed between the Institute, the central Air Staff, the Procurement Executive and the aircraft industry was most productive, helping to ensure that the needs of the human operator were fully recognised and met as far as possible in air systems. The influence of the RAF IAM in ensuring that physiological requirements were understood and met in the development and operational use of aircraft predominated in the thirty or so years between 1960 and the early 1990s. Furthermore the Institute was intimately involved in the development and evaluation of the compromises between the ideal and what was practical, which is probably one of the most important aspects of applied science. The Institute of Naval Medicine and the Royal Naval Physiological Research Laboratory were also successful in these respects.

The recognition of the value of the applied physiological and medical support provided by the Royal Naval and Royal Air Force Institutes increased progressively over the twenty years following the Second World War. By the early 1960s the two research Institutes and the Royal Naval Physiological Laboratory were being consulted on and were providing inputs to virtually all the new weapon systems and items being developed to enhance the performance and safety of the operators. Success in ensuring that sound physiological principles were incorporated at the design stage and that adequate physiological assessments were performed during development and prior to production was achieved in most weapon systems and life support and protective equipment including submarine escape, the crew environment in nuclear powered submarines, fast jet aircraft especially the Harrier, Phantom, Tornado and most recently Eurofighter, and the NBC protection aircrew.

The increasing recognition of the effects of the subtleties of human performance upon the effectiveness of weapon systems led to the use of psychological techniques and to the birth of ergonomics with its emphasis upon the man-machine interface. The second World War had seen the beginnings of aviation psychology and the development of this branch of science was nurtured and supported within the RAF Institute of Aviation Medicine. A parallel but somewhat slower recognition occurred in the Army which culminated in the formation of the Army Personnel Research Establishment with a significant cadre of psychologists in the early 1970s.

The very high standard of applied research in human physiology attained by the personnel research establishments required well trained and motivated investigators, a good proportion of whom had indepth experience of working in a particular area. The enlightened attitude of the Royal Air Force, and to a lesser extent the Royal Navy, in establishing medical research specialities in the physiology based sciences was undoubtedly a very important factor in the success of the RAF IAM and INM in providing high quality applied research in a timely manner. The great success of the RAF IAM in attracting and retaining high quality staff was due to the offer of a long career in research and to the provision of time and encouragement to conduct academic/basic research as well as to conduct high quality applied research. The roles which both RAF IAM and INM had in educating others, both within the Ministry of Defence and civilians in aviation and diving medicine respectively, also enhanced their attractiveness to potential staff. Both establishments were the only organisations in the United Kingdom providing training in aviation medicine (IAM) and diving and submarine medicine (INM) respectively. The long course in aviation medicine established by IAM in 1967 rapidly gained world wide recognition with over half the students coming from overseas.

The single service arrangement of applied physiological and psychological research gave rise within the Ministry of Defence to concerns in the 1960s that there might be duplication of research in the three personnel research establishments and an uneven distribution of the total effort available. A working party comprising twelve academics chaired by Professor Krohn of Birmingham university was established in 1964 to review 'personnel and anti-personnel research' in MOD. The report of this Working Party in 1967 found that whilst applied physiological and medical research should remain single service relating to sea, land and air weapon systems and operations there would be merit in establishing a tri-service human factors (applied psychology) institute. The latter recommendation was not implemented. More formal reviews of the work of INM, APRE and RAF IAM by their respective directors were established in order to ensure that duplication of research did not occur. A further internal MOD review in the early 1980s concluded that there was no significant duplication of research effort between the three services. The changes in the organisation of the Ministry of Defence and reductions in the Defence budget which occurred in the late 1980s and early 1990s led in 1991 to the Chief Scientific Adviser setting up a review of the organisation of Human Factors research defence wide including the feasibility of establishing a single centre of excellence. This review which reported in 1992 recommended the setting up of a tri-service Human Factors Establishment to include applied physiological research. This recommendation was implemented with the formation of the Centre for Human Sciences as a division of the Defence Evaluation and Research Agency in 1994. The Centre was formed at DERA Farnborough by the amalgamation of old Army Personnel Research Establishment in its entirety, the civilian research staff of the RAF Institute of Aviation but without the Royal Air Force Specialists in Aviation Medicine, and a small research component of the Institute of Naval Medicine. It is too early and inappropriate to consider here whether in the long term these changes will enhance the applied physiological research support to the armed forces and UK industry.

Acknowledgement

It is a pleasure to acknowledge my indebtedness and thanks to Surgeon Rear Admiral Frank Golden OBE (retired) and Major General Joe Crowdy CB (retired) and to the colleagues with whom I served at the RAF Institute of Aviation Medicine who so willingly provided contributions to this chapter.

1. A.E. Boycott, D.C. Damant and J.S. Haldane, 'The prevention of compressor air illness', *Journal of Hygiene*, 8 (1907): 342–444.
2. T.M. Gibson, M.H. Harrison, *Into Thin Air. A History of Aviation Medicine in the RAF* (Robert Hale, 1984).
3. The Talbot Committee, *Report on Naval Lifesaving*, Admiralty Papers (1946).
4. Medical Research Committee, Committee on the Care of Shipwrecked Personnel, *A Guide to the Preservation of Life at Sea after Shipwreck*, War Memorandum No. 8 (HMSO, 1943).
5. F.E. Smith, *Survival at Sea*, Royal Naval Personnel Research Committee Report No. SS 1/76 (Medical Research Council, 1976).
6. The Talbot Committee (n. 3 above).
7. W.R. Keatinge, *Survival in Cold Water* (Blackwell, 1969).
8. G.R. Hervey and R.A. McCrance, 'The effects of carbohydrate and seawater on the metabolism of men without food or sufficient water', *Proc. Roy. Soc. B.*, 139 (1952): 527–45.
9. G.R. Hervey, 'Drinking sea water', *Lancet*, 1 (1957): 533–4.
10. W.R. Keatinge, *Survival in Cold Water* (n. 7 above).
11. F.StC. Golden, G.R Hervey and M.J. Tipton, 'Circum-rescue collapse: collapse, sometimes fatal, associated with rescue of immersion victims', *Journal of the Royal Naval Medical Service*, 157 (1991): 629–32.
12. M.J. Tipton, 'The initial responses to cold water immersion in man', *Clinical Sciences*, 77 (1989): 581–8.
13. S. Miles, *Underwater Medicine* (Staples Press, 1962).
14. H.V. Hempleman, *A New Theoretical Basis for the Calculation of Decompression Tables*, Royal Naval Personnel Research Committee Report No. UPS 131 (Medical Research Council, 1952).
15. H.V. Hempleman, *Medical Code of Practice for Work in Compressed Air*, Construction Industry Research and Information Association Report No. 44, 1st edition (1973).
16. E.E.P. Barnard, 'Fundamental studies in decompressions from steady state exposures', in *Proceedings of the Fifth Symposium Underwater Physiology*, edited by C.J. Lambertsen (Fed. Am. Soc. Exp. Biol, 1976).

17. P.B. Bennett and J.C. Rostain, 'The high pressure nervous syndrome', *The Physiology and Medicine of Diving*, edited by P.B. Bennett and D.H. Elliott (Saunders, 1993): 194–237.
18. Flag Officer Submarines, *Condensed History of Submarine Escape in the Royal Navy* (Fort Blockhouse, Gosport, 1993).
19. K.W. Donald, W.M. Davidson and W.O. Shelford, 'Submarine escape breathing air', *Journal of Hygiene*, 46 (1948).
20. S.L. Cowan, W. Wilson-Dickson and H.C. Wright, *Respiratory and Other Responses of Human Subjects During Rapid Decompression—Preliminary Experiments Using a Hot Wire Recorder*, Royal Naval Personnel Research Committee Report No. UPS/82 (Medical Research Council, 1947).
21. L.D. Hamlyn, M.H. Parsons, *A Report on Submarine Escape Trials: Upshot 1*, Royal Naval Personnel Research Committee Report No. UPS/229 (Medical Research Council, 1962).
22. E.E.P. Barnard, W.G. Eaton, *Experiments in Submarine Escape*, Royal Naval Personnel Research Committee Report No. UPS/241 (Medical Research Council, 1965).
23. M.R. Todd, *Upshot V. Report to Flag Officer Submarines* (Fort Blockhouse, Gosport, 1971).
24. D.H. Elliott, J.A.B. Harrison and E.E.P. Barnard, 'Clinical and radiological features of 88 cases of decompression barotrauma', in *Proceedings of the Sixth Symposium on Underwater Physiology*, edited by C.J. Lambertsen (Fed. Am. Soc. Exp. Biol., 1978): 527–36.
25. A.H. Bebb, H.N.V. Temperley and J.S.P. Rawlins, *Underwater Blast: experiments and researches by British investigators*, Report No. R81 401 (Admiralty Marine Technology Establishment, Gosport, 1981).
26. J.S.P. Rawlins, 'Trials of underwater blast from the Martin-Baker Type IV ejection seat gun', *Journal of the Royal Naval Medical Service*, 48 (1962): 57–62.
27. Royal Naval Environmental Research Working Party, *1st Report 1972–1977* (Institute of Naval Medicine, Gosport, 1977).
28. C.C. Ungley, G.D. Channell and R.L. Richards, 'The immersion foot syndrome', *Br. J. Surg.*, 33 (1945): 17–31.
29. F.StC. Golden, 'The Falklands experience', in *Medical Operational Problems in a Cold Environment*, Joint UK/US Casualty Care Workshop (Institute of Naval Medicine, Gosport, 1984): 90–4.
30. B. McArdle, W. Dunham, H.E. Holling, W.S.S. Ladell, J.W. Scott, M.L. Thomson and J.S. Weiner, *The Prediction of the Physiological Effects of Warm and Hot Environments*, Report No. RMP 47/391 (Medical Research Council, 1946).
31. K.D. Duncan, J.R. Allan, D.T. Beeston, J.P. Crowdy, M.F. Haisman, D.W.A. Peters and L. Zurick, *The Effect of an Artificial Acclimatisation Technique on Infantry Performance in a Hot Climate*, Report No. 14/63 (Army Operational Research Establishment, Byfleet, 1963).
32. R.H. Fox, R. Goldsmith, D.J. Kidd and H.W. Lewis, 'Acclimatisation to threat in man by controlled elevation of body temperature', *J. Physiol.*, 166 (1963): 530–47.
33. C.Y. Gooderson, *Anthropometric Survey of the British Army. Clothed Anthropometry and Increments for Three Clothing Assemblies*, Report No. AR22/77 (Army Personnel Research Establishment, Farnborough, 1977).
34. R.J. Whitney, *Fighting Vehicle Research Capsule Investigation*, Report APRC 69/FVI, Army Personnel Research Committee (Medical Research Council, 1969).
35. C.Q. Large, *The APRE Human Factors Unit (MVEE)*, Report No. 72 (Army Personnel Research Establishment, Farnborough, 1972).
36. J.P. Crowdy, *Drinking Water Requirements in Hot Climates*, Report No. 1/68 (Army Personnel Research Establishment, Farnborough, 1968).
37. D.E. Worsley, A.F. Amor, D.A. Ramsay, W.P. Hughes and N.E. Ince, *Physiological Trials of Cold Weather Clothing and Equipment. Honky Tonk I Norway 1974*, Report No. 16/74 (Army Personnel Research Establishment, Farnborough, 1974).
38. APRE/CDE/DNBCS, *Assessment of the Effects of Wearing Full NBC Protective Clothing and Equipment on the Military Performance of all Arms (Exercise Jeremiah)*, Report No. 1/73 (Army Personnel Research Establishment, Farnborough, 1973).
39. A.F. Amor, D.E. Worsley, *Energy Cost of Wearing Multi-layer Clothing*, Report No. 18/73 (Army Personnel Research Establishment, Farnborough, 1973).
40. J.P. Crowdy, *Combat Nutrition—the Effects of a Restricted Diet on the Performance of Hard and Prolonged Physical Work*, Report No. 2/71 (Army Personnel Research Establishment, Farnborough, 1971).
41. J.V.G.A. Durnin, J. Wormersley, 'Body fat assessed from total body density and its estimation from skin fold thickness: measurements on 481 men and women aged 16 to 72 years', *Br. J. Nutr.*, 32 (1974): 77–97.
42. M.F. Haisman, D.W.A. Peters, *Assessment of Fitness for Prolonged Physical Exertion Phase III. Investigation of the Relationship between Fitness as measured by Physiological Tests and the Performance of Athletic and Military Tasks*, Report No. 8/68 (Army Personnel Research Establishment, Farnborough, 1968).
43. S.H. Cole, *The Vibration Environment in Armoured Fighting Vehicles, with Reference to the Human Occupant*, Report No. 21/76 (Army Personnel Research Establishment, Farnborough, 1976).

44. J.M. Rylands, *Intelligibility Assessment of Warrior Integrated Noise Reduction System Headsets*, Report GO R013 (Army Personnel Research Establishment, Farnborough, 1990).
45. H.J. Jacobs, A.I. Karstens, *Physiological Evaluation of the Partial Pressure Suit*, Report No. MCR EXD-696–104H (Air Material Command, Wright Field, Dayton, Ohio, 1948).
46. R.C. London, H.L. Roxburgh, *Emergency Partial Pressure Suit. Part I Narrative*, Report No. 814 (Flying Personnel Research Committee, Ministry of Defence, UK, 1953).
47. H.L. Roxburgh, P. Howard, W.H. Dainty and R.L. Holmes, *A High Pressure Breathing System for Short Endurance Fighter Aircraft*, Report No. 842 (Flying Personnel Research Committee, Ministry of Defence, UK, 1953).
48. H.L. Roxburgh, J. Ernsting and I.G. Badger, *The Pressure Jerkin*, Report No. 948 (Flying Personnel Research Committee, Ministry of Defence, UK, 1956).
49. I.J. Badger, J. Ernsting and D.J. Parry, *The Arm Jerkin G-suit Combination*, Report No. 979 (Flying Personnel Research Committee, Ministry of Defence, UK, 1956).
50. J. Ernsting, I.D. Green, R.E. Nagle and P.R. Wagner, 'High altitude protection from pressure breathing mask with trunk and lower limb counterpressure', *J. Aviat. Med.*, 31 (1960): 40–6.
51. P.R. Wagner, *The Altitude Performance of the Type B Full Pressure Suit*, Report No. 1191 (Flying Personnel Research Committee, Ministry of Defence, UK, 1962).
52. D.M. Denison *et al.*, 'Complex Reaction Times at Simulated Altitudes of 5000 and 8000 ft.', *Aerospace Medicine*, 37 (1960): 1010–3.
53. J. Ernsting, 'Prevention of hypoxia—acceptable compromises', *Aviat. Space Environ. Med.*, 49 (1978): 495–502.
54. A.N. Nicholson and J. Ernsting, 'Neurological sequalae of prolonged decompression', *Aerospace Medicine*, 38 (1967): 389–94.
55. J. Ernsting, 'Physiological requirements for oxygen systems of high performance aircraft', 17th Meeting of Aerospace Medical and Life Support Systems Working Party of the Air Standardisation Co-ordination Committee Ministry of Defence, UK, *Symposium on Biomedical and Biophysical Aspects of Oxygen Systems* (1976).
56. J. Ernsting and R.L. Miller, 'Advanced oxygen systems for aircraft', *Agardograph*, 286 (1996): 21–33.
57. J. Ernsting and R.L. Miller, 'Advanced oxygen systems for aircraft' (n. 56 above): 54–5.
58. J. Ernsting, *The Inflight Assessment of the Harrier GR5 Oxygen System in Hunter XL563 (July 1985 – March 1986)*, Report No. 544 (Royal Air Force Institute of Aviation Medicine, Farnborough, 1986).
59. A.N. Nicholson, *Anti G Suits—Research and Development (1962–1966)*, IAM Report No. 346 (Royal Air Force Institute of Aviation Medicine, Farnborough, 1965).
60. H.P. Ruffell Smith, F.G Cumming, R.S. Waambeck and R.T. Robinson, *Flight Trials of the Prone Position Meteor, Final Report*, Report No. 943 (Flying Personnel Research Committee, Ministry of Defence, UK, 1955).
61. D.H. Glaister, 'The effects of gravity and acceleration on the lung', *Agardograph*, 113 (1970).
62. R.J. Crossley, R.R. Burton, S.D. Leverett, E.D. Michaelson, S.J. Shubbrooks, *Human Physiological Responses to High, Sustained, +Gz Acceleration*, IAM Report No. 537 (Royal Air Force Institute of Aviation Medicine, Farnborough, 1973).
63. A.R.J. Prior, G.J. Cresswell, *Flight Trial of an Enhanced G. Protection System in Hawk XX327*, IAM Report 678 (Royal Air Force Institute of Aviation Medicine, Farnborough, 1989).
64. A.R.J. Prior, *Physiological Aspects of an Enhanced G Protection System*, 11th Ann. Meeting IUPS Commission on Gravitational Physiology (Lyon, France, 1989).
65. W.K. Stewart, *Ejection of Pilots from Aircraft. A Review of the Applied Physiology*, Report No. 671 (Flying Personnel Research Committee, Ministry of Defence, UK, 1946).
66. F. Latham, 'A study in body ballistics. Seat ejection', *Proc. Roy. Soc. B.*, 147 (1957): 121–39.
67. S. Sharman, *Sir James Martin* (Haynes, 1996).
68. A.N. Nicholson, 'Sleep patterns of an airline pilot operating world-wide east–west routes', *Aerospace Medicine*, 41 (1970): 626–32.
69. J.S. Howitt, J.S. Balkwill, T.C.D. Whiteside and P.D.G.V. Whittingham, *A Preliminary Study of Flight Deck Work Loads in Civil Air Transport Aircraft*, Report No. 1240 (Flying Personnel Research Committee, Ministry of Defence, UK, 1965).
70. A.N. Nicholson, *Neurosciences and Aviation Medicine* (International Academy of Aviation and Space Medicine, Auckland, New Zealand, 1998).
71. M.B. Spencer, B.M. Stone, A.S. Rogers and A.N. Nicholson, 'Circadian rhythmicity and sleep of aircrew during polar schedules', *Aviat. Space Environ. Med.*, 62 (1991): 3–13.
72. A.N. Nicholson, T. Roth and B.M. Stone, 'Hypnotics and aircrew', *Aviat. Space Environ. Med.*, 56 (1985): 299–303.
73. J.A. Baird, P.K.L. Coles and A.N. Nicholson, 'Human factors and air operations in the South Atlantic Campaign', *J. Roy. Soc. Med.*, 76 (1983): 933–7.
74. T.G. Dobie, *Air Sickness in Aircrew*, Report No. 177 (Advisory Group Aerospace Research and Development, Neuilly-sur-Seine, France, 1974).

75. M. Bagshaw, J.R.R. Stott, 'The desensitisation of chronically motion sick aircrew in the Royal Air Force', *Aviat. Space Environ. Med.*, 56 (1985): 1144–51.
76. J.P. Reason, 'Motion sickness adaptation: a neural mismatch model', *J. Roy. Soc. Med.*, 71 (1978): 819–29.
77. G. Melvill Jones, *A Study of Current Problems Associated with Disorientation in Man-Controlled Flight*, Report No. 1021 (Flying Personnel Research Committee, Ministry of Defence, UK, 1957).
78. P.J. O'Connor, 'Differential diagnosis of disorientation in flying', *Aerospace Medicine*, 38 (1967): 1155–60.
79. A.J. Benson, *Orientation/Disorientation Training of Flying Personnel. A Working Group Report*, Report No. 625 (Advisory Group for Aerospace Research and Development, Neuilly-sur-Seine, France, 1974).
80. International Organisation for Standardisation, *Guide for the Evaluation of Human Exposure to Whole-body Vibration*, ISO 2631 (1974).
81. British Standards Institution, *Guide to the Evaluation of Human Exposure to Whole-body Mechanical Vibration and Repeated Shock*, BS 6841 (London, 1987).
82. A.J. Benson, G.R. Barnes, 'Vision during angular oscillation: the dynamic interaction of visual and vestibular mechanisms', *Aviat. Space Environ. Med.*, 49 (1978): 340–45.
83. T.C.D. Whiteside, *Problems of Vision in Flight at High Altitude* (Butterworth, 1957).
84. T.C.D. Whiteside, *Eye Protection by Visors*, Report No. 750 (Flying Personnel Research Committee, Ministry of Defence, UK, 1950).
85. D.H. Brennan, J.M. Moon, *Service Trial of a Dual Visor Assembly*, AEG Report 415 (Royal Air Force Institute of Aviation Medicine, Farnborough, 1976).
86. T.C.D. Whiteside, *The Observation and Luminance of a Nuclear Explosion*, Report No. 1075 (Flying Personnel Research Committee, Ministry of Defence, UK, 1960).
87. R.G. Borland, J. Marshall, D.H. Brennan and A.N. Nicholson, 'Safety with lasers', *Proc. Roy. Soc. Med.*, 66 (1973): 841–6.
88. R.E. Simpson, *Proposals for a Chemical Defence Hood for Use by Military Aircrew (the Aircrew Respirator NBC No. 3)*, Technical Report 73071 (Royal Aircraft Establishment, 1973).
89. R.E. Simpson, *Proposals for a Revised Form of Aircrew NBC Respirator (Aircrew Respirator NBC No. 5)*, Letter Report YSH/107/06/RES dated 25 February 1976 (Royal Aircraft Establishment, 1976).
90. J. Ernsting, A.W. Cresswell, A.J.F. Macmillan, R.E. Simpson and B.C. Short, *United Kingdom Aircrew Chemical Defence Assemblies*, Conference Proceedings 264 (Advisory Group for Aerospace Research and Development, Neuilly-sur-Seine, France, 1979).

CHAPTER 13

Civil Spinoff From the Defence Research Establishments

GRAHAM SPINARDI

INTRODUCTION

The defence research establishments (DREs) have been at the heart of much of the technological activity sponsored by the British government.[1] Similarly, defence production has been a major component of Britain's manufacturing output. Taken together, it would be surprising if this level of defence-related activity had not had some effects on civil science and technology. Put simplistically, these effects could be seen as beneficial, in supporting basic scientific research and providing commercial technological spinoff, or deleterious, in skewing the aims and outputs of science and technology and in diverting manufacturing from commercial markets.

Whatever the analysis, the fact is that shrinking defence budgets have put pressure on the level of R&D supported by the military. During the 1980s it became explicit Ministry of Defence (MoD) policy that the DREs should focus more narrowly on their defence mission. More recently, the end of the Cold War has led defence R&D budgets to be further reduced, and perceptions (and organisational change in the shape of DERA's creation) have shifted again to some encouragement of relations with civil technology. A better understanding of the role of such military R&D in UK civil science and technology is thus extremely timely. Two policy issues assume particular importance in the current climate of concern over economic competitiveness. First, if the work of the DREs has underpinned basic science and technology in the UK, then reductions in this role may need to be compensated for by other civil mechanisms. Second, the ways in which the DREs contribute to civil innovation need to be scrutinised and enhanced to the extent that is possible within their overall defence mission.

The relationship between the DREs and civil industrial innovation is thus the main concern of this chapter. The related issue of how the DREs have influenced scientific activities in the UK cannot be adequately dealt with here. Whereas the influence of defence on science has come under increasing scrutiny elsewhere—especially in the USA—this remains an area underresearched by historians of British science.[2] Chapter 11 does, however, add greatly to our knowledge of the role of defence-sponsored work in scientific disciplines related to CBW.

The bulk of this chapter comprises three case studies of the influence of the defence establishments and their work on civil technology. The cases of liquid crystals, carbon fibre and high-integrity computing show a range of outcomes in differing technological and industrial contexts. These cases are preceded by two sections setting out the background: first, to the concept of spinoff, and then to the history of institutional mechanisms for promoting such spinoff.

THE SPINOFF DEBATE

It is widely acknowledged that scientific and technological developments sponsored through defence funding produce some civil spinoff.[3] Much of the evidence, however, is anecdotal, focusing on a few successful cases. The aim here is to look at the range of ways in which UK DREs have influenced civil innovation, and at the differing outcomes, with particular reference to three detailed case studies: liquid crystals, carbon fibre, and high-integrity computing. For some, even the use of the term spinoff is controversial, and so some clarification is necessary. The public promotion of the idea of spinoff has clearly been a political tool to gain support for programmes, most notably in the USA with NASA, and also more recently with the 'Star Wars' Strategic Defense Initiative (SDI) project. 'Spinoff' is used here simply to describe the transfer of inventions from the DREs to civil firms and their subsequent commercial exploitation. It is not suggested that this is either common or very successful in outcome, only that it does sometimes occur. Nor is it implied that the spinoff process involves the simple transfer of an invention or gadget which can be readily sold into civil markets: again, all the evidence shows that a significant level of adaptation to the requirements of civil markets is involved.

Consideration of the processes of technology transfer and innovation required for successful spinoff indicates two main outlets for DRE inventions, depending on whether the company involved is a traditional defence contractor or not. Traditionally, the DREs have had good working relationships with their extramural contractors. High levels of knowledge flow between the DREs and these firms enable technology transfer to be potentially very effective. However, as most of these firms are geared towards working in defence markets, their interest in, and aptitude for, civil commercialisation is often limited.

Typically, the other mode of spinoff involves the transfer of DRE inventions to non-defence companies outside the normal relationships of the establishments. In some cases this transfer has happened passively; in others it has been the result of active efforts by organisations such as the National Research Development Corporation (NRDC) or Defence Technology Enterprises (DTE), or by especially motivated DRE inventors.

Although such spinoff into the wider civil industry of the country typically entails greater awareness of the needs of civil markets, there are other handicaps that may limit its success. Specifically, many of the inventions generated by the DREs will have a strong focus on defence missions, a focus which makes them unsuitable, at least in their current state, for civil applications. Moreover, compared to technologies developed within existing DRE–firm relationships, a much greater effort of technology transfer will be required to recreate the invention in its new commercial environment.

However, whether the DRE origins of an invention make it uniquely difficult to adapt to civil markets is debatable. All technology transfer from a research environment—be it DRE, civil laboratory or university—to a commercial setting requires substantial further work to produce a viable product. Inventions from the DREs are no different in this regard. Nor do they differ in being equally vulnerable to the general problems that have been associated with industrial innovation in the UK—in particular, a lack of long-term investment in technological developments.

THE INSTITUTIONAL ARRANGEMENTS OF SPINOFF

The first formal institution established to promote technology transfer was the National Research Development Corporation. Stemming in part from concern over the UK's failure to benefit fully from its own inventions, the NRDC was set up in 1949 to ensure that such inventions were at least patented (so that overseas exploitation would result in licence royalties), and ideally to encourage their exploitation by UK firms.[4]

Typically, but not invariably, the NRDC would handle the licensing of inventions from the DREs which were seen to have civil applications (though much was excluded because work in collaboration with industry usually involved the assignment of intellectual property rights to the firm). Examples from Farnborough include structural materials such as carbon fibre and glass and asbestos plastic composites.[5] The SERL at Baldock and the RRE at Malvern (which were combined into the RSRE in 1976) were the originators of many developments in semiconductor technology. As well as the work done on lasers (see Chapter 10), these included the development of the Malvern Correlator (a laser device for measuring gas or liquid flows), licensed to Malvern Instruments Ltd, and much work on the properties of semiconductor materials such as gallium phosphide/arsenide, and on methods for producing them, such as the Malvern crystal-growth systems which were licensed to Metals Research Ltd.[6]

In 1964 the Wilson government established the Ministry of Technology (MinTech) to 'guide and stimulate a major national effort to bring advanced technology and new processes into British industry'.[7] As well as taking over administrative responsibility for NRDC (although NRDC remained, as before, an independent public corporation), MinTech also sought to encourage the application of the DRE's resources to civil technological uses.[8] For example, in 1967 MinTech set up an Electronic Materials Unit within the Royal Radar Establishment at Malvern. This was intended to supply newly available semiconductor materials for use by industry, universities or other government research establishments, thus avoiding the necessity for each to produce their own.[9]

In some cases MinTech went beyond simply encouraging spinoff, as staff at the establishments were encouraged to become involved in working on civil technologies and socially beneficial projects. This was particularly noticeable at the Atomic Weapons Research Establishment at Aldermaston, where projects were carried out on biomedical applications, including artificial limbs; on computerised engineering; and on rationalised production methods.[10]

With the demise of MinTech after the 1970 election that brought the Conservative party back into power, the NRDC was left as the central apparatus for promoting technology transfer to the civil sector. In 1981 the NRDC was merged with the National Enterprise Board to form the British Technology Group (BTG), although its activities continued much as before.

The next effort geared specifically towards enhancing spinoff from the DREs did not occur until the mid-1980s, when in response to a plea from Prime Minister Thatcher, Defence Technology Enterprises (DTE) was set up. A consortium of eight organisations, including BTG and seven financial institutions, DTE set out to adopt a more proactive approach to discovering and exploiting inventions from the DREs than what was seen as NRDC/BTG's previously passive role. Instead, DTE's 'ferrets' were to scour the DREs in search of unexploited inventions, which would be added to a database and circulated to DTE Associate Members.

DTE had only limited success and had effectively ceased its ferreting function by the end of 1990.[11] The fundamental problem underlying DTE's demise was the short-term expectations of the shareholders. Without long-term financial backing DTE had little chance (without extraordinary luck) of becoming profitable, given the (often underestimated) effort needed to turn promising laboratory developments into commercially viable products. Moreover, most of the more promising inventions generated within the DREs were already being handled by 'the usual channels'—through the existing industrial contacts stemming from extramural contracts.[12] One of the factors undermining the confidence of DTE's shareholders was the impending reorganisation of the DREs into a Defence Research Agency. It was feared that the more commercial attitudes implied by this arrangement would prejudice the access of DTE to DRE inventions. Whether this would have been the case or not, the advent of the DRA did lead to changes in relation to technology transfer, reversing to some extent the *laissez faire* mood of the 1980s. Within the DRA, and DERA, the commercial freedom of the new setup seems to have encouraged a return to viewing the DREs' technological assets as a resource which should be exploited for commercial benefit, in addition to serving the primary function of supporting UK defence requirements.

MALVERN AND LIQUID CRYSTALS

Discussion of spinoff has typically focused on technologies developed through defence R&D which have then found significant civil usage and commercial success. A number of examples of this type of spinoff can be attributed to the DREs, but by far the most successful in generating licensing royalties has been the work of the former RSRE at Malvern on liquid crystals. However, despite the licensing revenue generated, UK companies have not become significant manufacturers of liquid crystal displays (LCDs). This case therefore highlights both strengths and weaknesses in the commercial development of innovations from UK defence R&D.

Liquid crystals have been known since 1888, and tens of thousands such materials

FIGURE 13.1: LIQUID CRYSTAL DISPLAYS PROCESSING LABORATORY (© CROWN COPYRIGHT, DERA)

have been synthesised since then (Fig. 13.1). They combine the properties of a liquid with orientational order similar to that usually found in solid crystals. The optical and electrical properties associated with this molecular orientation give liquid crystals their potential for display technology. The typical liquid crystal display relies on the 'twisted nematic' effect, which utilises the polarising properties of the material. A thin layer of the liquid crystal is held between two polarised glass plates, whose inner surfaces are coated with a transparent conductor. In the absence of an electrical field, the liquid crystal rotates the plane of polarised light. If the enclosing plates are crossed polarisers, then light is transmitted. When an electric field is applied to the liquid crystal, the polarising effect is removed and so light is not transmitted, giving the dark image which, for example, makes up the numerals on a digital watch display.[13]

RSRE's interest in liquid crystal displays stems from 1967, but even before this advances in semiconductor electronics had led to a perception that the standard cathode-ray tube (CRT) display constituted something of a bottleneck. However, although developments in electronics offered smaller size and lower power requirements which made the CRT look outdated, they also kept staff at RRE (as it was then) so busy as to rule out any speculative work on alternative display technologies. This changed in 1967, following a visit to RRE by the then Minister of State for Technology, John Stonehouse, in March. During his discussions with the Director,

Dr (now Sir) George MacFarlane, the issue of financial returns from inventions came up, and MacFarlane stated that the UK was paying royalties to RCA for CRT technology which amounted to more than the development costs of Concorde. The next day the minister telephoned MacFarlane to suggest that the UK should instigate a programme to develop a solid-state alternative to the CRT.[14]

This was initially greeted with scepticism at RRE, but a working party was set up to investigate the possibilities. The first draft of the working party's report was ready in October 1969, and recommended work on eight main topics, not including liquid crystals. However, the final version, issued in December, was substantially altered by Cyril Hilsum, the head of RRE's Display Section, to emphasise liquid crystals, as he later recalled:

> the original 'final' version recommended a large programme on ferro-electric ceramics, and one small-scale exploratory contract on liquid crystals. I was uneasy as I read it, feeling that there was something wrong, and almost doodling, replaced the words 'ferro-electric ceramics' everywhere by 'liquid crystals'. In some curious way it now had the right feel, and I had the final version issued. Three years had passed since the Stonehouse visit, and at last we knew what we wanted to do.[15]

Hilsum had been impressed by the knowledge of liquid crystals demonstrated by George Gray of Hull University when he attended a meeting on liquid crystals held on 1 October 1968.[16] Gray and his team had been carrying out academic research on the properties of liquid crystals for some 20 years. With the decision to pursue liquid crystals, RRE arranged CVD[17] funding for Hull University, starting in October 1970.

By this time, developments elsewhere had influenced RRE's views as to which types of liquid crystal appeared to offer most promise. The initial potential for displays had come from the properties of negative liquid crystals, which turn across an electric field and block light transmission by dynamic scattering. An alternative approach was offered by the discovery of positive or nematic liquid crystals, which in their normal state rotate or twist the polarisation of light, and a device in which a further rotation of polarisation allowed light transmission. When activated by an electric charge such positive liquid crystals cease to rotate light polarisation, with the effect that such a device—known as a twisted nematic cell—will not transmit light.

This team continued its search for suitable materials that were stable and worked over the desired temperature range. Two years of eliminating unsuitable materials eventually, in August 1972, led to the synthesis of cyanobiphenyls. By mixing various of these compounds it was possible to obtain a reasonably wide range of operating temperatures, and the general performance and stability of the materials was good. Realising the commercial potential of this, RRE immediately started to mark consortium documents 'Commercial-in-confidence' and to use codewords to describe the materials, a patent for which was filed on 9 November.[18] Demand for the materials now exceeded Hull's production capability, and so an industrial supplier was chosen. Although lacking any previous experience with liquid crystals, BDH Chemicals was chosen because of other successful work they had done for RRE.[19]

So far, the biphenyl materials provided the desired optical properties and stability, but not the necessary range of operating temperatures. The properties could be varied by combining different biphenyls in eutectic mixtures, and one important advance made by Peter Raynes at Malvern was a theoretical means for predicting the properties of such mixtures based on a knowledge of the component materials.

In November 1973, about a year after accepting the liquid crystal work, BDH announced the commercial availability of four liquid crystals. Of these, one known as E4, with an operating range of 4–61°C, was thought to be the most useful for display devices.[20] Then, in August 1974, a mixture of biphenyls named E7 was created which had properties close to requirements RRE had been given by watch manufacturers. These included an operating range of –9–59°C.

Despite some initial resistance—owing to device manufacturers having invested in processes specific to other materials, and rumours that biphenyls were carcinogenic—by 1977 BDH had become the world's largest liquid crystal producer.[21] Biphenyls produced under licence to BDH, Merck and Hoffman LaRoche were at one point earning the MoD royalties of about £0.5 million per year. In 1979 this success was marked by a Queen's Award for Technological Achievement, which was conferred jointly on RSRE, BDH and Hull University.

Liquid crystal technology can be divided into two types of commercial product: the liquid crystal materials themselves, and devices that use them. Commercialisation of the materials developed under the aegis of CVD has been one of the MoD's greatest, best-publicised success stories, with royalties of some £27 million earned up to 1994.[22] Commercialisation of liquid crystal devices, on the other hand, has been mainly undertaken outside the UK, especially by Japanese companies.

In contrast to its success with the development and production of liquid crystal materials, the UK failed to exploit devices that utilised them. A typical instance is the 'supertwist' display invented by Peter Raynes at RSRE in 1982, and thereafter widely used in laptop computers. GEC was funded to develop prototypes of the technology, but did not take commercial advantage of this pioneering collaboration. Instead, GEC concentrated on collaborating with Malvern on liquid crystal displays for defence use.

With the advent of the DRA, the 1990s have seen another phase in the liquid crystal work at Malvern, as full advantage has been taken of the operating freedom encouraged by the new setup. Most significantly, licensing arrangements were sought with Japanese companies that were exploiting work done by the RSRE in the 1970s, but for which Japanese patents had only recently been established. These licence agreements have generated about another £15 million. Malvern also sponsored key work on active matrix display technology at Strathclyde University, but inadvertent publication meant that this could not be patented.

This licensing activity in Japan led in turn to closer relationships with some of the Japanese display manufacturers, and to DRA Malvern (as it is now called) gaining contracts to carry out research for these companies. Malvern's liquid crystal team has thus grown significantly since the inception of the DRA, providing a UK core of excellence which can both fulfil its role to advise the MoD on relevant technological issues and collaborate with academia and industry. A pragmatic view has been taken that, on the whole, this will not be UK industry, and that significant exploitation of liquid crystal display technology is unlikely to be undertaken by UK companies.

THE ROYAL AIRCRAFT ESTABLISHMENT (RAE) AT FARNBOROUGH AND CARBON FIBRE

Public interest in the exploitation of carbon fibre arose in March 1968, when it became known that the Rolls-Royce RB-211 engine had been chosen by a number of US airlines to power the Lockheed 1011 Tri-jet airliner. The engine's outstanding performance was mainly attributed to the use of Hyfil, a carbon-fibre-reinforced composite material, in its front fan and low-pressure compressor blades. On 17 October 1968, a House of Commons subcommittee took evidence on the question of how best to ensure British exploitation of this indigenous invention. Further hearings were then held, with a substantial report published in February 1969. Prophetically, the report asked: 'How then is the nation to reap the maximum benefit without it becoming yet another British invention to be exploited more successfully or more fully overseas?'[23]

Farnborough's breakthrough in carbon fibre development in 1963 came as others, including Rolls-Royce at Derby and Japanese researchers, were also carrying out similar work. The potentially high strength and stiffness of carbon fibre had already been recognised and, to some extent, realised. Farnborough's advance was to develop a production process which not only achieved much higher strength and stiffness than Shindo had achieved in Japan, but also did so in a much more practical manner than that used by Rolls-Royce.[24]

Although known since the last century (Joseph Swan and Thomas Edison carbonised various fibres in making their electric lamp filament), the high theoretical strength and stiffness of carbon fibre could only be approached with the discovery of a method of manufacturing a highly orientated crystalline fibre.[25] The process developed at Farnborough was patented in the names of Johnson, Phillips and Watt as British Patent 1,110,791 (published 24 April 1968), and involved the controlled carbonisation of polyacrylonitrile (PAN) manmade fibre.

Obtaining the benefits of carbon fibre's properties requires their incorporation into a composite material, forming a matrix with materials such as a synthetic resin, a metal, ceramic or glass. In the late 1960s the composite that appeared to offer the highest specific strength and specific stiffness for practical applications was carbon-fibre-reinforced plastic (CFRP). This was seen as superseding existing composite materials, particularly glass-reinforced plastics (GRP), for some applications.

The Farnborough inventors bridged this disciplinary divide, as Leslie Phillips worked in the Plastics Technology Section. The main research effort on developing carbon fibres, however, was done by a small team comprising Bill Watt, Bill Johnson and Roger Moreton of the High Temperature Materials Laboratory (Fig. 13.2). The key contribution credited to Phillips at this stage was his suggestion that Orlon

FIGURE 13.2: CARBON FIBRE TESTING AT FARNBOROUGH (© CROWN COPYRIGHT, DERA)

(a proprietary polyacrilonitrile fibre) be used as the precursor in carbon fibre production.[26]

This search for a way of making a stiff, strong fibre was motivated by a general view that such properties would be useful in aerospace applications, rather than by any specific military requirements.[27] As such properties had already been observed in very short graphite 'whiskers', the potential of carbon fibres seemed clear.

Finding a suitable precursor fibre was only part of the Farnborough work. Two other crucial aspects were specified in the first Farnborough patent. The precursor needed to be preoxidised before being carbonised in a furnace, and this needed to be done at tension to prevent shrinkage (Fig. 13.3). Initial batches were wrapped around frames made out of Meccano. The encouraging properties of the carbon fibre thus produced led to a scaling up of production, from about 5 grams each batch to 50 grams (with an immediate benefit to the local Farnborough toyshop).[28]

In accordance with the normal practice of the times, the exploitation of carbon fibre was handled by the NRDC. NRDC established licences with three firms that already had considerable contact with the work at RAE: Courtaulds, Morgan Crucible and Rolls-Royce. The Atomic Energy Research Establishment at Harwell,

FIGURE 13.3: INTERIOR VIEW OF AN OXIDISING OVEN USED IN THE EARLY DEVELOPMENT OF CARBON FIBRE AT THE ROYAL AIRCRAFT ESTABLISHMENT (RAE), FARNBOROUGH. THE OVEN CONTAINS POLYACRILONITRILE IN VARIOUS STAGES OF OXIDATION. (SCIENCE MUSEUM INV. NO. 1985–1349, SCIENCE MUSEUM/SSPL)

which had experience in high-temperature graphite technology, was also brought in to develop larger-scale production than was possible at RAE.

Farnborough's breakthrough had come as Roger Prescott at Rolls-Royce was separately working on ways of making carbon fibres from a polyacrilonitrile precursor. By the end of 1964 Rolls-Royce was making usable carbon fibre, albeit in very small quantities. At the same time, Rolls-Royce was planning to try to break into the US market for large aeroengines, and carbon fibre was seen as offering the kind of technical edge to facilitate this. The company had already been using glass fibre for gas turbine blades, and it decided to 'go all out to make carbon fibre reinforced blades on the basis that we knew that glass fibre reinforced blades were a practical proposition from the previous eight years' experience'.[29]

Rolls-Royce's own process was used to make carbon fibre which was incorporated into compressor blades used in their Conway engine, which was used in the BOAC VC-10 in the late 1960s, accumulating as much as 5000 hours' flying time each. However, once Farnborough's superior process had been patented, Rolls-Royce dropped its own approach and licensed the RAE's production method.

Farnborough's approach differed in two main ways from that developed at Rolls-Royce.[30] First, although Prescott had discovered the importance of stretching to achieve molecular orientation, he did not use the preoxidation stage that stabilised this orientation prior to carbonisation. Second, Farnborough was using one-and-a-half-denier special acrylic material supplied by Courtaulds, which could be processed much faster than the larger-diameter precursor used by Rolls-Royce. In practice, these differences meant that Rolls-Royce was taking a much longer time to make batches of carbon fibre which were probably of lower quality.

Whereas the other two UK licensees of the technology were concerned with the manufacture of carbon fibres for sale, Rolls-Royce was only interested in making a composite material for use in aerospace components for its engines. 'It became apparent during 1966 that the only way the Company could meet the exacting material requirements of its engineers, within the requisite time scale and the necessary degree of commercial security, was to develop and build its own plant. Courtaulds supplied the crucial precursor material in a special form to meet the Rolls-Royce specification'.[31]

Over the period 1967–69 Rolls-Royce developed its carbon-fibre production facility from a batch process capable of making 1 tonne per year to a continuous-process plant with a capacity of 25 tonnes per year.[32] However, Rolls-Royce's expectations for the use of carbon fibre in the RB-211 were dashed by the results of the 'dead chicken' impact tests. The engine's prototype carbon-fibre compressor blades were simply not robust enough to survive bird hits, and the material had to be discarded for this application. What had seemed a great British technological triumph in 1968 quickly became a disaster as Rolls-Royce's ambitious schedule for the RB-211 was endangered. Indeed, Rolls-Royce's problems became so great that in 1971 the company was eventually nationalised by the Conservative government of Edward Heath.

The outcome of all this was that the RB-211 used titanium instead of carbon fibre. Nevertheless, it was an inherently good design in other ways, and was successful in enabling Rolls-Royce to achieve its strategic objective of breaking into the American large airliner business. As part of the rationalisation that came with nationalisation of the company, the carbon fibre production plant was sold off to form Bristol Composites.

Despite withdrawing from carbon fibre production, Rolls-Royce continued to be interested in the use of carbon fibre in its aeroengines. The first result of this was the use in 1979 of carbon fibre for aeroengine pod doors. The choice of Japanese-sourced carbon fibre for this—formed into a preimpregnated sheet by Ciba-Geigy at Cambridge—meant the loss of a prestigious customer for the remaining large UK producer of carbon fibre, Courtaulds.

In 1963 Courtaulds had been asked by RAE to supply organic fibres suitable for conversion to high-quality carbon fibres, and started to supply an acrylic fibre manufactured specifically to meet RAE's requirement. Rolls-Royce's requirements for the RB-211 engine led Courtaulds to build a plant, which subsequently also supplied some of this precursor to Morganite, with some being used internally. In 1965 Courtaulds also started to convert its own special acrylic fibre into carbon fibres using the Farnborough process. From this Courtaulds developed its own conversion process and by the end of 1968 had filed 52 patent applications in the UK 'on the production of special acrylic fibre, on the production of carbon fibres and on the use of carbon fibre'.[33]

By 1968 Courtaulds was producing carbon fibres in a plant running 24 hours a day, and selling them on the open market in the UK and USA. On 3 February 1969 an agreement was announced between Courtaulds Ltd and Hercules Inc. of Wilmington, Delaware, whereby 'Hercules is granted the exclusive right to sell "Grafil", Courtaulds carbon fibre, in the USA and an option to manufacture there at a later date'.[34] Hercules was a pioneer in the application of fibre technology in the aerospace industry. The agreement provided for the two-way exchange of information if the option to manufacture was taken up.

Courtaulds continued to manufacture carbon fibre throughout the 1970s and 1980s, developing two main types of market: aerospace and sports equipment (such as golf-club shafts). During this period Courtaulds made many improvements in the production process, significantly increasing the speed of production (one of the main reasons why carbon fibre is expensive), as well as the quality of the product. In the latter activity, Farnborough's expertise was extremely helpful. One particular example is typical of the way that Farnborough was able to provide complementary expertise to assist industry. Courtaulds suddenly found that the quality of its carbon fibre had dropped significantly and, unable to detect the reason, sent a sample to Farnborough. The analytical techniques developed there were able to find the flaws causing the drop in quality and to identify their chemical composition.

Despite this close collaboration between Farnborough and Courtaulds, the feeling persisted that the quality improvements that stemmed from the work during the 1970s of Roger Moreton and others at the RAE were not taken full advantage of. In part, this seems to have stemmed from the fact that the PAN precursor was Courtelle and, as the manufacturer of this material, it was not easy for Courtaulds (and for people associated with the development of Courtelle) to accept criticism of its quality as regards contamination with impurities.

Courtaulds' position as a manufacturer of the precursor material thus proved both a strength and a weakness. Low cost and ready availability were potential advantages, but the water-based inorganic process used to make Courtelle was susceptible to impurities which did not affect the organic process used by other carbon fibre manufacturers. Courtaulds was also considered by some to be unresponsive to customer requirements. For example, overseas companies were prepared to provide bobbins of carbon fibre of a precise length, something composite manufacturers considered important. A crucial failure in this regard was Courtaulds' unwillingness to provide a sample to Rolls-Royce because the type of carbon fibre requested was not one that Courtaulds' plant was set up to produce. Courtaulds considered it uneconomic to set up a production run simply to make the sample for Rolls-Royce, but the outcome was that Rolls-Royce became instead an important user of Japanese-sourced carbon fibre.

Courtaulds' exit from carbon fibre manufacture was largely due to management decisions, which first greatly expanded investment in carbon fibre production during the 1980s (including building a production plant in California), and then lost faith when insufficient returns were generated. Courtaulds decided to pull out of carbon fibre production in 1991, though ironically, the one surviving UK-based carbon fibre manufacturer continues to thrive making fibre based on Courtaulds' precursor. RK Carbon Fibres Ltd, based in Inverness (now owned by the German SGL Carbon group), has concentrated on producing carbon fibre for low-grade industrial applications, and thus does not need to compete at the high quality levels reached by overseas manufacturers. Although the process used is essentially that developed at Farnborough, the company has had little other direct benefit from Farnborough's expertise (except indirectly: ex-Courtaulds' staff recruited by RK Carbon had worked closely with Farnborough).

After a gap during the 1980s, when work at Farnborough was limited to trying to improve the interfacial bonding between the fibres and the resin, and thus composite properties, interest in carbon fibres themselves was revived in the early 1990s. The increasing use of carbon fibre composites as structural materials in aerospace applications led to a desire for improved compression strength. This could be done either by improving the properties of the resin used to make the composite, or by increasing the stability of the fibres. The obvious way to improve fibre stability was to increase diameter, but this was limited in practice by the diffusion time required to process the fibres. The solution was to achieve the same-diameter fibres with a hollow centre, and to do this Farnborough has worked with Strathclyde University on the production of a hollow PAN precursor fibre. When Courtaulds decided to pull out of carbon fibre production, Farnborough acquired one of Courtaulds' processing lines and again became the main focus of carbon-fibre work in the UK. However, the lack of UK carbon fibre manufacturers raises doubts about the industrial development of Farnborough's hollow-fibre work, despite the setting up of a Structural Materials Dual Use Technology Centre (DUTC) on 12 April 1994.

MALVERN AND INFORMATION TECHNOLOGY: THE CASE OF HIGH-INTEGRITY COMPUTING

The work of the DREs, both in-house and extramural, has maintained a core of UK technological expertise in many fields. One aspect of this is that the DREs—especially the former RSRE—have acted as independent arbiters of collaborative projects involving a number of firms (and also often academia). Thus RSRE staff were prominent in the Alvey programme, and have coordinated several industry consortia geared towards precompetitive research (for example the gallium arsenide consortium).[35] Although much of this work may have little direct military orientation, defence sponsorship can play a major role in the setting of standards in new technologies.

The area described here is that of the integrity of computer systems, especially as regards the reliability of software used in safety- and security-critical applications. Malvern's work on systems integrity has found military application in the Eurofighter 2000, the ship helicopter operating limits instrumentation system (SHOLIS), and the EH101 helicopter, as well as a number of secure information systems. It has also, however, played a major role in sponsoring work and setting standards in the field, contributing both to the development of the field in academia and to its application in civil uses.

Concerns about the integrity of computer systems—especially in the context of maintaining the security of information held by networked systems—first began to gain attention in the UK in the mid-1970s. A second concern—ensuring that systems reliant on computers were safe—arose as more and more weapons systems came to be computer controlled.

Malvern's work on high-integrity computing falls into two main categories, both of which have increasingly involved championing the use of formal methods—the attempt to find precise mathematical representations for software. So far, verification of software—to demonstrate that it does what it is supposed to do—has been most influential, with the development of tools for static analysis. Second, RSRE has had a long record of work on well-structured languages, and more recently on attempts to enable formal specification of software. RSRE was also a central actor in the initial formulation of Def-Stan 00-55, a standard which seeks to set out guidelines for the use of both these approaches, so as to ensure high-integrity software in safety- and security-critical applications.

Computer systems are usually tested dynamically by running them in their intended manner of use. Bugs may then be discovered and corrected, although—as personal computer users are all too well aware—much debugging of commercial products is in effect done by the consumer after the computer system has been bought. Most of these bugs are caused by software errors, although hardware errors do also occur—most famously that discovered in 1994 in Intel's new Pentium chip—and RSRE has also been at the forefront of the use of formal methods in chip verification.[36]

Static analysis is one solution to the problem of not being able to test software in all its possible states of use. Two main approaches are involved: flow analysis looks at the structure of the program in terms of data and information flows, checking, for example, that variables are set before they are used; semantic analysis seeks to provide a deeper form of analysis by looking at the transformation of logical expressions by the program, and potentially allowing proof of correctness.

Work on static analysis at RSRE was started by Bob Phillips, superintendent of one of the computing divisions, and championed thereafter most strongly by John Cullyer. Initially, the work was 'sold' as being of use to security-critical applications, and RSRE's main outside links were with the Communications Electronic Security Group (CESG) of GCHQ. Cullyer's enthusiasm for these software analysis techniques led him to seek their dissemination to a wider industrial audience.

Some of the work on static analysis was being carried out under extramural funding by Bernard Carré at the University of Southampton, and, with support from RSRE, he started to run training courses in static analysis in the late 1970s. This had two important, though not entirely planned, consequences. First, the process of running the courses led the techniques developed at Southampton to be refined through use, 'so they actually brought these tools up to a much more production status than would have happened if it hadn't been for the course'.[37] Second, the courses fostered a community of knowledgeable users (for example within British Aerospace), who would thereafter make use of the technology.

A powerful demonstration of the utility of static analysis was provided when these techniques were used between 1982 and 1985 to check existing military avionic software. A significant number of deviations from specification were discovered, including some 'which would have had direct and observable effects on an aircraft's performance'.[38]

The interest generated in industry by Carré's courses led to demand for the static analysis tools to be made commercially available. This led Carré to form a company, Program Validation Limited, in 1983 to market the analysis tools as SPADE (Southampton Program Analysis and Development Environment), as well as continuing to run the courses. At the same time, however, RSRE had been continuing its in-house work on static analysis—along broadly similar lines to the work at Southampton, but with specific differences in, for example, the graph-reduction algorithms used. As it seemed by no means certain that the Southampton commercial venture would go ahead, or be successful if it did, it was decided that the RSRE work should also be made available for commercial exploitation. Known as MALPAS (Malvern Program Analysis Suite), this was seen as one of the few successes of the ill-fated defence technology enterprises.[39]

SPADE and MALPAS are successful commercial products which are the direct results of work at, and sponsored by, RSRE. Although the static analysis method they use is seen as a peculiarly British approach, which has not been widely taken up elsewhere, some export sales have also been achieved. At the heart of this modest success has been the drive and enthusiasm of product champions such as Cullyer and Carré. Significantly, what they did was not simply to develop a product they believed in, but also to create a market for this product, through the RSRE-supported static analysis courses run at Southampton University. A classic failing of many inventions is that their very newness means they do not fit into existing social and technological frameworks. Here, although perhaps inadvertently, these courses helped create a constituency of users for the technology, ensuring that it was not too far ahead of its time commercially.

However, the ease of static analysis and the credibility of its results is generally thought to depend to some extent on the language used to write the software being analysed.[40] Well structured languages are considered more amenable to static analysis. Indeed, concern over the nature of programming languages has been at the heart of RSRE's work on computer systems since the earliest such systems were developed. It is therefore not surprising that much of RSRE's computing efforts over the years have gone into the rationalisation of programming languages. Three phases of this are of particular interest. First, right from the start of the computer age RSRE was concerned with the promotion of what were seen (at the establishment) as well structured languages, notably ALGOL-68 and CORAL-66. A second phase grew out of the work on static analysis, with the development of an ADA subset, SPARK, specifically geared towards enhancing the credibility of such analysis. Finally, RSRE has also promoted the use of the Z formal specification language.

The first phase of RSRE's involvement in languages was marked by a characteristic mixture of technical achievement coupled with limited impact outside of defence. Initially RSRE was working with its own 'home-built' computers, with what was then the Mathematics department providing a computing service to other RSRE scientists. Only when RSRE started to use commercial computers—the first was an ICL 1902—did its work on languages begin to have wider influence.

Most influential was RSRE's development of the first compiler for ALGOL-68. This was taken up and distributed by ICL, and thus became widely used in UK universities. Ultimately, however, there was too little penetration into the commercial world, and ALGOL would lose out to the dominance of C, a language considered at RSRE to be much more of a compromise.

A similar fate would also await CORAL, the first UK high-level programming language designed specifically for computer systems embedded in military (and other) equipment and functioning in real time. Although high-level languages became widespread in scientific and commercial computing in the late 1950s and early 1960s, real-time computers—much more limited in their capacity, and also in the tasks they performed—were still typically programmed in lower-level assembly language. During the 1960s, however, both the power of real-time computers and the complexity of their computational tasks, increased considerably. Interest grew in developing a high-level language specifically designed for programming such systems. Following the development of such a language—JOVIAL—in the USA, RSRE set to work developing a language for the UK: CORAL.

CORAL was based mainly on compiler technology developed at RSRE. Its acceptance as a Ministry of Defence standard was steered by Philip Woodward, then head of RSRE's Mathematics Division. Again, however, although much military software was written using CORAL, and still remains in use, industry's preference was to revert to using C. CORAL nevertheless provided the crucial link to the second phase of work on languages, in which concerns went beyond simply having a well structured language to a specific focus on languages amenable to verification techniques, such as static analysis. The static analysis work involved working with intermediate languages developed at RSRE, and it was also possible to translate directly from assembly code (widely used for many small safety-critical military applications) to these intermediate languages.

The most recent development in programming language prompted by RSRE is a subset of ADA—the language chosen as the standard for use in US and UK defence procurement—known as SPARK. SPARK was a direct consequence of the work on static analysis supported by RSRE. Given the belief (at least at RSRE) that the credibility of the results of static analysis depended to some extent on the programming language used to write the software, it was decided that a Pascal subset of ADA would be desirable for safety-critical applications. Extramural funding from RSRE (and initially GCHQ) again supported this work, which was done by Carré and Program Validation Limited.

Finally, RSRE has also sponsored work on the practical implementation of a formal specification language known as Z. Such languages are seen as providing a mathematically rigorous way of writing down what the software is supposed to do, which can then be translated into a programming language, such as SPARK. RSRE has 'tried to support the standardisation of Z', including working with ICL on tools for verifying the compliance of a SPARK program to a Z specification.[41]

RSRE has thus been at the heart of many UK developments in safety- and security-critical computing. Static analysis tools have proved a modest commercial success, finding some applications outside defence (albeit often in similar sectors, such as nuclear power). Despite arguments in favour of wider application, formal methods remains a niche approach to software engineering, seen by many as costly and overelaborate. Despite its apparent generic nature, it thus remains a typical 'gold-plated' defence technology, being seen as overspecified and overexpensive for many civil uses.

It was for this reason that RSRE's attempt to standardise high-integrity software engineering caused alarm in the industry. As the driving force behind Def-Stan 00-55, RSRE played a major role in the development of what was seen as a potentially influential standard for the production of safety- and security-critical software:

> 00-55 is more than just a defence standard. It is a regulatory framework which may produce a sea-change in the software engineering field over the next decade.[42]

Although far from being the only player, RSRE was one of the major influences in initiating and developing Def-Stan 00-55. As procurement policy shifted during the 1980s towards a more 'hands-off' approach, the argument was made that it would be desirable to have a standard that contractors would need to meet in writing software for critical defence systems.

Def-Stan 00-55 differed from other similar standards for safety-critical computing in the extent to which it required the use of 'formal methods', both in specification and verification of the software, in addition to the usual software engineering good practice. This emphasis on formal mathematical techniques resulted in considerable opposition to 00-55 within the UK software industry. In essence, this opposition stemmed from concern that 00-55 was too demanding, requiring skills which were not widely available (most software engineers do not have mathematical backgrounds), and was likely to increase software development times and costs unnecessarily.

Industry's fierce criticism of earlier drafts of 00-55 has led to some compromise over its demands. Its influence has not been as great as some hoped (or feared), as the civil uptake of high-integrity computing approaches remains very limited.[43] As in many areas of information technology, RSRE's influence has been considerable in the development of high integrity computing in the UK without necessarily having major commercial significance.[44]

CONCLUSIONS: DEFENCE R&D, INNOVATION AND THE ECONOMY

It is generally believed that innovation is a key to economic competitiveness, and that this in part depends on the extent to which scientific and technical developments are commercialised. Defence R&D—at or in conjunction with the DREs—has been at the heart of post-1945 UK science and technology. The importance of the DREs stems not just from the scale of defence R&D funding, but also from the state-of-the-art expertise of the DREs in many areas of high technology. However, the UK as a whole—with some honourable exceptions—appears not to have made the most commercial and economic gain from its scientific and technical excellence.

The examples detailed above—and many others described elsewhere in this book—convey something of the breadth and depth of defence-sponsored work on science and technology in the UK. Most of this work, not surprisingly, has had defence missions as its focus, with limited potential for civil exploitation. Technology transfer into industry has been widespread, but almost exclusively to the defence industry and of technology geared towards military applications. Inevitably, the social arrangements that have favoured successful technology transfer have also meant that many of the recipients—orientated towards the particular requirements of the defence market—have not been well placed to carry out the significant amounts of further innovatory activities required for success in civil markets.

Where there has been significant potential for spinoff into the civil market, success has depended on the wider range of factors that have shaped the innovatory performance of UK firms in the postwar period. This is strongly exemplified by the story of carbon fibre, where Farnborough's work was successfully taken up by UK industry, but ultimately market dominance was achieved by overseas firms.

A similar tale of mixed commercial success resulted from the work done at the Services Electronics Research Laboratory (SERL) at Baldock, and later at Malvern, on gallium semiconductor materials. The very successful development by English Electric Valve Company of image-intensifying gallium-arsenide photocathodes, for which EEV and Malvern were jointly awarded a Queens Award for Technological Achievement in 1987, was a short-lived success, as EEV later ceased manufacturing the photocathodes. A more sustained success was the exploitation of semiconductor crystal-growth technologies by Metals Research Ltd (which later became Cambridge Instruments Ltd), which made the UK a major supplier of crystal-pulling machines (Fig. 13.4). On the other hand, however, the original work at Baldock on light-emitting diodes (LEDs) did not lead to any UK firm capturing a share of the civil LED market.

Overall, exploitation of the gallium semiconductor work done by the DREs points up the same limitations as in the liquid crystals case. Whereas UK firms continued to do well in chemicals (such as the liquid crystal materials) and in developing electronic products for niche markets (not just defence, but also, for example, 'high-

FIGURE 13.4: CZOCHRALSKI CRYSTAL PULLER (© CROWN COPYRIGHT, DERA)

end' audio equipment), there has been a conspicuous lack of success in the production of mass-market semiconductor devices in the UK. The British electronics giants (now either taken over by GEC or under overseas ownership) became expert in manufacturing for the defence market, both domestic and export. They did not, however, put the same investment into large production plants for LCDs or for computer chips. Where such plants do exist in the UK, they are now almost entirely the result of inward investment by firms from Japan, the USA, Korea and elsewhere. It has been argued that military funding of UK semiconductor R&D undermined its commercial potential.[45] What does seem clear is that a secure domestic government market in the form of the MoD limited the incentive for UK electronics companies to invest in mass-market commercialisation.

The harsh truth is that most inventions, from whatever source, do not lead to commercially successful innovation. Most patented inventions are not licensed at all, and of those that are, only a small proportion make a significant return on investment. It may be true that 'There is absolutely top-class research, world-winning research [at Malvern], not being fully exploited in the United Kingdom.'[46] It does not necessarily follow that all such research can, or should be, exploited by UK firms. In many cases, as in Malvern's current collaboration with Japanese firms on liquid crystals, it may be more realistic to work with overseas firms that are able to make the necessary investment to exploit the technology.

Nor should 'inventions' be seen as the only way in which the DREs can benefit UK industrial performance. Sharing of facilities has obviously been very important in many instances, notably for the major aerospace companies Rolls-Royce and British Aerospace.[47] Perhaps the most important example of this is the access that the British aerospace industry has had to wind tunnels and other facilities and expertise at the RAE, Farnborough, and other related establishments. The continuing success of UK companies in civil aviation manufacturing may be in part due to this assistance.

In evidence to the House of Lords Select Committee on Science and Technology, British Aerospace and Rolls-Royce noted the importance that several defence establishment facilities had had to the UK aerospace industry. These were referred to as 'unique national facilities which a company would not want, in fact, could not afford, to replace or to duplicate'.[48] In the case of Rolls-Royce these are mainly the facilities of the former National Gas Turbine Establishment at Pyestock, and included a jet-noise facility which was the only one of its type in Europe, an altitude test facility, and other facilities geared towards compressors, combustion and turbines.[49] British Aerospace had benefited most from access to wind tunnels:

> The Defence Research Agency owns a number of quite unique wind tunnels, unique in so far as the United Kingdom is concerned and unique in some cases as far as Europe is concerned. We make sporadic use of these facilities, and I think that that is a reflection of the way in which computational methods are overtaking wind tunnel test methods. That is not to say that we should do away with the need for wind tunnels, but there is less use being made of wind tunnels these days and they are used more for confirmatory testing of the results of the computations. They are nevertheless still essential to the whole process of designing commercially viable and defence capable aircraft and we certainly see the continued availability of the Defence Research Agency wind tunnels as being very important for our future competitivity.[50]

Malvern has provided similar infrastructural assistance in other areas, such as semiconductor research by, for example, supplying samples of materials to companies before they were widely available. Probably even more important has been the opportunity for industry to access a broad range of relevant knowledge, either through working on extramural contracts or by participating in consortia—such as that on gallium arsenide run by Malvern.[51] The establishment of the dual-use technology centres continues this approach, although some now worry that DERA's financial motivation could make them less of a neutral collaborator and more of a competitor.

Further improvements to industry's exploitation of DERA's technological resources rest on overcoming a critical paradox—that the strongest links, and therefore the best potential for technology transfer, is between the DREs and traditional defence companies, which have proved notoriously bad at, or uninterested in, civil exploitation.

1. For figures on the proportion of UK government R&D funded for defence, see POST, *Relationships Between Defence and Civil Science and Technology* (May 1991): 3, 9. For an account of the UK's defence orientation, see D. Edgerton, 'Liberal militarism and the British state', *New Left Review*, 185 (1991): 138–69.
2. See, above all, P.L. Galison, *Image and Logic: a material culture of microphysics* (University of Chicago Press, 1997); and, also, the American contributions in *National Military Establishments and the Advancement of Science and Technology*, edited by P. Forman and J.M. Sánchez-Ron (Kluwer Academic Publishers, 1996).
3. For a number of examples of 'spin-off' from the former RSRE at Malvern, see J.F. Barnes and B.R. Holman, 'The transfer of defence research on electronic materials to the civil field', in *Technology in the 1990s: the promise of advanced materials*, edited by E. Mondros and A. Kelly, *Philosophical Transactions of the Royal Society of London* 322 (1987): 335–46.
4. On the origins of the NRDC, see S.T. Keith, 'Inventions, patents and commercial development from governmentally financed research in Great Britain: the origins of the National Research Development Corporation', *Minerva*, 19, no. 1 (1981): 93–122.
5. See 'Plastic structures', *NRDC Bulletin*, 2 (July 1954): 12–15.
6. RSRE and Metals Research Ltd received a joint Queen's Award for Technological Achievement in 1979 for the work on the Malvern crystal growth systems. See also 'Electronic materials', *NRDC Bulletin*, 37 (April 1971): 19–20; G.R. Blackwell, 'Green light for LEDs', *NRDC Bulletin*, 43 (Autumn 1975): 3–8; A.G. Woodruff, 'The Malvern correlator', *NRDC Bulletin*, 42 (Spring 1975): 30–7; N.A. Ratcliff, 'White-hot technology of boron nitride', *NRDC Bulletin*, 46 (Summer 1977): 7–9.
7. Ministry of Technology, *The Ministry of Technology* (HMSO, 1967).
8. Initially MinTech had responsibility for the Department of Scientific and Industrial Research (DSIR) research stations, the NRDC, and the Atomic Energy Authority, which included the establishments at Harwell and Aldermaston. Initial technological sponsorship by MinTech focused on four industries: computers, electronics, telecommunications, and machine tools.
9. 'Electronic materials' (see n. 6 above).
10. See R. Coopey, 'Restructuring civil and military science and technology: the Ministry of Technology in the 1960s', in *Defence Science and Technology*, edited by R. Coopey, M. Uttley and G. Spinardi (Harwood, 1993): 65–84.
11. The agreement between MoD and DTE was not renewed when it came up for review in 1991, but DTE continues to operate in exploiting the existing licence arrangements.
12. A fuller account of the DTE episode is G. Spinardi, 'Defence Technology Enterprises: a case study in technology transfer', *Science and Public Policy*, 19 (August 1992): 198–206.
13. This description is drawn from E.P. Raynes, 'Cyanobiphenyl liquid crystals: an MoD success story', *Journal of Naval Science*, 7, no. 1 (January 1981): 40–6.
14. This account is based on C. Hilsum, 'The anatomy of a discovery: biphenyl liquid crystals', *TS* (6 March 1984): 2. I am grateful to Professor Damien McDonnell of DERA for providing me with a copy of this.
15. C. Hilsum, 'The anatomy of a discovery' (n. 14 above): 5.
16. C. Hilsum, 'The anatomy of a discovery' (n. 14 above): 3–4.
17. CVD was instigated by the Admiralty in 1938 to enhance the development of valves for military communication uses through collaboration with industry. For the early part of its history, CVD stood for Coordination of Valve Development, but this was later revised to Components, Valves and Devices.
18. C. Hilsum, 'The anatomy of a discovery' (n. 14 above): 11.
19. C. Hilsum, 'The anatomy of a discovery' (n. 14 above): 11.
20. C. Hilsum, 'The anatomy of a discovery' (n. 14 above): 13.
21. C. Hilsum, 'The anatomy of a discovery' (n. 14 above): 15.
22. Deputy Under Secretary of State Michael Bell, *Minutes of Evidence Taken Before the Select Committee on Science and Technology* (Sub-committee II Defence Research Agency) House of Lords, Session 1993–94, House of Lords Paper 24-II, par. 61 (24 February 1994): 18.
23. House of Commons Select Committee on Science and Technology, *Carbon Fibres Report*, 157, par. 36 (1969): xiii.
24. R. Moreton, W. Watt and W. Johnson, 'Carbon fibres of high strength and high breaking strain', *Nature*, 213 (18 February 1967): 690–1.
25. House of Commons Select Committee on Science and Technology, par. 1, v (n. 23 above).
26. W. Watt, *Biographical Memoirs of Fellows of the Royal Society*, 33 (1987): 651.
27. Interview with R. Moreton, 6 February 1997.
28. Interview with R. Moreton (n. 27 above).
29. House of Commons Select Committee on Science and Technology, par. 20, ix (n. 23 above).
30. Interview with J. Johnson, 8 April 1997.
31. R.J. Moore, *Memorandum by Rolls-Royce Ltd*, House of Commons, 157, par. 7 (1969): 32.

32. D.J. Thorne, 'Manufacture of carbon fibre from PAN', in *Strong Fibres, Vol. 1: Handbook of Composites*, edited by W. Watt and B. V. Perov (Elsevier Science Publishers, 1985): 475–94, at 476.
33. R.J. Moore, *Memorandum by Rolls-Royce Ltd* (n. 31 above): 43.
34. Press notice, House of Commons, 157 (1969): 112.
35. D.T.J. Hurle, 'National collaborative research into gallium arsenide', *Electronic Engineering*, 57 (June 1985): 149–53.
36. RSRE's most famous (some might say infamous) work in hardware verification was the VIPER chip, where doubts about the degree to which the VIPER chip in question was actually formally proven, and claims that it had been 'oversold', made this programme controversial. See D. MacKenzie, 'The fangs of the VIPER', *Nature*, 352 (1994): 467–8.
37. Interview with C.T. Sennett, 15 August 1996.
38. R. Dettmer, 'Making software safer', *IEE Review* (September 1988).
39. 'Case history: the licensing of the MALPAS verification and validation software', *DTE Spotlight*, 1, no. 1 (February 1986): 5–6.
40. There is apparently a split here between the companies which market SPADE and MALPAS, with the former (in agreement with RSRE) supporting the view that only programs written in well-structured languages can be analysed with any credibility, while the latter takes the view that benefits can be gained from using static analysis on any programs.
41. Interview with Sennett (n. 37 above).
42. Margaret Tierney, 'The Evolution of Def Stan 00-55: A socio-history of a design standard for safety-critical software', in Social Dimensions of Systems Engineering, edited by Paul Quintas (London, Ellis Horwood, 1993): 111–43.
43. G. Cleland and D. MacKenzie, 'The industrial uptake of formal methods in computer science: an analysis and a policy proposal', *Science and Public Policy*, 22, no. 6 (December 1995): 369–82.
44. One project involving DTI is building on this work on 'computer security, where there is top-class work taking place both at Malvern and indeed at GCHQ, where the DTI recognised there was a need to make computer security skills available into, for example, the banking sectors, and hence thought it would make sense to work jointly with what was then RSRE, and an important initiative emerged', House of Lords Paper 24-II (n. 22 above): 35.
45. K. Dickson, 'The influence of Ministry of Defence funding on semiconductor research and development in the United Kingdom', *Research Policy*, 12 (1983): 113–20.
46. R. Foster, *DTI*, House of Lords Paper 24-II, par. 127 (n. 22 above): 35.
47. T.H. Kerr, 'The role of the Research Establishments in the developing world of aerospace', *Aeronautical Journal*, 86 (1982): 359–69. Interestingly, the strong links between military and civil aerospace technology have not been mirrored in naval technology.
48. S. Miller, Director of Engineering and Technology, Rolls-Royce plc., *Minutes of Evidence Taken Before the Select Committee on Science and Technology* (Sub-committee II Defence Research Agency) House of Lords, Session 1993–94, House of Lords Paper 24-vi, par. 372 (24 February 1994): 110.
49. P.C. Ruffles, Director of Engineering, Rolls-Royce Aerospace Group, *Minutes of Evidence Taken Before the Select Committee on Science and Technology* (n. 48 above). See also the MoD memorandum: 231–2.
50. M. Peters, Director of Technical Planning, British Aerospace plc., *Minutes of Evidence Taken Before the Select Committee on Science and Technology*, par. 377 (n. 48 above): 111. He was also concerned that the DRA's more 'commercial approach might have implications for the long-term availability of major test facilities. Wind tunnels in particular are a cause of some concern. We are worried that perhaps short-term commercial pressures might lead to the closure of some of these facilities and we believe that this would be a considerable loss to the nation as a whole,' 107, par. 359: 107.
51. D.T.J. Hurle, 'National collaborative research into gallium arsenide' (n. 35 above).

CHAPTER 14

Government Management of Defence Research Since the Second World War

STEPHEN ROBINSON

INTRODUCTION

Although the roots of defence research go back as far as war itself, the period since 1945 is one of special interest. Not only was a huge research platform built up during the Second World War to exploit the flowering of physical science, but thanks to the Cold War, there has subsequently been a remarkably coherent evolution of a large military research programme. The study of such a long period of sustained effort is likely to throw light on some of the perennial problems of applied research management, notably on the direction of research investments, on quality in research programmes, on the effective utilisation of new knowledge, and (in the case of defence) on the civil impact of military innovation.

The last item immediately throws up intriguing questions. It is almost a cliché that defence companies have difficulties in civil markets, and that the 'only civil product to come out of the space race is the non-stick frying pan'. But who can ignore the impact of atomic energy; gas turbines, rockets and space; computers, communications and the Internet; and surveillance, signal processing and imaging; and doubt the overwhelming influence of defence research on society in recent times. The contradiction relates to the great difference between the nature of knowledge and of its application. The link between research and innovation is far from straightforward, and the fact that knowledge usually costs less to acquire than to apply does not mean that it can be acquired more quickly. The source of knowledge is often lost in the mists of time, but its application, when needed, can rarely be fast enough. Thus, like so many ants, scientists build the knowledge base bit by bit, while automated design and production facilities deliver complex artefacts to the public with a product lifecycle as short as three months.

Research presents interesting and extreme management challenges. Like all exploration it is a lonely, self-driven and obsessive activity which benefits enormously from urgency and clear objectives. Research is competitive, but whereas cross-fertilisation of ideas between colleagues is important, direct duplication is anathema. Although obviously labour intensive, the capital needs of research are high and its measuring instruments quickly become outdated. With such characteristics it is not surprising that in research, as in the arts, 'directors' and 'leaders', rather than 'managers', preside and, much to the chagrin of those wanting immediate solutions, many scientists scorn the title 'research manager' as a contradiction in terms.

Opportunities for exploiting science are normally identified by key individuals working at the leading edge, and many defence establishments grew up *ad hoc* during and just after the war as technology, with obvious application in key military fields, was identified. Thus in 1945 there existed, among others, separate laboratories for aircraft, gas turbines, rockets, air defence, communications, military vehicles, bridge building, guns, gun laying, ships, ships' weapons, submarines, atomic weapons, chemical weapons, biological weapons and electronic components.

These establishments, with origins within the Armed Services and numerous government departments, were typically commanded by military officers and administered by the civil service. They were enlarged and strengthened during the war, notably by university graduates drafted in from the Scientific and Technical Register, which had been established by Sir Henry Tizard and Dr John Cockcroft in 1939. United by a common aim, the organisation proved to be well suited to effective research and development, and the momentum injected then has hardly been dissipated by the friction of intervening years.

Thus the history of the management of the defence research programme since 1945 inevitably revolves round the coordination and rationalisation of a large and disparate set of establishments. Bearing in mind the number of government departments and civil service staff classes involved, this rationalisation (especially in the 1960s) has been a considerable management achievement in its own right. The research programme has therefore been conducted against a background of rationalisation and evolving government administration policy, as well as emerging technologies and the changing military priorities of the Cold War. This chapter will address defence research management from three points of view. First, the administration of the establishments will be discussed, tracing the development and evolution of staffing and financial management policies, from military-style 'Units' of the war to the commercialised 'Agencies' of today. Second, research programming methodology will be examined. Achievements and weaknesses, in research management terms, will be identified. Third, some aspects of links with industry will be reviewed. To exemplify the role of the government R&D establishments and the impact of research management, brief historical case studies of innovation, associated with the Harrier aircraft and the coordination of electronic component development, will be included in the second and third parts, respectively.

PERSONNEL, ADMINISTRATION AND FINANCE

The personnel and administration systems used for defence research have been those of the public sector at large, and in this respect the consequences of the 1968 report[1] of the Committee on the Civil Service, led by Lord Fulton, have been most significant. The Committee sought to unify the 1400 staff classes across the service, with a view to making it more professional and opening up the most senior positions to specialists. Unification has gradually taken place and now covers the seven most senior grades, but the higher ranks of policy making have remained dominated by the 'intelligent generalists', who were so stoutly defended by Lord Simey on the original committee. Lord Fulton also foresaw the formation of Agencies, which were articulated and defined in the 1980s by Sir Robin Ibbs. Until the Defence Research Agency was formed in 1991, the most significant change was the move towards more financial accountability, which proceeded slowly and was driven by wider considerations than the need for more effective research. This is not to imply that public sector regimes have been inappropriate. Indeed, knowledge is power not money, and applied research, in many ways, sits more comfortably in government than under the short-term financial pressures of industry, or even in the universities.

Academics have developed the scholastic practices of the humanities to progress basic scientific research, through a productive link with teaching, but have been less successful with applied research, which must deliver all the knowledge required by the market. Thus the Defence Evaluation and Research Agency has the opportunity to address the neglected distinction between pure and applied research management; a territory outside both government administration and commercial venture, which large companies, and such institutions as the Lawrence Livermore Laboratories in the USA, are only now beginning to explore.

Public affairs are run largely by civil servants in the administrative and executive classes, and as government research expanded between the wars, more and more scientists were recruited. In 1945 arrangements were formalised and the science class was created, with separate career streams for scientists, scientific assistants and experimental staff. This was very much a matching arrangement, in which experimentalists largely carried out the programmes planned by scientists just as executives largely implemented the policy formulated by administrators. All these classes were involved in the defence research and development establishments. Typically, graduates would join the science class as Scientific Officers at grade 10, and some would aspire to become a Permanent Secretary at grade 1. One defence scientist, known to his friends as 'Fulton Man', has actually achieved this rank, albeit not in the Ministry of Defence. As direct military involvement in the R&D establishments waned after 1945, there was a tussle for authority between the senior administrator (the Secretary) and the senior scientist (the Superintendent), who often had similar ranks at about grade 6. This was resolved in favour of the scientists, as Superintendents became first Chief Superintendents and then Directors at ranks as high as grade 2. However, the Director and the Secretary continued to have separate reporting lines to the Ministry of Defence and, in spite of the seniority difference, research establishments remained very much dual-headed organisations: the Director was responsible for the research programme and the Secretary looked after staff, finance and general services. The arrangement worked remarkably well, not least because scientists were familiar with similar structures in the universities, and because strong establishment loyalty united Director and Secretary to defuse most potential conflicts.

However desirable the inflation of science grades might have been to attract the best staff and to give scientists essential access to defence policy debate at all but the highest level, it was of doubtful benefit to the research programme. Modern business thinking and the information revolution have recently led to fewer levels in the management hierarchies of business, but in research it has always been obvious that a flat management structure was important, and that those with ability and maturing experience should remain in direct contact with the experimental programme. The abolition of experimental and scientific assistant staff grades in the 1970s (following Lord Fulton) and the introduction of performance pay and overlapping grade salary scales in the 1980s were half-hearted attempts to address the dilution of accountability and programme commitment, associated with an extended management structure. It was not until the formation of the Defence Evaluation and Research Agency in the 1990s that it was possible to organise and reward research leaders, project managers and department heads in a more flexible and pragmatic way.

Thus new and apparently more appropriate arrangements have largely replaced the old grading structure, and evidence of improved research output is keenly awaited. Although very inflexible, grading systems have some merit in research, for stability and security are needed in the face of constant technical disappointment, and even when it is accepted that seemingly lucky discoveries follow ability, endeavour and imagination, research achievement cannot be the only determinant of reward. Moreover, the individual merit (IM) promotion scheme, introduced in 1946, was derived from the grading system and has served research well. Prior to the amalgamation of scientific and experimental grades, principal scientific officer (grade 7) was the career grade in the scientific civil service. Most programmes were led at this level, but it was possible to gain merit promotion to grades 6, 5 and even 4, on the basis of a review, carried out by independent assessors, within the civil service personal promotion system. Over the years, the IM scheme has enabled first-class scientists to remain closely associated with important programmes, and to provide the core of excellence so important both to national defence and to international knowledge exchange. Nevertheless, the scheme has suffered from two flaws. The panel of independent assessors has included world-class scientists more familiar with pure than applied work, and the five-year review, stipulated to retain the new rank, has sometimes enhanced the predisposition of staff to remain in a particular field against the run of changing defence priorities. Apocryphal stories relating to IM promotion abound, including that of the candidate who failed for being involved in research 'too closely associated with defence matters'!

With the exception of a short period in the mid-1980s, when an economic boom corresponded with low morale from earlier defence cuts, the Civil Service Commission has regularly succeeded in recruiting many of the best scientists from British universities, and in most cases very satisfactory careers have been pursued to retirement within the MoD. Staff have frequently progressed to high positions in the universities and industry, but the lack of mobility in the middle ranks has been a distinct failing. Extramural research contracts have been used to offset any programme weaknesses owing to the absence of specialists on short-term contracts, but the low level of short- and long-term staff exchanges with industry has undoubtedly impeded technology transfer and added to the problems of defence equipment procurement. Many efforts have been made to reverse growing introspection and stimulate mobility and more extrovert attitudes but, sadly, pension rules, more restrictive intellectual property policy and tighter procurement practices have conspired to make matters worse.

Defence research costs are recovered from Vote 2, the Treasury allocation to the Ministry of Defence for equipment, and, in common with all government expenditure, are subject to cash limits within each tax year. Thus research has enjoyed the luxury of a fixed income stream, with costs controlled by rigid adherence to staffing targets, and balancing capital investments filling the gap between staff costs and the funds available. Although such an arrangement might be expected to starve capital expenditure in the cause of staff security, this has only been partly true. Staff often leave unexpectedly, and security clearance and the procedures needed to preserve the high probity standards of the Civil Service Commission have made recruitment a lengthy process. In consequence, establishments have tended to run below their staff complement and there has been pressure to spend surplus cash on equipment at the end of each tax year. Such funds would otherwise not only be lost, but future budgets would be put at risk in the constant struggle to control public expenditure. Thus, surprisingly, surplus cash creates a crisis, but one which can be exploited to circumvent complex capital expenditure procedures and allow laboratories to acquire the latest test equipment, so necessary for research but so difficult to justify on normal investment criteria. By these means, and at the expense of staff numbers, defence scientists have achieved a clear advantage over their colleagues in industry and the universities, for whom lack of access to the most advanced test equipment is a running sore. Of course, the absence of capital accounting, with its depreciation costs and notional rents, has made investment in fixed assets very difficult, and the contrast between government and industry accommodation has been stark. However, when normal capital accounting practices are used all research assets, other than buildings, are written off quickly. In consequence, it is dangerously easier to prepare an investment case for a building than for a piece of research equipment. The depreciation account can only be spent once, and now that the Defence Evaluation and Research Agency has adopted commercial accounting practices it must beware the grain of truth in Professor Northcote Parkinson's old tongue-in-cheek dictum that the 'excellence of research is inversely proportional to the quality of the premises in which it is carried out'!

Since 1945 the defence R&D establishments have been subject to almost continuous financial cuts, and of the 25 major laboratories reviewed in a report on government R&D[2] prepared by Sir Solly (subsequently Lord) Zuckerman in 1961, only four remained when the Defence Research Agency was formed in 1991. More recently, organisations not considered by Zuckerman, notably the Chemical and Biological Defence Establishment and the Defence Operational Analysis Establishment, together with the main military test and evaluation units, have been added to form the Defence Evaluation and Research Agency as it is today. Annual reductions have varied from Treasury-imposed efficiency savings (3 per cent) to more severe measures (10 per cent or more) prompted by financial crisis or defence review. However, in all this change, which has involved massive staff reductions of more than 80 per cent, research has been well protected. Costs, and therefore staff reductions, have been accommodated by withdrawal from equipment design and development, by improvement in direct to indirect staff ratios, which still lag behind those in industrial laboratories, and by the reduction in experimental hardware activity as software-based research has expanded. In consequence, the number of research workers has not suffered unduly, and there were many more staff in the scientific career stream in 1991 than in 1961 (then about 2000). The surprising rise in scientific staff numbers is undoubtedly explained by the incorporation of scientific assistant and experimental officer grades into the scientific career stream during the period. Nevertheless, it is clear that development has borne the brunt of establishment cuts. Moreover, the expenditure ratio between internal and contracted research has been defended as a matter of policy, and, at about 2:1, has remained largely unchanged. In reality, the lack of capital costs in Treasury accounting has diluted the incentive to reduce real-estate holdings, and establishment mergers have been due as much to a demand for improved efficiency and the need to address defence requirements at a higher systems level, as to reduced expenditure. Thus although the Defence Research Agency inherited only four establishments in 1991, many times that number of significant laboratory sites were involved.

The cost profiles of government-sponsored civil and defence R&D programmes have always been quite different, and this difference, after 50 years, still confuses the debate on the most appropriate balance between civil and defence expenditure. Government accepts the role of supporting most basic and strategic research in the civil sector, whereas in defence little basic work is done and strategic research must be approved in London on the basis of its relevance to military needs. Zuckerman estimated in 1961 that, although more than six times as much was spent by the government on defence R&D as on civil R&D, the ratios between basic/strategic research, applied research and development were 1:19:80 for defence expenditure of £242m, but 40:45:15 for civil expenditure of £40m (education and atomic energy excluded). The cost profile contrast is even greater now, and the profile for the civil sector in 1995 was about 55:40:5. By normal civil industry standards defence research costs are a high fraction of development costs (20 per cent), and R&D costs as a fraction of production costs are even higher (25 per cent). However, against the peacetime aim of 'preparedness', such proportions do not seem unreasonable. The defence equipment budget has been about half the total defence vote, which, although subject to fluctuation from economic pressures and changes in the perceived threat, has fallen fairly steadily by a few percentage points over the second half of the twentieth century, to about 3 per cent of the gross national product (GNP) following the end of the Cold War. The GNP has increased considerably in real terms, and the draconian cuts which are such a regular feature of equipment programmes have been due as much to the unexpected escalation of costs, as to dramatic changes in defence policy.

PROGRAMMING RESEARCH

Until the 1990s, when research began to be contracted and to become identified by the budget source, the Ministry of Defence had found it helpful to separate its research investments into five categories. Category 1 was concerned with technical support for equipment procurement, and would not normally be considered as research. Category 2 covered research directly backing agreed equipment programmes, and, for example, could include the construction of military demonstration equipment. Much work in these categories was funded directly from purchasing departments, and although often subcontracted to industry, relied heavily for guidance on government defence establishment scientists with mature experience. Many equipment procurement programme successes can be traced to direct funding of substantial Category 2 work in industry from the headquarters purchasing department, but sadly Category 2 work has more often been squeezed between the aspirations of scientists to do longer-term work and lack of priority within the appropriate equipment budget. Category 3 work was concerned with equipment effectiveness studies, and was linked with the much wider area of operational analysis and intelligence studies conducted outside the research and equipment sector. The main research programmes were carried out under Category 4, where applied research for headquarters (Deputy Chief of the Defence Staff (Systems) in recent times and single Service Chief Scientific Advisers previously) directly supported the definition of the staff requirements on which future equipment development and production would be based. The level of procurement risk depended crucially on the quality of this work, and the effective transfer of technology from government research establishments to industry was vital. Companies need a foundation of research as well as design experience, and a significant fraction of Category 4 work was contracted to industry to form a bridge for technology transfer and to complement their privately funded programmes. Category 5 research was strategic and the programme was approved annually by the Deputy Under-Secretary (Science and Technology) at the Ministry of Defence. Here, the military relevance of new scientific trends and technologies was explored, and here joint programmes with universities, including staff exchanges, were to be found. Ideally, Category 5 work would react with operational analysis to identify important possibilities for specific applied research. It has been an undoubted strength of the government defence laboratories that Categories 3, 4 and 5 work could be closely coordinated under common management. However, this advantage has not been exploited as vigorously as it might have been.

In practice, research, more than most activities, is programmed in detail by those doing the work, within constraints imposed from above. For example, the constraints could relate to narrow applications, as in Category 1 work performed in support of the purchase of army boots, or broad, as in the Category 5 case of the development of new analytical tools for fluid dynamics. The method of defining the constraints presents great difficulty, and for this reason has always been in a constant state of flux. Over the years, however, two aspects have remained common: the need for military staff to interpret user requirements and set programme constraints with the help of scientific advice, and the attenuation of unwanted projects by prioritising candidate programmes within fixed resource allocations. Defence research programmes have always been approved by the military customer, and MoD methods to some extent inspired the customer–contractor principle made famous by Lord Rothschild in his 1971 report[3] on the organisation and management of government R&D. His proposals were aimed mainly at applied research in the civil sector, notably that sponsored by the Research Councils in their own laboratories. However, the proposal was enthusiastically adopted by government, and since then it has been firm MoD policy to align funds ever more closely with the current perception of military needs. Lord Rothschild was well aware that overzealous customer control would stultify research progress, and he proposed that a surcharge of 10 per cent should be available to scientists for discretionary work. It is a cause of concern that although the MoD allows such cost recoveries in industry, it has not seen them as appropriate in its own laboratories.

After the war scientific advisers became quite powerful. Each Armed Service had one at Director General level (grade 3). They and their staff supervised laboratory programmes through the annual review of the main fields (such as, for example, Electronics and Guided Weapons), and managed the extramural programme directly. However, these scientific staffing organisations were increasingly seen as out of touch with fast-moving science, and by the 1980s tri-Service military operational requirements staff were negotiating programmes directly with the R&D establishments, with the help of experts on short-term postings. The laboratories, in turn, initiated research contracts with industry and the universities as required. This trend increased the independence and reduced the responsiveness of the establishments. It prompted the introduction of package management, which coincidentally suited the agency regime introduced in 1991. The formation of an agency to conduct government intramural defence research was seen as a way of improving administrative flexibility and research programme responsiveness. Thus advantage was taken of the general recommendation by Sir Robin Ibbs that all services provided by government (both internal and to the public) would be best delivered on an 'arm's-length' commercial basis. In the new arrangements, package managers (normally scientists) manage a set of research contracts on behalf of military budget holders, responsible for research in support of a 'package' of military functions.

Package management tends to lead to well defined and therefore small projects, and to resource fragmentation. This limits flexibility and curtails the option of closing by attrition those programmes seen by the customer to have low priority. Such loose management is in any case out of keeping with the more macho styles of today, but it did have the advantage of allowing the best aspects of poorly funded programmes to survive by natural selection, for as long as possible. Low priority today can be high priority tomorrow. Thus when the Director of the Royal Radar Establishment wrote solemnly in the 1958 annual review that 'work on infra-red has ceased as there is no operational requirement to detect such wavelengths'[4], the irony was that in no time at all the work to detect objects by their natural heat emissions (infrared radiation) had blossomed into one of the most successful programmes ever sponsored by government, with numerous Queen's Awards and a MacRobert Award to its credit. By conferring full night-fighting capability, infrared became the major military technology discriminator in the Cold War, and was most effective in the Gulf conflict. Whether the Director concerned thought he had closed the programme, or knew that he had not, has never been revealed.

The key to good applied research is that high-calibre research staff should be able to steer their work by intimate knowledge of users and their potential needs, and transfer technology effectively to those who will apply it to meet such needs. The challenge to management is to lead, and to deploy capital, human resources and advice in a way that sharpens rather than blunts this process. It is a challenge which is rarely met as well as might be hoped, but it is fair to claim that performance within the MoD has compared favourably over the years with that in similar organisations overseas.

Research excellence is sought by scientists and managers alike, not just because discoveries can only be made once, but also because leading scientists, with key information, are welcomed by colleagues throughout the world. They have the enormous advantage of being able to multiply their knowledge, by exchange, many times over. However, early access to information can confer great military and commercial advantage. Thus government experts are constantly involved in information exchange, but must strike a balance between maximising their own knowledge, avoiding possible benefit to potential enemies, compromising the security of allies and respecting legitimate industrial commercial interests. An astonishingly complex set of defence science networks has grown up within NATO over the last 50 years. Many nets are opened, some may become dormant, but few are closed. Bilateral and multilateral exchanges take place between countries at varying levels of secrecy. They are driven by local political and commercial concerns, as well as the general merit of open research and programme cooperation. In particular, the technical links built up with the USA during the Second World War have remained very strong, and The Technical Cooperation Program (TTCP), with representation from Australia, Canada, New Zealand, the USA and the UK, has survived for 50 years. In spite of the acronym being mischievously misconstrued as 'The Travelling Cocktail Party', there have been major benefits in the stimulation of ideas, the encouragement of joint programmes, the organisation of trials, the exchange of data and, above all, the identification of those expensive nugatory channels of research that should be avoided.

'A week is a long time in politics', and although general trends may be clear, actual events are often unpredictable, as the Falklands and Gulf conflicts have shown. At the same time, the cost and sophistication of equipment have escalated, and the interval between a weapon system and its replacement has increased. Thirty years ago a fortunate scientist could apply the wisdom he gained in the research and design of one generation of equipment to the next. Now he is lucky if he sees one design through to completion, and even if he should survive to its successor, there is a likelihood that rapid advances in technology will have passed him by. With military officers in post for only two years and industry under increasing pressure to improve its return on capital employed, there is a real sense in which defence research programmes maintain the technical knowledge archive, and provide that continuity of technical expertise so necessary for quick reaction in war and good commercial judgement in peace. Moreover, the internationalisation of arms supply, the increasing integration of systems and the introduction of automation at higher and higher levels of military command, have all added to the problem of providing a rapid technical response. Growing difficulties in the maintenance of appropriate industrial capacity and the deep understanding of ever more complex military systems have long been recognised. Increasingly, these are the overriding considerations in formulating policy for research programmes and in the career development of staff.

INNOVATION IN THE HARRIER AIRCRAFT PROGRAMME

Most defence equipment innovation is characterised by a long gestation period within which changes in personnel and the market environment provide an ever-moving backdrop influencing priorities and direction. Moreover, each significant equipment or operational event alters for ever the evolution base for future activity. The history of the Harrier short takeoff and vertical landing (STOVL) aircraft, and its subsequent influence on the European Fighter Aircraft and Joint Strike Fighter programmes, is typical and exemplifies many defence research management issues.

By 1950, it was clear that, owing partly to the sophistication of electronics and weapons, the cost of military aircraft was growing at an unsustainable rate. Moreover, it was also clear that the long concrete runways required for heavy, fast aircraft would be vulnerable to highly accurate ballistic missiles. Thus the lightweight fighter programme was born. A competition was sponsored by Colonel O'Driscoll, the American head of the NATO Mutual Weapons Development Program, in Paris, and the Fiat G 91 was preferred to the Folland Gnat, which could not land on grass. The G 91 used the lightweight Bristol Orpheus engine, which was put into large-scale production in both Germany and Italy. It had nearly 5000 lb thrust, and a thrust to weight ratio of more than 6:1. O'Driscoll believed that short takeoff aircraft would be essential in Europe, and persuaded the Bristol Engine Company to take an interest in the ideas of a French engineer called Wibault. With financial support from the Rockefeller Foundation, he was proposing an aeroplane with vectored thrust, called the 'ground attack gyropter'. Four centrifugal compressors were to be set horizontally at right-angles to the engine, and the thrust direction controlled by rotating the casing and radial exit port of each compressor about its axis. Expecting problems from weight and reliability, Bristol were not impressed, but O'Driscoll persisted and, with 75 per cent funding from the Mutual Weapons Development Programme, Bristol added rotatable nozzles to the established Orpheus engine.[5,6] First, a low-pressure compressor with two nozzles, which could continuously alter the direction of thrust from the vertical to the horizontal, was added to the front of the engine. Later, to give more stability, the hot turbine exit gases were redirected by a similar pair of nozzles added at the rear. Gyroscopic effects were minimised by making the low- and high-pressure compressor/turbine systems rotate in opposite directions, and attitude control stability was assisted by ensuring that the thrust vectors converged on the centre of gravity of the host vehicle. Full development of

FIGURE 14.1: THE PEGASUS GAS TURBINE USED BY THE HARRIER VERTICAL TAKEOFF AIRCRAFT, SHOWING TWO OF THE FOUR THRUST VECTORING NOZZLES ADDED TO THE ORIGINAL ORPHEUS ENGINE (ROLLS-ROYCE PLC, BRISTOL)

the engine, now called the Pegasus, started in 1958 (Fig. 14.1). It won the first MacRobert Award, which was established in 1969 as the premier UK engineering innovation prize, and has since been successively developed to give thrust levels well in excess of 20,000 lb.

The first jet-propelled vertical takeoff demonstration in the UK was the tethered flight of the Rolls-Royce 'Flying Bedstead' at the Royal Aircraft Establishment in the early 1950s (Fig. 14.2). This was soon followed in 1954 by the Short Brothers' SC1 research aircraft. The SC1 had four Rolls-Royce RB-108 engines to give vertical lift, and a fifth for forward movement. Observing this work, and that on the Pegasus engine, Sir Sidney Camm in 1957 proposed the Hawker Kestrel (P1127), and it was this initiative that accelerated full development of the engine. The Kestrel (Fig. 14.3) demonstrated the feasibility of STOVL operations and was the direct forerunner of the Royal Air Force Harrier, the Royal Navy Sea Harrier and the US Marine Corps AV8B.

The Royal Air Force has always been enthusiastic for STOVL aircraft and commissioned a squadron of Kestrels, even though their weapons-carrying capacity was poor. It took nearly a decade to develop the engine and to cleverly optimise thrust vector control during takeoff, so that the aircraft could lift credible operational loads. Moreover, the logistic problem of working away from main bases dismayed the Air Force, and an interest in vertical takeoff aircraft would not have helped the Navy in their long rearguard action to keep large aircraft carriers. Also, to make matters

FIGURE 14.2: THE ROLLS-ROYCE 'FLYING BEDSTEAD' IN TETHERED FLIGHT. THE FIRST JET SUSTAINED VERTICAL LIFT VEHICLE USED AT THE ROYAL AIRCRAFT ESTABLISHMENT, BEDFORD, TO STUDY STABILITY CONTROL IN THE 1950S (© CROWN COPYRIGHT, DERA)

worse, the American, German and UK high-level government panel, set up in the 1960s to coordinate STOVL aircraft requirements, had grandiose ambitions for supersonic fighters and high-load-carrying freighters, which were far from current reality. However, the panel disbanded in disagreement, and in 1966 the Navy finally accepted that cost and vulnerability to missile attack made large carriers untenable. This was the turning point. Air-defence guided weapons were not good enough to let the Navy dispense with fighter cover, and the Harrier was seen as an essential complement to surveillance helicopters if the UK was to continue to maintain a deep-sea surface fleet. Similarly, the US Marines did not operate large aircraft carriers and were keen to add a version of the Harrier to their helicopter ships.

Of the many—a dozen or more—serious STOVL programmes worldwide, only the derivatives of the Kestrel have survived into operational service. This can be attributed to very close collaboration between research at the Royal Aircraft Establishment at Bedford and the aircraft contractor British Aerospace at Warton. Clearly, the development of a vehicle with such a complex potential flight envelope is a hazardous business, and an incremental research approach has been pioneered at Bedford in which the very best test pilots collaborate with the scientists on the parallel

FIGURE 14.3: THE KESTREL AIRCRAFT UNDERGOING APPROACH AND VERTICAL LANDING TRIALS IN A WOODED CLEARING, AT THE ROYAL AIRCRAFT ESTABLISHMENT, BEDFORD, IN THE 1960S (© CROWN COPYRIGHT, DERA)

development of a research flight simulator and an experimental aircraft. Thus small steps with acceptable risks can be demonstrated on the simulator before being confirmed in real flight. Feedback from flight measurements then helps to evolve the design of the simulator without loss of its assured integrity. This approach is particularly appropriate when seeking to provide the pilot with computer-derived assistance (so-called 'fly by wire'), and with the help of a twin-seat aircraft, where one pilot is computer aided and the other is not, substantial progress has been made towards flying possible higher-performance descendants of the Harrier more simply.

Thus it was that a few squadrons of Royal Air Force and Royal Navy Harriers were available at the start of the Falklands war in 1982. Whether the war could have been fought at all without the Harrier is debatable, but it was certainly the only air cover available to the Navy. Moreover, the Harrier acquitted itself well, and was able to outmanoeuvre and defeat apparently superior aircraft. Victory in the air battle not only secured its position with the Navy and US Marines, but also stimulated interest in the wider use of vectored thrust, notably in supersonic agile fighters. Consequently, by the 1990s the combined UK government and industry defence science community had established a vital role in flight-control systems research for the proposed Anglo-American Joint Strike Fighter.

Modern fighters need a very high-performance intercept radar to engage low-flying targets at night, in bad weather and beyond visual range. Such radars must operate in the presence of interference and very strong reflections from the ground and, with severe size and weight constraints, require the most advanced radar technology. Such a radar, compatible with the latest air-to-air missiles, was needed by the Harrier as it came of age as a frontline fighter in the Falklands War.

Airborne intercept (AI) radars are manufactured in relatively large numbers and, although pioneered in Britain, by the end of the Second World War most production was taking place in the USA. However, Ferranti, casting around for a product to replace its successful gyro-stabilised gunsight, invested in AI and won contracts for the Lightning fighter and the Buccaneer. These successes led Ferranti to be chosen, along with GEC, as the preferred AI suppliers, and to share a continuous stream of research contracts, both from the R&D establishments and the MOD's purchasing department for airborne electronics. Thus, in spite of disappointment when the TSR-2 was cancelled, and when US equipment manufactured in Germany was chosen for the Phantom, Ferranti, with the help of government scientists, was well placed to pick up information, through The Technical Cooperation Program, on the latest medium-pulse-rate coherent radar research published in the USA. In this type of AI equipment a continuous wave is generated to excite a multimode pulsed transmitter, and moving targets can be identified in the presence of strong ground echoes by using fast integrated circuits to analyse the frequency spectrum, as well the time delay, of reflected signals. Although pioneered for interference suppression, these techniques conferred for the first time the important operational advantage of determining the track of a target as soon as it is detected. Design principles were consolidated and extended in a significant set of MoD flight trials using modules from a proposed Ferranti demonstration radar called Blue Falcon, which was never completely built. By including this so-called 'supporting technology programme' in its fixed-price production contract, Ferranti, at considerable commercial risk, was able to ensure that key design and performance features were proved by demonstration before design commitments were made. The urgency of the Sea Harrier requirement, and the effectiveness of government–industry collaboration in the supporting technology flight trials, enabled British industry to take a half-generation lead in AI radar for the first time since 1945. In consequence, a further enhanced Ferranti (now GEC Marconi Avionics) radar, called the ECR90, was able to beat off US/German competition and win the AI contract for the European Fighter Aircraft (EFA) in 1991. This radar represents the ultimate system using a mechanically scanned antenna, and research is now being concentrated on the use of electronic beam scanning for the future.

It is worth summarising the research management characteristics common to the history of the Harrier and its radar. First, it is clear that the source of invention and of basic research knowledge is relatively unimportant, compared with an appreciation of what can be achieved with current knowledge and a continuous readiness to meet unpredicted market requirements. An important aim of defence research management in peacetime must be to ensure that the research programme maintains maximum relevance over long periods at minimum cost.

Second, the importance of demonstration equipment can hardly be overstated. Not only will unknown information needed for development be uncovered, but a consolidated knowledge platform will have been established for the better direction of future research. However, the choice of what to demonstrate and when is critical, as failure will be expensive and success may encourage the Armed Services to adopt that particular weapon configuration, whether or not it is the most appropriate that could be made available. A successful demonstration programme requires close coordination between research, industrial and weapons-procurement staff. Research staff are the best judges of the available knowledge, but industrialists have the widest insight and are most likely to identify the best specification for demonstration equipment, and the optimum timing for its construction. At the same time, a visionary budget holder in the government purchasing office is essential if a product well matched to the military requirement is to be obtained in the most timely and cost-effective way. Many, including the National Audit Office, have extolled the virtues of 'demonstrators' as an excellent way of reducing risk, and making a net saving of time and money in procurement. The value of demonstration equipment is increasingly recognised, but such equipment is expensive and remains vulnerable to the argument deployed by the Treasury (and others) that, if the military requirement is clear, then a demonstrator is not needed, and if it is not then the equipment, and therefore its demonstrator, is not justified.

INDUSTRY AND TECHNOLOGY TRANSFER

In prewar Britain, as in second-century Rome, the rich pickings of Empire, and commitments to govern abroad, conspired to undermine wealth creation at home, and by 1939 manufacturing industry had been in relative decline for well over 50 years. Thus the growth in the government research and development establishments during the war must be seen against the background that Germany and the USA were ahead in most sectors, but that Britain, at least in the early days, had a more urgent and direct view of what was needed. The establishments were first and foremost development and trials units, with strong military involvement, working as close to the front as to the frontiers of science, and producing drawings and successful prototypes for subsequent manufacture in the UK and the USA. For example, in the early 1930s the USA had a better grounding than Britain in both radar and operational analysis, but by 1939 these subjects were being pioneered in Britain, where seeds were sown for much of the postwar military activity in the USA. In particular, the American group formed within the Telecommunications Research Establishment at Malvern helped found the Lincoln Laboratory, near Boston, which now leads radar research worldwide. Moreover, the superior production engineering capability and industrial capacity of the USA swung into action, and by the end of the war many radar applications pioneered with distinction in the UK were already largely a US industrial preserve. Other technique areas with similar histories are to be found in aeroengines, electronic intelligence and submarine warfare.

Thus, in 1945, and in spite of heroic wartime production achievements, British industry was in a far from enviable position. Mechanical engineering firms were fragmented, and there were too many shipyards, airframe and engine companies.

However, with deeper roots they fared better than those in newer industries, such as electronics, which found themselves with weak design facilities and fast-moving technology passing them by. The establishments, on the other hand, had a wealth of military and technical knowledge, but needed to develop their research skills and to pass their development and design capacity to industry as quickly as possible. The electronics industry was reasonably strong in components, but in equipment its position immediately after the war is exemplified by comparing the capability of the defence arms of Philips in the Netherlands and in England. After regrouping in the late 1940s, and in spite of German occupation, the Dutch factories were well ahead of their British counterparts in subjects where there was direct equivalence, such as military radio and optoelectronics. Only when the knowledge and skills base of the government R&D establishments was added could any sort of UK superiority be claimed. Moreover, with government establishments taking an even-handed approach to competing UK firms, Dutch knowhow was very significant in helping Philips UK to build a strong position in its home defence market during the 1950s and 1960s. It has been claimed that concentration of research resources in military R&D establishments since the Second World War has contributed to a decline in the competitiveness of the British electronics industry. However, the Philips example suggests that the problem originated much earlier, and was only exposed when Empire preference declined in the 1970s and global markets developed in the 1980s. In fact, the surviving companies have largely closed the productivity gap with companies overseas, at least in Europe, and those firms best able to benefit from defence applied research have been the most successful.

On the whole, the transition to peace was not well handled and the high world standing of both the defence industry and the research establishments in the 1990s speaks volumes for the momentum derived from wartime knowledge, for the merit of a user-driven research strategy, and above all for a united desire, born of military pride and not seen elsewhere in the economy, to struggle forward in the face of flawed managerial direction. So far, and in the end, a common perception of what is in the national interest has united military, government and industrial staffs to counter such policy weakness and the managerial failures that have arisen.

There have been three areas of particular importance where policy weakness and management failure has been evident. These are the slow withdrawal of government from development and design, the abuse of power naturally accruing to research embedded in a monopoly customer, and the lack of efficient coordination of R&D investments by the three Armed Services.

TRANSFER TO INDUSTRY OF DESIGN, SYSTEM INTEGRATION AND PROJECT MANAGEMENT

None of these subjects is easy. It is difficult to recruit scientists while laying off engineers, and much of industry was only too happy to accept the low-risk option of manufacturing equipment to government drawings. The implications of the Ministry of Defence not withdrawing quickly from development and design were brought to a head during the surge of activity on antiaircraft guided weapons in the

1950s. Drifting in decline after 1945, the R&D establishments were spurred into action when the technical promise of replacing expensive piloted aircraft with guided missiles coincided with the escalation of East–West confrontation at the time of the Korean war. Each service had its own programme. Missiles and radars were designed separately, industry manufactured under contract at subsystem level, and little attention was given to command and control, even though Battle of Britain achievements were close in corporate memory. Much of the technical work was excellent, and some of the systems—notably Bloodhound, Rapier and Seawolf—have seen long service and successful foreign sales. But management control was poor, and horrendous overspends led to the famous report[7] by the Steering Group on Development Cost Estimating chaired by W.G. Downey in 1966, which outlined good project management practice in a commendable way that has stood the test of time.

The Downey report had a major impact on equipment procurement procedures, but the transfer of development and design to industry remained slow. Lord Strathcona, in a report[8] recommending the response research establishments should make to defence cuts in 1980, was prompted to observe that 'Most scope [for staff reduction] exists in the tasks of design, development, project support and post design services, but even here care has to be taken to protect important current projects, utilise existing capital facilities and make a realistic appraisal of industrial capabilities'. There was a general understanding that the management of equipment development projects should be transferred to industry, but indicators of the weakness with which the policy was pursued are that the Downey report largely assumed 'hands-on' control by MoD staff, and that the driver for Strathcona nearly 15 years later was not policy implementation but shortage of funds. It had not been sufficiently clearly understood that in war, time is the essence and that research and design will overlap within development, but that in peace, when the priority is to minimise cost, research may help, but must not be allowed to distract design. Special procedures for the control of design and its links with research are therefore needed.

Thus, not only was industry slow to build project management capability, but overseas marketing was impeded and the important defence need to integrate equipment at a higher systems level was delayed. Lack of clarity concerning the proper development and project management responsibility of industry may also have contributed to the incredible concept of the 1980s that, by more rigorous contracts, a monopoly customer could increase competition and reduce technical risk at the same time. Sadly for managers, it is all too easy to fail in the implementation of good policy, but virtually impossible to succeed when it is bad.

The project management transfer process could finally be said to have been completed, after 40 years, in the mid-1980s, when a systems management contract for the torpedo programme was let to GEC. Against this background, and in fear of backsliding, it is not surprising that industry was as keen to see development and production excluded from the Defence Research Agency Framework Document in 1991, as it had been to see the design and production of aircraft banned at the Royal Aircraft Establishment more than 70 years before.

USER-BASED RESEARCH RESPONSIBILITIES AND THE ABUSE OF RESEARCH POWER

The policy decision to maintain a strong defence research and equipment evaluation activity within government was wise. Not only do both benefit from close association with the military user, but the need to minimise expensive duplication and to make all relevant knowledge available to suppliers is facilitated. Moreover, with global competition becoming more intense, companies are less able to support research and see intellectual property more and more as a commodity to be obtained from the community at large. However, considerable power is conferred on research when it is embedded in the customer's organisation, and management has the responsibility to ensure that such power is not abused. With budgets to hand, research staff have been able to lead military requirements by technical demonstration, and dictate the direction of industrial research through extramural contracts. The combination is dangerous unless the scientists concerned have good judgement of production practicality, of commercial realities and of operational effectiveness. This has not always been the case. Consequently, well-meaning, excellent and articulate scientific staff have sometimes presented management with problems they have not even been able to recognise, let alone solve, and valuable industrial resources have been misused. As might be expected, this phenomenon has been most evident in areas such as data processing, where civil technology has been overtaking that more familiar within the defence community. However, it can also occur in pure defence technology as, for example, happened when scientists defended their hitherto unrivalled rifled tank gun design technique against the much superior smooth-bore approach that had overtaken it. With the formation of the Defence Evaluation and Research Agency, budgets have been transferred to military staff and the Pathfinder scheme, in which industry is encouraged to propose collaborative research programmes, has been introduced. These changes put the direction of research on a broader basis, which is to be welcomed, but have yet to be proved effective.

Although this problem may, on occasion, have led to suboptimal defence procurement, the exploitation record of defence research has been generally good. Queen's Awards for Technology are given when commercial success in industry has derived from highly original scientific work, and, bearing in mind the small fraction of defence establishment activity which is appropriate, it is highly commendable that, between 1976 and 1996, government defence laboratories (20 awards) have shared in awards three times as often as Research Council laboratories (7 awards), and ten times as often as university departments (2 awards). On balance, British industry has certainly gained considerable commercial advantage from knowledge discovered by government scientists for defence purposes.

FAILURE OF INTER-SERVICE PROGRAMME COORDINATION

In August 1939, observing that the Army and Air Force were working together on radar, at Bawdsey Manor, and that the Navy had hardly started, Robert Watson Watt proposed that radar R&D should be carried out on a tri-Service basis. On 1 September the Air Ministry moved its scientists to Dundee, and on the same day the War Office moved its staff to Christchurch, and at a policy meeting on

5 September the Admiralty refused to become involved. Agreement on a compromise proposal—that all research should be conducted together—was stalled until August 1940, and by then it was too late so that nothing was done. Turf wars between the Armed Services are common, and urgency in war can justify research and weapon design proceeding side by side.

However, in time of peace, with budgets on a tri-Service basis and the work moving towards generic research, there is little doubt that the research establishments have been too parochial. It is of course true that equipment needs to be developed with a particular environment in mind, but it is also true that the use of common subsystems can reduce costs, and that separate applied research streams will often duplicate the same work and fail to identify common ground. Nor can the argument be accepted that duplication leads to healthy competition and increases the probability of success. Valuable insights can be achieved by the competitive interaction of individuals working on complementary research programmes, but it is a feature of parallel work at separate locations that scientists hardly communicate at all. Moreover, 'focus' and 'viable levels of resource' are watchwords of applied research, and dispersal is almost always a recipe for fragmentation and failure.

As development moved to industry after the war the inefficiencies of uncoordinated research were exposed, and the practice of some firms being strongly linked to a particular Armed Service was an obvious distortion of the market. Further, when industrial rationalisation gathered pace in the 1960s, companies found themselves putting in the same research proposal to three (sometimes more) research establishments, and matters were demonstrably unsatisfactory. Electronics, being generic and of increasing importance to all three Services, was the main problem area, and matters improved somewhat when the principal electronics laboratories were merged at Malvern.

Essentially, each R&D establishment Director had his own fiefdom. Even the introduction of a common reporting line to the Controller, Establishments and Research (CER), within the tri-Service Procurement Executive, set up in 1971, following the report[9] by Derek Rayner on defence procurement, did not help significantly. Thus it came as no surprise that, when the Defence Research Agency was established with a single Chief Executive, its new structure was based on application sectors rather than on geographical locations and single Service loyalties.

ELECTRONIC COMPONENT DEVELOPMENT

The Navy and its establishments have been the most insular, but it is to them that the credit must go for the managed coordination of R&D in electronic components, easily the first programme area in which full tri-Service integration was attempted. The use of components is an effective way of applying technical knowledge, and quantity is the economic driver for modern production. Thus component sophistication and standardisation dominate equipment design, and the defence need for small quantities of extremely specialised equipment presents a difficulty. Nowhere is this difficulty greater than in electronics, where, for example, a few mobile radars may each require a single, unique megawatt component for transmission, and a special processor with a million identical microwatt elements to analyse target reflections received. It is not surprising, therefore, that many defence research management and policy issues are encapsulated in the history of the coordinated development of components. The headquarters directorate concerned was always known as CVD, and although when first proposed in 1938 this stood for the Coordinated Valve Department, this was later changed to Coordinated Valve Development, and later still to Components, Valves and Devices.

Recognising the importance of science to the needs of the Navy, the Admiralty set up a Department of Scientific Research at the end of the First World War, and laboratories to develop special valves for communications soon followed at Portsmouth in the early 1920s. This was the base from which the Admiralty Signal Establishment grew in 1941, and which in turn led to the first tri-Service research establishment, the Services Electronics Research Laboratory (SERL), being formed at Baldock in 1945.[10] The aim was to centralise research for CVD, which had already been formed to meet demands by the Navy for a coordinated approach to the purchase of electronic valves from industry. SERL was located in an unused electronic valve plant, built as part of the dispersal of the Cosmos Manufacturing Company from London in the face of the blitz. This brought benefits from close links between scientists and local 'production-literate' technological support, and the subsequent record in transferring government research on electronic components to production in industry has been second to none. The Navy led in the early days, and both CVD and its laboratory were staffed by members of the Royal Naval Scientific Service, which was later merged with the scientific branch of the civil service. Much later, in 1976, SERL joined the Signals Research and Development Establishment (Army Communications) and the Royal Radar Establishment to form the Royal Signals and Radar Establishment, at Malvern.

Although after 1945 the development of all components was passed quickly to industry, CVD acted as sponsor and adopted a structured approach to ensure that standardised advanced components were available to the Armed Forces.[11] Each year CVD invited research proposals from industry in areas identified by the whole scientific community as having potential defence importance. With access to MoD research, research contractors were able to acquire and expand knowledge and thus develop enabling technologies on which later production could be based. When an actual weapons requirement was approved, CVD would invite tenders for an experimental valve (VX) stage 'A' development to demonstrate the feasibility of any special component required. This would be followed by stage 'B', to achieve 'design approval of manufacturability', and often initial government production orders, so that equipment contractors could be supplied with the latest standard components.

There is no doubt that CVD was effective, and that most of the problems facing the successful application of research were addressed in an exemplary way. Wide-ranging debate identified relevant promising areas, applied research was conducted by the most technically competitive supplier, without stifling novelty, and research knowledge acquired was made freely available to those with a legitimate interest. Then, by presenting advanced components to equipment manufacturers, standardisation across component families could be achieved, with minimum sacrifice of performance. CVD presided over some extraordinary achievements in British electronics, and competitive research positions in such subjects as gallium arsenide devices, thermal detectors, image intensifiers, lasers, liquid crystals for displays and high-power microwave tubes have been maintained with credit, against overwhelming US investments, long after most civil colleagues had left the field.[12]

Why then was CVD dissolved in 1992? It was certainly to some extent redundant, in that all research was by then run on a tri-Service basis, and with schemes such as Pathfinder it could be argued that the best of its practices were being deployed more widely. However, it is also clear that CVD had run its course and had failed to adapt adequately to three significant changes in the environment.

First, the practice of purchasing parts of systems separately proved incompatible with the tighter procurement procedures of the 1980s. Contractors were all too often able to justify delay and excess expenditure by real or imaginary shortcomings in government-furnished supplies, and when, in consequence, government ceased to recommend the components to be used in defence equipment, uncertainty increased and it was difficult for CVD to justify new development investments. Thus, first stage B and then stage A programmes withered, leaving research on components exposed, with little differentiation from other extramural research activity. Second, American components could be purchased with increasing ease as the lead taken by the UK before the USA entered the Second World War was progressively eroded. Such foreign purchases undermined the cost-effectiveness of CVD development. Finally, the very standardisation that CVD sought was outflanked when civil technology, in such areas as communications and computing, outpaced defence technology and delivered world-standard components at very low cost.

In so far as the demise of CVD indicates a loss of national ability to compete at the highest technical level, the message is a sombre one, for a glance inside any personal computer will show that the capability of systems is increasingly represented by the sophistication of the components. In future conflicts a protagonist with access to superior components, for example for target acquisition, intelligence gathering or communications, is likely to prevail. The loss of stage A component development programmes is probably the most serious consequence of the closure of CVD. Research and quality work can be and is done elsewhere, and production is rightly part of procurement. However, unless stage A has demonstrated the feasibility of the most advanced components, the level of risk associated with their adoption will be unacceptable to equipment procurement programmes, and new weapons systems will not be as effective as they might otherwise have been.

CONCLUSIONS

Although management theories and fashionable styles developed to address particular problems abound, management remains a largely pragmatic activity, a game with uncertain rules, played on an uneven field, in an unfair world. Performance assessment means little except in terms of the success of that which is managed. It is often unclear and subjective, especially in the short term. However, 50 years is not short term, even for research, and an overall assessment of government defence research management should be attempted.

The defence (including aerospace) and pharmaceutical industries share the distinction of being the most significant indigenous British manufacturing sectors with world standing. It cannot be a coincidence that both benefit from large applied research activities, set outside the market but focused strongly on it. Further, in time of war the defence knowledge base has not been found wanting. The technical community in government and industry has achieved a speed of effective response in recent conflicts, which has rivalled the better-publicised achievements of the Second World War (see Chapter 7).[12] Examples, such as the deployment of helicopter-borne surveillance radar in the Falklands (1982), and infrared/optical targeting systems in the Gulf (1990), could be quoted.

Nor should the staff-development role of research be forgotten. Major laboratories at GEC, Plessey, Philips, EMI and ITT grew out of the research establishments after 1945, and there has been a steady stream of technical directors and professors from defence research in subsequent years. In the 1990s, among those who spent their formative years in defence research, were directors at Hewlett-Packard, GEC-Marconi, EMI and Nortel, as well as three professors who have recently headed the prestigious Department of Electrical Engineering at University College, London.

However, whereas there have been outstanding research achievements, the assessment 'Could do better' is also in order, for peacetime defence equipment procurement continues to suffer from costly delays, attributable to technical risk. Either the pure tone of good advice is not heard above bureaucratic noise in the Procurement Executive of the MoD, or contractors do not have the necessary knowledge when it is required. The Defence Evaluation and Research Agency leads defence research and cannot duck responsibility for either of these problems. The onus is on it, not only to give advice which is both convincing and correct, but also to ensure that technical knowledge necessary for the defence of the realm is available, where and when it is required. Applied research workers in the defence community should feel involved and strongly committed to successful equipment procurement, for it is written: 'the Lord said unto Cain, Where is Abel thy brother? and he said, I know not: Am I my brother's keeper?' After all their endeavours, it would be sad if the Defence Evaluation and Research Agency survived, only to carry the mark of Cain into the next millennium.

Two other difficult, but legitimate questions, are often asked: 'Has the work been cost effective?' and 'Could the skilled research manpower have been better deployed elsewhere?' The first requires valuation of human life, and the second is hypothetical. Both matters merit discussion, but lie outside the scope of this work.

Looking to the future, the Defence Evaluation and Research Agency faces many management challenges. Harsh financial discipline in a period of declining defence expenditure will put strains on programme flexibility, that progenitor of research excellence, and agency status has already exposed the difficulty of rigorous customer–contractor relationships in applied research. However, the user interface has been, and will always be, vital, and the discipline of closer bonds and more openness may be no bad thing. Positioning the role of the Defence Evaluation and Research Agency within Europe is likely to be the most fundamental issue in the near future, as industrial rationalisation changes the technology transfer role. When playing the strong UK defence card in European negotiations, government and industry should not forget that the Defence Evaluation and Research Agency is a jewel in their crown. In a world of increasing technical sophistication, where industry will be less and less able to carry out training and applied research at the level demanded by the market, it may well prove to be a model of the type of integrated staff development and applied research institute that society requires.

1. [Great Britain, Committee on the Civil Service. Chairman Lord Fulton] *The Civil Service. Vol.1, Report of the Committee 1966–68*, Cmnd 3638 (HMSO, 1968).
2. [Great Britain, Minister for Science, Committee on the Management and Control of Research and Development. Chairman Solly Zuckerman] *Report of the Committee on the Management and Control of Research and Development* (HMSO, 1961).
3. *A Framework for Government Research and Development*, Cmnd 4814 (HMSO, 1971).
4. David A. Smith, *Thermal Imaging—A Transfer of Technology* (Manor Park Press, 1988).
5. S. Hooker, 'History of the Pegasus Vectored Thrust Engine', *Journal of Aircraft*, 18 (1981): 322–6.
6. G.M. Lewis, *Pegasus Vectored-Thrust Turbofan Engine 1960* (International Historic Engineering Landmark, A.S.Mech.E and I.Mech.E., 1993).
7. [Great Britain Ministry of Technology Steering Group on Development Cost Estimating. Chairman W.G. Downey] *Report of the Steering Group on Development Cost Estimating* (HMSO, 1969).
8. Lord Strathcona, *Steering Group on the Research and Development Establishments Document* (June 1980).
9. [Great Britain, Civil Service Department. Chairman D.G.Rayner], *Government Organisation for Defence Procurement and Civil Aerospace*, Cmnd 4641 (HMSO, 1971).
10. G.P. Wright, *Services Electronics Research Laboratory, 1945–1976* (Defence Research Information Centre, 1986).
11. K. Dickson, 'The influence of Ministry of Defence funding on semiconductor research and development in the United Kingdom', *Research Policy*, 12 (1983): 113–20.
12. Richard Coopey, Matthew Uttley and Graham Spinardi, *Defence Science and Technology—Adjusting to Change* (Harwood Academic Publishers, 1993).

Index